Pitman Research Notes in Mathematics Series

Submission of proposals for consideration
Suggestions for publication, in the form of outlines and representative samples, are invited by the Editorial Board for assessment. Intending authors should approach one of the main editors or another member of the Editorial Board, citing the relevant AMS subject classifications. Alternatively, outlines may be sent directly to the publisher's offices. Refereeing is by members of the board and other mathematical authorities in the topic concerned, throughout the world.

Preparation of accepted manuscripts
On acceptance of a proposal, the publisher will supply full instructions for the preparation of manuscripts in a form suitable for direct photo-lithographic reproduction. Specially printed grid sheets can be provided and a contribution is offered by the publisher towards the cost of typing. Word processor output, subject to the publisher's approval, is also acceptable.

Illustrations should be prepared by the authors, ready for direct reproduction without further improvement. The use of hand-drawn symbols should be avoided wherever possible, in order to maintain maximum clarity of the text.

The publisher will be pleased to give any guidance necessary during the preparation of a typescript, and will be happy to answer any queries.

Important note
In order to avoid later retyping, intending authors are strongly urged not to begin final preparation of a typescript before receiving the publisher's guidelines. In this way it is hoped to preserve the uniform appearance of the series.

Longman Scientific & Technical
Longman House
Burnt Mill
Harlow, Essex, CM20 2JE
UK
(Telephone (0279) 426721)

Titles in this series. A full list is available from the publisher on request.

51 Subnormal operators
J B Conway
52 Wave propagation in viscoelastic media
F Mainardi
53 Nonlinear partial differential equations and their applications: Collège de France Seminar. Volume I
H Brezis and J L Lions
54 Geometry of Coxeter groups
H Hiller
55 Cusps of Gauss mappings
T Banchoff, T Gaffney and C McCrory
56 An approach to algebraic K-theory
A J Berrick
57 Convex analysis and optimization
J-P Aubin and R B Vintner
58 Convex analysis with applications in the differentiation of convex functions
J R Giles
59 Weak and variational methods for moving boundary problems
C M Elliott and J R Ockendon
60 Nonlinear partial differential equations and their applications: Collège de France Seminar. Volume II
H Brezis and J L Lions
61 Singular Systems of differential equations II
S L Campbell
62 Rates of convergence in the central limit theorem
Peter Hall
63 Solution of differential equations by means of one-parameter groups
J M Hill
64 Hankel operators on Hilbert Space
S C Power
65 Schrödinger-type operators with continuous spectra
M S P Eastham and H Kalf
66 Recent applications of generalized inverses
S L Campbell
67 Riesz and Fredholm theory in Banach algebra
B A Barnes, G J Murphy, M R F Smyth and T T West
68 Evolution equations and their applications
K Kappel and W Schappacher
69 Generalized solutions of Hamilton–Jacobi equations
P L Lions
70 Nonlinear partial differential equations and their applications: Collège de France Seminar. Volume III
H Brezis and J L Lions
71 Spectral theory and wave operators for the Schrödinger equation
A M Berthier
72 Approximation of Hilbert space operators I
D A Herrero
73 Vector valued Nevanlinna theory
H J W Ziegler
74 Instability, nonexistence and weighted energy methods in fluid dynamics and related theories
B Straughan
75 Local bifurcation and symmetry
A Vanderbauwhede
76 Clifford analysis
F Brackx, R Delanghe and F Sommen
77 Nonlinear equivalence, reduction of PDEs to ODEs and fast convergent numerical methods
E E Rosinger
78 Free boundary problems, theory and applications. Volume I
A Fasano and M Primicerio
79 Free boundary problems, theory and applications. Volume II
A Fasano and M Primicerio
80 Symplectic geometry
A Crumeyrolle and J Grifone
81 An algorithmic analysis of a communication model with retransmission of flawed messages
D M Lucantoni
82 Geometric games and their applications
W H Ruckle
83 Additive groups of rings
S Feigelstock
84 Nonlinear partial differential equations and their applications: Collège de France Seminar. Volume IV
H Brezis and J L Lions
85 Multiplicative functionals on topological algebras
T Husain
86 Hamilton–Jacobi equations in Hilbert spaces
V Barbu and G Da Prato
87 Harmonic maps with symmetry, harmonic morphisms and deformations of metric
P Baird
88 Similarity solutions of nonlinear partial differential equations
L Dresner
89 Contributions to nonlinear partial differential equations
C Bardos, A Damlamian, J I Díaz and J Hernández
90 Banach and Hilbert spaces of vector-valued functions
J Burbea and P Masani
91 Control and observation of neutral systems
D Salamon
92 Banach bundles, Banach modules and automorphisms of C^*-algebras
M J Dupré and R M Gillette
93 Nonlinear partial differential equations and their applications: Collège de France Seminar. Volume V
H Brezis and J L Lions
94 Computer algebra in applied mathematics: an introduction to MACSYMA
R H Rand
95 Advances in nonlinear waves. Volume I
L Debnath
96 FC-groups
M J Tomkinson
97 Topics in relaxation and ellipsoidal methods
M Akgül
98 Analogue of the group algebra for topological semigroups
H Dzinotyiweyi
99 Stochastic functional differential equations
S E A Mohammed

100 Optimal control of variational inequalities
V Barbu
101 Partial differential equations and dynamical systems
W E Fitzgibbon III
102 Approximation of Hilbert space operators Volume II
C Apostol, L A Fialkow, D A Herrero and D Voiculescu
103 Nondiscrete induction and iterative processes
V Ptak and F-A Potra
104 Analytic functions – growth aspects
O P Juneja and G P Kapoor
105 Theory of Tikhonov regularization for Fredholm equations of the first kind
C W Groetsch
106 Nonlinear partial differential equations and free boundaries. Volume I
J I Díaz
107 Tight and taut immersions of manifolds
T E Cecil and P J Ryan
108 A layering method for viscous, incompressible L_p flows occupying R^n
A Douglis and E B Fabes
109 Nonlinear partial differential equations and their applications: Collège de France Seminar. Volume VI
H Brezis and J L Lions
110 Finite generalized quadrangles
S E Payne and J A Thas
111 Advances in nonlinear waves. Volume II
L Debnath
112 Topics in several complex variables
E Ramírez de Arellano and D Sundararaman
113 Differential equations, flow invariance and applications
N H Pavel
114 Geometrical combinatorics
F C Holroyd and R J Wilson
115 Generators of strongly continuous semigroups
J A van Casteren
116 Growth of algebras and Gelfand–Kirillov dimension
G R Krause and T H Lenagan
117 Theory of bases and cones
P K Kamthan and M Gupta
118 Linear groups and permutations
A R Camina and E A Whelan
119 General Wiener–Hopf factorization methods
F-O Speck
120 Free boundary problems: applications and theory. Volume III
A Bossavit, A Damlamian and M Fremond
121 Free boundary problems: applications and theory. Volume IV
A Bossavit, A Damlamian and M Fremond
122 Nonlinear partial differential equations and their applications: Collège de France Seminar. Volume VII
H Brezis and J L Lions
123 Geometric methods in operator algebras
H Araki and E G Effros
124 Infinite dimensional analysis–stochastic processes
S Albeverio
125 Ennio de Giorgi Colloquium
P Krée
126 Almost-periodic functions in abstract spaces
S Zaidman
127 Nonlinear variational problems
A Marino, L Modica, S Spagnolo and M Degliovanni
128 Second-order systems of partial differential equations in the plane
L K Hua, W Lin and C-Q Wu
129 Asymptotics of high-order ordinary differential equations
R B Paris and A D Wood
130 Stochastic differential equations
R Wu
131 Differential geometry
L A Cordero
132 Nonlinear differential equations
J K Hale and P Martinez-Amores
133 Approximation theory and applications
S P Singh
134 Near-rings and their links with groups
J D P Meldrum
135 Estimating eigenvalues with *a posteriori/a priori* inequalities
J R Kuttler and V G Sigillito
136 Regular semigroups as extensions
F J Pastijn and M Petrich
137 Representations of rank one Lie groups
D H Collingwood
138 Fractional calculus
G F Roach and A C McBride
139 Hamilton's principle in continuum mechanics
A Bedford
140 Numerical analysis
D F Griffiths and G A Watson
141 Semigroups, theory and applications. Volume I
H Brezis, M G Crandall and F Kappel
142 Distribution theorems of L-functions
D Joyner
143 Recent developments in structured continua
D De Kee and P Kaloni
144 Functional analysis and two-point differential operators
J Locker
145 Numerical methods for partial differential equations
S I Hariharan and T H Moulden
146 Completely bounded maps and dilations
V I Paulsen
147 Harmonic analysis on the Heisenberg nilpotent Lie group
W Schempp
148 Contributions to modern calculus of variations
L Cesari
149 Nonlinear parabolic equations: qualitative properties of solutions
L Boccardo and A Tesei
150 From local times to global geometry, control and physics
K D Elworthy

151 A stochastic maximum principle for optimal control of diffusions
U G Haussmann
152 Semigroups, theory and applications. Volume II
H Brezis, M G Crandall and F Kappel
153 A general theory of integration in function spaces
P Muldowney
154 Oakland Conference on partial differential equations and applied mathematics
L R Bragg and J W Dettman
155 Contributions to nonlinear partial differential equations. Volume II
J I Díaz and P L Lions
156 Semigroups of linear operators: an introduction
A C McBride
157 Ordinary and partial differential equations
B D Sleeman and R J Jarvis
158 Hyperbolic equations
F Colombini and M K V Murthy
159 Linear topologies on a ring: an overview
J S Golan
160 Dynamical systems and bifurcation theory
M I Camacho, M J Pacifico and F Takens
161 Branched coverings and algebraic functions
M Namba
162 Perturbation bounds for matrix eigenvalues
R Bhatia
163 Defect minimization in operator equations: theory and applications
R Reemtsen
164 Multidimensional Brownian excursions and potential theory
K Burdzy
165 Viscosity solutions and optimal control
R J Elliott
166 Nonlinear partial differential equations and their applications: Collège de France Seminar. Volume VIII
H Brezis and J L Lions
167 Theory and applications of inverse problems
H Haario
168 Energy stability and convection
G P Galdi and B Straughan
169 Additive groups of rings. Volume II
S Feigelstock
170 Numerical analysis 1987
D F Griffiths and G A Watson
171 Surveys of some recent results in operator theory. Volume I
J B Conway and B B Morrel
172 Amenable Banach algebras
J-P Pier
173 Pseudo-orbits of contact forms
A Bahri
174 Poisson algebras and Poisson manifolds
K H Bhaskara and K Viswanath
175 Maximum principles and eigenvalue problems in partial differential equations
P W Schaefer
176 Mathematical analysis of nonlinear, dynamic processes
K U Grusa
177 Cordes' two-parameter spectral representation theory
D F McGhee and R H Picard
178 Equivariant K-theory for proper actions
N C Phillips
179 Elliptic operators, topology and asymptotic methods
J Roe
180 Nonlinear evolution equations
J K Engelbrecht, V E Fridman and E N Pelinovski
181 Nonlinear partial differential equations and their applications: Collège de France Seminar. Volume IX
H Brezis and J L Lions
182 Critical points at infinity in some variational problems
A Bahri
183 Recent developments in hyperbolic equations
L Cattabriga, F Colombini, M K V Murthy and S Spagnolo
184 Optimization and identification of systems governed by evolution equations on Banach space
N U Ahmed
185 Free boundary problems: theory and applications. Volume I
K H Hoffmann and J Sprekels
186 Free boundary problems: theory and applications. Volume II
K H Hoffmann and J Sprekels
187 An introduction to intersection homology theory
F Kirwan
188 Derivatives, nuclei and dimensions on the frame of torsion theories
J S Golan and H Simmons
189 Theory of reproducing kernels and its applications
S Saitoh
190 Volterra integrodifferential equations in Banach spaces and applications
G Da Prato and M Iannelli
191 Nest algebras
K R Davidson
192 Surveys of some recent results in operator theory. Volume II
J B Conway and B B Morrel
193 Nonlinear variational problems. Volume II
A Marino and M K V Murthy
194 Stochastic processes with multidimensional parameter
M E Dozzi
195 Prestressed bodies
D Iesan
196 Hilbert space approach to some classical transforms
R H Picard
197 Stochastic calculus in application
J R Norris
198 Radical theory
B J Gardner
199 The C^*-algebras of a class of solvable Lie groups
X Wang
200 Stochastic analysis, path integration and dynamics
K D Elworthy and J C Zambrini

201 Riemannian geometry and holonomy groups
S Salamon
202 Strong asymptotics for extremal errors and polynomials associated with Erdös type weights
D S Lubinsky
203 Optimal control of diffusion processes
V S Borkar
204 Rings, modules and radicals
B J Gardner
205 Two-parameter eigenvalue problems in ordinary differential equations
M Faierman
206 Distributions and analytic functions
R D Carmichael and D Mitrovic
207 Semicontinuity, relaxation and integral representation in the calculus of variations
G Buttazzo
208 Recent advances in nonlinear elliptic and parabolic problems
P Bénilan, M Chipot, L Evans and M Pierre
209 Model completions, ring representations and the topology of the Pierce sheaf
A Carson
210 Retarded dynamical systems
G Stepan
211 Function spaces, differential operators and nonlinear analysis
L Paivarinta
212 Analytic function theory of one complex variable
C C Yang, Y Komatu and K Niino
213 Elements of stability of visco-elastic fluids
J Dunwoody
214 Jordan decomposition of generalized vector measures
K D Schmidt
215 A mathematical analysis of bending of plates with transverse shear deformation
C Constanda
216 Ordinary and partial differential equations. Volume II
B D Sleeman and R J Jarvis
217 Hilbert modules over function algebras
R G Douglas and V I Paulsen
218 Graph colourings
R Wilson and R Nelson
219 Hardy-type inequalities
A Kufner and B Opic
220 Nonlinear partial differential equations and their applications: Collège de France Seminar. Volume X
H Brezis and J L Lions
221 Workshop on dynamical systems
E Shiels and Z Coelho
222 Geometry and analysis in nonlinear dynamics
H W Broer and F Takens
223 Fluid dynamical aspects of combustion theory
M Onofri and A Tesei
224 Approximation of Hilbert space operators. Volume I. 2nd edition
D Herrero
225 Operator theory: proceedings of the 1988 GPOTS–Wabash conference
J B Conway and B B Morrel
226 Local cohomology and localization
J L Bueso Montero, B Torrecillas Jover and A Verschoren
227 Nonlinear waves and dissipative effects
D Fusco and A Jeffrey
228 Numerical analysis 1989
D F Griffiths and G A Watson
229 Recent developments in structured continua. Volume III
D De Kee and P Kaloni
230 Boolean methods in interpolation and approximation
F J Delvos and W Schempp
231 Further advances in twistor theory. Volume I
L J Mason and L P Hughston
232 Further advances in twistor theory. Volume II
L J Mason and L P Hughston
233 Geometry in the neighborhood of invariant manifolds of maps and flows and linearization
U Kirchgraber and K Palmer
234 Quantales and their applications
K I Rosenthal
235 Integral equations and inverse problems
V Petkov and R Lazarov
236 Pseudo-differential operators
S R Simanca
237 A functional analytic approach to statistical experiments
I M Bomze
238 Quantum mechanics, algebras and distributions
D Dubin and M Hennings
239 Hamilton flows and evolution semigroups
J Gzyl
240 Topics in controlled Markov chains
V S Borkar
241 Invariant manifold theory for hydrodynamic transition
S Sritharan
242 Lectures on the spectrum of $L^2(\Gamma\backslash G)$
F L Williams
243 Progress in variational methods in Hamiltonian systems and elliptic equations
M Girardi, M Matzeu and F Pacella
244 Optimization and nonlinear analysis
A Ioffe, M Marcus and S Reich
245 Inverse problems and imaging
G F Roach
246 Semigroup theory with applications to systems and control
N U Ahmed
247 Periodic-parabolic boundary value problems and positivity
P Hess
248 Distributions and pseudo-differential operators
S Zaidman
249 Progress in partial differential equations: the Metz surveys
M Chipot and J Saint Jean Paulin
250 Differential equations and control theory
V Barbu

251 Stability of stochastic differential equations with respect to semimartingales
X Mao
252 Fixed point theory and applications
J Baillon and M Théra
253 Nonlinear hyperbolic equations and field theory
M K V Murthy and S Spagnolo
254 Ordinary and partial differential equations. Volume III
B D Sleeman and R J Jarvis
255 Harmonic maps into homogeneous spaces
M Black
256 Boundary value and initial value problems in complex analysis: studies in complex analysis and its applications to PDEs 1
R Kühnau and W Tutschke
257 Geometric function theory and applications of complex analysis in mechanics: studies in complex analysis and its applications to PDEs 2
R Kühnau and W Tutschke
258 The development of statistics: recent contributions from China
X R Chen, K T Fang and C C Yang
259 Multiplication of distributions and applications to partial differential equations
M Oberguggenberger
260 Numerical analysis 1991
D F Griffiths and G A Watson
261 Schur's algorithm and several applications
M Bakonyi and T Constantinescu
262 Partial differential equations with complex analysis
H Begehr and A Jeffrey
263 Partial differential equations with real analysis
H Begehr and A Jeffrey
264 Solvability and bifurcations of nonlinear equations
P Drábek
265 Orientational averaging in mechanics of solids
A Lagzdins, V Tamuzs, G Teters and A Kregers
266 Progress in partial differential equations: elliptic and parabolic problems
C Bandle, J Bemelmans, M Chipot, M Grüter and J Saint Jean Paulin
267 Progress in partial differential equations: calculus of variations, applications
C Bandle, J Bemelmans, M Chipot, M Grüter and J Saint Jean Paulin
268 Stochastic partial differential equations and applications
G Da Prato and L Tubaro
269 Partial differential equations and related subjects
M Miranda
270 Operator algebras and topology
W B Arveson, A S Mishchenko, M Putinar, M A Rieffel and S Stratila
271 Operator algebras and operator theory
W B Arveson, A S Mishchenko, M Putinar, M A Rieffel and S Stratila
272 Ordinary and delay differential equations
J Wiener and J K Hale
273 Partial differential equations
J Wiener and J K Hale
274 Mathematical topics in fluid mechanics
J F Rodrigues and A Sequeira
275 Green functions for second order parabolic integro-differential problems
M G Garroni and J F Menaldi
276 Riemann waves and their applications
M W Kalinowski
277 Banach C(K)-modules and operators preserving disjointness
Y A Abramovich, E L Arenson and A K Kitover
278 Limit algebras: an introduction to subalgebras of C*-algebras
S C Power
279 Abstract evolution equations, periodic problems and applications
D Daners and P Koch Medina
280 Emerging applications in free boundary problems
J Chadam and H Rasmussen
281 Free boundary problems involving solids
J Chadam and H Rasmussen
282 Free boundary problems in fluid flow with applications
J Chadam and H Rasmussen
283 Asymptotic problems in probability theory: stochastic models and diffusions on fractals
K D Elworthy and N Ikeda
284 Asymptotic problems in probability theory: Wiener functionals and asymptotics
K D Elworthy and N Ikeda
285 Dynamical systems
R Bamon, R Labarca, J Lewowicz and J Palis

George A Anastassiou

Memphis State University, USA

Moments in probability and approximation theory

Copublished in the United States with
John Wiley & Sons, Inc., New York

Longman Scientific & Technical
Longman Group UK Limited
Longman House, Burnt Mill, Harlow
Essex CM20 2JE, England
and Associated companies throughout the world.

Copublished in the United States with
John Wiley & Sons Inc., 605 Third Avenue, New York, NY 10158

First published 1993

AMS Subject Classification: 41, 60

ISSN 0269-3674

ISBN 0 582 22770 4

British Library Cataloguing in Publication Data

A catalogue record for this book is
available from the British Library

Library of Congress Cataloging-in-Publication Data

A catalog record for this book is available

Printed and bound in Great Britain
by Biddles Ltd, Guildford and King's Lynn

To

Angela

TABLE OF CONTENTS

PAGE

CHAPTER ONE

A Preview 1

I) On Chapter 2 2

II) On Chapter 3 4

III) On Chapter 4 6

IV) On Chapter 5 11

V) On Chapter 6 12

VI) On Chapter 7 14

VII) On Chapter 8 19

VIII) On Chapter 9 24

IX) On Chapter 10 26

X) On Chapter 11 29

XI) On Chapter 12 31

XII) On Chapter 13 33

XIII) On Chapter 14 35

XIV) On Chapter 15 36

CHAPTER TWO

Geometric Moment Theory

2.1) Methods of Optimal Distance and Optimal Ratio 38

2.2) Convex Moment Methods 59

CHAPTER THREE

Moment Problems Of Kantorovich Type And Kantorovich Radius

3.1) Moment Problems of Kantorovich Type 68

3.2) Kantorovich Radius 73

CHAPTER FOUR

Moment Problems Related to c–Rounding Proportions

4.1) Moment Problems Related to c–Rounding Proportions Subject to One Moment Condition 80

4.2) Moment Problems Related to c–Rounding Proportions Subject to Two Moment Conditions 90

4.3) Moment Problems Related to Jefferson-Rounding Proportions Subject to Two Moment Conditions 100

4.4) Moment Problems Related to Adams Rule of Rounding Subject to Two Moment Conditions 107

4.5) Moment Problems Related to Jefferson and Adams rules of Rounding Subject to One Moment Condition 112

CHAPTER FIVE

The Levy Radius

5.1) The Levy Radius of a Set of Probability Measures Satisfying Moment Conditions Involving $\{t, t^2\}$ 123

5.2) The Levy Radius of a Set of Probability Measures Satisfying Two Moment Conditions Involving a Tchebycheff System 129

CHAPTER SIX

The Prokhorov Radius

6.1) The Prokhorov Radius of a Set of Probability Measures Satisfying Moment Conditions involving $\{t, t^2\}$ 150

6.2) The Trigonometric Prokhorov Radius 170

PAGE

CHAPTER SEVEN

Probability Measures, Positive Linear Operators and Korovkin Type Inequalities

7.1) Introduction 199

7.2) Optimal Korovkin Type Inequalities 212

7.3) Nearly Optimal Korovkin Inequalities 223

7.4) Multivariate Korovkin Type Inequalities 234

CHAPTER EIGHT

Optimal Korovkin Type Inequalities Under Convexity

8.1) On the Degree of Weak Convergence of a Sequence of Finite Measures to the Unit Measure Under Convexity 242

8.2) On the Rate of Weak Convergence of Convex Type Finite Measures to the Unit Measure 259

8.3) On the Smooth Rate of Weak Convergence of Convex Type Finite Measures to the Unit Measure 278

CHAPTER NINE

Optimal Korovkin Type Inequalities for Convolution Type Operators

9.1) Sharp Inequalities for Convolution Operators 296

9.2) Sharp Inequalities for Non-positive Generalized Convolution Operators 303

CHAPTER TEN

10.1) Optimal Korovkin Type Inequalities for Positive Linear Stochastic Operators ...313

PAGE

CHAPTER ELEVEN

11.1) Optimal Korovkin Type Inequalities for Positive Linear Operators Using an Extended Complete Tchebycheff System 332

CHAPTER TWELVE

12.1) A General "K-Attained" Inequality Related to the Weak Convergence of Probability Measures to the Unit Measure 355

CHAPTER THIRTEEN

13.1) A General Stochastic Inequality Involving Basic Moments 370

CHAPTER FOURTEEN

14.1) Miscellaneous Sharp Inequalities and Korovkin-type Convergence Theorems Involving Sequences of Basic Moments 379

CHAPTER FIFTEEN

15.1) A Discrete Stochastic Korovkin Type Convergence Theorem 386

Index 392

List of Symbols 393

References 396

Preface

The trend of using Probability methods in other areas of Mathematics in recent years has become larger and larger. This approach has the advantage of producing short, elegant proofs and going deeper into the subject under investigation. To mention some successful examples, see Probabilistic Number Theory and the blends of Probability with Functional Analysis, Harmonic Analysis and Graph Theory. This book is a research monograph giving an account to *author's work of the last eleven years* in the field of applications of Geometric Moment Theory to Probability (Moment Theory) and especially Approximation Theory.

More precisely, here we apply Kemperman's Geometric approach to Moment Theory to study the degree of approximation and rate of convergence of sequences of Positive Linear Operators (pointwise convergence to the Unit operator) and Probability (finite) measures (weak convergence to the Unit measure). These are motivated by the famous Korovkin convergence theorems. Our approach is quantitative, most of the produced inequalities are attained i.e. they are sharp.

Using the same methods we solve several moment problems which have applications to the Mass–transportation Problem, Political Science and Queueing Theory. This monograph is the first of this kind and its main characteristic is that the established results by the Geometric Moment methods are optimal. This book can be used for doing further research, but also in seminars of Probability, Approximation Theory, Numerical Analysis and Statistics. The author wants to thank J. H. B. Kemperman (Rutgers University) for introducing him into the subject and for having interesting and profitable discussions. Also wants to thank O. Shisha (U.R.I) and S. T. Rachev (U.C.S.B) for having interesting discussions on related matters. This eleven year research body was conducted at University of Rochester (1981–84), U. R. I. (1984–86) and Memphis State University (1986–1992).

The author also wants to thank Shannon Diamond and Kate MacDougall for a heroic typing job.

Memphis
April 8, 1992
George A. Anastassiou

CHAPTER ONE

A PREVIEW

This monograph among others studies in great detail the overlaps of Geometric Moment Theory and Approximation Theory and also solves basic important Moment Problems. In particular studies the wide applications of the methods of Optimal Distance and Optimal Ratio from the Geometric Moment Theory of Kemperman (1968), (1971), see Chapter 2. These are into the theory of pointwise convergence of Positive Linear Operators to the Unit Operator and into the theory of weak convergence of Probability (finite) measures to the unit (Dirac) measure (Approximation Theory): a quantitative approach, where we establish best upper bounds leading to sharp/attained inequalities, giving us the optimal rate of convergence of above entities. The last are contained in Chapters 5–10. See also in the References several related Anastassiou publications.

In Chapters 3, 4 we solve some very important moment problems of Kantorovich type and related to c–Rounding Proportions by the use of the method of Optimal Distance. These have tremendous applications to the Theory of Mass-transportation when we have only partial knowledge of the marginals (see also the forthcoming related book of S. T. Rachev: Mass-transportation Problems, Part 3). Also §6.2 has applications in Queueing Theory.

In Chapters 11–15 are involved again basic moments of probability measures and have to do greatly with the convergences of Positive Linear Operators and Probability Measures in other related ways. The field of activity, within Approximation Theory, of Geometric Moment Theory has been greatly motivated and at large determined by the famous Korovkin convergence theorems. This Monograph is unique of its kind, as it is easily seen from the above and especially what follows in detail. It is based mainly on the related publications and some unpublished material of the author. If not much related material is included from other authors papers and books, it is because to the best of author's knowledge such material does not exist in the spirit of applications of Geometric Moment Theory.

The main purpose of this Monograph is to initiate and encourage further research in applications of Moment Theory into Approximation Theory in an optimal–dynamic way where further improvement is hardly expected! This monograph also contains a lot of applications and examples related to the presented results.

More specifically, next we present very briefly parts of the contents of all the chapters to follow. For the convenience of the reader sometimes this exposition is presented in a simplified form.

I. ON CHAPTER 2: ABOUT GEOMETRIC MOMENT THEORY.

(§1) Main Problem.

Let $g_1, \ldots, g_n$ and h be given real-valued Borel measurable functions on a fixed measurable space $X := (X, A)$. We would like to find the best upper and lower bounds on

$$\mu(h) := \int_X h(t)\mu(dt),$$

given that μ is a probability measure on X with prescribed moments

$$\mu(g_i) := \int g_i(t)\mu(dt) = y_i, i = 1, \ldots, n. \tag{1}$$

Here we assume μ is such that

$$\int_X |g_i|\mu(dt) < +\infty, i = 1, \ldots, n$$

and

$$\int_X |h|\mu(dt) < +\infty.$$

For $y := (y_1, \ldots, y_n) \in \mathbb{R}^n$ as in (1), we consider the optimal quantities

$$L(y) := L(y|h) := \inf_{\mu} \mu(h),$$

$$U(y) := U(y|h) := \sup_{\mu} \mu(h).$$

If $h := \mathcal{X}_S$ is the characteristic function of a given measurable set S of X, then we write

$$L(y|\mathcal{X}_S) := L_S(y), U(y|\mathcal{X}_S) := U_S(y).$$

Consider $g : X \longrightarrow \mathbb{R}^n$ such that $g(t) := (g_1(t), \ldots, g_n(t))$.

The Method Of Optimal Distance

Call

$$M := \operatorname*{conv}_{t \in X} (g_1(t), \ldots, g_n(t), h(t)).$$

Then $L(y|h)$ is equal to the *smallest* distance between $(y_1, \ldots, y_n, 0)$ and $(y_1, \ldots, y_n, z) \in \bar{M}$. Also $U(y|h)$ is equal to the *largest* distance between $(y_1, \ldots, y_n, 0)$ and $(y_1, \ldots, y_n, z) \in \bar{M}$.

The Method Of Optimal Ratio

We would like to find

$$L_S(y) := \inf \mu(S)$$

and

$$U_S(y) := \sup \mu(S),$$

over all probability measures μ such that

$$\mu(g_i) = y_i, i = 1, \ldots, n.$$

Set $S' := X - S$. Call $W_S := \overline{convg(S)}$, $W_{S'} := \overline{convg(S')}$ and $W := \overline{convg(X)}$.

Finding $L_S(y)$

1) Pick a boundary point z of W and *"draw"* through z a hyperplane H of support to W.

2) Determine the hyperplane H' parallel to H which supports $W_{S'}$ as well as possible, and on the same side as H supports W.

3) Denote

$$A_d := W \cap H = W_S \cap H \quad \text{and}$$

$$B_d := W_{S'} \cap H'.$$

Given that $H' \neq H$, set $G_d := \overline{\text{conv}}(A_d \cup B_d)$. Then we have that

$$L_S(y) = \frac{\Delta(y)}{\Delta},$$

for each $y \in \text{int}(V)$ such that $y \in G_d$. Here, $\Delta(y)$ is the distance from y to H' and Δ is the distance between the distinct parallel hyperplanes H, H'.

In an analogous way we find $U_S(y)$.

(§2) The Convex Moment Problem

Definition 1. *Let $s \geq 1$ be a fixed natural number and let $x_0 \in \mathbb{R}$ be fixed. By $m_s(x_o)$ we denote the set of probability measures μ on $\mathbb{R}$ such that the associated cumulative distribution function F possesses an $(s-1)-th$ derivative $F^{(s-1)}(x)$ over $(x_0, +\infty)$ and furthermore $(-1)^s F^{(s-1)}(x)$ is convex in $(x_0, +\infty)$.*

The Problem

Let $g_i, i = 1, \ldots, n; h$ are Borel measurable functions from $\mathbb{R}$ into itself. These are assumed to be locally integrable on $[x_0, +\infty)$ relative to Lebesque measure. Consider $\mu \in m_s(x_o), s \geq 1$ such that

$$\mu(|g_i|) := \int_{\mathbb{R}} |g_i(t)|\mu(dt) < +\infty, i = 1, \ldots, n$$

and

$$\mu(|h|) := \int_{\mathbb{R}} |h(t)|\mu(dt) < +\infty.$$

Let $c := (c_1, \ldots, c_n) \in \mathbb{R}^n$ be such that

$$\mu(g_i) = c_i, i = 1, \ldots, n, \mu \in m_s(x_o).$$

We would like to find

$$L(c) := \inf_{\mu} \mu(h)$$

and

$$U(c) := \sup_{\mu} \mu(h),$$

where μ is as above described.

Here, the method will be to transform the above convex moment problem into an ordinary one usually handled as in (§1).

II. ON CHAPTER 3: KANTOROVICH TYPE MOMENT PROBLEMS AND RADIUS

(§3) We face and solve the following problem.

Let $S := [a, b] \times [c, d] \subset \mathbb{R}^2$ and $\varphi(x, y) = |x - y|^p, p \geq 1$. Suppose $(\alpha, \beta) \in S$, then find

$$U := U(\varphi, \alpha, \beta) := \sup_{\mu} \int_S |x - y|^p \mu(dx, dy)$$

subject to

$$\int_S x\mu(dx, dy) = \alpha, \int_S y\mu(dx, dy) = \beta,$$

where μ is a probability measure on S. If we change the domain S the results change greatly:

In the following theorem we consider stripes in $\mathbb{R}^2$: for $b, \gamma > 0$,

$$S_1^b := \{(x, y) : y = x + b', \quad \text{where } 0 \leq b' \leq b\},$$

$$S_2^\gamma := \{(x, y) : y = x - \gamma', \quad \text{where } 0 \leq \gamma' \leq \gamma\},$$

$$S^{b,\gamma} := S_1^b \cup S_2^\gamma.$$

Theorem 1. *Assume again that $p \geq 1$.*

(i) If $S = S_1^b, (\alpha,\beta) \in S$, then

$$U := U(\varphi,\alpha,\beta) = b^{p-1}(\beta-\alpha),$$

where $\varphi(x,y) := |x-y|^p$ and

$$U := \sup_\mu \int_S |x-y|^p \mu(dx,dy)$$

such that

$$\int_S x\mu(dx,dy = \alpha, \int_S y\mu(dx,dy) = \beta;$$

μ is a probability measure on S.

(ii) If $S = S_2^\gamma, (\alpha,\beta) \in S$, then

$$U := U(\varphi,\alpha,\beta) = \gamma^{p-1}(\alpha-\beta).$$

(iii) If $S = S^{b,\gamma}, (\alpha,\beta) \in S$, then

$$U = \frac{(b^p-\gamma^p)(\beta-\alpha-b) + b^p(b+\gamma)}{b+\gamma}$$

When $b = \gamma$ in the above stripes i.e. they are symmetric we solve the same problem for $0 < p \leq 1$, there we also find

$$L := \inf_\mu \int_S |x-y|^p \mu(dx,dy).$$

(§4) The Kantorovich Radius

For given $x_0 \in \mathbb{R}, \alpha > 0, p \in \mathbb{R}, q > 0 (p^2 \leq q), -\infty \leq a < b \leq +\infty$ we find the *Kantorovich radius*

$$K := K(x_0,\alpha,p,q,a,b) := \sup\{E|X-x_0|^\alpha : X \in [a,b] a.s., EX = p, EX^2 = q\},$$

where X stands for a random variable and E for the expectation operator.

We meet various cases:

Theorem 2. (Case (A): $\alpha \geq 2, -\infty < a < b < +\infty$)

Suppose X's as above take values on $[a,b]$. Let $x_0 := (a+b)/2, a \leq p \leq b$ and $0 \leq q \leq b^2 + (a+b)(p-b)$. Then

$$K \leq (\frac{b-a}{2})^{\alpha-2} \cdot [q - p(a+b) + \frac{(a+b)^2}{4}].$$

Moreover, if there exist $\lambda_1, \lambda_2 \geq 0$, $\lambda_1 + \lambda_2 \leq 1$ such that

$$p = (\frac{a+b}{2}) + (\frac{b-a}{2})(\lambda_1 - \lambda_2)$$

and

$$q = \frac{(a+b)^2}{4} + (\frac{b^2-a^2}{2})(\lambda_1 - \lambda_2) + \frac{(b-a)^2}{4}(\lambda_1 + \lambda_2),$$

then

$$K = \left(\frac{b-a}{2}\right)^{\alpha-2} \cdot [q - p(a+b) + \frac{(a+b)^2}{4}].$$

Another important case is

Theorem 3. (Case (B): $0 < \alpha \leq 2, a = -\infty, b = +\infty$).
For any $x_0 \in \mathbb{R}, p \in \mathbb{R}, q > 0, p^2 \leq q$, the Kantorovich radius K takes the following value

$$K = K(x_0; \alpha; p; q) = (q - 2x_0 p + x_0^2)^{\alpha/2}.$$

E.t.c. we meet almost all other cases of K.

III. ON CHAPTER 4: MOMENT PROBLEMS RELATED TO c–ROUNDING PROPORTIONS

(§5) Solutions Subject to One Moment Condition

Definition 2. Let $0 \leq c \leq 1$ fixed: for any $x \geq 0$,

$$[x]_c := \begin{cases} m, & \text{if } m \leq x < m + c \\ m+1, & \text{if } m + c \leq x < m + 1, \end{cases}$$

where $m \in \mathbb{N} \cup \{0\}$.

Related Moment Problem

Find

$$\sup_{\mu} \int_A [t]_c \mu(dt),$$

where μ is a probability measure on $A := [0, a]$ or $[0, \infty)$ and

$$\inf_{\mu} \int_A [t]_c \mu(dt),$$

subject to the given moment condition

$$\int_A t^r \mu(dt) = d_r, r > 0, d_r > 0.$$

Here X stands for a random variable. We obtain the following results.

Theorem 4. *Let* $0 < c < 1, r > 0, d > 0$ *and*

$$U := U_{[\cdot]_c}(r,d) := \sup\{E[X]_c : X \geq 0 \; a.s.(EX^r)^{1/r} = d\}.$$

(I) If $0 < r < 1$, then $U = +\infty$.

(II) If $r \geq 1$ and $0 \leq d \leq c$, then $U = \frac{d^r}{c^r}$.

(III) Suppose $r \geq 1$. Define $k \in \mathbb{N}$ by $k - 1 + c \leq d < k + c$. Then

$$U = k + \frac{d^r - (k-1+c)^r}{(k+c)^r - (k-1+c)^r} \leq 1 - c + d.$$

Its counterpart has as follows:

Theorem 5. *Let* $c \in (0,1), r > 0, d > 0$ *and*

$$L = L_{[\cdot]_c}(r,d) := \inf\{E[X]_c : X \geq 0 \; a.s., \; (EX^r)^{1/r} = d\}.$$

(I) If $r > 1$, then $L = 0$.

(II) If $0 < r \leq 1, 0 < d \leq c$, then $L = 0$.

(III) If $r = 1, c \leq d < +\infty$, then $L = d - c$.

(IV) If $0 < r \leq 1$ define $k \in \mathbb{N} \cup \{0\}$ by $k + c \leq d < k + 1 + c$. Then

$$L = k + \frac{d^r - (k+c)^r}{(k+1+c)^r - (k+c)^r}.$$

(§6) Solutions subject to Two Moment Conditions

Related Moment Problem

Find

$$\sup_\mu \int_A [t]_c \mu(dt),$$

where μ is a probability measure on $A := [0,a]$ or $[0,+\infty)$ and

$$\inf_\mu \int_A [t]_c \mu(dt),$$

subject to the given two moment conditions

$$\int_A t\mu(dt) = d_1, d_1 > 0$$

and

$$\int_A t^r \mu(dt) = d_r, r > 0, d_r > 0.$$

Theorem 6. *Let* $0 < c < 1, 0 < r \neq 1, d_1 > 0, d_r > 0$ *and*

$$U = U_{[\cdot]_c}(r, d_1, d_r) := \sup\{E[X]_c : X \geq 0 \; a.s., \; EX = d_1, EX^r = d_r\}.$$

(I) Suppose $r > 1$.

(i) Let $0 < d_1 \leq c$ and $c^r \leq d_r < +\infty$. Then $U = d_1/c$.

(ii) Suppose that there exist $\lambda_1, \lambda_2 \geq 0$ with $\lambda_1 + \lambda_2 \leq 1$ and such that $d_1 = \lambda_1 c, d_r = (\lambda_1 + \lambda_2)c^r$. Then $U = d_1/c$.

(iii) Suppose there exists integer $k \geq 0$ such that $k + c \leq d_1 < k + 1 + c$ and $(k+c)^r + [(k+1+c)^r - (k+c)^r](d_1 - k - c) \leq d_r < +\infty$. Then $U = d_1 + 1 - c$.

(II) Suppose $0 < r < 1$.

(i) Let $c \leq d_1 < +\infty, 0 < d_r \leq c^r$. Then $U = d_r c^{-r}$.

(ii) Suppose there exist $\lambda_1 \geq 0, \lambda_2 \geq 0$, such that $\lambda_1 + \lambda_2 \leq 1$ and $d_1 = (\lambda_1 + \lambda_2)c, d_r = \lambda_2 c^r$. Then $U = d_r c^{-r}$.

(iii) Suppose that there exists an integer $k \geq 0$ with $k + c \leq d_1 < k + 1 + c$ and such that $c^r < d_r \leq (k+c)^r + [(k+1+c)^r - (k+c)^r](d_1 - k - c)$. Then $U = d_1 + 1 - c$.

And its counterpart has as follows:

Theorem 7. *Let* $0 < c < 1, d_1 > 0, d_r > 0, 0 < r \neq 1$. *Denote*

$$L := L_{[\cdot]_c}(r, d_1, d_r) := \inf\{E[X]_c : X \geq 0 \; a.s., \; EX = d_1, EX^r = d_r\}.$$

(I) Suppose one of the following three conditions holds:

(i) there exists t_1, t_2, λ such that $0 \leq t_1 \leq t_2 \leq c, 0 \leq \lambda \leq 1$, such that

$$d_1 = (1-\lambda)t_1 + \lambda t_2, d_r = (1-\lambda)t_1^r + \lambda t_2^r;$$

(ii) $r > 1$ and there exist $\lambda_1, \lambda_2 \geq 0, \lambda_1 + \lambda_2 \leq 1$ such that $d_1 = \lambda_1 c, d_r = (\lambda_1 + \lambda_2)c^r$;

(iii) $r > 1$ and $0 < d_1 \leq c$ and $d_r \geq c^r$.

Then $L = 0$.

(II) Suppose there exists integer k such that $k + c \leq d_1 < k + 1 + c$ and either when $r > 1$ we have $d_r \geq d_{r,k} := (k+c)^r + ((k+1+c)^r - (k+c)^r)(d_1 - k - c)$; or when $0 < r < 1$ we have $d_{r,k} \geq d_r \geq c^r$. Then $L = d_1 - c$.

(III) Let $0 < r < 1$, and either there exist $\lambda_1 \geq 0, \lambda_2 \geq 0, \lambda_1 + \lambda_2 \leq 1$ such that $d_1 = (\lambda_1 + \lambda_2)c, d_r = \lambda_2 c^r$, or $d_1 \geq c$ and $0 < d_r \leq c^r$. Then $L = (c^{r-1}d_1 - d_r) \cdot c^{1-r}$.

(§7) Moment Problems Related to Jefferson-Rounding Proportions Subject to Two Moment Conditions.

In the case of conventional (Jefferson) rounding or MYZ–rounding $[x] = [x]_1$, $x \geq 0$; the integral part of x, we solve the following:

Moment Problem

Find

$$\sup_{\mu} \int_A [t]\mu(dt),$$

where μ is a probability measure on $A := [0, a]$ or $[0, +\infty)$ and

$$\inf_{\mu} \int_A [t]\mu(dt),$$

subject to the given two moment conditions

$$\int_A t\mu(dt) = d_1, d_1 > 0$$

and

$$\int_A t^r \mu(dt) = d_r, r > 0, d_r > 0. \tag{2}$$

The same problem is solved completely when is subject only to second moment condition in (2).

(§8) Moment Problems Related to Adams Rule of Rounding Subject to Two Moment Conditions.

In the case of Adams rule of rounding we have $[x]_0 = \lceil x \rceil; x \geq 0 (c = 0)$, where $\lceil \cdot \rceil$ is the *ceiling* of the number.

We Solve the following

Moment Problem

Find

$$\sup_{\mu} \int_A \lceil t \rceil \mu(dt)$$

where μ is a probability measure on $A = [0, a]$ or $[0, +\infty)$ and

$$\inf_{\mu} \int_A \lceil t \rceil \mu(dt)$$

subject to the given two moment conditions

$$\int_A t\mu(dt) = d_1, d_1 > 0,$$

and

$$\int_A t^r \mu(dt) = d_r, r > 0, d_r > 0. \tag{2}$$

The same problem is solved completely when is subject only to the second moment condition in (2).
We have
Theorem 8. *Let* $0 < r \neq 1, d_1, d_r > 0$. *Denote*

$$U := U_{\lceil\cdot\rceil}(r, d_1, d_r) = \sup\{E\lceil X\rceil : X \geq 0 \ a.s., \ EX = d_1, EX^r = d_r\}.$$

Denote $k := [d_1]$ *and*

$$\Delta_r = k^r + [(k+1)^r - k^r] \cdot (d_1 - k).$$

Suppose that either
(I) $r > 1, d_r \geq \Delta_r$, or
(II) $0 < r < 1, d_r \leq \Delta_r$.
Then $U = d_1 + 1$.

A counterpart has as follows:

Theorem 9. *Let* $0 < r \neq 1, d_1, d_r > 0, a > 0$. *Denote*

$$L := L_{\lceil\cdot\rceil}(a, r, d_1, d_r) = \inf\{E\lceil X\rceil : 0 \leq X \leq a \ a.s. \ \text{ and } \ EX = d_1, EX^r = d_r\}.$$

Call $\theta := \lceil a\rceil$, *the ceiling of* a.
(I) Suppose $0 < a \leq 1$ and $d_1 = \lambda a, d_r = \lambda a^r$ for some $\lambda \in [0,1]$. Then $L = d_1/a$.
(II) Let $a > 1$ and suppose $d_1 = \lambda_1(\theta - 1) + \lambda_2 a, d_r = \lambda_1(\theta-1)^r + \lambda_2 a^r$ for some $\lambda_1, \lambda_2 \geq 0, \lambda_1 + \lambda_2 \leq 1$. Then $L = \frac{(a^r - \theta(\theta-1)^{r-1})d_1 + (\theta-a)d_r}{a^r - a(\theta-a)^{r-1}}$.
(III) Suppose one of the following holds:

(i) $r > 1$ and for some $k \in \{0, 1, \ldots, \theta-2\}, k \leq d_1 < k+1, \Delta_r := k^r + [(k+1)^r - k^r] \cdot (d_1 - k) \leq d_r \leq (\theta-1)^{r-1} d_1$;

(ii) $0 < r < 1$ and for some $k \in \{0, 1, \ldots, \theta-2\}, k \leq d_1 < k+1, (\theta-1)^{r-1} d_1 \leq d_r \leq \Delta_r$.

Then $L = d_1$.

IV. ON CHAPTER 5: THE LEVY RADIUS

(§9) Subject to Moment Conditions Involving $\{t, t^2\}$.

Here we consider probability measures μ on $\mathbb{R}$ such that both $\int |t| d\mu, \int t^2 d\mu$ are finite. We consider

$$M(\epsilon_1, \epsilon_2) := \{\mu : | \int t^j d\mu - s_0^j| \le \epsilon_j, j = 1, 2\},$$

where s_0 is a given point in $\mathbb{R}$, also $0 < \epsilon_j \le 1; j = 1, 2$. We would like to measure the *"size"* of $M(\epsilon_1, \epsilon_2)$. Since weak convergence of probability measures is of great importance and their standard weak topology is well described by the *Levy distance* d_L, it is natural to define the *Levy radius* for $M(\epsilon_1, \epsilon_2)$ as

$$D := \sup_{\mu \in M(\epsilon_1, \epsilon_2)} d_L(\mu, \delta_{s_0}),$$

where δ_{s_0} is the unit measure at s_0.

We prove the following algorithm for the calculation of D.

Theorem 10. *Let $s_0 \ge 0, \epsilon_1, \epsilon_2 > 0$ and $\epsilon_1, \epsilon_2 \le 1$. Let $M(\epsilon_1, \epsilon_2)$ denote the set of all probability measures μ on $\mathbb{R}$ such that*

$$|y_1 - s_0| \le \epsilon_1, \quad |y_2 - s_0^2| \le \epsilon_2,$$

where $y_j := \int t^j \mu(dt), j = 1, 2$.

The quantity $D := \sup_{\mu \in M(\epsilon_1, \epsilon_2)} d_L(\mu, \delta_{s_0})$ is calculated as follows:

Define r_1 as follows: If $s_0 \le (1 - \epsilon_2)/2$. then $r_1 \in (0, 1]$ is the unique root of g in $(0, 1]$, where

$$g(x) := x^3 - 2s_0 x^2 + s_0^2 x - (s_0^2 + \epsilon_2).$$

If $s_0 > (1 - \epsilon_2)/2$, then set $r_1 := 1$.

Case (i). If $\epsilon_1 > s_0 + (s_0^2 + \epsilon_2)^{1/2}$ then $D = r_1$.

Case (ii). If $\epsilon_1 \le s_0 + (s_0^2 + \epsilon_2)^{1/2}$, let $r_2 \in (0, 1]$ be the smallest root of the cubic polynomial

$$F(x) := x^3 - 2\epsilon_1 x^2 + (2\epsilon_1 s_0 + \epsilon_2)x - (2\epsilon_1 s_0 + \epsilon_2 - \epsilon_1^2)$$

satisfying both

$$0 \le x \le r_1 \quad \text{and} \quad x^2 - s_0 x + (s_0 - \epsilon_1) \ge 0.$$

If r_2 exists, then $D = r_2$. If not, then $D = r_1$.

As a special case we have

Corollary 1. *Let $s_0 = 0$ in the above. If $\epsilon_1^{1/2} \geq \epsilon_2^{1/3}$ then $D = \epsilon_2^{1/3}$. If $\epsilon_1^{1/2} < \epsilon_2^{1/3}$ then D equals the unique root of*

$$x^3 - 2\epsilon_1 x^2 + \epsilon_2 x - \epsilon_2 + \epsilon_1^2 = 0$$

in $[\epsilon_1^{1/2}, \epsilon_2^{1/3}]$.

We can generalize, namely Levy Radius:

(§10) Subject to Moment Conditions Involving a Tchebycheff System

Here we consider probability measures μ on $\mathbb{R}$, defined on a $\sigma-$algebra containing all the singletons, such that both $\int |g_1| d\mu, \int |g_2| d\mu$ are finite, where $\{1, g_1, g_2\}$ is a Tchebycheff system of continuous functions from $\mathbb{R}$ into $\mathbb{R}$.

We consider

$$M(\epsilon_1, \epsilon_2) := \{\mu : |\int g_j d\mu - g_j(s_0)| \leq \epsilon_j, j = 1, 2\},$$

where s_0 is a given point in $\mathbb{R}$, also $0 < \epsilon_j \leq 1; j = 1, 2$.

We would like again to measure the *"size"* of $M(\epsilon_1, \epsilon_2)$ with the help of Levy distance d_L.

It is natural to define again the Levy radius for $M(\epsilon_1, \epsilon_2)$ as:

$$D = \sup d_L(\mu, \delta_{s_0}), \mu \in M(\epsilon_1, \epsilon_2).$$

Note that $D = D(\epsilon_1, \epsilon_2)$ and as $\epsilon_1, \epsilon_2 \longrightarrow 0$ we get that $D \longrightarrow 0$, thus $d_L(\mu, \delta_{s_0}) \longrightarrow 0$, consequently, $\mu \underset{\longrightarrow}{\longrightarrow} \delta_{s_0}$ weakly. Given that we know D then we know the rate of weak convergence of μ to δ_{s_0}.

Under mild assumptions on $\{1, g_1, g_2\}$, by the use of geometric moment method of optimal ratio, we find theoretically the magnitude of D and then by the use of Newton-Puiseux diagram, we give asymptotic expansions for D when ϵ_1, ϵ_2 are small.

V. ON CHAPTER 6: THE PROKHOROV RADIUS

(§11) The Prokhorov Radius Subject to Moment Conditions Involving $\{t, t^2\}$

Here we consider probability measures μ on $\mathbb{R}$ such that both $\int |t| d\mu, \int t^2 d\mu$ are finite. We consider

$$M(\epsilon_1, \epsilon_2) = \left\{\mu : \left|\int t^j d\mu - \alpha^j\right| \leq \epsilon_j, j = 1, 2\right\}$$

where α is a given point in $\mathbb{R}$, also $0 < \epsilon_j < 1, j = 1, 2$ and $0 < \epsilon_2 + 2|\alpha|\epsilon_1 < 1$. We would like to measure the *"size"* of $M(\epsilon_1, \epsilon_2)$ to be given by a simple formula involving only $\epsilon_1, \epsilon_2, \alpha$.

Since weak convergence of probability measures is of great importance and their standard weak topology is also described by the Prokhorov distance p it is natural to define the *Prokhorov radius* for $M(\epsilon_1, \epsilon_2)$ as

$$D := \sup_{\mu \in M(\epsilon_1, \epsilon_2)} p(\mu, \delta_\alpha)$$

where δ_α is the Dirac measure at α. Clearly $D \longrightarrow 0$ as $\epsilon_1, \epsilon_2 \longrightarrow 0$, giving us that $\mu \xrightarrow{\longrightarrow} \delta_\alpha$ weakly. The knowledge of D gives the rate of weak convergence of μ to δ_α.

Here we calculate the exact value of D.

Theorem 11.

i) If $\alpha = 0$ we get $D = \epsilon_2^{1/3}$.

ii) If $|\alpha| \geq 1$ we get $D = (\epsilon_2 + 2|\alpha|\epsilon_1)^{1/3}$.

iii) For $0 < |\alpha| < 1$ and sufficiently small ϵ_1, ϵ_2 we obtain again

$$D = (\epsilon_2 + 2|\alpha|\epsilon_1)^{1/3}.$$

(§12) The Trigometric Prokhorov Radius

Here we consider probability measures μ on $[0, 2\pi]$. We consider

$$M(\epsilon) := \{\mu : |\int cost\ d\mu - cos\ \alpha| \leq \epsilon, |\int sint\ d\mu - sin\ \alpha| \leq \epsilon\},$$

where $\alpha \in [1, 2\pi - 1]$ be given and $0 < \epsilon < \frac{1}{\sqrt{2}} \cdot (1 - cos\ 1)$.

We would like to measure the *"size"* of $M(\epsilon)$ to be given by a simple equation involving only ϵ, α.

It is natural to define the *Trigonometric Prokhorov Radius* for $M(\epsilon)$ as

$$D := \sup_{\mu \in M(\epsilon)} p(\mu, \delta_\alpha)$$

We find the exact value of D.

Theorem 12. *Let $\alpha \in [1, 2\pi - 1]$ be fixed and $0 < \epsilon < \frac{1}{\sqrt{2}} \cdot (1 - cos\ 1)$. Then $D = r_0$, which is the unique solution of the equation*

$$r_0 - r_0 \cdot cos\ r_0 = \epsilon \cdot (|cos\ \alpha| + |sin\ \alpha|); 0 < r_0 < 1.$$

VI. ON CHAPTER 7: PROBABILITY MEASURES - POSITIVE LINEAR OPERATORS AND KOROVKIN TYPE INEQUALITIES

Here we find estimates to

$$|\int_M f d\mu - f(x_0)|,$$

where $f \in C^n(M)(n \in \mathbb{Z}^+)$ in the Frechet sense, M is a non-empty convex compact subset of real normed vector space $(V, \|\cdot\|), x_0 \in M$ is fixed and μ is a finite measure on M. When $n \geq 1, (V, \|\cdot\|)$ is assumed to be a Banach space.

(§13) Optimal Results

We start with a best upper bound general case.

Theorem 13. *Let μ be a finite measure of mass m on the non-empty convex and compact subset M of the real normed vector space $(V, \|\cdot\|)$. Consider $x_0 \in M$ and $C(x_0) > 0$ such that $0 \leq \| x - x_0 \| \leq C(x_0)$ for all $x \in M$. Assume that*

$$\left(\int_M \| x - x_0 \|^r \cdot \mu(dx)\right)^{1/r} := D_r(x_0),$$

where $r > 0$ and $D_r(x_0) > 0$ are given. For the existence of μ we suppose that $D_r^r(x_0) \leq m \cdot C^r(x_0)$. Let $f : M \longrightarrow \mathbb{R}$ be such that

$$|f(x) - f(y)| \leq w \quad \text{if} \quad \| x - y \| \leq h, \quad x, y \in M$$

where $w, h > 0$ are given. Then the best possible constant $K(x_0) = K(m, r, D_r(x_0), h, w, f(x_0))$ in the inequality

$$\left|\int_M f d\mu - f(x_0)\right| \leq |m - 1| \cdot |f(x_0)| + mK(x_0)$$

is given as follows (independently of $f(x_0)$). Set

$$n(x_0) := \lceil \frac{C(x_0)}{h} \rceil, \quad k(x_0) := \lceil \frac{D_r(x_0)}{h \cdot m^{1/r}} \rceil.$$

It is $1 \leq k(x_0) \leq n(x_0)$ because $D_r^r(x_0) \leq mC^r(x_0)$.

(i) $K(x_0) = n(x_0)w$ when $k(x_0) = n(x_0)$, i.e., when $D_r(x_0)/m^{1/r} > C(x_0) - h$.

(ii) $K(x_0) = [1 + \frac{1}{m} \cdot (\frac{D_r(x_0)}{h}) \cdot^r (n(x_0) - 1)^{1-r}] \cdot w$ when $r \leq 1$ and $k(x_0) < n(x_0)$.

(iii) $K(x_0) = (k(x_0) + \theta_{k(x_0)}) \cdot w \leq [1 + D_r(x_0)/(hm^{1/r})] \cdot w$ when $r \geq 1$ and $k(x_0) < n(x_0)$. Here

$$\theta_{k(x_0)} := [(D_r^r(x_0)/m) - (k(x_0) - 1)^r \cdot h^r]/[k^r(x_0)h^r$$

$$-(k(x_0)-1)^r \cdot h^r].$$

Let $h > 0, x \in \mathbb{R}$, we consider the function

$$\phi_n(x) = \int_0^{|x|} \lceil \frac{t}{h} \rceil \frac{(|x|-t)^{n-1}}{(n-1)!} dt, \quad n \in \mathbb{N},$$

where $\phi_0(x) := \lceil \frac{|x|}{h} \rceil$.

Next we present a sharp/attained inequality.

Theorem 14. *Let μ be a finite measure on $[a,b] \subset \mathbb{R}, 0 \in (a,b)$ and $|a| \leq b$. Put*

$$c_k := \int t^k \mu(dt), \quad k = 0,1,\ldots,n : \quad d_n := \left(\int |t|^n \mu(dt) \right)^{1/n}.$$

Let $f \in C^n([a,b])$ be such that

$$|f^{(n)}(s) - f^{(n)}(t)| \leq w \quad \text{if} \quad a \leq s, t \leq b, \quad \text{and } |s-t| \leq h \tag{3}$$

where w, h are given positive numbers.

Then we have the upper bound

$$\left| \int f d\mu - f(0) \right| \leq |f(0)| \cdot |c_0 - 1| + \sum_{k=1}^{n} \frac{|f^{(k)}(0)|}{k!} \cdot |c_k| + w\phi_n(b) \cdot \left(\frac{d_n}{b} \right)^n. \tag{4}$$

The above inequality is in a certain sense attained by the measure μ with masses $[c_0-(d_n/b)^n]$ and $(d_n/b)^n$ at 0 and b, respectively, and when, moreover, the optimal function is

$$\tilde{f} := \begin{cases} w\phi_n, & \text{on } [0,b]; \\ 0, & \text{on } [a,0] \end{cases}$$

Namely, the latter is the limit of a sequence of functions f having continuous nth derivatives satisfying (3) and $f^{(k)}(0) = 0 (k = 0, \ldots, n)$ and such that the difference of the two sides of (4) tends to 0. In fact, $\lim_{N \longrightarrow +\infty} f_{nN}(t) = \tilde{f}(t)$, where $(a \leq t \leq b)$

$$f_{nN}(t) := w \int_0^t \left(\int_0^{t_1} \cdots \left(\int_0^{t_{n-1}} f_{0N}(t_n) dt_n \right) \cdots \right) dt_1.$$

Here, for $k = 0, 1, \ldots, \lceil b/h \rceil - 1$ and $N \geq 1$ f_{0N} is the continuous function defined

$$f_{0N}(t) := \begin{cases} 0, & \text{if } a \leq t < 0; \\ \frac{Nwt}{2h} + kw(1 - \frac{N}{2}), & \text{if } kh \leq t \leq (k + \frac{2}{N})h; \\ (k+1)w, & \text{if } (k + \frac{2}{N})h < t \leq (k+1)h; \\ \lceil b/h \rceil w, & \text{if } (\lceil b/h \rceil - 1 + \frac{2}{N})h < t \leq b. \end{cases}$$

Observe that $f_{0N}(t)$ fulfills (3) and further

$$\lim_{N\longrightarrow+\infty} f_{0N}(t)=\begin{cases}\lceil t/h\rceil w, & t\in[0,b];\\ 0, & t\in[a,0].\end{cases}$$

The Riesz representation theorem is the bridge between probability (finite) measures and positive linear functionals/positive linear operators. We obtain:

Corollary 2. *Consider the positive linear operator*

$$L: C^n([a,b])\longrightarrow C([a,b]),\quad n\in\mathbb{N}.$$

Let

$$c_k(x):=L((t-x)^k,x),\ \ k=0,1,\ldots,n;$$
$$d_n(x):=[L(|t-x|^n,x)]^{1/n.},$$
$$c(x):=\max(x-a,b-x)(c(x)\geq(b-a)/2).$$

Let $f\in C^n([a,b])$ such that $\omega_1(f^{(n)},h)\leq w$, where w,h are fixed positive numbers, $0<h<b-a,\omega_1$ is the first modulus of continuity. Then we have the upper bound

$$|L(f,x)-f(x)|\leq|f(x)|\cdot|c_0(x)-1|+\sum_{k=1}^{n}\frac{|f^{(k)}(x)|}{k!}\cdot|c_k(x)|+R_n.$$

Here

$$R_n:=w\phi_n(c(x))\cdot\left(\frac{d_n(x)}{c(x)}\right)^n=\frac{w}{n!}\cdot\Theta_n(h/c(x))\cdot d_n^n(x),$$

where $\Theta_n(h/u):=n!\phi_n(u)/u^n$.

The above inequality is sharp. Analogous to the last theorem, it is in a certain sense attained by $w\phi_n((t-x)_+)$ and a measure μ_x supported by $\{x,b\}$ when $x-a\leq b-x$, also attained by $w\phi_n((x-t)_+)$ and a measure μ_x supported by $\{x,a\}$ when $x-a\geq b-x$: in each case with masses $c_0(x)-(d_n(x)/c(x))^n$ and $(d_n(x)/c(x))^n$, respectively.

(§14) Almost Optimal Inequalities

Theorem 15. *Let μ be a measure of mass $m>0$ on $[a,b]$ and $x_0\in[a,b]$ fixed. Let n be a fixed positive integer and put*

$$h:=\left[\int_{[a,b]}|t-x_0|^{n+1}\mu(dt)\right]^{1/(n+1)}.$$

Suppose $f\in C^n([a,b])$ satisfies

$$|f^{(n)}(s)-f^{(n)}(t)|\leq w\ \ \text{if}\ \ a\leq s,t\leq b,\ \text{and}\ |s-t|\leq h,$$

where w is a given positive number. Then

$$\left|\int_{[a,b]} f d\mu - f(x_0)\right| \leq |f(x_0)| \cdot |m-1| + \sum_{k=1}^{n} \frac{|f^{(k)}(x_0)|}{k!} \cdot \left|\int_{[a,b]} (t-x_0)^k \mu(dt)\right|$$

$$+\frac{wh^n}{n!} \cdot (m^{1/(n+1)} + 1/(n+1)).$$

Proposition 1. *Let $x_0 \in [a,b]$ and the measure μ on $[a,b]$ of mass $m > 0$ satisfy the moment conditions*

$$\int_{[a,b]} (t-x_0)\mu(dt) := 0 \quad \text{and} \quad d_2(x_0) := \left(\int_{[a,b]} (t-x_0)^2 \mu(dt)\right)^{1/2}.$$

Consider $r > 0$ and $f \in C^1([a,b])$. Then

$$\left|\int_{[a,b]} f d\mu - f(x_0)\right| - |f(x_0)| \cdot |m-1| \leq$$

$$\begin{cases} \frac{1}{8r} \cdot (2+\sqrt{m}r)^2 \cdot \omega_1(f', r d_2(x_0)) \cdot d_2(x_0), & \text{if } r \leq 2/\sqrt{m}; \\ \sqrt{m}\omega_1(f', r d_2(x_0)) \cdot d_2(x_0), & \text{if } r > 2/\sqrt{m}. \end{cases}$$

When $r = 1/2$, in a similar way we obtain:

Theorem 16. *Let the random variable Y have distribution μ, $E(Y) := x_0$, and $Var(Y) := \sigma^2$. Consider $f \in C^1_B(\mathbb{R})$. Then*

$$|Ef(Y) - f(x_0)| = \left|\int f d\mu - f(x_0)\right| \leq (1.5625) \cdot \omega_1(f', \frac{1}{2}\sigma) \cdot \sigma.$$

Application

Let B_n be the classical Bernstein polynomials, we have:

Corollary 3. *For any $f \in C^1([0,1])$ consider $(B_n f)(t) := \sum_{k=0}^{n} f(k/n)\binom{n}{k} t^k (1-t)^{n-k}, t \in [0,1]$. Then*

$$|(B_n f)(t) - f(t)| \leq (1.5625) \cdot \omega_1\left(f', \frac{1}{2} \cdot \sqrt{\frac{t(1-t)}{n}}\right) \cdot \sqrt{\frac{t(1-t)}{n}}$$

$$\leq \left(\frac{0.78125}{\sqrt{n}}\right) \cdot \omega_1\left(f', \frac{1}{4\sqrt{n}}\right), n \in \mathbb{N}.$$

(§15) Multivariate Korovkin Type Inequalities

We need the following:

Lemma 1. *Take Q a compact and convex subset of $\mathbb{R}^k$ and let $\overrightarrow{x_0} := (x_{01}, \ldots, x_{0k}) \in Q$ be fixed and let μ be a probability measure on Q. Let $f \in C^n(Q)$ and suppose that each nth partial derivative $f_\alpha = \partial^\alpha f / \partial x^\alpha$, where $\alpha := (\alpha_1, \ldots, \alpha_k), \alpha_i \in Z^+, i = 1, \ldots, k$ and $|\alpha| := \sum_{i=1}^k \alpha_i = n$, has, relative to Q and the l_1- norm $\|\cdot\|$, a modulus of continuity $\omega_1(f_\alpha, h) \leq w$. Here h and w are fixed positive numbers. Then*

$$\left|\int_Q f d\mu - f(\overrightarrow{x_0})\right| \leq \left|\sum_{j=0}^n \frac{1}{j!} \cdot \int_Q g_{\overrightarrow{x}}^{(j)}(0)\mu(d\overrightarrow{x})\right|$$

$$+w \cdot \int_Q \phi_n(\|\overrightarrow{x} - \overrightarrow{x_0}\|)\mu(d\overrightarrow{x}),$$

where $g_{\overrightarrow{x}}(t) := f(\overrightarrow{x_0} + t(\overrightarrow{x} - \overrightarrow{x_0})), t \geq 0$.

Here comes a main result.

Theorem 17. *Take Q of the form $Q := \{\overrightarrow{x} \in \mathbb{R}^k : \|\overrightarrow{x}\| \leq 1\}$, where $\|\cdot\|$ the l_1-norm in $\mathbb{R}^k$, and let $\overrightarrow{x_0} := (x_{01}, \ldots, x_{ok}) \in Q$ be fixed. Let the measure μ satisfy $\mu(Q) = 1$ and $\int_Q \|\overrightarrow{x} - \overrightarrow{x_0}\| \mu(d\overrightarrow{x}) := d^*$. Also, let $f \in C^n(Q)$ and suppose that each of its nth partial derivatives has a modulus of continuity $\omega_1(f_\alpha, h) \leq w$, where h and w are fixed positive numbers. Then we have the estimate*

$$\left|\int_Q f d\mu - f(\overrightarrow{x_0})\right| \leq \left|\sum_{j=1}^n \frac{1}{j!} \cdot \int_Q g_{\overrightarrow{x}}^{(j)}(0)\mu(d\overrightarrow{x})\right|$$

$$+w \cdot d^* \cdot \phi_n(1 + \|\overrightarrow{x_0}\|)/(1 + \|\overrightarrow{x_0}\|)$$

where $g_{\overrightarrow{x}}(t) := f(\overrightarrow{x_0} + t \cdot (\overrightarrow{x} - \overrightarrow{x_0})), t \geq 0$.

Remark 1. For $n = 1$, this yields the inequality $(x_0 \in Q)$:

$$\left|\int_Q f d\mu - f(\overrightarrow{x_0})\right| \leq \left|\sum_{i=1}^k \frac{\partial f}{\partial x_i}(\overrightarrow{x_0}) \cdot \int_Q (x_i - x_{0i})\mu(d\overrightarrow{x})\right|$$

$$+w \cdot d^* \cdot \phi_1(1 + \|\overrightarrow{x_0}\|)/(1 + \|\overrightarrow{x_0}\|).$$

VII ON CHAPTER 8:

OPTIMAL KOROVKIN TYPE INEQUALITIES AND CONVEXITY

(§16) On weak convergence of finite measures when functions are of convex type.

One main result follows:

Theorem 18. *Let $r > 0, \mu$ a finite measure of mass m on an interval $[a,b], x_0 \in (a,b)$. Set $c(x_0) = \max(x_0 - a, b - x_0)$ and*

$$\left(\int |t - x_0|^r \mu(dt)\right)^{1/r} = d_r(x_0),$$

and assume $d_r(x_0) > 0$. In order that μ exist, we also assume that $d_r^r(x_0) \leq m \cdot (c(x_0))^r$. Next consider $f : [a,b] \longrightarrow \mathbb{R}$ for which $|f(t) - f(x_0)|$ is convex in t and

$$|f(s) - f(t)| \leq w \quad \text{when} \quad s,t \in [a,b]; \ |s - t| \leq h.$$

Here $0 < h \leq \min(x_0 - a, b - x_0)$ and $w > 0$ are fixed.

A best upper bound is given by

$$\left|\int f d\mu - f(x_0)\right| - |m - 1| \cdot |f(x_0)| \leq \begin{cases} w m^{1-(1/r)} \left(\frac{d_r(x_0)}{h}\right), & r \geq 1, \\ w (c(x_0))^{1-r} \frac{d_r^r(x_0)}{h}, & r \leq 1. \end{cases} \tag{5}$$

Remark 2. When $m = 1$, (5) implies

$$\left|\int f d\mu - f(x_0)\right| \leq \begin{cases} w\left(\frac{d_r(x_0)}{h}\right), & r \geq 1, \\ w (c(x_0))^{1-r} \frac{d_r^r(x_0)}{h}, & r \leq 1. \end{cases} \tag{6}$$

If $w = \omega_1(f,h)$ the modulus of continuity of f in $[a,b]$, and $r \geq 1$, (6) becomes

$$\left|\int f d\mu - f(x_0)\right| \leq \omega_1(f,h) \frac{d_r(x_0)}{h}. \tag{7}$$

Note that inequality (7) is sharp when $r = 1$, namely, equality is attained by $f(t) = |t - x_0|$ where both sides are $d_1(x_0)$.

For differentiable functions we get:

Theorem 19. *Let $r > 0, \mu$ a finite measure on $[a,b] \subseteq \mathbb{R}, x_0 \in (a,b)$ and $c(x_0) = \max(x_0 - a, b - x_0)$. Set*

$$\begin{aligned} c_k(x_0) &= \int (t - x_0)^k \mu(dt), \quad k = 0,1,\ldots,n; \\ d_r(x_0) &= \left(\int |t - x_0|^r \cdot \mu(dt)\right)^{1/r}. \end{aligned}$$

Let $f \in C^n[a,b], n \geq 1$, *and assume* $|f^{(n)}(t) - f^{(n)}(x_0)|$ *is convex in* t *and*

$$|f^{(n)}(s) - f^{(n)}(t)| \leq w \quad \text{if} \quad s,t \in [a,b] \quad \text{and} \quad |s-t| \leq h.$$

Here $0 < h \leq \min(x_0 - a, b - x_0)$ *and* $w > 0$ *are fixed.*

Then

$$E(x_0) := \left| \int f d\mu - f(x_0) \right| - |f(x_0)| \cdot |c_0(x_0) - 1| - \sum_{k=1}^{n} \frac{|f^{(k)}(x_0)|}{k!} \cdot |c_k(x_0)|$$

$$\leq \begin{cases} \frac{w}{h(n+1)!} \cdot d_r^{n+1}(x_0) \cdot c_0(x_0)^{1-((n+1)/r)}, & r \geq n+1 \\ \frac{w}{h(n+1)!} \cdot d_r^r(x_0) \cdot (c(x_0))^{(n+1)-r}, & r \leq n+1. \end{cases}$$

Remark 3. When $r = n+1$ and $w = \omega_1(f^{(n)}, h)$,

$$E(x_0) \leq \frac{\omega_1(f^{(n)}, h)}{h(n+1)!} \cdot d_{n+1}^{n+1}(x_0), \tag{8}$$

Inequality (8) is sharp, namely equality is attained.

(§17) On Weak Convergence Of Convex Type Finite Measures.

Main Result

Theorem 20. *Let* μ *be a finite measure of mass* m *on the non-empty convex and compact subset* M *of the real normed vector space* $(V, \|\cdot\|)$. *Consider* $x_0 \in M$ *and* $C(x_0) > 0$ *such that* $0 \leq \| x - x_0 \| \leq C(x_0)$ *for all* $x \in M$. *Let* $\tau(x) = \| x - x_0 \|$ *and let the induced probability measure* $\rho = m^{-1}\mu \circ \tau^{-1}$ *on* $[0, C(x_0)]$. *Assume that the cumulative distribution function corresponding to* ρ *is concave in* $(0, C(x_0)]$. *Suppose further that*

$$\left(\int_M \| x - x_0 \|^r \mu(dx) \right)^{1/r} = D_r(x_0),$$

where $0 < r \leq 1$ *or* $r \geq (ln2/ln1.5) \simeq 1.7095114$ *and* $D_r(x_0) > 0$ *are given, with*

$$D_r^r(x_0) \leq m(r+1)^{-1} C^r(x_0).$$

Consider $h > 0$ *as given. In the case of* $r \geq 1.7095114$ *we would assume that*

$$\beta(\beta+1)\left(\frac{r+1}{2r}\right) h[1 - \left(\frac{\beta-1}{\beta+1}\right)^{(r/(r+1))}] \leq C(x_0),$$

where $\beta = \lceil C(x_0)/h \rceil - 1$; $\lceil \cdot \rceil$ *is the ceiling of the number.*

Let $f : M \longrightarrow \mathbb{R}$ be such that

$$|f(x) - f(y)| \leq w \quad \textit{if} \quad \| x - y \| \leq h, \quad x, y \in M,$$

where $w > 0$ is given.

Then the best possible constant $K(x_0) = K(m, r, D_r(x_0), h, w, f(x_0))$ in the inequality

$$\left| \int_M f d\mu - f(x_0) \right| \leq |m-1||f(x_0)| + mK(x_0) \tag{9}$$

is given as follows (independently of $f(x_0)$).

Let $\phi_1(u) = \int_0^u \lceil z/h \rceil d\, z, u > 0$,

(i)
$$K(x_0) = [1 + \left(\frac{\phi_1(C(x_0)) - C(x_0)}{(C(x_0))^{r+1}} \right) \cdot \left(\frac{r+1}{m} \right) \cdot D_r^r(x_0)] \cdot w,$$

when $0 < r \leq 1$;

(i)
$$K(x_0) = \psi((r+1) \cdot m^{-1} \cdot D_r^r(x_0)),$$

when $r \geq 1.7095114$. Here $\psi(y)$ is defined in $[0, C^r(x_0)]$ as follows: For $k = 1, \ldots, \beta$ consider the numbers

$$y_{1k} := (k-1)^{(r/(r+1))} \cdot k^r \cdot (k+1)^{(r^2/(r+1))}$$
$$\times h^r \cdot \left(\frac{r+1}{2r} \right)^r \cdot \left[1 - \left(\frac{k-1}{k+1} \right)^{(r/(r+1))} \right]^r$$

and

$$y_{2k} := k^r \cdot (k+1)^r \cdot h^r \cdot \left(\frac{r+1}{2r} \right)^r \cdot \left[1 - \left(\frac{k-1}{k+1} \right)^{(r/(r+1))} \right]^r.$$

Hence

$$\psi(y) := \begin{cases} w \cdot \left[\left(\frac{y_{21} - y}{y_{21}} \right) + y \cdot \frac{\phi_1(y_{21}^{1/r})}{y_{21}^{(1+1/r)}} \right]; \quad 0 \leq y \leq y_{21}, \\ w \cdot \left[\left(\frac{y_{2k} - y}{y_{2k} - y_{1k}} \right) \cdot \frac{\phi_1((y_{1k}^{1/r})}{y_{1k}^{1/r}} + \left(\frac{y - y_{1k}}{y_{2k} - y_{1k}} \right) \cdot \frac{\phi_1(y_{2k}^{1/r})}{y_{2k}^{1/r}} \right]; \\ y_{1k} \leq y \leq y_{2k}, k = 2, \ldots, \beta, \\ w \cdot \frac{\phi_1(y^{1/r})}{y^{1/r}}, \quad \textit{elsewhere.} \end{cases}$$

Remark 4 When $r \geq 1.7095114$ and $(r+1) \cdot m^{-1} \cdot D_r^r(x_0) \geq y_{21}$, inequality (9) is attained by $w \cdot \lceil \| x - x_0 \| / h \rceil$ and a measure μ such that the cumulative distribution function corresponding to $\rho = m^{-1} \mu \circ \tau^{-1}$ is concave in $(0, C(x_0)]$.

The measure ρ is the weak*–homeomorphic image of an at most two–points–supported probability measure on $[y_{21}^{1/r}, C(x_0)]$.

When f fulfills a Lipschitz type condition the best upper bounds we get are simpler.

Theorem 21. *Let μ be a finite measure of mass m on the non-empty connected and compact subset M of the real normed vector space $(V, \|\cdot\|)$. Consider $x_0 \in M$ and $C(x_0) > 0$ such that $0 \leq \| x - x_0 \| \leq C(x_0)$ for all $x \in M$. Let $\tau(x) = \| x - x_0 \|$ and let the induced probability measure $\rho = m^{-1}\mu \circ \tau^{-1}$ on $[0, C(x_0)]$. Assume that the cumulative distribution function F corresponding to ρ possesses an $(s-1)$th derivative $F^{(s-1)}(x) (s \geq 1)$ throughout the interval $(0, C(x_0)]$ and that further $(-1)^s F^{(s-1)}(x)$ is convex in $(0, C(x_0)]$. Suppose also that*

$$\left(\int_M \| x - x_0 \|^r \mu(dx)\right)^{1/r} = D_r(x_0) > 0, \quad r > 0$$

with

$$D_r^r(x_0) \leq \binom{r+s}{s}^{-1} \cdot m \cdot C^r(x_0).$$

Let $f : M \longrightarrow \mathbb{R}$ be such that

$$|f(x) - f(y)| \leq k \| x - y \|^\alpha, \quad (k > 0, 0 < \alpha \leq 1).$$

Then we find the following best upper bound,

$$\left|\int_M f d\mu - f(x_0)\right| - |m - 1| \cdot |f(x_0)|$$

$$\leq \begin{cases} k\binom{\alpha+s}{s}^{-1} \cdot \binom{r+s}{s} \cdot C(x_0)^{\alpha-r} D_r^r(x_0), & \alpha \geq r; \\ k\binom{\alpha+s}{s}^{-1} \cdot \binom{r+s}{s}^{\alpha/r} \cdot D_r^\alpha(x_0) \cdot m^{1-(\alpha/r)}, & \alpha \leq r. \end{cases}$$

The last inequality is attained.

Application

Definition 3. Let the sequence of real random variables $\{X_{nt}\}, n \geq 1, t \in \mathbb{R}$, have the exponential distribution with probability density function

$$f_{nt}(x) := \begin{cases} \epsilon_n^{-1} \cdot e^{-\epsilon_n^{-1} \cdot (x-t)}, & t \leq x \leq +\infty \\ 0, & -\infty \leq x < t, \end{cases}$$

where $\epsilon_n > 0$. For $f \in C_B(\mathbb{R})$ we define the operator

$$L_n(f, t) := \epsilon_n^{-1} \cdot e^{\epsilon_n^{-1} t} \cdot \int_t^{+\infty} f(x) \cdot e^{-\epsilon_n^{-1} \cdot x} dx, \quad \text{for all } t \in \mathbb{R},$$

called exponential.

Theorem 22. *Let $\{L_n\}, n \geq 1$, be a sequence of exponential operators and $f \in C_B(\mathbb{R})$. Then*

$$\| L_n f - f \|_\infty \leq \min\Big\{(1.4268781) \cdot w_1(f, \sqrt{2}\epsilon_n);$$

$$(2.2679492) \cdot w_1\left(f, \frac{\sqrt{2}}{2}\epsilon_n\right)\Big\}.$$

(§18) On weak convergence of higher order convex type finite measures

Using convex moment theory methods we get

Theorem 23. *Let μ be a finite measure on $[a,b] \subset \mathbb{R}$ such that $\mu([a,b]) = c_0 > 0$. Consider $x_0 \in [a,b]$ and $0 \leq |x - x_0| \leq C(x_0) = \max(b - x_0, x_0 - a)$, all x in $[a,b]$. Let $\tau(x) = |x - x_0|$ and the induced probability measure $\rho = c_0^{-1}\mu \circ \tau^{-1}$ on $[0, C(x_0)]$.*

Assume that the corresponding to ρ cumulative distribution function F possesses an $(s-1)$st derivative $F^{(s-1)}(x)(s \geq 1)$ throughout the interval $(0, C(x_0)]$ and that further $(-1)^s F^{(s-1)}(x)$ is convex in $(0, C(x_0)]$.

Suppose also that

$$\left(\int_{[a,b]} |x - x_0|^n \mu(dx)\right)^{1/n} = d_n(x_0) > 0, \quad n \geq 1$$

with

$$d_n^n(x_0) \leq \binom{n+s}{s}^{-1} \cdot c_0 \cdot C^n(x_0).$$

Call

$$\int_{[a,b]} (x - x_0)^k \mu(dx) = c_k(x_0), \quad k = 1, \ldots, n.$$

Let $f \in C^n([a,b])$ such that the first modulus of continuity $\omega_1(f^{(n)}, h) \leq w$, where w, h are fixed positive numbers, $0 < h \leq b - a$.

Then we have the upper bounds

$$\left|\int_{[a,b]} \left(f(x) - \sum_{k=0}^{n} \frac{f^{(k)}(x_0)}{k!} \cdot (x - x_0)^k\right) \mu(dx)\right| \leq U(x_0)$$

and

$$\left|\int_{[a,b]} f d\mu - f(x_0)\right|$$

$$\leq |f(x_0)| \cdot |c_0 - 1| + \sum_{k=1}^{n} \frac{|f^{(k)}(x_0)|}{k!} \cdot |c_k(x_0)| + U(x_0),$$

where

$$U(x_0) := w \cdot \left(\prod_{j=1}^{s} (n+j) \right) \cdot \left(\frac{d_n(x_0)}{C(x_0)} \right)^n \cdot \frac{\phi_{n+s}(C(x_0))}{C^s(x_0)}.$$

The last two inequalities are sharp, namely are attained.

(VIII) ON CHAPTER 9:

OPTIMAL KOROVKIN TYPE INEQUALITIES FOR CONVOLUTION TYPE OPERATORS

(§19) Sharp inequalities for convolution operators.

Here these operators are of the form:

Definition 4. Let $f \in C([-2a, 2a]), a > 0$, and let μ be a probability measure on $[-a, a]$. For every $x \in [-a, a]$, we put

$$L(f, x) := \int_{-a}^{a} ([f(x+y) + f(x-y)]/2)\mu(dy).$$

We call L a positive linear convolution operator from $C([-2a, 2a])$ into $C([-a, a])$.

Define

$$c_{2\rho} := \int_{-a}^{a} y^{2\rho} \mu(dy), \rho \in \mathbb{N}.$$

Here ω_2 stands for the second modulus of smoothness. We present the following result:

Proposition 2. *Let μ be a probability measure on $[-a, a], a > 0$, and define $d_r := [\int_{-a}^{a} |y|^r \mu(dy)]^{1/r}, r > 0$. Let $n \geq 2$ be even, and consider all $f \in C^n([-2a, 2a])$ such that*

$$\omega_2(f^{(n)}, |y|) \leq 2A|y|^\alpha, \quad 0 \leq |y| \leq a, 0 < \alpha \leq 2, A > 0.$$

Then, for $x_0 \in [-a, a]$,

$$\left| L(f, x_0) - f(x_0) - \sum_{\rho=1}^{n/2} \frac{f^{(2\rho)}(x_0)}{(2\rho)!} \cdot c_{2\rho} \right|$$

$$\leq A d_{n+2}^{n+\alpha} / [(\alpha+1)(\alpha+2)\dots(\alpha+n)].$$

The last inequality is attained (therefore is sharp), when $x_0 = 0$, by the function

$$f_*(y) := A|y|^{\alpha+n} / [(\alpha+1)(\alpha+2)\dots(\alpha+n)]$$

and the associated probability measure μ having mass 1/2 at the two points $\pm d_{n+2}$.

The next result assumes that $f \in C^n([-2a, 2a]), (a > 0)$, and that $\omega_2(f^{(n)}, |t|) \le g(t)$ $(0 \le |t| \le a)$, where g is a given, arbitrary, bounded, even positive function which is Borel measurable. We put

$$\hat{G}_n(y) := \int_0^y g(t)(y-t)^{n-1}/(n-1)!dt.$$

Theorem 24. *Let ψ be a function on $[0, a]$ such that $\psi(0) = 0$, which is continuous and strictly increasing.*

Let a probability measure μ exist on $[-a, a]$ with

$$\psi^{-1}\left(\int_{-a}^{a} \psi(|y|)\mu(dy)\right) = d.$$

Suppose ($n \ge 2$ even) that

$$\mathcal{H}_n(u) := \hat{G}_n(\psi^{-1}(u))$$

is concave on $[0, \psi(a)]$. Then, for every $x_0 \in [-a, a]$,

$$E(x_0) := \left| L(f, x_0) - f(x_0) - \sum_{\rho=1}^{n/2} \frac{f^{(2\rho)}(x_0)}{(2\rho)!} \cdot c_{2\rho} \right| \le \frac{1}{2} \cdot \hat{G}_n(d).$$

The last inequality is sharp, more precisely it is attained.

(§20) Sharp Inequalities for Non-positive Generalized Convolution Operators

These operators are described as follows:

Definition 5. Let $(\mu_N)_{N \in \mathbb{N}}$ be a sequence of probability measures on $[a, b] \subset \mathbb{R}$. Here $o \in (a, b)$ and $|a| \le b$.

For $r \in \mathbb{N}$ and $n \in \mathbb{Z}_+$ we put

$$\alpha_j := \begin{cases} (-1)^{r-j} \cdot \binom{r}{j} j^{-n}, & j = 1, \ldots, r \\ 1 - \sum_{j=1}^{r} (-1)^{r-j} \cdot \binom{r}{j} j^{-n}, & j = 0. \end{cases}$$

Let $[\alpha, \beta] \subset \mathbb{R}$ such that

$$[a, b] \subset \left[\frac{\alpha}{r}, \frac{\beta}{r}\right].$$

Let $(L_n)_{N \in N}$ be the sequence of linear operators from $C^n([\alpha, \beta])$ into $C([\alpha - ra, \beta - rb])$ defined by

$$(L_N f)(x_0) := \int_{[a,b]} \left(\sum_{j=0}^{r} \alpha_j \cdot f(x_0 + j \cdot y) \right) \cdot \mu_N(dy),$$

for all $f \in C^n([\alpha, \beta])$, all $x_0 \in [\alpha - ra, \beta - rb]$. These operators are of convolution type and in general are not positive.

Let ω_r be the rth modulus of smoothness. Next comes the main result:

Theorem 25. *Here we consider $n \in \mathbb{N}$. Let*

$$\delta_k = \sum_{j=0}^{r} \alpha_j \cdot j^k, \quad k = 1, \ldots, n$$

and

$$c_{kN} = \int_{[a,b]} y^k \mu_N(dy), \quad all \quad N \in \mathbb{N}.$$

Also, consider the even function

$$G_n(y) = \int_0^{|y|} \frac{(|y| - t)^{n-1}}{(n-1)!} \cdot \omega_r(f^{(n)}, t) \cdot dt,$$

all $y \in [a, b]$; $f \in C^n([\alpha, \beta])$. Then

$$\begin{aligned} |(L_N f)(x_0) - f(x_0)| \ \leq\ & \sum_{k=1}^{n} \frac{|f^{(k)}(x_0)|}{k!} \cdot |\delta_k| \cdot |c_{kN}| \\ & + \int_{[a,b]} G_n(y) \mu_N(dy), \quad all \quad x_0 \in [\alpha - ra, \beta - rb]. \end{aligned} \tag{10}$$

At $x_0 = 0$ inequality (10) is attained by $f(x) = x^{r+n}$ and a probability measure μ_N supported by $[0, b]$. I.e. inequality (10) is sharp.

(IX) ON CHAPTER 10: OPTIMAL KOROVKIN TYPE INEQUALITIES FOR POSITIVE LINEAR STOCHASTIC OPERATORS

Let $(\underline{\Omega}, A, P)$ denote a probability space and $L^1(\underline{\Omega}, A, P)$ be the space of all real-valued random variables $Y = Y(\omega)$ with

$$\int_{\underline{\Omega}} |Y(\omega)| P(d\omega) < \infty.$$

Let $X = X(t, \omega)$ denote a stochastic process with index set $[a, b] \subset \mathbb{R}$ and real state space $(\mathbb{R}, \mathcal{B})$, where $\mathcal{B}$ is the $\sigma-$field of Borel subsets of $\mathbb{R}$. Here $C([a, b])$ is the space of continuous real–valued functions on $[a, b]$ and $B([a, b])$ is the space of bounded real- -valued functions on $[a,b]$. Also $C_{\underline{O}}([a,b]) := C([a,b], L^1(\underline{O}, A, P))$ is the space of L^1–continuous stochastic processes in t and

$$B_{\underline{O}}([a,b]) := \left\{ X : \sup_{t\in[a,b]} \int_{\underline{O}} |X(t,\omega)| P(d\omega) < \infty \right\};$$

obviously $C_{\underline{O}}[a,b]) \subset B_{\underline{O}}([a,b])$. For $n \in \mathbb{N}$ consider the subspace $C^n_{\underline{O}}([a,b]) := \{X$: there exists $X^{(k)}(t,\omega) \in C_{\underline{O}}([a,b])$ and it is continuous in t for each $\omega \in \underline{O}$, $k = 0,1,\ldots,n\}$, i.e., for every $\omega \in \underline{O}$ we have $X(t,\omega) \in C^n([a,b])$. $(EX)(t) = \int_{\underline{O}} X(t,\omega)P(d\omega)$ is the expectation operator, defined for all $t \in [a,b]$.

Consider the linear operator

$$T : C_{\underline{O}}([a,b] \hookrightarrow B_{\underline{O}}([a,b]).$$

If $X \in C_{\underline{O}}([a,b])$ is nonnegative and TX, too, then T is called positive. If $ET = TE$, then T is called E–commutative.

One main result is presented next:

Theorem 26. *Consider the positive E–commutative linear operator*

$$T : C_{\underline{O}}([a,b]) \hookrightarrow B_{\underline{O}}([a,b]).$$

Let $t_0 \in [a,b]$ and

$$\begin{aligned} c(t_0) &= \max(t_0 - a, b - t_0), \\ d_n(t_0) &= ((T(|t - t_0|^n))(t_0))^{1/n}, \quad n \in \mathbb{N}. \end{aligned}$$

Let $X \in C^n_{\underline{O}}([a,b])$ such that $\omega_1(EX^{(n)}, h) \le w$, where ω_1 is the first modulus of continuity and w, h are fixed positive numbers, $0 < h \le b - a$. Then we get the upper bound

$$|(E(TX))(t_0) - (EX)(t_0)| \le |(EX)(t_0)| \cdot |(T(1))(t_0) - 1|$$

$$+ \sum_{k=1}^{n} \frac{|(EX^{(k)})(t_o)|}{k!} \cdot |(T(t - t_0)^k)(t_0)| + w \cdot \phi_n(c(t_0)) \cdot \left(\frac{d_n(t_0)}{c(t_0)} \right)^n .$$

The last inequality is sharp.

A simpler variation of the last theorem has as follows.

Theorem 27. *Consider the positive E– commutative linear operator*

$$T : C_{\underline{O}}([a,b]) \hookrightarrow B_{\underline{O}}([a,b]).$$

Let $T(1)(t_0) = m_{t_0}, t_0 \in [a,b]$ *and*

$$[m_{t_0}^{-1} \cdot (T(|t-t_0|^{n+1}))(t_0)]^{1/n+1} = d_{n+1}(t_0) > 0, \quad n \in \mathbb{N}.$$

Let $X \in C_{\underline{o}}^n([a,b])$ *such that* $\omega_1(EX^{(n)}, rd_{n+1}(t_0)) \leq w$, *where* r, w *are given positive numbers. Then*

$$\begin{aligned} |(E(TX))(t_0) - (EX)(t_0)| &\leq |(EX)(t_0)| \cdot |m_{t_0} - 1| \\ &+ \sum_{k=1}^{n} \frac{|(EX^{(k)})(t_0)|}{k!} \cdot |(T(t-t_0)^k)(t_0)| \\ &+ \frac{wm_{t_0}}{rn!} \cdot \left[\frac{nr^2}{8} + \frac{r}{2} + \frac{1}{n+1}\right] \cdot d_{n+1}^n(t_0). \end{aligned}$$

The last inequality is proved to be sharp.

We have also multidimensional similar results.

Application

Convergence of Multidimensional Stochastic Bernstein Operators

Let $\mathcal{Q}$ be the following compact convex subset of $(\mathbb{R}^k, \|\cdot\|_{l_1}), k \geq 1$,

$$\mathcal{Q} = \{(z_1, \ldots, z_k) \in \mathbb{R}^k : 0 \leq z_i \leq 1 \text{ for each } i \text{ and } \sum_{i=1}^{k} z_i \leq 1\}.$$

Given $N \in \mathbb{N}$ the Nth multidimensional stochastic Bernstein operator B_N maps $X \in C_{\underline{o}}(\mathcal{Q})$ onto the stochastic process

$$(B_N X)(z_1, \ldots, z_k, \omega) := \sum_{(m_1,\ldots,m_k)\in M} X\left(\frac{m_1}{N}, \ldots, \frac{m_k}{N}, \omega\right) \cdot F_{(m_1,\ldots,m_k)}(z_1, \ldots, z_k),$$

where summation is taken over the set $M = \{(m_1, \ldots, m_k) : m_i \in \mathbb{Z}_+$ for each i and $\sum_{i=1}^k m_i \leq N\}$ and

$$\begin{aligned} F_{(m_1,\ldots,m_k)}(z_1, \ldots, z_k) : &= \frac{N!}{(\prod_{i=1}^k m_i!) \cdot (N - \sum_{i=1}^k m_i)!} \\ &\times \left(\prod_{i=1}^{k} z_i^{m_i}\right) \cdot \left(1 - \sum_{i=1}^{k} z_i\right)^{N - \sum_{i=1}^k m_i} \end{aligned}$$

is defined for each $(z_1, \ldots, z_k) \in \mathcal{Q}$.

As a related result we give

Theorem 28. *Let $X \in C^1_{\underline{o}}(Q)$. Call*

$$w := \max\{\omega_1\left(EX_\alpha, \frac{k}{4\sqrt{N}}\right); \quad \text{all } \alpha \text{ such that } |\alpha| = 1 \ \}.$$

Then

$$\sup_{x_0 \in Q} |(E(B_N X))(x_0) - (EX)(x_0)| \le 0.78125 \cdot w \cdot \frac{k}{\sqrt{N}}, \quad \text{all} \quad N \in \mathbb{N}.$$

Here X_α stands for a mixed partial of X of order one and $C^1_{\underline{o}}(Q)$ is defined similarly as $C^1_{\underline{o}}([a,b])$.

(X) ON CHAPTER 11: OPTIMAL KOROVKIN TYPE INEQUALITIES FOR POSITIVE LINEAR OPERATORS USING AN EXTENDED COMPLETE TCHEBYCHEFF SYSTEM (E.C.T.).

Let the functions $f, u_0, u_1, \ldots, u_n \in C^{n+1}([a,b]), n \ge 0$, and the Wronskians

$$W_i(x) := W[u_0(x), u_1(x), \ldots, u_i(x)], i = 0, 1, \ldots, n.$$

Here we assume that all $W_i(x)$ are positive throughout $[a,b]$. Consider the linear differential operator of order $i \ge 0$

$$L_i f(x) := \frac{W[u_0(x), u_1(x), \ldots, u_{i-1}(x), f(x)]}{W_{i-1}(x)},$$

$i = 1, \ldots, n+1$; $L_0 f(x) := f(x)$, all $x \in [a,b]$. Here $W[u_0(x), u_1(x), \ldots, u_{i-1}(x), f(x)]$ denotes the Wronskian of $u_0, u_1, \ldots, u_{i-1}, f$.

Consider also the functions

$$g_i(x,t) := \frac{1}{W_i(x)} \cdot \begin{vmatrix} u_0(t) & u_1(t) & \ldots & u_i(t) \\ u_0'(t) & u_1'(t) & \ldots & u_i'(t) \\ \vdots & \vdots & & \\ u_0^{(i-1)}(t) & u_1^{(i-1)}(t) & \ldots & u_i^{(i-1)}(t) \\ u_0(x) & u_1(x) & \ldots & u_i(x) \end{vmatrix},$$

$i = 1, 2, \ldots, n$; $g_0(x,t) := 1$, all $x, t \in [a,b]$. Denote

$$N_n(x,t) := \int_t^x g_n(x,s)ds, \quad n \ge 0.$$

Our main result follows:

Theorem 29. *Let μ be a finite measure of mass m on $[a,b] \subset \mathbb{R}$ and t a fixed point in (a,b), such that*

$$\int_{[a,b]} |x-t|\mu(dx) = d > 0.$$

Let the functions $f(x), u_0(x), u_1(x), \ldots, u_n(x)$ belong to $C^{n+1}([a,b]), n \geq 0$, and let the Wronskians $W_0(x), W_1(x), \ldots, W_n(x)$ be positive throughout $[a,b]$ (i.e. $\{u_i\}_{i=0}^n$ is an E.C.T. system).

Assume that $u_0(x) = c > 0$ and $u_1(x)$ is a concave function for $x \leq t$ and a convex function for $x \geq t$. Define

$$\tilde{G}_n(x,t) := \left| \int_t^x g_n(x,s) \left\lceil \frac{|s-t|}{h} \right\rceil ds \right|, \quad x,t \in [a,b],$$

where $0 < h \leq b-a$ is given and $\lceil \cdot \rceil$ is the ceiling of the number; $n \geq 0$. Assume that the first modulus of continuity $\omega_1(L_{n+1}f, h) \leq w$, where $w > 0$ is given.

Consider the error function

$$E_n(x,t) := f(x) - f(t) - \sum_{i=1}^n L_i f(t) \cdot g_i(x,t) - L_{n+1}f(t) \cdot N_n(x,t).$$

Then we have the upper bounds

$$\int_{a,b]} |E_n(x,t)|\mu(dx) \leq w \cdot \max\left\{ \frac{\tilde{G}_n(b,t)}{b-t}, \frac{\tilde{G}_n(a,t)}{t-a} \right\} \cdot d,$$

and

$$\begin{aligned} \left| \int_{[a,b]} f d\mu - f(t) \right| &\leq |m-1| \cdot |f(t)| + \sum_{i=1}^n |L_i f(t)| \cdot \left| \int_{[a,b]} g_i(x,t)\mu(dx) \right| \\ &+ |L_{n+1}f(t)| \cdot \left| \int_{[a,b]} N_n(x,t)\mu(dx) \right| \\ &+ w \cdot \max\left\{ \frac{\tilde{G}_n(b,t)}{b-t}, \frac{\tilde{G}_n(a,t)}{t-a} \right\} \cdot d. \end{aligned}$$

The last two inequalities are proved to be attained i.e. they are sharp.

A simpler variation of the last theorem follows:

Theorem 30. *Let μ be a finite measure of mass m on $[a,b] \subset \mathbb{R}$ and t a fixed point in $[a,b]$, such that*

$$\left[\int_{[a,b]} |x-t|^{n+2}\mu(dx)\right]^{1/(n+2)} = h,$$

where $0 < h \leq b-a$ *is given;* $n \geq 0$.

Let the functions $f(x), u_0(x), u_1(x), \ldots, u_n(x)$ belong to $C^{n+1}([a,b])$ and let the Wronskians $W_0(x), W_1(x), \ldots, W_n(x)$ be positive throughout $[a,b]$. Assume that the first modulus of continuity $\omega_1(L_{n+1}f, h) \leq w$, where $w > 0$ is given.

Consider the error function

$$E_n(x,t) := f(x) - \sum_{i=0}^{n} L_i f(t) \cdot g_i(x,t) - L_{n+1}f(t) \cdot N_n(x,t).$$

Then we have the upper bounds

$$\int_{[a,b]} |E_n(x,t)|\mu(dx) \leq$$

$$w \cdot [m^{1/(n+2)} + 1] \cdot \left[\int_{[a,b]} |N_n(x,t)|^{(n+2/n+1)}\mu(dx)\right]^{(n+1/n+2)}$$

and

$$\left|\int_{[a,b]} f d\mu - f(t)\right| \leq |f(t)| \cdot |m-1|$$

$$+\sum_{i=1}^{n} |L_i f(t)| \cdot \left|\int_{[a,b]} g_i(x,t)\mu(dx)\right| + |L_{n+1}f(t)| \cdot \left|\int_{[a,b]} N_n(x,t)\mu(dx)\right|$$

$$+w \cdot [m^{1/(n+2)} + 1] \cdot \left[\int_{[a,b]} |N_n(x,t)|^{(n+2/n+1)} \cdot \mu(dx)\right]^{(n+1/n+2)}.$$

ON CHAPTER 12: A GENERAL "K–ATTAINED" INEQUALITY FOR THE WEAK CONVERGENCE OF PROBABILITY MEASURES.

Let the functions $g, u_0, u_1, \ldots, u_n \in C^{n+1}([a,b]), n \geq 0, [a,b] \subset \mathbb{R}$ such that $u_0(x) = c > 0$ and $u_1(x)$ is a concave function for $x \leq t$ and a convex function for $x \geq t$, where t is a fixed point in (a,b).

Consider the Wronskians

$$W_i(x) := W(u_0(x), u_1(x), \ldots, u_i(x)), i = 0, 1, \ldots, n$$

and assume that all are positive everywhere on $[a,b]$.

Let the linear differential operator of order $i \geq 0$

$$L_i g(x) := \frac{W[u_0(x), u_1(x), \ldots, u_{i-1}(x), g(x)]}{W_{i-1}(x)}$$

$i = 1, \ldots, n+1$; $L_0 g(x) := g(x)$ all $x \in [a,b]$. Here $W[u_0(x), u_1(x), \ldots, u_{i-1}(x), g(x)]$ denotes the Wronskian of $u_0, u_1, \ldots, u_{i-1}, g$.

Consider also the functions

$$g_i(x,t) := \frac{1}{W_i(x)} \cdot \begin{vmatrix} u_0(t) & u_1(t) & \ldots & u_i(t) \\ u_0'(t) & u_1'(t) & \ldots & u_i'(t) \\ \vdots & & & \\ u_0^{(i-1)}(t) & u_1^{(i-1)}(t) & \ldots & u_i^{(i-1)}(t) \\ u_0(x) & u_1(x) & \ldots & u_i(x) \end{vmatrix},$$

$i = 1, 2, \ldots, n$; $g_0(x,t) := 1$, all $x, t \in [a,b]$.

Here we denote by

$$\begin{aligned} N_n(x,t) : &= \int_t^x g_n(x,s)ds, \\ \tilde{N}_n(x,t) : &= |N_n(x,t)|; \quad n \geq 0. \end{aligned}$$

Next we would like to introduce the following measure of smoothness.

Definition 6. Let $f \in C([a,b])$. Let $n \geq 0$ and $t \in (a,b)$ be fixed. We define the *generalized reduced K– functional* as:

$$\tilde{K}_{n+1}^{(t)}(f,h) := K(f,h) := K(h) := \inf_g (\| f - g \|_\infty + h \cdot \| L_{n+1} g \|_\infty), h \geq 0.$$

Here g ranges through the functions $g \in C^{n+1}([a,b])$ satisfying

$$L_i g(t) = 0, \quad i = 1, \ldots, n,$$

where $\| \cdot \|_\infty$ denotes the sup norm.

Theorem 31. *The generalized reduced K– functional has the following properties: (i) Subadditive in terms of f. And in terms of h is (ii) continuous, (iii) nonnegative, (iv) monotonely increasing, (v) concave, hence (vi) subadditive. (vii)* $\tilde{K}_{n+1}^{(t)}(f,0) = 0$.

Our main result here follows:

Theorem 32. *Let μ be a probability measure on $[a,b] \subset \mathbb{R}$ such that*

$$\int_{[a,b]} |x-t|\mu(dx) = d_1(t) > 0,$$

where t is a fixed point of (a,b).

Let the functions $u_0(x), u_1(x), \ldots, u_n(x)$ belong to $C^{n+1}([a,b]), n \geq 0$, and let the Wronskians $W_0(x), W_1(x), \ldots, W_n(x)$ be positive throughout $[a,b]$ (i.e. $\{u_i\}_{i=0}^n$ is an E.C.T. system).

Assume that $u_0(x) = c > 0$ and $u_1(x)$ is a concave function for $x \leq t$ and a convex function for $x \geq t$.

Define

$$\Lambda_n(t) := \max\left\{\frac{\tilde{N}_n(b,t)}{b-t}, \frac{\tilde{N}_n(a,t)}{t-a}\right\}.$$

Consider $f \in C([a,b])$. Then

$$\left|\int_{[a,b]} f d\mu - f(t)\right| \leq 2 \cdot \tilde{K}_{n+1}^{(t)}\left[f; \frac{\Lambda_n(t)\cdot d_1(t)}{2}\right], \quad n \geq 0.$$

For special choices of $d_1(t)$ the last inequality is attained *(i.e., it is sharp)* by the function $N_n(x,t)$ and the probability measure μ_0 described as follows:

(i) If $\frac{\tilde{N}_n(b,t)}{b-t} \geq \frac{\tilde{N}_n(a,t)}{t-a}$ and $d_1(t) \leq b-t$, then the optimal probability measure μ_0 is supported at $\{t,b\}$ with masses $[1-\frac{d_1(t)}{b-t}], [\frac{d_1(t)}{b-t}]$, respectively.

(ii) If $\frac{\tilde{N}_n(b,t)}{b-t} \leq \frac{\tilde{N}_n(a,t)}{t-a}$ and $d_1(t) \leq t-a$, then the optimal probability measure μ_0 is supported at $\{t,a\}$ with masses $[1-\frac{d_1(t)}{t-a}], [\frac{d_1(t)}{t-a}]$, respectively.

(XII) ON CHAPTER 13: A GENERAL STOCHASTIC INEQUALITY

Let $(\underline{0}, \mathcal{F}, P)$ be an arbitrary probability space and $(X_n)_{n\in\mathbb{N}} : \underline{0} \longrightarrow \mathbb{R}$ be a sequence of real, independent, not necessarily identically distributed random variables with distribution functions F_{X_n}, and $S_n = \sum_{i=1}^{n} X_i$.

Here denote $C_B := C_B(\mathbb{R})$, the space of bounded continuous functions from $\mathbb{R}$ into itself. For $r \in \mathbb{N}, 0 < a \leq r$ an important class of functions is

$$\text{Lip}\,(a, r, C_B) := \{f \in C_B : \omega_r(t, f, C_B) \leq L_f \cdot t^a\},$$

where ω_r is the rth modulus of smoothness on $\mathbb{R}$.

If Z is any real random variable the operator $V_Z : C_B \longrightarrow C_B$, used by Trotter, is defined by

$$(V_Z f)(y) := \int f(x+y) dF_Z(x), \quad y \in \mathbb{R}.$$

Let $(X_i)_{i \in \mathbb{N}}$ be a sequence of real independent $r.v.$'s, the corresponding limiting $r.v. X$ is assumed to be φ–decomposable into the independent components $Z_{i,n}, (1 \leq i \leq n)$, $n \in \mathbb{N}$ (define $\varphi : \mathbb{N} \longrightarrow \mathbb{R}^+$ such that $\varphi(n) = 0(1)$ as $n \longrightarrow +\infty$, then X is called φ–decomposable if $F_X = F_{\varphi(n) \cdot \sum_{i=1}^{n} Z_{i,n}}$, for all $n \in \mathbb{N}$).

We put $G_{i,n}(x) := (F_{X_i} - F_{Z_{i,n}})(x)$,

$$\mu_{j,i,n} := \int_{\mathbb{R}} x^j \cdot dG_{i,n}(x), V_{j,i,n} := \int_{\mathbb{R}} |x|^j \cdot d|G_{i,n}(x)|,$$

where $|G_{i,n}(x)|$ denotes the total variation of $G_{i,n}(x)$ over $(-\infty, x]$. Here we assume that the $s \leq r, r \in \mathbb{N}$ moments of $G_{i,n}(x)$ are finite.

Also, we consider the $r.v.$ $T_n := \varphi(n) \cdot (\sum_{i=1}^{n} X_i)$.

Next comes the main result

Theorem 33. *Let $r \in \mathbb{N}$ and*

$$V_{r,i,n} := \int_{\mathbb{R}} |x|^r \cdot d|G_{i,n}(x)| < +\infty; \quad i, n \in \mathbb{N}.$$

Further, assume that there are constants G_j such that

$$\varphi(n)^{j-r} \cdot \left(\sum_{i=1}^{n} |\mu_{j,i,n}| \right) \leq G_j \cdot \left(\sum_{i=1}^{n} V_{r,i,n} \right), \quad (0 \leq j \leq r-1).$$

Then for any $f \in Lip\ (r, r, C_B)$ we obtain

$$\| V_{T_n} f - V_X f \|_\infty \leq \left(C_f + \frac{2 \cdot L_f}{r!} \right) \cdot (\varphi(n))^r \cdot \left(\sum_{i=1}^{n} V_{r,i,n} \right),$$

where $C_f := \sum_{j=0}^{r-1} G_j \cdot \frac{\|f^{(j)}\|_\infty}{j!}$ and L_f is the Lipschitz constant of $f^{(r-1)}$.

(XIII) ON CHAPTER 14: MISCELLANEOUS OPTIMAL INEQUALITIES AND KOROVKIN-TYPE CONVERGENCE RESULTS INVOLVING BASIC MOMENTS

Lemma 2. *Let $u, l \geq 2, u, l \in \mathbb{N}$, then there exists a positive constant $C(k,l) \geq (k^2(k^2-1))/(l^2(l^2-1))$ such that*

$$[k^2 \cdot (1 - \cos t) - (1 - \cos kt)] \leq C(k,l).$$

$$[l^2 \cdot (1 - \cos t) - (1 - \cos lt)], \quad all \quad t \in [0, \pi].$$

The last inequality for small k, l is sharp.

Definition 7. *Let μ be a probability measure on $[0, \pi]$. Its Fourier-Stieltjes coefficients are defined by*

$$p_k := \int_0^\pi \cos kt \mu(dt) \quad (k = 0, 1, 2, \ldots).$$

We need

Lemma 3. *Let μ be a probability measure on $[0, \pi]$ with Fourier-Stieltjes coefficients $p_k, k \in \mathbb{Z}^+$ and $p_1 \neq 1$.*

$$\left[k^2 - \left(\frac{1-p_k}{1-p_1}\right)\right] \leq C(k,l) \cdot \left[l^2 - \left(\frac{1-p_l}{1-p_1}\right)\right],$$

where $k, l \geq 2, k, l \in \mathbb{N}$.

Next comes one main result

Theorem 34. *Let $l \in \mathbb{N}, l \geq 2$. If $\{\mu_n\}_{n \in \mathbb{N}}$ is a sequence of probability measures on $[0, \pi]$ with Fourier-Stieltjes coefficients p_{kn} such that $p_{1n} \neq 1$ and $\lim_{n \longrightarrow \infty}((1-p_{ln})/(1-p_{1n})) = l^2$, then $\lim_{n \longrightarrow \infty}((1-p_{kn})/(1-p_{1n})) = k^2$ for all $k \geq 2, k \in \mathbb{N}$.*

Also we consider

Definition 8. *Let $\{\mu_n\}_{n \in \mathbb{N}}$ be a sequence of probability measures on $\mathbb{R}$ such that the following integrals exist:*

$$\tilde{p}_{kn} := \int_{\mathbb{R}} Coshkt\mu_n(dt) \quad (k \in \mathbb{R}).$$

We shall call the numbers $\tilde{p}_{k,n}$ hyperbolic coefficients of the measure μ_n.

Here we have another main result:

Theorem 35. *Let $\{\mu_n\}_{n\in\mathbb{N}}$ be a sequence of probability measures on $\mathbb{R}$ with $\tilde{p}_{k,n} < +\infty$ and $\tilde{p}_{1,n} \neq 1$. Let $k > 1$ and suppose that*

$$\lim_{n\longrightarrow\infty} \frac{\tilde{p}_{k,n} - 1}{\tilde{p}_{1,n} - 1} = k^2,$$

then

$$\lim_{n\longrightarrow\infty} \frac{\tilde{p}_{l,n} - 1}{\tilde{p}_{1,n} - 1} = l^2 \quad \textit{for all} \quad 1 < l \leq k.$$

ON CHAPTER 15: A DISCRETE STOCHASTIC KOROVKIN THEOREM

Here is our result

Theorem 36. *Let $(\underline{0}, A, \tau)$ be a probability space and $P = \{t_1, \ldots, t_j, \ldots\}$ be a countable set of cardinality ≥ 2. Consider the space of stochastic processes with real state space*

$$B_{\underline{0}}(P) := \{X : \sup_{t\epsilon P} \int_{\underline{0}} |X(t,\omega)|\tau(d\omega) < +\infty\}$$

and the space

$$B(P) := \{f : P \longrightarrow \mathbb{R} : \quad \| f \|_\infty < +\infty\},$$

where

$$\| f \|_\infty := \sup_{t\epsilon P} |f(t)|; B(P) \subset B_{\underline{0}}(P).$$

Let $T_n : B_{\underline{0}}(P) \longrightarrow B_{\underline{0}}(P)$ be a sequence of positive linear operators that are E–commutative, i.e.

$$(E(T_n X))(t,\omega) = (T_n(EX))(t,\omega), \ \textit{for all} \ (t,\omega)\epsilon P \times \underline{0}$$

where E is the expectation operator.

Also assume that $(T_n 1)(t,\omega) = 1$, for all $(t,\omega)\epsilon P \times \underline{0}$. For

$$\{X_1(t,\omega), \ldots, X_k(t,\omega)\} \subset B_{\underline{0}}(P)$$

assume that

$$\lim_{n\longrightarrow\infty} E[(T_n X_i)(t_j,\omega) - X_i(t_j,\omega)] = 0,$$

for all $t_j\epsilon P$ and all $i = 1, \ldots, k$.

In order that

$$\lim_{n\longrightarrow\infty} E[(T_n X)(t_j,\omega) - X(t_n,\omega)] = 0,$$

for all $t_j \epsilon P$ and all $X \epsilon B_{\underline{0}}(P)$, it is enough to assume that for each $t_j \epsilon P$ there are real constants $\beta_1, \ldots, \beta_k$ such that

$$\sum_{i=1}^{k} \beta_i \cdot E[X_i(t,\omega) - X_i(t_j,\omega)] \geq 1, \quad \text{for all } t \epsilon P - \{t_j\}.$$

Comment

For the convenience of the reader, we have made a serious effort in preparing the chapters to follow in a way to be self-contained and independent among themselves as much as possible. However we feel that there exists a continuation and logical sequence.

CHAPTER TWO

GEOMETRIC MOMENT THEORY

2.1 Methods Of Optimal Distance And Optimal Ratio

(§1) The first part in this section-tool comes from Kemperman (1968).

Let $g_1, \ldots, g_n$ and h be given real-valued Borel measurable functions on a fixed measurable space $X := (X, A)$. We would like to find the best upper and lower bound on

$$\mu(h) := \int_X h(t)\mu(dt),$$

given that μ is a probability measure on X with prescribed moments

$$\int g_i(t)\mu(dt) = y_i, i = 1, \ldots, n.$$

Here we assume μ such that

$$\int_X |g_i|\mu(dt) < +\infty, i = 1, \ldots, n$$

and

$$\int_X |h|\mu(dt) < +\infty.$$

For each $y := (y_1, \ldots, y_n) \in \mathbb{R}^h$, consider the optimal quantities

$$L(y) := L(y|h) := \inf_\mu \mu(h),$$

$$U(y) := U(y|h) := \sup_\mu \mu(h),$$

where μ is a probability measure as above with

$$\mu(g_i) = y_i, i = 1, \ldots, n.$$

If there is not such probability measure μ we set $L(y) := +\infty, U(y) := -\infty$.

If $h := \mathcal{X}_S$ the characteristic function of a given measurable set S of X, then we agree to write

$$L(y|\mathcal{X}_S) := L_S(y), U(y|\mathcal{X}_S) := U_S(y).$$

Hence, $L_S(y) \leq \mu(S) \leq U_S(y)$. Consider $g : X \longrightarrow \mathbb{R}^n$ such that $g(t) := (g_1(t), \ldots, g_n(t))$. Set also $g_0(t) := 1$, all $t \in X$. Here we present Kemperman's geometric methods for solving the above main moment problems which were motivated by Markov (1884), Riesz (1911) and Krein (1951). Kempermans's results here will be presented without proofs, the interested reader could consult for more in Kemperman(1968). The advantage of geometric method is that many times is simple and immediate giving us the optimal quantities L, U in a closed-numerical form, on the top of this is very elegant. Here the σ-field A contains all subsets of X.

The next result comes from Richter (1957), Rogosinsky (1958, p.4) and Mulholland and Rogers (1958).

Theorem 2.1.1. *Let $f_1, \ldots, f_N$ be given real- valued Borel measurable functions on a measurable space Ω (such as $g_1, \ldots, g_n$ and h on X). Let μ be a probability measure on Ω such that each f_i is integrable with respect to μ. Then there exists a probability measure μ' of finite support on Ω (i.e. having non-zero mass only at a finite number of points) satisfying*

$$\int_\Omega f_i(t)\mu(dt) = \int_\Omega f_i(t)\mu'(dt),$$

all $i = 1, \ldots, N$.

One can even achieve that the support of μ' has at most $N+1$ points. So from now on in this section we can talk only about finite supported probability measures.

Call

$$V := \text{conv}\, g(X),$$

where $g(X) := \{z \in \mathbb{R}^n : z = g(t) \text{ for some } t \in X\}$ is a curve in $\mathbb{R}^n$ (if $X = [a,b] \subset \mathbb{R}$ or if $X = [a,b] \times [c,d] \subset \mathbb{R}^2$).

Let $S \subset X$, and let $M^+(S)$ denote the set of all probability measures on X whose support is finite and contained in S.

Lemma 2.1.1. *Given $y \in \mathbb{R}^n$, then $y \in V$ iff $\exists \mu \in M^+(X)$ such that*

$$\mu(g) = y \quad (i.e.\ \mu(g_i) := \int_X g_i(t)\mu(dt) = y_i, i = 1, \ldots, n).$$

Hence $L(y|h) < +\infty$ iff $y \in V$ (see that by Theorem 2.1.1

$$L(y|h) = \inf\{\mu(h) : \mu \in M^+(X), \mu(g) = y\}$$

and

$$U(y|h) = \sup\{\mu(h) : \mu \in M^+(X), \mu(g) = y\}).$$

Easily one can see that

$$L(y) := L(y|h)$$

is a convex function on V, i.e.

$$L(\lambda y' + (1-\lambda)y'') \leq \lambda L(y') + (1-\lambda)L(y''),$$

whenever $0 \leq \lambda \leq 1$ and $y', y'' \in V$. Also $U(y) := U(y|h) = -L(y|-h)$ is a concave function on V.

After some considerations the following three properties are clearly equivalent:

(i) int(V):= interior of $V \neq \phi$;

(ii) $g(X)$ is not the subset of any hyperphane in $\mathbb{R}^n$;

(iii) $1, g_1, g_2, \ldots, g_n$ are linearly independent on X.

From now on we assume that $1, g_1, \ldots, g_n$ are linearly independent, i.e. int $(V) \neq \phi$.

Let D^* denote the set of all $(n+1)$–types of real numbers $d^* := (d_0, d_1, \ldots, d_n)$ satisfying

$$h(t) \geq d_0 + \sum_{i=1}^{n} d_i g_i(t), \quad \text{all} \quad t \in X. \tag{2.1.1}$$

Theorem 2.1.2 *For each $y \in$ int (V) we have that*

$$L(y|h) = \sup\{d_0 + \sum_{i=1}^{n} d_i y_i : d^* = (d_0, \ldots, d_n) \in D^*\}. \tag{2.1.2}$$

Given that $L(y|h) > -\infty$, the supremum in (2.1.2) is even assumed by some $d^* \in D^*$. If $L(y|h)$ is finite in int (V) then for almost all $y \in$ int(V) the supremum in (2.1.2) is assumed by a unique $d^* \in D^*$. Thus $L(y|h) < +\infty$ in int(V) iff $D^* \neq \phi$. Note that $y := (y_1, \ldots, y_n) \in$ int $(V) \subset \mathbb{R}^n$ iff $d_0 + \sum_{i=1}^n d_i y_i > 0$ for each choice of the real constants d_i not all zero such that $d_0 + \sum_{i=1}^n d_i g_i(t) \geq 0$, all $t \in X$. (The last statement comes e.g. from Kemperman (1965), p.573.)

If h is bounded then $D^* \neq \phi$, trivially.

Theorem 2.1.3. *Let $d^* \in D^*$ be fixed and set*

$$B(d^*) := \{z = g(t) : d_0 + \sum_{i=1}^{n} d_i g_i(t) = h(t), t \in X\}. \tag{2.1.3}$$

Then for each point

$$y \in conv\, B(d^*) \tag{2.1.4}$$

the quantity $L(y|h)$ *is gotten as follows. Set*

$$y = \sum_{j=1}^{m} p_j g(t_j)$$

with

$$g(t_j) \in B(d^*),$$

and

$$p_j \geq 0, \quad \sum_{j=1}^{m} p_j = 1. \tag{2.1.5}$$

Then

$$L(y|h) = \sum_{j=1}^{m} p_j h(t_j) = d_0 + \sum_{i=1}^{n} d_i y_i \tag{2.1.6}$$

Theorem 2.1.4. *Let* $y \in int(V)$ *be fixed. Then the following are equivalent:*

(i) $\exists \mu \in M^+(X)$ *such that* $\mu(g) = y$ *and* $\mu(h) = L(y|h)$, *i.e. infimum is attained.*

(ii) $\exists d^* \in D^*$ *satisfying* (2.1.4).

Furthermore for almost all $y \in int(V)$ *there exists at most one* $d^* \in D^*$ *satisfying (2.1.4).*

In many situations the above infimum is not attained so that Theorem 2.1.3 is not applicable. The next theorem has more applications. For that set

$$\eta(z) := \underline{lim}_{\delta \to 0} \inf_t \{h(t) : t \in X, |g(t) - z| < \delta\}. \tag{2.1.7}$$

If $\epsilon \geq 0$ and $d^* \in D^*$, define

$$C_\epsilon(d^*) := \{z \in \overline{g(T)} : 0 \leq \eta(z) - \sum_{i=0}^{n} d_i z_i \leq \epsilon\}, (z_0 = 1) \tag{2.1.8}$$

and

$$G(d^*) := \cap_{N=1}^{\infty} \overline{conv} C_{1/N}(d^*) \tag{2.1.9}$$

It is easily proved that $C_\epsilon(d^*)$ and $G(d^*)$ are closed, furthermore $B(d^*) \subset C_0(d^*) \subset C_\epsilon(d^*)$, where $B(d^*)$ is defined by (2.1.3).

Theorem 2.1.5. *Let* $y \in int(V)$ *be fixed.*

(i) *Let* $d^* \in D^*$ *be such that* $y \in G(d^*)$. *Then*

$$L(y|h) = d_0 + d_1 y_1 + \ldots + d_n y_n. \tag{2.1.10}$$

(ii) *Assume that* g *is bounded. Then there exists* $d^* \in D^*$ *satisfying*

$$y \in conv\, C_0(d^*) \subset G(d^*)$$

and

$$L(y|h) = d_0 + d_1 y_1 + \ldots + d_n y_n.$$

(iii) *We further obtain, whether or not* g *is bounded, that for almost all* $y \in int(V)$ *there exists at most one* $d^* \in D^*$ *satisfying* $y \in G(d^*)$.

(§2). The next part of this section comes from the Kemperman (2) (1983) "Moment Theory," class notes, University of Rochester. Our aim here is to present explicitly the moment method of optimal distance which is purely a geometric method

Let again $g_1, \ldots, g_n, h$ be real-valued Borel measurable functions on $X := (X, A)$ and

$$A_n := \{y = (y_1, \ldots, y_n) \in \mathbb{R}^n : \int_X g_i(t)\mu(dt) = y_i, i = 1, \ldots, n \tag{2.1.11}$$

holds for at least one finite or infinite (nonnegative) measure μ on $X\}$.

Here g_i is assumed to be μ–integrable, all $i = 1, \ldots, n$. A_n is called the *moment space* corresponding to g_i and their integrals are called *moments.*

Theorem 2.1.6. *The set* A_n *coincides with the convex cone convc* $(g(X))$ *generated by the points* $g(t) := (g_1(t), \ldots, g_n(t)), t \in X$.

Theorem 2.1.7. *Let* $y = (y_1, \ldots, y_n) \in \mathbb{R}^n$ *be given. Then the following statements are equivalent:*

(i) $y \in A_n$;

(ii) $\exists\mu$ *measure on* X *having a finite support and satisfying (2.1.11) moment conditions;*

(iii) $\exists\mu$ *measure on* X *having a finite support consisting of at most* n *points and satisfying (2.1.11) moment conditions;*

(iv) $y \in convc(g(X))$.

Next we fix the *moment point* $y = (y_1, \ldots, y_n) \in A_n$. Let $M[y]$ denote the set of all measures (finite or infinite) satisfying (2.1.11) moment conditions and let $M_f[y]$ consist of all $\mu \in M[y]$ which are of finite support. From Theorem 2.1.7 follows that $M_f[y] \neq \phi$.

Let again $h : X \longrightarrow \mathbb{R}$ be given measurable function. We are interested in finding

$$U(y|h) = \sup\{\mu(h) : \mu \in M[y]; \mu(|h|) < +\infty\};$$

and

$$L(y|h) = \inf\{\mu(h) : \mu \in M[y]; \mu(|h|) < +\infty\}. \tag{2.1.12}$$

(Here, $\mu(h) = \int_X h(t)\mu(dt)$).

It is clear that

$$U(y|h) = \sup\{y_{n+1} : (y_1, \ldots, y_n, y_{n+1}) \in A_{n+1}\}$$

and

$$L(y|h) = \inf\{y_{n+1} : (y_1, \ldots, y_n, y_{n+1}) \in A_{n+1}\}. \tag{2.1.13}$$

Here A_{n+1} denotes the moment space corresponding to the $n+1$ functions $g_1(t), \ldots, g_n(t), h(t)$.

Again here we consider that $g_1(t), \ldots, g_n(t)$ are linearly independent, equivalently by $A_n = conv c(g(X))$ has a non-empty interior. Obviously $A_{n+1} = \underset{t \in X}{conv c}\,(g_1(t), \ldots, g_n(t), h(t))$ is not always a closed set. Therefore $(y_1, \ldots, y_n, U(y|h)), (y_1, \ldots, y_n, L(y|h))$ may be boundary points of A_{n+1}. Namely, $U(y|h)$ is equal to the *largest* distance between $(y_1, \ldots, y_n, 0)$ and $(y_1, \ldots, y_n, z) \in \overline{A_{n+1}}$, also $L(y|h)$ is equal to the *smallest* distance between $(y_1, \ldots, y_n, 0)$ and $(y_1, \ldots, y_n, z) \in \overline{A_{n+1}}$.

The above constitute the very essence *of the moment method of optimal distance.*

Theorem 2.1.8. *Let $y = (y_1, \ldots, y_n)$ is a given point of int(A_n). Then*

$$U(y|h) = \inf_c \left(\sum_{i=1}^{n} c_i y_i\right), \tag{2.1.14}$$

where $c = (c_1, \ldots, c_n)$ runs through the closed and convex set C of all points in $\mathbb{R}^n$ such that

$$h(t) \leq \sum_{i=1}^{n} c_i g_i(t), \quad \text{all } t \in X. \tag{2.1.15}$$

In particular, $U(y) = +\infty$ iff $C = \phi$. Furthermore, the infimum in (2.1.14) is attained if $U(y)$ is finite.

Remark 2.1.1. Often we are only interested in probability measures. Then we choose

$$g_1(t) \equiv 1, \quad \text{all } t \in X; y_1 = 1. \tag{2.1.16}$$

In that case one proves that

$$y := (1, y_2, \ldots, y_n) \in A_n \text{ iff } \bar{y} := (y_2, \ldots, y_n) \in$$

$$B_{n-1} := \underset{t\in X}{conv}\,(g_2(t), \ldots, g_n(t)) \subset \mathbb{R}^{n-1}.$$

And

$$(1, y_1, \ldots, y_n) \in \text{int}\,(A_n) \text{ iff } (y_2, \ldots, y_n) \in \text{int}\,(B_{n-1}).$$

Set

$$\overline{U}(\bar{y}|h) := \sup\{\mu(h) : \mu(g_1) = 1; \mu(g_i) = y_i, i = 2, \ldots, n\} \tag{2.1.17}$$

Therefore Theorem 2.1.8 specializes to

Theorem 2.1.9 *Let $\bar{y} \in \mathbb{R}^{n-1}$ be given such that $\bar{y} = (y_2, \ldots, y_n) \in int(B_{n-1})$. Then*

$$\bar{U}(\bar{y}|h) = \inf_{c\in C}(c_1 + c_2y_2 + \ldots + c_ny_n), \tag{2.1.18}$$

where c runs through the set C of n-tuples $c = (c_1, \ldots, c_n)$ such that

$$h(t) \leq c_1 + c_2g_2(t) + \ldots + c_ng_n(t), \quad \text{all } t \in X. \tag{2.1.19}$$

In particular, $\bar{U}(\bar{y}|h) = +\infty$ iff $C = \phi$. If $\bar{U}(\bar{y}|h) < +\infty$, then the above infimum (2.1.18) is attained.

Remark 2.1.2. Method of optimal distance for probability measures.

Consider $g_i; i = 1, \ldots, n, h$ etc. as in (§1) of this section. Call $M := \underset{t\in X}{conv}\,(g_1(t), \ldots, g_n(t), h(t))$. Then $L(y|h)$ is equal to the *smallest* distance between $(y_1, \ldots, y_n, 0)$ and $(y_1, \ldots, y_n, z) \in \bar{M}$. Also, $U(y|h)$ is equal to the *largest* distance between $(y_1, \ldots, y_n, 0)$ and $(y_1, \ldots, y_n, z) \in \bar{M}$. This method will be applied a lot in the following chapters. The main difficulty of the last method is to fully determine $\bar{M}$. This is possible in $\mathbb{R}^2$ (almost always), some times possible in $\mathbb{R}^3$ and almost impossible in $\mathbb{R}^n, n \geq 4$. In general it is not all that easy to study the convex hull of the graph of a real-valued function $y = f(x), x \in [a, b]$. One can prove that

$$F^+(u) = \inf_p(pu + \sup_t(f(t) - pt)),$$

$$\tag{2.1.20}$$

and

$$F^-(u) = \sup_p(pu + \inf_t(f(t) - pt)).$$

Here F^+, F^- represent the upper and lower boundaries, respectively, of the convex hull of the graph of $\{(x, f(x))\}$. And of course it is not always easy to give F^+, F^- in a closed form.

Let K be a simply connected subset of $\mathbb{R}^n$ and $g : K \longrightarrow \mathbb{R}$ i.e. $z = g(x_1, \ldots, x_n)$. Then one can generalize and determine the upper H^+ and lower H^- envelopes of the closed convex hull of $(x_1, \ldots, x_n, z)$. We have

$$H^+(u_1, \ldots, u_n) = \inf_{(\alpha_1,\ldots,\alpha_n)} \{(\alpha_i u_1 + \ldots + \alpha_n u_n)$$

$$+ \sup_{(x_1,\ldots,x_n)} (g(x_1, \ldots, x_n) - (\alpha_1 x_1 + \ldots + \alpha_n x_n))\} \tag{2.1.21}$$

and

$$H^-(u_1, \ldots, u_n) = \sup_{(\alpha_1,\ldots,\alpha_n)} \{(\alpha_1 u_1 + \ldots + \alpha_n u_n)$$

$$+ \inf_{(x_1,\ldots,x_n)} (g(x_1, \ldots, x_n) - (\alpha_1 x_1 + \ldots + \alpha_n x_n))\}.$$

Example 2.1.1. Let μ denote probability measures on $[0, a], a > 0$, i.e. $\mu([0, a]) = 1$. Fix $0 < d < a$. Find

$$U := \sup_{\mu} \int_{[0,c]} t^2 \mu(dt) \quad \text{and} \quad L := \inf_{\mu} \int_{[0,c]} t^2 \mu(dt), \tag{2.1.22}$$

Knowing that

$$\int_{[0,c]} t \mu(dt) = d. \tag{2.1.23}$$

Consider the graph

$$G := \{(t, t^2) : 0 \leq t \leq a\}.$$

Call $M := \overline{convG} = \text{conv } G$.

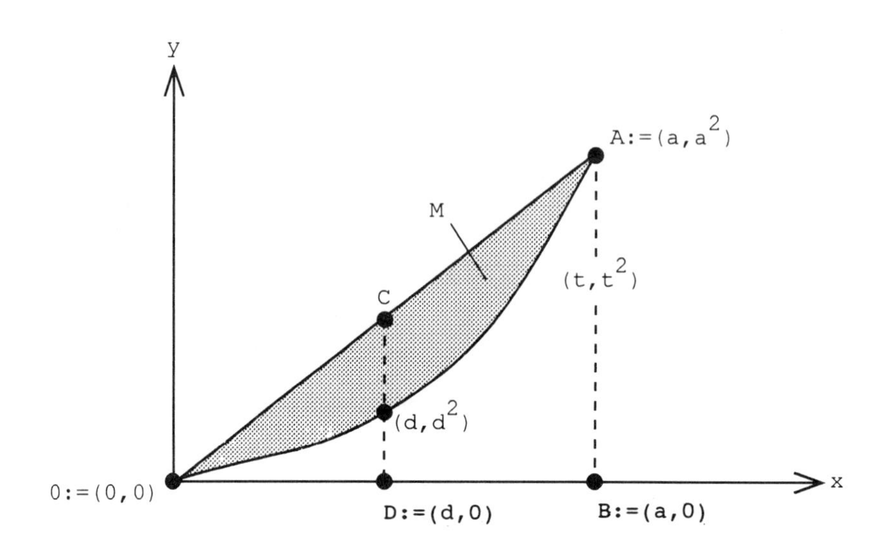

Figure 2.1.1: $M = \text{conv}G$

A direct application of the optimal distance method gives us

$$L = d^2 \quad \text{(an optimal measure } \mu \text{ is supported at } d \text{ with mass 1)} \tag{2.1.24}$$

and

$$U = \overline{DC}.$$

Note that triangle $\overset{\triangle}{ODC}$ is similar to $\overset{\triangle}{OBA}$. Therefore

$$\frac{\overline{DC}}{\overline{BA}} = \frac{\overline{OD}}{\overline{OB}} \quad i.e.$$

$$\frac{U}{a^2} = \frac{d}{a}, \quad \text{or } U = \tfrac{d}{a} \cdot a^2.$$

Hence

$$U = d \cdot a. \tag{2.1.25}$$

An optimal measure μ is supported at 0 and a with masses $(1 - \frac{d}{a})$ and $\frac{d}{a}$, respectively.

(§3) We continue from (§1), again the presented material is taken from Kemperman (1968).

Here we treat the special case of $h = \mathcal{X}_S$, where $\mathcal{X}_S$ is the characteristic function of a subset S of X. Call $S' = X - S$ i.e. $S' \cup S = X$ and $S' \cap S = \phi$. Assume that both $S, S' \neq \phi$. To remind $g(t) = (g_1(t), \ldots, g_n(t), t \in X$. Consider the convex sets

$$V_S = \text{convg } (S), \quad V_{S'} = \text{convg } (S') \quad V = \text{convg } (X)$$

and

$$W_S = \overline{V_S}, W_{S'} = \overline{S'}, W = \overline{V}. \tag{2.1.26}$$

Denote $L(y|\mathcal{X}_S) = L_S(y)$, which is the smallest mass in S for a probability measure μ on X with $\mu(g) = y$. Clearly,

$$L_S(y) = 0 \text{ if } y \in V_{S'}. \tag{2.1.27}$$

Also, one can easily see that

$$L_S(y) = 0, \quad \text{if } y \in W_{S'}, y \in \text{int}(V). \tag{2.1.28}$$

Thus, we only consider the case of

$$y \in \text{ int } (V) = \text{int}(W); y \notin W_{S'} \tag{2.1.29}$$

Here the function η, see (2.1.7), collapses to

$$\begin{aligned} \eta(z) &= 0, \quad \text{if} \quad z \in \overline{g(S')}, \\ \eta(z) &= 1, \quad \text{if} \quad z \in \overline{g(S)}, z \notin \overline{g(S')}. \end{aligned} \tag{2.1.30}$$

So $d^* \in D^*$ iff

$$d_0 + \sum_{i=1}^{n} d_i z_i \leq 1,$$

all $z \in \overline{g(X)}$, hence for all $z \in W$, and

$$d_0 + \sum_{i=1}^{n} d_i z_i \leq 0, \text{ all } z \in \overline{g(S')}, \text{ hence for all } z \in W_{S'} \tag{2.1.31}$$

Suppose momentarily that not all of $d_1, \ldots, d_n$ are zero, and introduce the distinct parallel hyperplanes in $\mathbb{R}^n$ defined as follows

$$H := H(d^*) := \{z : d_0 + \sum_{i=1}^{n} d_i z_i = 1\}; \tag{2.1.32}$$

$$H' := H'(d^*) := \{z : d_0 + \sum_{i=1}^{n} d_i z_i = 0\} (z := (z_1, \ldots, z_n) \in \mathbb{R}^n).$$

Condition (2.1.31) implies that H supports all of W and H' supports all of $W_{S'}$, on the same side as H supports W (so that H' is in between H and $W_{S'}$).

We call such a pair of *distinct* hyperplanes H, H' an *admissible* pair. Given a pair of hyperplanes as above described there exists a unique $(n+1)-$ tuple $d^* = (d_1, \ldots, d_n)$, not all the components zero, such that H, H' are given from (2.1.32). Furthermore (2.1.31) is true i.e. $d^* \in D^*$.

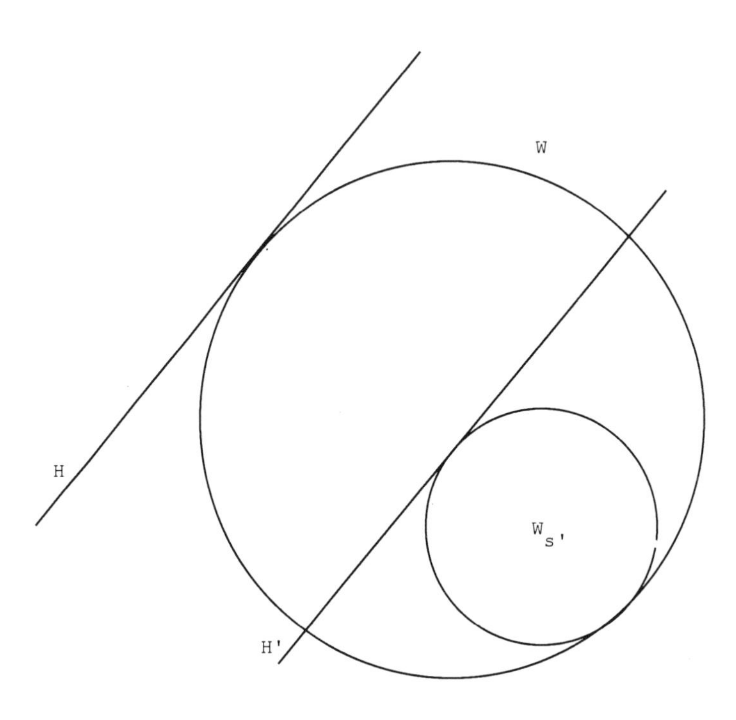

Figure 2.1.2: admissible H, H' for $L_S(y)$

Theorem 2.1.10. (i) *Given a pair of admissible hyperplanes* $H := H(d^*)$ *and* $H' := H'(d^*)$. *Set*

$$G(d^*) := \overline{conv}[(H \cap W_S) \cup (H' \cap W_{S'})] \tag{2.1.33}$$

Then for each $y \in G(d^*)$ *with* $y \in int(V)$ *we have*

$$L_S(y) = d_0 + \sum_{i=1}^{n} d_i y_i = \frac{\Delta(y)}{\Delta}. \tag{2.1.34}$$

Here, $\Delta(y)$ *denotes the distance from* y *to* H' *and* Δ *denotes the distance between the parallel hyperplanes* H *and* H'.

(ii) *For almost all* $y \in int(V)$ *there exists at most one admissible pair* $H(d^*), H'(d^*)$ *so that* $y \in G(d^*)$.

(iii) *Assume that* g *is bounded. Then*

$$G(d^*) = conv[(H \cap \overline{g(S)}) \cup (H' \cap \overline{g(S')})]. \tag{2.1.35}$$

(iv) *Assume again* g *bounded. For all* $y \in int(V) = int(W), y \notin W_{S'}$ *there exists at least one admissible pair* $H(d^*), H'(d^*)$ *such that* $y \in G(d^*)$, *(in this case again* $L_S(y) = d_0 + \sum_{i=1}^{n} d_i y_i = \frac{\Delta(y)}{\Delta}$ *as in (2.1.34)).*

The content of above Theorem 2.1.10, especially (2.1.34), is the very essence of the *optimal ratio* method for the infimum case. Similarly is treated the supremum case.

Non-interior Points of a Moment Space

Lemma 2.1.2. *Let* $K \neq \phi$ *be a convex subset of* $\mathbb{R}^m, m \geq 1$ *and let* $y \in \mathbb{R}^m$. *Then each of the next three properties defines the set* K_y.

(i) *If* y *is an extreme point of* K *then* $K_y = \{y\}$. *Otherwise,* K_y *is the union of all closed line segments* $[y', y'']$ *which are entirely contained in* K *and have* y *as an interior point.*

(ii) K_y *is the set of those* $z, z' \in K$ *which occur in some representation of* y *of the form*

$$y = \lambda z + (1 - \lambda) z', \quad 0 < \lambda < 1. \tag{2.1.36}$$

(iii) K_y *is the largest convex subset of* K *such that*

$$y \in int_K(K_y) \tag{2.1.37}$$

Denote by $m^+ := m^+(X)$ the set of all probability measures on X such that $\mu(|g_i|) < +\infty$, $i = 1, \ldots, n$ and $\mu(|h|) < +\infty$. Consider $g(t) := (g_1(t), \ldots, g_n(t)), t \in X$, that is $g : X \longrightarrow \mathbb{R}^n$ is a measurable function and $V := convg(X)$. Set

$$X^y := g^{-1}(V_y), \qquad \Gamma^y := g(X) \cap V_y. \tag{2.1.38}$$

Here V_y is defined as in Lemma 2.1.2 and K is replaced by V.

Theorem 2.1.11. *Let $y \in V$ be fixed. Then each probability measure $\mu \in m^+(X)$ such that $\mu(g) \in V_y$ is concentrated on the measurable subset X^y of X; the converse is clear. Also it holds*

$$V_y = conv(\Gamma^y). \tag{2.1.39}$$

Remark 2.1.3. Theorem 2.1.11 implies the following: Let $S \subset X$ and $y \in V_y$ be fixed. Then by this theorem all involved measures are concentrated on X^y. Here $X^y \subset X$ properly, if $y \notin \text{int}_F(V)$ (where F is the smallest translate of a linear subspace of $\mathbb{R}^n$ containing V).

Let g_y be the restriction of g on X^y. Hence, the range of g_y equals Γ^y. Therefore by (2.1.39), the role of V is taken over by V_y. From (2.1.37), $y \in int_F(V_y)$, thus, *we go back to the "original" and desired situation*, where now X is replaced by X^y, g replaced g_y and V replaced by V_y.

A Special Case Which Is Important

Let $S \subset X$ and let again

$$L_S(y) = \inf\{\mu(S) : \mu \in M^+(X), \mu(g) = y\}.$$

Assume S and $S' = X - S$ are not empty and $g_0(= 1), g_1, g_2, \ldots, g_n$ are linearly independent so that $\text{int}V \neq \phi, V := \text{convg}(X)$. V_S, W_S are defined as in (2.1.26). Let $y \in V$ then there exists $\mu \in M^+(X)$ such that $\mu(g) = y$. Separating the support of μ into points $t_i \in S$ and $t_i \in S'$, we get

$$y = \lambda u + (1 - \lambda)u' \tag{2.1.40}$$

where

$$u \in V_S, u' \in V_{S'}, 0 \leq \lambda \leq 1. \tag{2.1.41}$$

Namely we have $\lambda = \mu(S)$.

Conversely each representation (2.1.40) of y fulfilling (2.1.41) comes in this way from at least one $\mu \in M^+(X)$ such that $\mu(g) = y$ and $\mu(S) = \lambda$.

If (2.1.40) and (2.1.41) hold, we say that y has a *$\lambda-$representation* and we name (u, u', λ) a *$V-$representation* of y. We call (u, u', λ) a *$W-$representation of y* if it satisfies (2.1.40) and

$$u \in W_S, u' \in W_{S'}, 0 \leq \lambda \leq 1 \tag{2.1.42}$$

Let $y \in \mathbb{R}^n$ be given. Further, $\inf_{(W)} \lambda$ and $\inf_{(V)} \lambda$ are taken over all λ for which we can find a (u, u', λ) which is a $W-$representation of y or a $V-$representation of y, respectively.

Lemma 2.1.3. *We obtain*

$$\underline{lim}_{y' \to y} L_S(y') \leq \; inf_{(W)}\lambda \leq \; inf_{(V)}\lambda = L_S(y). \tag{2.1.43}$$

Furthermore, if g is bounded then the first equality sign holds. If $y \in int(V)$ then all equality signs hold. In a similar may we get

$$U_S(y) = \; sup_{(V)}\lambda \leq \; sup_{(W)}\lambda \leq \overline{lim}_{y' \to y} U_S(y'). \tag{2.1.44}$$

Lemma 2.1.4. *If $y \in V_{S'}$ or $y \in W_{S'} \cap int(V)$ then $L_S(y) = 0$. Conversely, if g is bounded and $L_S(y) = 0$ then $y \in W_{S'}$. Similarly for $U_S(y) = 1 - L_{S'}(y)$. Hence, if g is bounded, $y \in int(V)$ then $U_S(y) = 1$ iff $y \in W_S$. At the end,*

$$0 \leq L_S(y) < U_S(y) \leq 1, \quad all \; y \in int(V), \tag{2.1.45}$$

unless $g(S)$ and $g(S')$ are located in distinct parallel hyperplanes.

Notation. Let $C \neq \phi$ be a subset of $\mathbb{R}^n$ and $d \in \mathbb{R}^n$. Set

$$\phi_d(C) := \sup_{y \in C} \sum_{i=1}^{n} d_i y_i \tag{2.1.46}$$

and

$$H_d(C) := \{x \in \mathbb{R}^n : \sum_{i=1}^{n} d_i x_i = \phi_d(C)\}. \tag{2.1.47}$$

We put

$$\phi_d := \phi_d(W_S), \phi_d' = \phi_d'(W_S)$$

and

$$H_d := H_d(W_S), H_d' := H_d(W_{S'}). \tag{2.1.48}$$

Therefore,

$$H_d := \{x \in \mathbb{R}^n : \sum_{i=1}^{n} d_i x_i = \phi_d\}$$

is the hyperplane supporting W_S in the direction d as well as possible (hence, also $g(S)$ and V_S). Similarly, H'_d is the hyperplane that supports the sets $g(S')$, $V_{S'}, W_{S'}$ as well as possible in the direction d.

Theorem 2.1.12. **(i)** *Let $y \in \mathbb{R}^n$ be given. Assume that $d \in \mathbb{R}^n$ and $\gamma \in \mathbb{R}$ are such that y admits a representation of the type*

$$y = \gamma u + (1-\gamma)u', 0 \leq \gamma \leq 1, \tag{2.1.49}$$

where

$$u \in H_d \cap V_S \text{ and } u' \in H'_d \cap V_{S'}. \tag{2.1.50}$$

Then

$$\gamma = L_S(y), \text{ if } \phi'_d < \phi_d < +\infty, \tag{2.1.51}$$

and

$$\gamma = U_S(y), \text{ if } \phi_d < \phi'_d < +\infty. \tag{2.1.52}$$

(ii) *If $y \in int(V)$, then (2.1.51) and (2.1.52) are still true when (2.1.50) is replaced by the weaker condition*

$$u \in H_d \cap W_S \quad \text{and} \quad u' \in H'_d \cap W_{S'}. \tag{2.1.53}$$

In the paragraph following (2.1.32) we introduced the notion of an admissible pair H, H' of hyperplanes in $\mathbb{R}^n$ (with respect to $S \subset X$). Using (2.1.48) the admissible pair H, H' is exactly the pair

$$H = H_d = H_d(W_S), H' = H'_d = H_d(W_{S'}),$$

where

$$\phi'_d < \phi_d < +\infty. \tag{2.1.54}$$

Proposition 2.1.1. *Let H, H' be an admissible pair of hyperplanes in $\mathbb{R}^n$ as in (2.1.54). Then for each point $y \in int(V)$ so that $y \in G_d$, where*

$$G_d := \overline{conv}[(H \cap W_S) \cup (H' \cap W_{S'})] \tag{2.1.55}$$

we have

$$L_S(y) = \frac{\Delta(y)}{\Delta}. \tag{2.1.56}$$

Here, $\Delta(y)$ is the distance from y to H' and Δ is the distance between the parallel hyperplanes H and H'. (The last is another expression of the optimal ratio method.)

Next let us list the following five statements related to a given $y \in V$.

(i) For some $\mu \in M^+(X)$ we have $\mu(g) = y$ and $\mu(S) = L_S(y)$.

(ii) From all V−representations (u, u', λ) of y there exists one for which λ is minimal.

(iii) For some $d \in \mathbb{R}^n$ we have

$$\phi_d' < \phi_d < +\infty \quad \text{and} \quad y \in conv[(H_d \cap V_S) \cup (H_d' \cap V_{S'})]. \tag{2.1.57}$$

(iv) From all $W-$ representations (u, u', λ) of y there exists one for which λ is minimal.

(v) For some $d \in \mathbb{R}^n$ we have

$$\phi_d' < \phi_d < +\infty \quad \text{and} \quad y \in conv[(H_d \cap W_S) \cup (H_{d'} \cap W_{S'})] \tag{2.1.58}$$

Notice that in (iii) and (v) the pair of hyperplanes H_d, H_d' is an admissible one.

Furthermore $L_S(y)$ can be given from either (2.1.51) or (2.1.56).

From Theorem 2.1.10 we obtain for almost all y, that if $\{H_d, H_d'\}$ in (v) exists then is unique.

Theorem 2.1.13. *Assume that*

$$y \in int(V), y \notin W_{S'} \tag{2.1.59}$$

Then the above properties (i), (ii), (iii) are equivalent. A sufficient condition for each of these is that both V and $V_{S'}$ are compact. Furthermore, the above properties (iv) and (v) are equivalent. A sufficient condition for each of them is that g is bounded.

Optimal Ratio Method Crystalized

Let $g_i, i = 1, \ldots, n$ real-valued Borel measurable functions on the measurable space $X := (X, A)$. Consider probability measures μ on X such that $\mu(|g_i|) < +\infty, i = 1, \ldots, n$ (here $\mu(|g_i|) := \int_X |g_i| d\mu$). Assume that $\mu(g_i) = y_i, i = 1, \ldots, n$, where $y := (y_1, \ldots, y_n) \in V := convg(X)$, with $g := (g_1, \ldots, g_n)$. Let $\phi \neq S \subset X$ and $S' := X - S \neq \phi$. We would like to find $L_S(y) := \inf \mu(S)$ and $U_S(y) := \sup \mu(S)$, over all probability measures μ such that $\mu(g_i) = y_i, i = 1, \ldots, n$ as above described.

Call $W_S := \overline{convg(S)}, W_{S'} = \overline{convg(S')}$ and $W := \overline{convg(X)}$.

(I) Finding $L_S(y)$

1) Pick a boundary point z of W and *"draw"* through z a hyperplane H of support to W.

2) Determine the hyperplane H' parallel to H which supports $W_{S'}$ as well as possible, and on the same side as H supports W. We are only interested in $H \neq H'$ in which case H' is between H and $W_{S'}$.

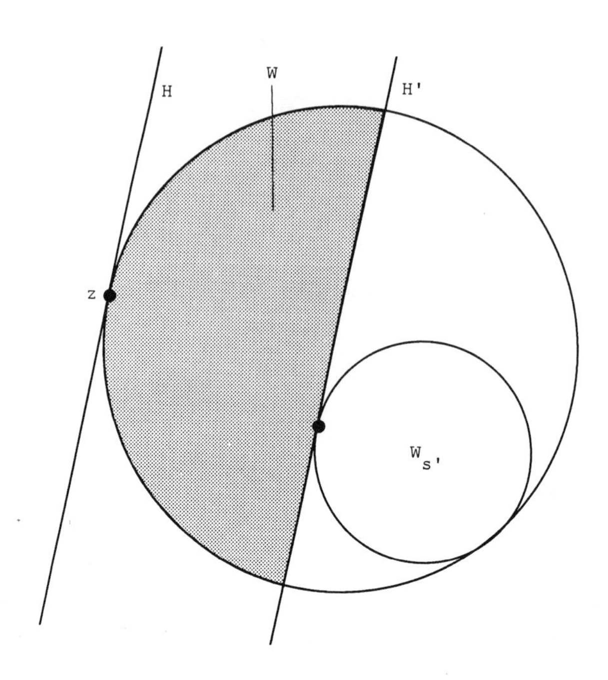

Figure 2.1.3: $L_S(y)$ values

Set

$$A_d := W \cap H = W_S \cap H$$

and

$$B_d := W_{S'} \cap H'.$$

Here, d stands for the common normal to the parallel hyperplanes H and H', pointing from H' to H. Hence, in the notation (2.1.47), we have

$$H = H_d = H_d(W), H' = H'_d = H_d(W_{S'} \tag{2.1.60}$$

and furthermore $\phi'_d < \phi_d < +\infty$.

3) Given that $H' \neq H$, set

$$G_d := \overline{conv}(A_d \cup B_d) \tag{2.1.61}$$

From Proposition 2.1.1 we find that

$$L_S(y) = \frac{\Delta(y)}{\Delta}, \quad \text{for each } y \in int(V) \quad \text{such that } y \in G_d. \tag{2.1.62}$$

Here, $\Delta(y)$ is the distance from y to H' and Δ is the distance between the distinct parallel hyperplanes H, H'.

Comment 2.1.1

Note that for each $y \in int(V)$ with $y \notin W_{S'}$, and whenever g is bounded, there always exist H, H' as above with $y \in G_d$, allowing us to obtain $L_S(y)$ from (2.1.62). Theorem 2.1.10 quarantees that for almost all y there exists at most one admissible pair $H_d, H_{d'}$ with $y \in G_d$.

The process to find $U_S(y) = 1 - L_{S'}(y)$ is very similar to the above in (I).

(II) Finding $U_S(y)$

1) Pick a boundary point z of W_S and *"draw"* through z a hyperplane H of support to W_S. Set $A_d := W_S \cap H$.

2) Determine the hyperplane H' parallel to H which supports $g(X)$ and hence W as well as possible, and on the same side as H supports W_S. We are interested only in $H' \neq H$ in which case H is between H' and W_S.

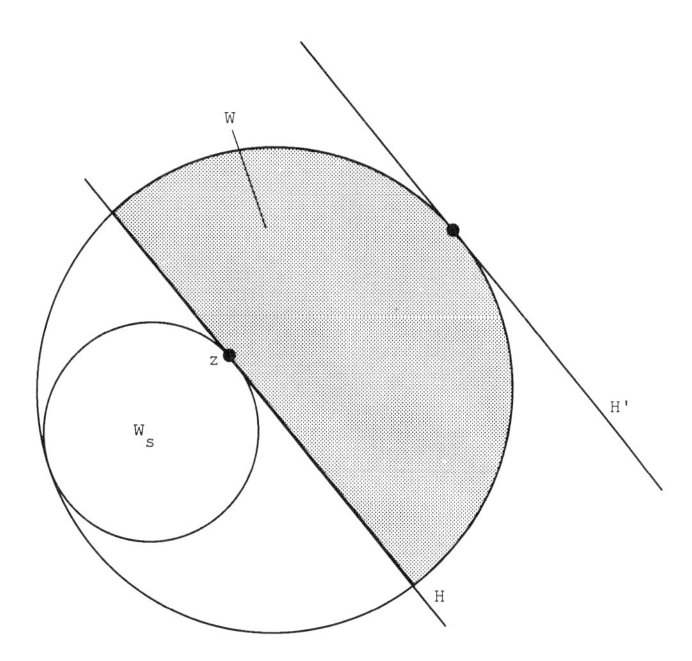

Figure 2.1.4: $U_S(y)$ values

Set $B_d := W \cap H' = W_{S'} \cap H'$. Let G_d as in (2.1.61). Then

$$U_S(y) = \frac{\Delta(y)}{\Delta}, \quad \text{for each } y \in int(V), \text{ where } y \in G_d, \tag{2.1.63}$$

assuming that H and H' are distinct. Here, $\Delta(y)$ and Δ are defined as in (2.1.62).

Comment 2.1.2

If $y \in W_S, y \in int(V)$ then $U_S(y) = 1$. If $y \notin W_S, y \in int(V)$ and g is bounded then there exist H, H' as above such that $U_S(y)$ can be gotten from (2.1.63). For almost all y $\{H, H'\}$ will be unique. Hence the different G_d's rarely intersect nontrivially.

Example 2.1.2.

Let $X := [\alpha, \beta]$ and $S := [a, b] \subset X$, such that $\alpha \neq a, \beta \neq b$. Assume $0 \in (a, b)$. Take $g_0(t) \equiv 1, g_1(t) = t, g_2(t) = t^2$. Set $g(t) := (g_1(t), g_2(t)) = (t, t^2)$.

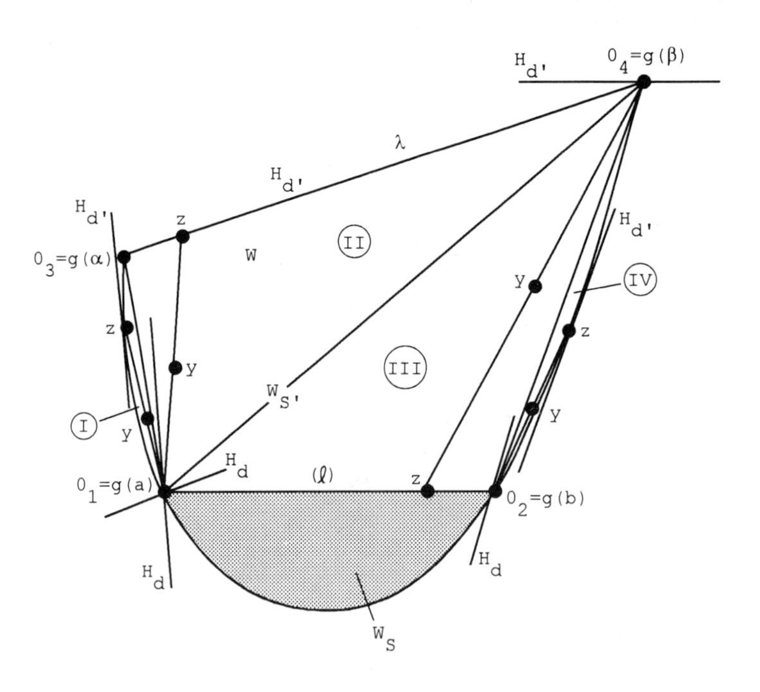

Figure 2.1.5: $g(t) = (t, t^2)$ for $U_S(y)$.

Here is assumed that λ and l intersect on the left. Denote $y := (y_1, y_2) \in intg(X)$. Assume that

$$\int_X t\mu(dt) = y_1, \int_X t^2\mu(dt) = y_2 \tag{2.1.64}$$

where μ is a probability measure on X.

(I) With respect to the last moment conditions (2.1.64) we would like to find

$$U_S(y) := \sup_{\mu} \mu(S).$$

Here we apply the method of optimal ratio in different cases. See Figure 2.1.5.

(i) If y belongs to region I we get

$$U_S(y) = \frac{\overline{yz}}{\overline{0_1 z}}. \tag{2.1.65}$$

(ii) If y belongs to region IV we have

$$U_S(y) = \frac{\overline{yz}}{\overline{0_2 z}}. \tag{2.1.66}$$

(iii) If y belongs to region II then z is on λ. Here $U_S(y)$ is given again by (2.1.65).

(iv) If y belongs to region III we obtain

$$U_S(y) = \frac{\overline{0_4 y}}{\overline{0_4 z}}. \tag{2.1.67}$$

Here z is on l.

(v) $U_S(y) = 1$ if $y \in W_S$.

(II) Next with respect to the moment condition (2.1.64) we would like to find

$$L_S(y) := \inf_{\mu} \mu(S).$$

Again here we apply the method of optimal ratio in different cases.
See Figure 2.1.6.

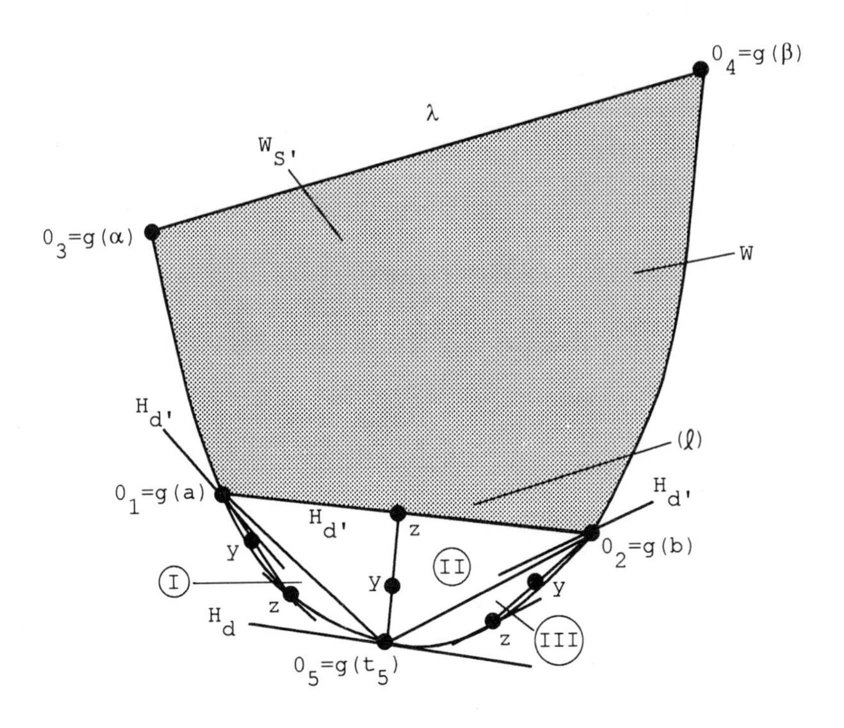

Figure 2.1.6: $g(t) = (t, t^2)$ for $L_S(y)$.

Let 0_5 denote the unique point on (t, t^2) from which passes a tangent line parallel to (l). This tangent supports $g(X)$ from below.

(i) If $y \in W_{S'}$, then $L_S(y) = 0$.

(ii) If y belongs to region I, then

$$L_S(y) = \frac{\overline{0_1 y}}{\overline{0_1 z}} \tag{2.1.68}$$

(iii) If y belongs to region II, then

$$L_S(y) = \frac{\overline{yz}}{\overline{0_5 z}} \tag{2.1.69}$$

Here z is on l.

(iv) If y belongs to region III, then

$$L_S(y) = \frac{\overline{y0_2}}{\overline{z0_2}}. \tag{2.1.70}$$

2.2. Convex Moment Methods

The material of this section-tool comes from Kemperman (1971).

Definition 2.2.1. Let $s \geq 1$ be a fixed natural number and let $x_0 \in \mathbb{R}$ be fixed. By $m_s(x_0)$ we denote the set of probability measures μ on $\mathbb{R}$ such that the associated cumulative distribution function (c.d.f) F possesses an $(s-1)$–th derivative $F^{(s-1)}(x)$ over $(x_0, +\infty)$ and furthermore $(-1)^s F^{(s-1)}(x)$ is convex in $(x_0, +\infty)$.
$m_0(x_0)$ denotes the class of all probability measures on $\mathbb{R}$.

Let $\mu \in m_s(x_0)$ with c.d.f. F. Since $(-1)^s F^{(s-1)}$ is convex in $(x_0, +\infty)$, then it is continuous there. One can easily see that

$$m_s(x_0) \subset m_{s-1}(x_0), (s \geq 2). \tag{2.2.1}$$

Thus,

$$(-1)^{j-1} F^{(j)}(x) \geq 0, \quad \text{all } x > x_0; j = 1, \ldots, s, \tag{2.2.2}$$

given that F corresponds to $\mu \in m_s(x_0)$. The set $m_s(x_0)$ is closed with respect to the convergence in distribution, therefore is compact in the weak*–topology.

A probability measure $\mu \in m_s(x_0)$ can be totally arbitrary on $(-\infty, x_0]$ and maybe have a positive mass at x_0; but on $(x_0 + \infty)$ it is always absolutely continuous with density $f(x) = F'(x+0)$. Since μ is also in $m_1(x_0)$ this density is even non-increasing.

Moment Problem 2.2.1

Let $g_i, i = 1, \ldots, n; h$ are Borel measurable functions from $\mathbb{R}$ into itself. These are assumed to be locally integrable on $[x_0, +\infty)$ relative to Lebesque measure. Consider $\mu \in m_s(x_0), s \geq 1$ such that

$$\mu(|g_i|) := \int_{\mathbb{R}} |g_i(t)| \mu(dt) < +\infty, i = 1, \ldots, n$$

and

$$\mu(|h|) := \int_{\mathbb{R}} |h(t)| \mu(dt) < +\infty. \tag{2.2.3}$$

Let $c := (c_1, \ldots, c_n) \in \mathbb{R}^n$ such that

$$\mu(g_i) = c_i, i = 1, \ldots, n, \mu \in m_s(x_0). \tag{2.2.4}$$

Here we would like to find

$$L(c) := \inf_{\mu} \mu(h)$$

(2.2.5)

and

$$U(c) := \sup_{\mu} \mu(h),$$

where μ is as in (2.2.3) and (2.2.4).

Here, our method will be to transform the above convex moment problem into an ordinary one, usually handled by section 2.1, as described earlier. For this we need:

Definition 2.2.2.

Consider here another copy of $(\mathbb{R}, \mathbb{B})$; $\mathbb{B}$ is the Borel σ–field, and further a given function $P(y, A)$ on $\mathbb{R} \times \mathbb{B}$.

Assume that for each fixed $y \in \mathbb{R}, P(y, \cdot)$ is a probability measure on $\mathbb{R}$ and for each fixed $A \in \mathbb{B}, P(\cdot, A)$ is a Borel-measurable real-valued function on $\mathbb{R}$. We call P a *Markov Kernel.* For each probability measure ν on $\mathbb{R}$, let $\mu := T\nu$ denote the probability measure on $\mathbb{R}$ given by

$$\mu(A) := (T\nu)(A) := \int_{\mathbb{R}} P(y, A)\nu(dy). \tag{2.2.6}$$

T is called a Markov transformation.

Next we Markov-transform the set $m_s(x_0)$.

To every $\mu \in m_S(x_0)$ (with associated c.d.f. F and density f on $x > x_0$), we assign the following nonnegative measures

$$\eta(A) := (-1)^s \cdot \int_A dF^{(s)}(u) \tag{2.2.7}$$

and

$$\nu(A) := \int_A \frac{(u - x_0)^s}{s!} \cdot \eta(du), \tag{2.2.8}$$

where A is a Borel subset of $(x_0, +\infty)$. We extend ν to a measure on all of $\mathbb{R}$ by

$$\nu(A) := \mu(A), \quad \text{all } A \subset (-\infty, x_0]. \tag{2.2.8$'$}$$

Here, the right hand derivative $F^{(s)}(x + 0)$ exists and is finite for all $x > x_0$ such that $(-1)^s \cdot F^{(s)}(x+0)$ is continuous to the right and non-decreasing. Furthermore, F is bounded, thus $F^{(s)}(+\infty) = 0$, and $(-1)^{(s)} \cdot F^{(s)}(x) \leq 0$ for all $x > x_0$ (the s–th derivative of F will be taken always as the right hand derivative). Because of $F^{(s)}(\infty) = 0$, the above measure η is finite on each $[x_0 + \epsilon, +\infty)$, where $\epsilon > 0$. Hence ν on $\mathbb{R}$ will be a probability measure. If μ is supported by an interval $(-\infty, x_1]$ then $F^{(s)}(x) = 0$ for all $x >$max(x_0, x_1). Therefore, μ is supported by $(-\infty, x_1]$ iff ν is supported by $(-\infty, x_1]$.

Lemma 2.2.1. Let $\mu \in m_s(x_0)$ and let η and ν as in (2.2.7) and (2.2.8). Then

$$(-1)^{s-1-j} \cdot F^{(s-j)}(x) = \int_x^{+\infty} \frac{(u-x)^j}{j!} \cdot \eta(du), \tag{2.2.9}$$

for all $x > x_0$ and all $j = 0, 1, \ldots, s-1$. In particular

$$f(x) = \int_x^{+\infty} \frac{(u-x)^{s-1}}{(s-1)!} \cdot \eta(du). \tag{2.2.10}$$

Hence,

$$\bar{F}(x) := 1 - F(x) = \int_x^{+\infty} \frac{(u-x)^s}{s!} \cdot \eta(du), \quad \text{if } x \geq x_0 \tag{2.2.11}$$

Especially,

$$\nu((x_0, +\infty)) = \int_{x_0}^{+\infty} \frac{(u-x_0)^s}{s!} \cdot \eta(du) = \bar{F}(x_0), \tag{2.2.12}$$

showing that ν is actually a probability measure.

(2.2.10) can be rewritten as

$$f(x) = \int_{\mathbb{R}} K_s(u,x)\nu(du), \quad \text{if } x > x_0, \tag{2.2.13}$$

where the kernel $K_s(u,x)$ is given by

$$K_s(u,x) := \begin{cases} \frac{s\cdot(u-x)^{s-1}}{(u-x_0)^s}, & \text{if } x_0 < x < u, \\ 0, & \text{elsewhere.} \end{cases} \tag{2.2.14}$$

Notice that $K_s(u,x) \geq 0$ and

$$\int_{\mathbb{R}} K_s(u,x)dx = 1, \quad \text{all } u > x_0. \tag{2.2.15}$$

In view of (2.2.8)′ and (2.2.13), to each probability measure ν on $\mathbb{R}$ we correspond another measure $\mu = T\nu$ defined the following way:

(i) if $A \subset (-\infty, x_0]$ then $\mu(A) = \nu(A)$;

(ii) if $A \subset (x_0, +\infty)$ then $\mu(A) = \int_A f(x)dx$;

where f is given by (2.2.13). Using (2.2.15), we get that μ is also a probability measure. Actually, this transformation T is exactly a Markov transformation

$$(T\nu)(A) = \int_{\mathbb{R}} P_s(u,A)\nu(du), \tag{2.2.16}$$

given that we take

$$P_s(u,A) := \begin{cases} \delta_u(A), & \text{if } u \le x_0; \\ \int_A K_s(u,x)dx, & \text{if } u > x_0. \end{cases} \quad (2.2.17)$$

Here δ_u is the unit (Dirac) measure at u. Denote $\mu_u := T\delta_u$, that is $\mu_u(A) = P_s(u,A)$. If $u \le x_0$ then $\mu_u = \delta_u$. If $u > x_0$ then μ_u is absolutely continuous (relative to Lebesque measure) having support $[x_0, u]$ and there a density $a(u-x)^{s-1}, (x_0 \le x \le u; a$ is a constant depending on $u)$. If $s = 0$ was allowed then take $P_0(u,A) = \delta_u(A)$ for all u, and here T is the identity transformation for measures and functions.

Theorem 2.2.1.

Let $x_0 \in \mathbb{R}$ and natural number $s \ge 1$ be fixed. Define the transformation $\mu = T\nu$ as in (2.2.16), (2.2.17) with $K_s(u,x)$ as in (2.2.14). Then the correspondence $\nu \longrightarrow \mu = T\nu$ is an (1-1) correspondence between the set m^* of all probability measures ν on $\mathbb{R}$ and the set $m_s(x_0)$ of all probability measures μ on $\mathbb{R}$ as in Definition 2.2.1. Therefore, $\mu \in m_s(x_0)$ iff its restriction to $(x_0, +\infty)$ is absolutely continuous with a density f admitting a representation as in (2.2.13) with ν as a nonnegative measure.

Corollary 2.2.1. Let $\mu \in m_s(x_0)$ and F is its corresponding distribution function then

$$\lim_{x \longrightarrow \infty} x^i \cdot F^{(i)}(x) = 0, (i = 1, \ldots, s), \quad (2.2.18)$$

and also

$$\lim_{x \downarrow x_0} (x - x_0)^i \cdot F^{(i)}(x) = 0, (i = 1, \ldots, s). \quad (2.2.19)$$

Remark 2.2.1

Theorem 2.2.1 verifies that each $\mu \in m_s(x_0)$ can be represented uniquely as a (convex) mixture

$$\mu = \int_{\mathbb{R}} \mu_u \nu(du)$$

of the special measures $\mu_u = T\delta_u$, where ν is a unique probability measure. It follows that $\mu_u (u \in \mathbb{R})$ are the extreme points of the convex set of probability measures $m_s(x_0)$. The above is related to the fact that $T : m^* \longrightarrow m_s(x_0)$ is $(1-1)$ and linear. Furthermore it is a homeomorphism given that m^* and $m_s(x_0)$ are endowed with the weak*$-$topology. Since T is (1-1) and m^* is compact, it is enough one to prove that T is continuous. Let $\phi : \mathbb{R} \longrightarrow \mathbb{R}$ be a bounded and continuous function. Introducing

$$\phi^*(u) := (T\phi)(u) := \int_{\mathbb{R}} \phi(x) \cdot P_s(u, dx), \quad (2.2.20)$$

we have

$$\int \phi d\mu = \int \phi^* d\nu.$$

One easily verifies that ϕ^* is bounded and continuous from $\mathbb{R}$ into itself. We also obtain (from (2.2.14) and (2.2.17)) that $\phi^*(u) = \phi(u)$, if $u \le x_0$; and

$$\begin{aligned}\phi^*(u) &= \int_{x_0}^{u} K_s(u,x)\cdot\phi(x)\cdot dx \\ &= \frac{s}{(u-x_0)^s}\int_{x_0}^{u}(u-x)^{s-1}\cdot\phi(x)dx \\ &= \int_0^1 \phi((1-t)u+tx_0)\cdot s\cdot t^{s-1}\cdot dt, \quad \text{if } u > x_0. \end{aligned} \tag{2.2.21}$$

Next we study more carefully the transformation $\phi^* = T\phi$ as given by (2.2.21). From now on, *let ϕ denote a measurable function on $[x_0,+\infty)$ which is locally integrable (relative to Lebesque measure)*, so the $\phi^*(u)$ is well-defined and finite for all $u \in \mathbb{R}$. If $u > x_0$ we have from (2.2.21) that

$$\frac{1}{s!}(u-x_0)^s\cdot\phi^*(u) = \frac{1}{(s-1)!}\cdot\int_{x_0}^{u}(u-x)^{s-1}\cdot\phi(x)\cdot dx. \tag{2.2.22}$$

Especially, if $r > -1$ we have for

$$\phi(u) := (u-x_0)^r \quad \text{that } \phi^*(u) = \tbinom{r+s}{s}^{-1}\cdot(u-x_0)^r, \tag{2.2.23}$$

for all $u > x_0$.

Here $r! := 1.2\ldots r(r\in\mathbb{R}; r > -1)$ and

$$\binom{r+s}{s} := \frac{(r+1)\ldots(r+s)}{s!} \tag{2.2.24}$$

It is interesting to see that the right hand side of (2.2.22) is equal to the s–fold integral $I^s\phi$, where

$$(I\psi)(u) := \int_{s_0}^{u}\psi(x)dx.$$

It follows that the right hand side of (2.2.22) has in $(x_0,+\infty)$ an $(s-1)$–th derivative which itself is absolutely continuous having almost everywhere a derivative equal to the locally integrable function $\phi(u)$. Also, $\phi^*(u)$ has in $(x_0,+\infty)$ an absolutely continuous $(s-1)$–th derivative such that

$$\begin{aligned}\phi(x) &= \frac{1}{s!}\cdot\left(\frac{d}{dx}\right)^s[(x-x_0)^s\cdot\phi^*(x)] \\ &= \sum_{j=0}^{s}\binom{s}{j}\cdot\frac{(x-x_0)^j}{j!}\cdot(\frac{d}{dx})^j\phi^*(x),\end{aligned} \tag{2.2.25}$$

for almost all $x > x_0$.

If we know ϕ, the differential equation (2.2.25) determines ϕ^* up to a function of the form

$$\sum_{j=1}^{s} a_j (x - x_0)^{-j},$$

where a_j are constants. Thus, (2.2.25) has at most one solution ϕ^* which is locally integrable at x_0.

Another conclusion from (2.2.22) is that for each $x_1 > x_0$ the function ϕ^* is more regular than ϕ by having there $(s-1)$ more derivatives; and s more derivatives if ϕ is continuous at x_1. At the point x_0, ϕ^* is at least regular as ϕ.

Remark 2.2.2.

In transferring moment problem 2.2.1 we observe the following: Let $\phi : \mathbb{R} \longrightarrow \mathbb{R}$ be a Borel measurable function which is locally integrable on $[x_0, +\infty)$ (relative to Lebesque measure) if ϕ is integrable relative to $\mu \in m_s(x_0), \mu = T\nu$, then $\phi^* = T\phi$ is integrable relative to $\nu \in m^*$. The converse is not always true. We also have $\int_{\mathbb{R}} \phi d\mu = \int \phi^* d\nu$, same as

$$\int_{\mathbb{R}} \phi d(T\nu) = \int (T\phi) d\nu.$$

For precisely, if $= \int_{\mathbb{R}} |\phi| d\mu < +\infty$ for $\mu \in m_s(x_0)$, then $\int_{\mathbb{R}} (T|\phi|) d\nu < +\infty$ for $\nu \in m^*$ (the corresponding probability measure on $\mathbb{R}$), the converse is also true. Because the restrictions of μ and ν to $(-\infty, x_0]$ coincide, we need only study the condition

$$\int_{x_0}^{+\infty} |\phi(x)| \mu(dx) = \int_{x_0}^{\infty} |\phi(x)| f(x) dx < +\infty$$

in comparison to the condition

$$\int_{x_0}^{+\infty} |\phi^*(u)| \nu(du) < +\infty.$$

To see the magnitude of ϕ^* on $(x_0, +\infty)$, observe that in (2.2.22) we have $|u - x| \leq |u - x_0|$. Therefore we obtain

$$|\phi^*(u)| \leq (T|\phi|)(u) \leq (\frac{s}{u - x_0}) \cdot \int_{x_0}^{u} |\phi(x)| dx, \tag{2.2.26}$$

true for all $u > x_0$.

Using (2.2.26) and Fubini's theorem we see that

$$\int_{x_0}^{a} |\phi(x) \cdot \log(x - x_0)| \cdot dx < +\infty$$

implies

$$\int_{x_0}^{a} |\phi^*(u)| \cdot du < +\infty. \tag{2.2.27}$$

If ϕ satisfies (2.2.27), which is a little stronger than integrability, then the differential equation (2.2.25) has a unique solution ϕ^* which is integrable near x_0. Also, from (2.2.26) we have that $(u - x_0) \cdot \phi^*(u) \longrightarrow 0$ as $u \downarrow x_0$. Furthermore, $T(|\phi|)$ is bounded in any neighborhood of x_0 where ϕ is bounded.

Solving Moment Problem 2.2.1.

Let T be the Markov transformation as described above. For each $\mu \in m_s(x_0)$ corresponds exactly one $\nu \in m^*$ such that $\mu = T\nu$. Call

$$g_i^* := Tg_i, i = 1, \ldots, n \quad \text{and } h^* := Th.$$

We have

$$\int_{\mathbb{R}} g_i^* d\nu = \int_{\mathbb{R}} g_i d\mu$$

and

$$\int_{\mathbb{R}} h^* d\nu = \int_{\mathbb{R}} h d\mu.$$

Notice that we get

$$\nu(g_i^*) := \int_{\mathbb{R}} g_i^* d\nu = c_i; i = 1, \ldots, n. \tag{2.2.28}$$

From (2.2.3) we get that

$$\int_{\mathbb{R}} T|g_i| d\nu < +\infty, i = 1, \ldots, n$$

and

$$\int_{\mathbb{R}} T|h| d\nu < +\infty. \tag{2.2.29}$$

Since T is a positive linear operator we obtain $|Tg_i| \leq T|g_i|, i = 1, \ldots, n$ and $|Th| \leq T|h|$ i.e.

$$\int_{\mathbb{R}} |g_i^*| d\nu < +\infty, i = 1, \ldots, n$$

and

$$\int_{\mathbb{R}} |h^*| d\nu < +\infty.$$

That is g_i^*, h^* are $\nu-$integrable.

Finally

$$L(c) = \inf_{\nu} \nu(h^*)$$

(2.2.30)

and

$$U(c) = \sup_{\nu} \nu(h^*),$$

where $\nu \in m^*$(probability measure on $\mathbb{R}$) such that (2.2.28) and (2.2.29) are true.

If in addition to the above we have that $g_i \geq 0, i = 1, \ldots, n$ and $h \geq 0$, then the solution to the moment problem (2.2.30) is given by implementing moment methods from Section 2.1.

Remark 2.2.3. Here we restrict our probability measures on $[0, +\infty)$ and we consider the case $x_0 = 0$. That is $\mu \in m_s(0), s \geq 1$, i.e. $(-1)^s F^{(s-1)}(x)$ is convex for all $x > 0$ but $\mu(\{0\}) = \nu(\{0\})$ can be positive, $\nu \in m^*$. We have

$$\phi^*(u) = su^{-s} \cdot \int_0^u (u-x)^{s-1} \cdot \phi(x) \cdot dx, u > 0. \tag{2.2.31}$$

Further $\phi^*(0) = \phi(0)$, $(\phi^* = T\phi)$. Especially if

$$\phi(x) = x^r \quad \text{then } \phi^*(u) = \binom{r+s}{s}^{-1} \cdot u^r, (r \geq 0). \tag{2.2.32}$$

Hence the moment

$$\alpha_r := \int_0^{+\infty} x^r \mu(dx) \tag{2.2.33}$$

is also expressed as

$$\alpha_r = \binom{r+s}{s}^{-1} \cdot \beta_r, \tag{2.2.34}$$

where

$$\beta_r := \int_0^{+\infty} u^r \nu(du). \tag{2.2.35}$$

Here remember that $T\nu = \mu$, where ν can be any probability measure on $[0, +\infty)$.

Remark 2.2.4. In here we restrict our probability measures on $[0, b]$, $b > 0$ and again we consider the case $x_0 = 0$. Let $\mu \in m_s(0)$ and

$$\int_{[0,b]} x^r \mu(dx) := \alpha_r, \tag{2.2.36}$$

where $s \geq 1, r > 0$ are fixed.

Also let ν be a probability measure on $[0, b]$ unrestricted, i.e. $\nu \in m^*$. Then $\beta_r = \binom{r+s}{s} d_r$, where

$$\beta_r := \int_{[0,b]} u^r \nu(du). \tag{2.2.37}$$

Let $h : [0,b] \longrightarrow \mathbb{R}_+$ be an integrable function with respect to Lebesque measure. Consider $\mu \in m_s(0)$ such that

$$\int_{[0,b]} h d\mu < +\infty \tag{2.2.38}$$

i.e.

$$\int_{[0,b]} h^* d\nu < +\infty, \nu \in m^*. \tag{2.2.39}$$

Here $h^* = Th, \mu = T\nu$ and

$$\int_{[0,b]} h d\mu = \int_{[0,b]} h^* d\nu.$$

Letting α_r be free, we have that the set of all possible $(\alpha_r, \mu(h)) = (\mu(x^r), \mu(h))$ coincides with the set of all

$$(\binom{r+s}{s}^{-1} \cdot \beta_r, \nu(h^*)) = (\binom{r+s}{s}^{-1} \cdot \nu(u^r), \nu(h^*)),$$

where μ as in (2.2.38) and ν as in (2.2.39), both probability measures on $[0,b]$. I.e. the set of all possible pairs $(\beta_r, \mu(h)) = (\beta_r, \nu(h^*))$ is precisely the convex hull of the curve

$$\Gamma := \{(u^r, h^*(u)) : 0 \leq u \leq b\}. \tag{2.2.40}$$

In order one to determine $L(\alpha_r)$ the infimum of all $\mu(h)$, where μ is as in (2.2.36) and (2.2.38), one must determine the lowest point in this convex hull which is on the vertical through $(\beta_r, 0)$. For $U(\alpha_r)$ the supremum of all $\mu(h), \mu$ as above, one must determine the highest point of above convex hull which is on the vertical through $(\beta_r, 0)$.

For more on the above see also section 2.1.

CHAPTER THREE

MOMENT PROBLEMS OF KANTOROVICH TYPE AND KANTOROVICH RADIUS

The content of this chapter is an extended version of the results contained in Anastassiou and Rachev (1992).

3.1 Moment Problems Of Kantorovich Type

Let $S = [a,b] \times [c,d] \subset \mathbb{R}^2$ and $\varphi(x,y) = |x-y|^p, p \geq 1$. Suppose $(\alpha, \beta) \in S$ and denote by

$$U := U(\varphi, \alpha, \beta) := \sup_{\mu} \int_S |x-y|^p \mu(dx, dy) \tag{3.1.1}$$

subject to

$$\int_S x\mu(dx, dy) = \alpha, \int_S y\mu(dx, dy) = \beta. \tag{3.1.2}$$

Theorem 3.1.1.

$$U = D\delta + T, \tag{3.1.3}$$

where

$$D := [|b-d|^p + |a-c|^p] - [|b-c]^p + |a-d|^p],$$
$$T := (1-B)|b-c|^p + (B+C-1)|a-c|^p + (1-C)|a-d|^p,$$

$$[B := \frac{b-\alpha}{b-a}, C := \frac{d-\beta}{d-c}], \delta := max(0, 1-B-C).$$

PROOF: From the convexity of φ, a simple checking shows that $D \leq 0$. From Proposition 1.4.6 and Lemma 1.4.7 (Alfsen (1971)) the optimal measure in (3.1.1) is carried by the four extreme points of S. From the given moment conditions one can determine the masses of three of the verteses of Sas a function of J, the mass of the fourth vertex of S. Optimization with respect to $J \in (0,1)$ of $\int_{\text{ext}(S)} \varphi dx$ gives (3.1.3), where ext stands for extreme points.

□

Remark 3.1.1. Since φ is convex on any $\Omega \subset \mathbb{R}^2$ then given (3.1.2)

$$\inf_{\mu} \int_{\Omega} \varphi d\mu = |\alpha - \beta|^p \tag{3.1.4}$$

Next, we extend Theorem 3.1.1 to non-bounded regions S. Namely, consider the following stripes in $\mathbb{R}^2$: for $b > 0$,

$$\begin{aligned} S_1^b : &= \{(x,y) : y = x + b', \text{ where } 0 \leq b' \leq b\}, \\ S_2^b : &= \{(x,y) : y = x - b', \text{ where } 0 \leq b' \leq b\}, \\ S^b : &= S_1^b \cup S_2^b. \end{aligned} \tag{3.1.5}$$

Theorem 3.1.2. *Assume that* $0 < p \leq 1$.

(i) If $S = S_1^b$ or S_2^b, $(\alpha, \beta) \in S$ then

$$U := U(\varphi, \alpha, \beta) = |\alpha - \beta|^p, \tag{3.1.6}$$

where

$$U := \sup_{\mu} \int_S |x - y|^p \mu(dx, dy) \tag{3.1.7}$$

such that

$$\int_S x\mu(dx, dy) = \alpha, \int_S y\mu(dx, dy) = \beta. \tag{3.1.8}$$

(ii) *If* $S = S^b$ *then* $U = b^p$.

(iii) *Let*

$$L := \inf_{\mu} \int_S |x - y|^p \mu(dx, dy) \tag{3.1.9}$$

subject to (3.1.8). *If*

$$S = S_1^b \text{ or } S = S_2^b \text{ or } S = S^b, (\alpha, \beta) \in S$$

then

$$L := L(\varphi; \alpha, \beta) = b^{p-1}|\alpha - \beta|. \tag{3.1.10}$$

Proof: We follow the Kemperman (1968) geometric approach to solve these moment problems (method of optimal distance).

(i) It is true because of concavity of

$$\varphi(x,y) = |x-y|^p, 0 < p \leq 1 \text{ in } S.$$

(ii) Here the upper envelope of the convex hull of $(x, y, |x-y|^p), (x,y) \in S^b$ is a plane stripe made out of the lines

$$(l_1): \quad \left.\begin{array}{c} x = t \\ y = b + t \\ z = b^p \end{array}\right\} \quad \text{and} \quad (l_2): \quad \left.\begin{array}{c} x = t \\ y = -b + t \\ z = b^p \end{array}\right\}$$

The plane through $(l_1), (l_2)$ has equation $z = b^p$.

Therefore

$$U = \sup_{\mu} \int_{S^b} |x-y|^p \mu(dx,dy) = b^p,$$

independently of α, β.

(iii) We consider the case of $S = S_1^b$, the cases of $S = S_2^b$ or $S = S^b$ are treated similarly.

Here the lower envelope of the convex hull of $(x, y, |x-y|^p), (x,y) \in S_1^b$ is a plane stripe generated by the lines

$$(l): \quad \left.\begin{array}{c} x = t \\ y = t \\ z = 0 \end{array}\right\} \quad \text{and} \quad (l_1): \quad \left\{\begin{array}{c} x = t \\ y = b + t \\ z = b^p \end{array}\right\}.$$

The plane through (l) and (l_1) has equation $z = b^{p-1}(y-x)$.

Therefore

$$L = \inf_{\mu} \int_{S_1^b} |x-y|^p \mu(dx,dy) = b^{p-1}(\beta - \alpha).$$

□

Remark 3.1.2. Note that as $b \longrightarrow +\infty$ then $S_1^b \longrightarrow \{(x,y) : x \leq y\}$ and $S_2^b \longrightarrow \{(x,y) : x \geq y\}$, half-planes in $\mathbb{R}^2$ and $S^b \longrightarrow \mathbb{R}^2$.

(i) Consider the moment problem

$$U_\infty := \sup_{\mu} \int_{\mathbb{R}^2} |x-y|^p \mu(dx,dy), 0 < p \leq 1 \tag{3.1.11}$$

subject to

$$\int_{\mathbb{R}^2} x d\mu = \alpha, \int_{\mathbb{R}^2} y d\mu = \beta; (\alpha, \beta) \in \mathbb{R}^2 \tag{3.1.12}$$

The

$$U_\infty = +\infty. \tag{3.1.13}$$

(ii) Also consider the moment problem

$$L_\infty := \inf_\mu \int_{\mathbb{R}^2} |s - y|^p \mu(dx, dy), 0 < p \leq 1 \tag{3.1.14}$$

subject to the moment conditions (3.1.12). Then

$$L_\infty = \begin{cases} 0, & 0 < p < 1 \\ |\alpha - \beta|, p = 1. \end{cases} \tag{3.1.15}$$

In the following theorem we consider the stripes in $\mathbb{R}^2$: for $b, \gamma > 0$,

$$\begin{aligned} S_1^b : &= \{(x, y) : y = x + b', \quad \text{where } 0 \leq b' \leq b\}, \\ S_2^\gamma : &= \{(x, y) : y = x - \gamma', \quad \text{where } 0 \leq \gamma' \leq \gamma\}, \\ S^{b,\gamma} &= S_1^b \cup S_2^\gamma. \end{aligned} \tag{3.1.16}$$

Theorem 3.1.3. *Assume that $p \geq 1$.*

(i) *If $S = S_1^b, (\alpha, \beta) \in S$ then*

$$U := U(\varphi, \alpha, \beta) = b^{p-1}(\beta - \alpha), \tag{3.1.17}$$

where U is given by (3.1.7) *and* (3.1.8).

(ii) *If $S = S_2^\gamma, (\alpha, \beta) \in S$ then*

$$U := U(\varphi, \alpha, \beta) = \gamma^{p-1}(\alpha - \beta). \tag{3.1.18}$$

(iii) *If $S = S^{b,\gamma}, (\alpha, \beta) \in S$ then*

$$U = \frac{(b^p - \gamma^p)(\beta - \alpha - b) + b^p(b + \gamma)}{b + \gamma}. \tag{3.1.19}$$

PROOF: Note here that $\varphi(x, y) = |x - y|^p$ is convex for $p \geq 1$.

(i) Here $S = S_1^b$ i.e. $\alpha \le \beta$. The upper envelope of the convex hull of $(x, y, |x - y|^p), (x, y) \in S_1^b$ is a plane stripe made out of the lines $(l_1), (l)$, where

$$(l_1): \quad \left.\begin{array}{c} x = t \\ y = b + t \\ z = b^p \end{array}\right\} \quad \text{and} \quad (l): \left.\begin{array}{c} x = t \\ y = t \\ z = 0 \end{array}\right\}.$$

The equation of the associated plane through (l_1) and (l) is

$$z = b^{p-1}(y - x), p \ge 1.$$

Note that μ is a probability measure on S. Thus by geometric moment theory (method of optimal distance) we get that $U = b^{p-1}(\beta - \alpha)$.

(ii) Acting similarly to (i) we obtain

$$U = \gamma^{p-1}(\alpha - \beta).$$

(iii) Here the upper envelope of the convex hull of $(x, y, |x - y|^p); p \ge 1, (x, y) \in S^{b,\gamma}$ is a plane stripe made out of the lines $(l_1), (l_2)$:

$$(l_1): \quad \left.\begin{array}{c} x = t \\ y = b + t \\ z = b^p \end{array}\right\} \quad \text{and} \quad (l_2): \left.\begin{array}{c} x = t \\ y = -\gamma + t \\ z = \gamma^p \end{array}\right\}$$

The plane through (l_1) and (l_2) has equation

$$(b^p - \gamma^p)x + (\gamma^p - b^p)(y - b) + (b + \gamma)(z - b^p) = 0.$$

$$i.e. \quad z = \frac{(b^p - \gamma^p)(y - x - b) + b^p(b + \gamma)}{b + \gamma}.$$

Here again μ is a probability measure on $S^{b,\gamma}$.

Hence by applying geometric moment theory (method of optimal distance) we obtain (3.1.19). □

Remark 3.1.3.

(i) $(\alpha, \beta) \in S^{b,\gamma}$ if $\beta = \alpha + \lambda$, where $-\gamma \le \lambda \le b$.

(ii) Given $\int\limits_{\mathbb{R}^2} x d\mu = \alpha$ and $\int\limits_{\mathbb{R}^2} y d\mu = \beta (\alpha \neq \beta)$, we get

$$U_\infty := \sup_\mu \int_{\mathbb{R}^2} |x-y|^p \mu(dx,dy) = +\infty, p > 1.$$

However, when $p = 1$ we have

$$U_\infty = U = |\alpha - \beta|,$$

independently of b, γ.

(iii) In case (iii) of Theorem 3.1.3 if $b = \gamma$, then $U = b^p$. Obviously there $U_\infty = +\infty, (p \geq 1)$.

3.2 Kantorovich Radius

For given $x_0 \in \mathbb{R}, \alpha > 0, p \in \mathbb{R}, q > 0 (p^2 \leq q), -\infty \leq a < b \leq +\infty$ find the *Kantorovich radius*

$$K := K(x_0; \alpha, p, q, a, b) \tag{3.2.1}$$

$$:= \sup\{E|X - x_0|^\alpha : X \in [a,b] a.s., EX = p, EX^2 = q\},$$

where X stands for a random variable.

Theorem 3.2.1. (Case (A): $\alpha \geq 2, -\infty < a < b < +\infty$). *Suppose X's in (3.2.1) take values on $[a,b]$. Let $x_0 = (a+b)/2, a \leq p \leq b$ and $0 \leq q \leq b^2 + (a+b)(p-b)$.*

Then

$$K \leq (\frac{b-a}{2})^{\alpha-2}[q - p(a+b) + \frac{(a+b)^2}{4}]. \tag{3.2.2}$$

Moreover, if there exist $\lambda_1, \lambda_2 \geq 0, \lambda_1 + \lambda_2 \leq 1$ such that

$$p = (\frac{a+b}{2}) + (\frac{b-a}{2})(\lambda_1 - \lambda_2)$$

and

$$q = \frac{(a+b)^2}{4} + (\frac{b^2-a^2}{2})(\lambda_1 - \lambda_2) + \frac{(b-a)^2}{4}(\lambda_1 + \lambda_2) \tag{3.2.3}$$

then

$$K = (\frac{b-a}{2})^{\alpha-2}[q - p(a+b) + \frac{(a+b^2}{4}]. \tag{3.2.4}$$

PROOF: By calling $Y := X - x_0$ we have $E(Y) = P$ and $E(Y^2) = Q$, where $P := p - x_0$ and $Q := q - 2px_0 + x_0^2$. Equivalently, we would like to find $K = K(Y,0) = \sup_Y E(|Y|^\alpha), \alpha \geq 2$ such that $EY = P$ and $EY^2 = Q$. To calculate K we use standard geometric moment theory from Kemperman (1968) (method of optimal distance). Here note that the graph of $(t^2, |t|^\alpha) = (z, z^{\alpha/2}), z \geq 0$. Also call $A := x_0 - a = b - x_0 > 0$, thus $|x - x_0| \leq A$, for all $x \in [a,b]$.

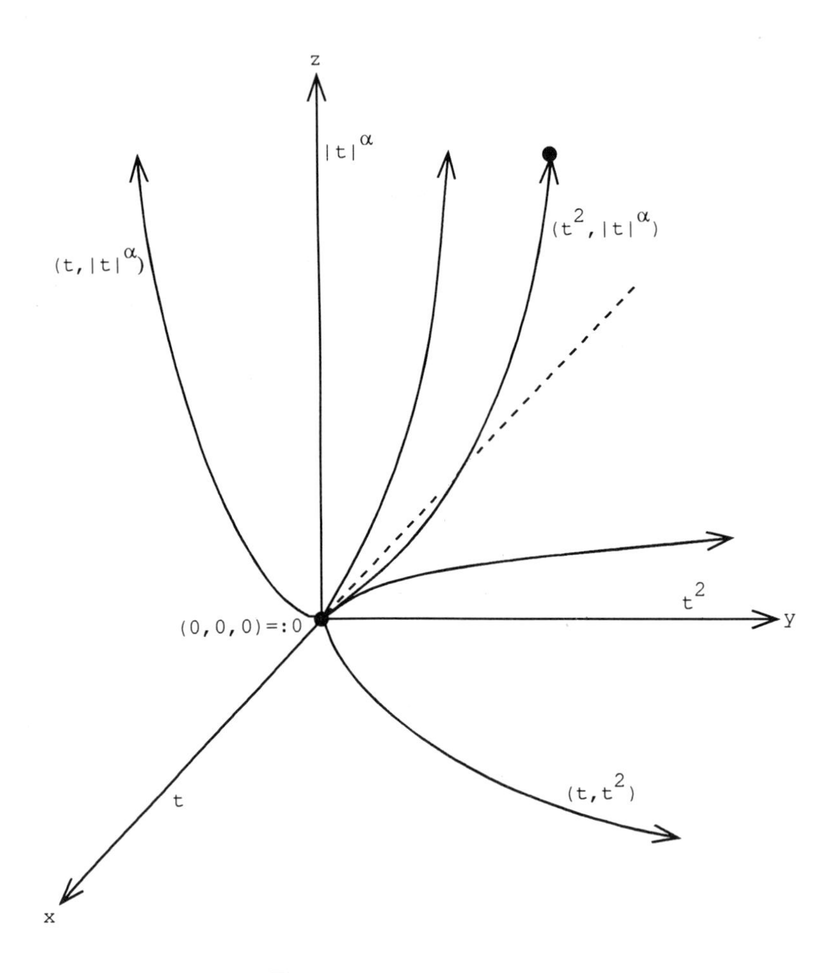

Figure 3.2.1, $\alpha \geq 2$.

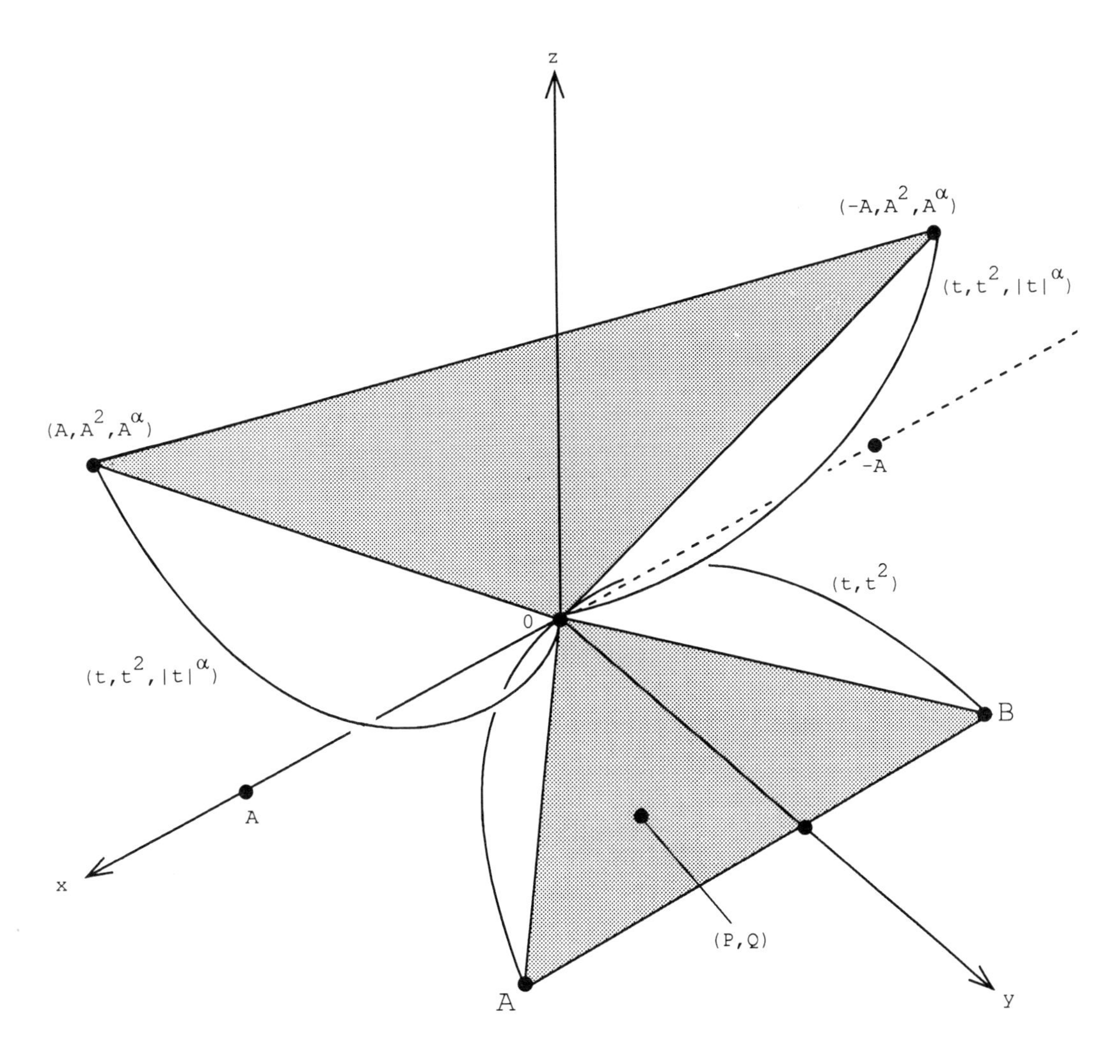

Figure 3.2.2, $(t, t^2, |t|^\alpha); \alpha \geq 2$.

Let $O := (0,0,0), \mathcal{A} = (A, A^2, 0), \mathcal{B} = (-A, A^2, 0)$. Then $(P,Q) \in \overset{\triangle}{\mathcal{OAB}}$ if there exist $\lambda_1, \lambda_2 \geq 0$ such that $\lambda_1 + \lambda_2 \leq 1$ with $P = A(\lambda_1 - \lambda_2)$ and $Q = A^2(\lambda_1 + \lambda_2)$ if there exist $\lambda_1, \lambda_2 \geq 0$ such that $\lambda_1 + \lambda_2 \leq 1$ with

$$p = (\frac{a+b}{2}) + (\frac{b-a}{2})(\lambda_1 - \lambda_2)$$

and

$$q = \frac{(a+b)^2}{4} + (\frac{b^2-a^2}{2})(\lambda_1 - \lambda_2) + \frac{(b-a^2)}{4}(\lambda_1 + \lambda_2).$$

When $(P,Q) \in \overset{\triangle}{OAB}$ then $P^2 \le Q$. According to moment theory, there exist $r.v.'s$ Y such that $EY = P, EY^2 = Q$ if the moment point (P,Q) is in the interior or boundary of the parabola (t,t^2) i.e. when $P^2 \le Q, -A \le P \le A$ and $0 \le Q \le A^2$. Equivalently, there exist $r.v.'sX$ such that $EX = p, EX^2 = q$ if $p^2 \le q, a \le p \le b$ and $0 \le q \le b^2 + (a+b)(p-b)$. The upper part of the convex hull of $(t,t^2,|t|^\alpha), \alpha \ge 2, -A \le t \le A$, which is above the triangle $\overset{\triangle}{OAB}$ is the triangle $\{(0,0,0),(A,A^2,A^\alpha),(-A,A^2,A^\alpha)\}$. The plane through the last triangle has equation $z = A^{\alpha-2}y$. Hence, $K = z$, where (P,Q,z) belongs to this plane. Therefore $K = A^{\alpha-2}Q$,

$$(A = \frac{b-a}{2} \text{ and } Q = q - p(a+b) + \tfrac{(a+b)^2}{4}).$$

Next one can observe that the curve $(t,t^2,|t|^\alpha)$ is below the plane $z = A^{\alpha-2}y$, for $-A \le t \le A$. Therefore the upper part of the convex hull of $(t,t^2|t|^\alpha)$ is below or on that plane.

Consequently

$$K \le A^{\alpha-2}Q$$

for all (P,Q) in or on the parabola (t,t^2) with $P^2 \le Q, -A \le P \le A$ and $0 \le Q \le A^2$.

Note that the optimal probability measure μ is supported at $\{-A,0,A\}$ with masses respectively

$$\{\frac{Q-P.A}{2A^2}, \frac{A^2-Q}{A^2}, \frac{P.A+Q}{2A^2}\}.$$

□

Theorem 3.2.2. (Case (B): $0 < \alpha \le 2, a = -\infty, b = +\infty$.)

For any $x_0 \in \mathbb{R}, p \in \mathbb{R}, q > 0, p^2 \le q$, the Kantorovich radius K admits the following value

$$K = K(x_0;\alpha;p;q) = (q - 2x_0p + x_0^2)^{\alpha/2} \tag{3.2.5}$$

Proof: Case $B1: 1 < \alpha \le 2$ (Case B2: $0 < \alpha \le 1$ as similar is omitted). Call $Y := X - x_0$, thus $EY = P, EY^2 = Q$, where $P := p - x_0, Q := q - 2x_0p + x_0^2$. We would like to find $K = \sup_Y E|Y|^\alpha, 1 < \alpha \le 2$ such that $EY = P, EY^2 = Q$. Here observe that $P^2 \le Q$ if $p^2 \le q$.

Let $V := \underset{t\in\mathbb{R}}{\text{conv}}(t,t^2,|t|^\alpha)$ and S be its upper part. This is formed by the line segments parallel to x-axis touching the curve $(t,t^2,|t|^\alpha), t \in \mathbb{R}$. The associated moment point on

(x, y)–plane is $(P, Q, 0)$. Thus $K = (\sqrt{Q})^\alpha, 1 < \alpha \leq 2$. The optimal probability measure μ is carried by $\{\sqrt{Q}, -\sqrt{Q}\}$ with massess $\{1-\lambda, \lambda\}, \lambda \in [0,1]$, respectively. □

Theorem 3.2.3. (Case (C): $0 < \alpha \leq 2, -\infty < a < b < +\infty$.)

For any $x_0 \in (a, b); p \in \mathbb{R}, p^2 \leq q$,

$$K = \sup\{E(|X - x_0|^\alpha) : X \in [a, b] a.s., EX = p, EX^2 = q\}.$$

Set $P = p - x_0, Q = q - 2x_0 p + x_0^2, A(x_0) = a - x_0,\ B(x_0) = b - x_0, C(x_0) = \min(-A(x_0), B(x_0))$.

(i)

$$\textit{Suppose } 0 \leq Q \leq C^2(x_0) \textit{ then } K = Q^{\alpha/2}. \tag{3.2.6}$$

(i) Suppose $Q \geq C^2(x_0), a \leq p \leq b$ *and*

$$(A(x_0) + B(x_0))P - Q - A(x_0)B(x_0) \geq 0. \tag{3.2.7}$$

Then

$$K \leq Q^{\alpha/2}. \tag{3.2.8}$$

Proof: (C_1) Case: $1 < \alpha \leq 2$ ((C_2) Case: $0 < \alpha \leq 1$ as similar is omitted.)

Call $Y := X - x_0$, then $E(Y) = P, EY^2 = Q$.

Hence, we would like to find $K = \sup_Y E(|Y|^\alpha), 1 < \alpha \leq 2$ given that $EY = P, EY^2 = Q$. Note that $P^2 \leq Q$ if $p^2 \leq q$.

(i) Suppose $0 \leq Q \leq C^2(x_0)$, then $|P| \leq C(x_0)$. Cal $V = \underset{A(x_0) \leq t \leq B(x_0)}{\text{conv}} (t, t^2, |t|^\alpha)$, $1 < \alpha \leq 2$.

Let S be the upper part of the convex hull V when $-C(x_0) \leq t \leq C(x_0)$. This is formed by line segments parallel to the x- axis touching $(t, t^2, |t|^\alpha)$. Our moment point on (xy)–plane is $(P, Q, 0)$ with $0 \leq Q \leq C^2(x_0)$. Thus $K = (\sqrt{Q})^\alpha, 1 < \alpha \leq 2$. Here we applied again the geometric moment method of optimal distance. The optimal associated probability measure μ is supported at $\{\sqrt{Q}, -\sqrt{Q}\}$ with masses $\{1-\lambda, \lambda\}, \lambda \in [0,1]$, respectively.

(ii) Suppose $Q \geq C^2(x_0)$ and

$$(A(x_0) + B(x_0))P - Q - A(x_0)B(x_0) \geq 0.$$

Hence, the moment point (P,Q) is below the line

$$y = (A(x_0) + B(x_0))x - A(x_0)B(x_0).$$

Therefore

$$K \leq (\sqrt{Q})^\alpha.$$

□

Theorem 3.2.4. (Case (D): $1 \leq \alpha \leq 2, -\infty < a < b \leq +\infty$).

For any $p \in \mathbb{R}, p^2 \leq q, a \leq p \leq b$, *set* $P = p - a, Q = q - 2ap + a^2, B = b - a$. *Suppose* $Q \leq BP$. *Then*

$$K := \sup\{E|X-a|^\alpha : X \in [a,b], EX = p, EX^2 = q\} = P^{2-\alpha}Q^{\alpha-1}. \tag{3.2.9}$$

PROOF: Call $Y = X - a$, then $E(Y) = P, E(Y^2) = Q$. Thus we want to find $K = \sup\limits_Y E(|Y|^\alpha), 1 \leq \alpha \leq 2$ given that $EY = P, E(Y^2) = Q$. Call $t := x - a$, then $0 \leq t \leq B$ when $a \leq x \leq b$. Note that $P^2 \leq Q$ if $p^2 \leq q$, also $0 \leq P \leq B$ if $a \leq p \leq b$. From $Q \leq BP$ we get $Q \leq B^2$. Call $V = \underset{0 \leq t \leq B}{\text{conv}} (t, t^2, t^\alpha)$, $1 \leq \alpha \leq 2$. Let S be the upper part of the convex hull V. This is formed by all line segments

$$\overline{(0,0,0),(t,t^2,t^\alpha)} \quad , 0 \leq t \leq B.$$

The line through $(0,0)$ and (B, B^2) is $y = Bx$. Since $Q \leq BP, (P,Q)$ is an admissible moment point. Call $0 := (0,0,0)$, $A := (P,Q,0)$, $M := (P,Q,z) \in S$, $K := (x_0, x_0^2, 0)$; $x_0 := Q/P$, $L := (x_0, x_0^2, x_0^\alpha) \in S$, such that $0, M, L$ are on the same line. Obviously, the triangles $\overset{\triangle}{OAM}$ and $\overset{\triangle}{OKL}$ are similar, therefore

$$\frac{\overline{0A}}{\overline{0K}} = \frac{\overline{AM}}{\overline{KL}} \quad \text{and} \quad \overline{AM} = \frac{\overline{OA}}{\overline{0K}}\overline{KL}.$$

Here

$$\overline{AM} = z, \overline{0A} = \sqrt{P^2 + Q^2}, \overline{0K} = \frac{Q}{P^2}\sqrt{P^2+Q^2} \quad \text{and} \quad \overline{KL} = \left(\frac{Q}{P}\right)^\alpha.$$

Hence $z = P^{2-\alpha}Q^{\alpha-1}$ and $K = z$. The associated optimal probability measure μ is supported at $\{0, \frac{Q}{P}\}$ with masses $(1 - \frac{P^2}{Q})$ and $(\frac{P^2}{Q})$ respectively. □

Theorem 3.2.5. (Case (E): $1 \leq \alpha \leq 2, -\infty \leq a < b < +\infty$).

For any $p \in \mathbb{R}, p^2 \leq q, a \leq p \leq b$, *set* $P = p - b, Q = q - 2bp + b^2, \Theta = a - b$. *Suppose* $Q \leq \Theta P$. *Then*

$$K := \sup\{E(|X - b|^\alpha) : X \in [a, b], EX = p, EX^2 = q\} = |P|^{2-\alpha} Q^{\alpha - 1}. \tag{3.2.10}$$

PROOF: As similar to Theorem (3.2.4) is omitted. □

Remark 3.2.1. Case $0 < \alpha \leq 1, X$ is a random variable that takes values on $\{[a, +\infty)$ or $[a, b]\}$ or on $\{(-\infty, b]$ or $[a, b]\}$ and x_0 is either a or b, respectively.

Then, using Hölder's inequality we get

$$\sup_X E(|X - x_0|^\alpha) \leq (\sqrt{Q})^\alpha, \tag{3.2.11}$$

where

$$Q := q - 2x_0 p + x_0^2; EX = p, EX^2 = q.$$

CHAPTER FOUR

MOMENT PROBLEMS RELATED TO $c-$ ROUNDING PROPORTIONS

The next is an extended version of results contained in Anastassiou and Rachev (1992).

4.1 Moment Problems Related To c- Rounding Proportions Subject To One Moment Condition.

Definition 4.1.1. *Let $0 \le c \le 1$ fixed: for any $x \ge 0$,*

$$[x]_c := \begin{cases} m & \text{if } \ m \le x < m + c \\ m+1, & \text{if } \ m + c \le x < m+1, \end{cases} \tag{4.1.1}$$

where $m \in \mathbb{N} \cup \{0\}$, or equivalently,

$$[x]_c = \begin{cases} 0, & \text{if } \ 0 \le x \le c, \\ m, & \text{if } \ m - 1 + c < x \le m + c, (m \in \mathbb{N}). \end{cases} \tag{4.1.2}$$

Moment Problem 4.1.1. Find

$$\sup_\mu \int_A [t]_c \mu(dt), \tag{4.1.3}$$

where μ is a probability measure on $A := [0, a]$ or $[0, \infty)$ and

$$\inf_\mu \int_A [t]_c \mu(dt), \tag{4.1.4}$$

subject to the given moment condition

$$\int_A t^r \mu(dt) = d_r, r > 0, d_r > 0. \tag{4.1.5}$$

The next four theorems deal with the solution of Moment Problem 4.1.1. In the sequel, the underlined probability space is assumed to be nonatomic and thus the space of laws of nonnegative random variables coincides with the space of all Borel probability measures on $\mathbb{R}_+$.

Theorem 4.1.1. *Let $c \in (0,1), r > 0, 0 < a < +\infty, d > 0,$ and*

$$U := U_{[\cdot]_c}(a, r, d) := \sup\{E[X]_c : 0 \le X \le a \ \ a.s., (EX^r)^{1/r} = d\}. \tag{4.1.6}$$

set $n := [a]$, integral part of a.

(I) *If* $n + c < a, n + c \le d \le a$, *then* $U = n + 1$.

(II) *If* $n + c \ge a, n - 1 + c \le d \le a$, *then* $U = n$.

(III) *If* $0 < a \le c$, *then* $U = 0$.

(IV) *If* $0 < r \le \frac{ln2}{ln(1+\frac{1}{c})}(< 1)$, $n + c < a$ *and* $0 \le d \le n + c$, *then* $U = (n + 1)d^r(n + c)^{-r}$.

(V) *If* $0 < r \le \frac{ln2}{ln(1+\frac{1}{c})}, n + c \ge a$ *and* $0 \le d \le n - 1 + c$, *then* $U = nd^r(n - 1 + c)^{-r}$.

(VI) *If* $r \ge 1$ *and* $0 \le d \le c$, *then* $U = d^r c^{-r}$.

(VII) *Suppose* $r \ge 1$. *If either*
(a) $n + c < a$ *and determine* $k \in \{1, \ldots, n\}$ *by* $k - 1 + c \le d < k + c$, *or*
(b) $n + c \ge a$ *and determine* $k \in \{1, \ldots, n - 1\}$ *by* $k - 1 + c \le d < k + c$. *Then*

$$U = k + \frac{d^r - (k - 1 + c)^r}{(k + c)^r - (k - 1 + c)^r} \le 1 - c + d.$$

PROOF: From geometric moment theory, the method of optimal distance, it follows that $U = \psi(d^r)$, where $A_1 = \{(u, \psi(u)) : 0 \le u \le a^r\}$ describes the upper boundary of the convex hull conv A_0 of the curve $A_0 = \{(t^r, [t]_c) : 0 \le t \le a\}$, see figure 4.1.1. Note that A_0 consists of the following parts:

Case (I) Assume that $n + c < a$.

(i) The closed interval $[0, c]$.

(ii) For $k = 1, \ldots, n$ the half open horizontal line segments $(P_k, Q_k]$, where $P_k = ((k - 1 + c)^r, k)$ and $Q_k = ((k + c)^r, k)$.

(iii) The half open non-empty horizontal line segment $(P_{n+1}, Q_*]$, where $P_{n+1} = ((n + c)^r, n + 1)$ and $Q_* = (a^r, n + 1)$.

Case (II) Assume that $n + c \ge a$.

(i) The closed interval $[0, c]$.

(ii) For $k = 1, \ldots, n - 1$ the half open horizontal line segments $(P_K, Q_k]$, where $P_k = ((k - 1 + c)^r, k)$ and $Q_k = ((k + c)^r, k)$.

(iii) The half open non-empty horizontal line segment $(P_n, Q_*]$, where $P_n = ((n-1+c)^r, n)$ and $Q_* = (a^r, n)$.

Parts (iii) of cases (I), (II) are parts of the upper boundary A_1 of conv A_0, yielding $U = n + 1; U = n$, respectively. Here we need $n + c \le d \le a; n - 1 + c \le d \le a$, respectively. The line segment from P_k to P_{k+1} has slope $\frac{1}{(k+c)^r - (k-1+c)^r}$ which is decreasing in k when $r \geq 1$ and increasing in k when $0 < r \leq 1$. Call $P_0 = (0,0)$. Slope $(\overline{P_0P_1}) = \frac{1}{c^r}$. Observe that slope $(\overline{P_0P_1}) \geq$ slope $(\overline{P_1P_2})$ for $r \geq 1$. When $0 < r \leq \frac{ln2}{ln(1+\frac{1}{c})}(< 1)$ we have slope $(\overline{P_0P_1}) \leq$ slope $(\overline{P_1P_2})$. Consequently, if $0 < r \leq \frac{ln2}{ln(1+\frac{1}{c})}$ then A_1 consists of the following two line segments:

Case (I) of $n + c < a$

$(\overline{P_0P_{n+1}})$ and $(\overline{P_{n+1}Q_*})$, where $P_0 = (0,0), P_{n+1} = ((n+c)^r, n+1), Q_* = (a^r, n+1)$. Here $0 \leq d \leq n + c$. Thus,

$$\frac{U}{n+1} = \frac{d^r}{(n+c)^r},$$

giving us

$$U = (n+1)\frac{d^r}{(n+c)^r}.$$

Case (II) of $n + c \geq a$

$(\overline{P_0P_n})$ and $(\overline{P_nQ_*})$, where $P_n = ((n-1+c)^r, n)$ and $Q_* = (a^r, n)$. Here $0 \leq d \leq n-1+c$. Thus

$$\frac{U}{n} = \frac{d^r}{(n-1+c)^r} \quad \text{i.e} \quad U = \frac{nd^r}{(n-1+c)^r}.$$

On the other hand, if $r \geq 1$, then A_1 is composed of the following line segments:

Case (I) of $n + c < a$

$(\overline{P_kP_{k+1}}), k = 0, 1, \ldots, n$ and $(\overline{P_{n+1}Q_*})$, where $P_k = ((k-1+c)^r, k), Q_* = (a^r, n+1)$.

Case (II) of $n + c \geq a$

$(\overline{P_kP_{k+1}}), k = 0, 1, \ldots, n-1$ and $(\overline{P_nQ_*})$, where $P_k = ((k-1+c)^r, k), Q_* = (a^r, n)$. In either case (I), (II) when $0 \leq d \leq c$ we get $U = d^r c^{-r}$. Note that the slope of $\overline{P_kP_{k+1}} = \frac{1}{(k+c)^r - (k-1+c)^r}$ which is decreasing in $k, r \geq 1$. The equation of the line through P_k, P_{k+1} has as follows

$$y - k = \frac{(x - (k-1+c)^r)}{[(k+c)^r - (k-1+c)^r]}.$$

Here $k - 1 + c \leq d < k + c$. Therefore

$$U = k + [\frac{d^r - (k-1+c)^r}{(k+c)^r - (k-1+c)^r}],$$

where in case (I), $k \in \{1, \ldots, n\}$; however in case (II), $k \in \{1, \ldots, n-1\}$. By the convexity of $x^r, r \geq 1; x \geq 0$ we obtain $U \leq (1-c) + d$. □

We extend Theorem 4.1.1 in the case $a = +\infty$ as follows.

Theorem 4.1.2. *Let* $0 < c < 1, r > 0, d > 0$ *and*

$$U := U_{[\cdot]_c}(r,d) := \sup\{E[X]_c : X \geq 0 \quad a.s., (EX^r)^{1/r} = d\}. \tag{4.1.7}$$

(I) *If* $0 < r < 1$, *then* $U = +\infty$.

(II) *If* $r \geq 1$ *and* $0 \leq d \leq c$, *then* $U = \frac{d^r}{c^r}$.

(III) *Suppose* $r \geq 1$. *Define* $k \in \mathbb{N}$ *by* $k - 1 + c \leq d < k + c$. *Then*

$$U = k + \frac{d^r - (k-1+c)^r}{(k+c)^r - (k-1+c)^r} \leq 1 - c + d.$$

PROOF: As similar and simpler of the proof of Theorem 4.1.1 is omitted. □

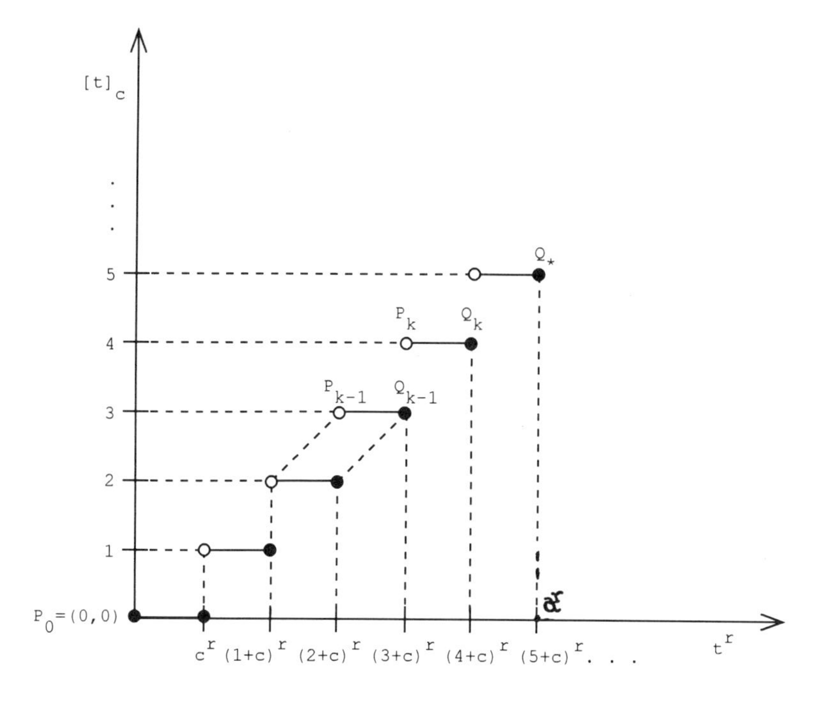

Figure 4.1.1, graph of $(t^r, [t]_c)$; $r > 0, 0 < c < 1, 0 \leq t \leq a$.

Theorem 4.1.3. *Let* $c \in (0,1), r > 0, 0 < a < +\infty, d > 0$ *and*

$$L := L_{[\cdot]_c}(a,r,d) := \inf\{E[X]_c : 0 \leq X \leq a \quad a.s, (EX^r)^{1/r} = d\}. \tag{4.1.8}$$

Set $n = [a]$.

(I) *If* $0 < d \leq c$, *then* $L = 0$.

(II) *If* $c < a \leq 1 + c$ *and* $c \leq d \leq a$ *then*

$$L = (d^r - c^r)/(a^r - c^r).$$

(III) *If* $0 < r \leq 1, n + c < a$ *and determine* $k \in \{0, 1, \ldots, n-1\}$ *by* $k + c \leq d < k + 1 + c$. *Then*

$$L = k + \frac{d^r - (k+c)^r}{(k+1+c)^r - (k+c)^r}.$$

(IV) *If* $0 < r \leq 1, n + c < a$ *and* $n + c \leq d \leq a$, *then*

$$L = n + \frac{d^r - (n+c)^r}{a^r - (n+c)^r}.$$

(V) *If* $0 < r \leq 1, n + c \geq a$ *and determine* $k \in \{0, 1, \ldots, n-2\}$ *by* $k + c \leq d < k + 1 + c$, *then*

$$L = k + (\frac{d^r - (k+c)^r}{(k+1+c)^r - (k+c)^r}).$$

(VI) *If* $0 < r \leq 1, n + c \geq a$ *and* $n - 1 + c \leq d \leq a$, *then*

$$L = (n-1) + (\frac{d^r - (n-1+c)^r}{a^r - (n-1+c)^r}).$$

From now on assume $r \geq 1$.

Case (A): $n + c < a$

(i) Assume that

$$\frac{1}{a^r - (n+c)^r} \geq \frac{n}{(n+c)^r - c^r}.$$

(i_1) If $c \leq d \leq n + c$, then

$$L = \frac{n(d^r - c^r)}{((n+c)^r - c^r)}.$$

(i_2) If $n + c \leq d \leq a$, then

$$L = n + [\frac{d^r - (n+c)^r}{a^r - (n+c)^r}].$$

(ii) Assume that

$$\frac{1}{a^r-(n+c)^r} \le \frac{n}{(n+c)^r-c^r} \quad \text{and } n \ge 1.$$

If $c \le d \le a$, then

$$L=(n+1)(\frac{d^r-c^r}{a^r-c^r}).$$

Case (B): $n+c \ge a, r \ge 1$

(i) Assume that

$$\frac{1}{a^r-(n-1+c)^r} \ge \frac{n-1}{(n-1+c)^r-c^r}$$

and $n \ge 1$.

(i_1) If $c \le d \le n-1+c$, then

$$L=\frac{(n-1)(d^r-c^r)}{((n-1+c)^r-c^r)}.$$

(i_2) If $n-1+c \le d \le a$, then

$$L=(n-1)+\frac{(d^r-(n-1+c)^r)}{(a^r-(n-1+c)^r)}.$$

(ii) Assume that

$$\frac{1}{a^r-(n-1+c)^r} \le \frac{n-1}{(n-1+c)^r-c^r}$$

and $n \ge 2$.
If $c \le d \le a$, then

$$L=\frac{n(d^r-c^r)}{(a^r-c^r)}.$$

The last is true also for $a \in \mathbb{N}-\{1\}$.

PROOF: Here we apply the geometric moment theory method of optimal distance. Let $Q_k := ((k+c)^r, k), k \in \mathbb{Z}_+; n=[a]$, see figure 4.1.1.

Case (A): $n+c<a$.

Here $k=0,1,\ldots,n$ and $Q_* := (a^r, n+1)$ is the last point of lower envelope of the convex hull of the graph $(t^r, [t])_c), 0 \le t \le a, r>0, c \in (0,1)$.

Case (B): $n + c \geq a$.

Here $k = 0, 1, \ldots, n-1$ and $Q_* := (a^r, n)$ is the last point of lower envelope of the above convex hull. Let $m_k :=$slope $(\overline{Q_k Q_{k+1}}) = ((k+1+c)^r - (k+c)^r)^{-1}$. If $r \geq 1$ we get that m_k are decreasing as k is increasing. And if $0 < r \leq 1$, then m_k are increasing in k. Obviously when $0 < d \leq c$ we have that $L = 0$.

Case of $0 < r \leq 1$.

Obviously in cases A, B we have that slope $(\overline{Q_n Q_*}) \geq$ all m_k, slope $(\overline{Q_{n-1} Q_*}) \geq$ all m_k, respectively. I.e. the lower envelope of the convex hull of $(t^r, [t]_c)$ is generated by $(\overline{P_0 Q_0})$, the $(\overline{Q_k Q_{k+1}})$'s and $(\overline{Q_n Q_*})$ or $(\overline{Q_{n-1} Q_*})$, respectively.

Case (A): $n + c < a, 0 < r \leq 1$.

Consider the line (l_k) through $Q_k, Q_{k+1}; k = 0, 1, \ldots, n-1$. It has equation

$$y = k + \left(\frac{x - (k+c)^r}{(k+1+c)^r - (k+c)^r}\right).$$

When $k + c \leq d < k + 1 + c$ we get

$$L = k + \left(\frac{d^r - (k+c)^r}{(k+1+c)^r - (k+c)^r}\right); k \in \{0, 1, \ldots, n-1\}.$$

Consider also the line (l) through Q_n and Q_*. This has equation

$$y = n + \frac{(x - (n+c)^r)}{(a^r - (n+c)^r)}.$$

When $n + c \leq d \leq a$, then

$$L = n + \left(\frac{d^r - (n+c)^r}{a^r - (n+c)^r}\right), n := [a].$$

Case (B): $n + c \geq a, 0 < r \leq 1$

Consider the line (l_k) through $Q_k, Q_{k+1}; k = 0, 1, \ldots, n-2$. It has equation

$$y = k + \left(\frac{x - (k+c)^r}{(k+1+c)^r - (k+c)^r}\right).$$

When $k + c \leq d < k + 1 + c$, then

$$L = k + \left(\frac{d^r - (k+c)^r}{(k+1+c)^r - (k+c)^r}\right), k \in \{0, 1, \ldots, n-2\}.$$

Let (l) be the line through Q_{n-1} and Q_*. It has equation

$$y = (n-1) + (\frac{x-(n-1+c)^r}{a^r-(n-1+c)^r}).$$

When $n-1+c \leq d \leq a$, then

$$L = (n-1) + (\frac{d^r-(n-1+c)^r}{a^r-(n-1+c)^r}).$$

(The last is true also when $a \in \mathbb{N}$.)

Case of $r \geq 1$. Here the slopes m_k are decreasing in k.

Case (A): $n+c < a$

Of interest are the points $P_0 = (0,0), Q_k = ((k+c)^r, k); k = 0,1,\ldots,n$ and $Q_* = (a^r, n+1); n := [a]$. Here denote by $Q_n := ((n+c)^r, n)$ and $Q_0 := (c^r, 0)$.

Subcase (i) of Case (A)

Assume that slope $(\overline{Q_n Q_*}) \geq$ slope $(\overline{Q_0 Q_n})$ i.e. assume that

$$\frac{1}{a^r-(n+c)^r} \geq \frac{n}{(n+c)^r-c^r}.$$

The line through (Q_0, Q_n) has equation $y = \frac{n(x-c^r)}{(n+c)^r-c^r}$. When $c \leq d \leq n+c$ we get

$$L = \frac{n(d^r-c^r)}{((n+c)^r-c^r)}.$$

The line through (Q_n, Q_*) has equation $y = n + \frac{(x-(n+c)^r)}{(a^r-(n+c)^r)}$. When $n+c \leq d \leq a$, we obtain

$$L = n + (\frac{d^r-(n+c)^r}{a^r-(n+c)^r}).$$

Subcase (ii) of Case (A)

Assume that slope $(\overline{Q_nQ_*}) \leq$ slope $\overline{(Q_0Q_n)}$ and $n \geq 1$. i.e. assume

$$\frac{1}{a^r - (n+c)^r} \leq \frac{n}{(n+c)^r - c^r}.$$

The line through (Q_0, Q_*) has equation

$$y = \frac{(n+1)(x - c^r)}{(a^r - c^r)}.$$

When $c \leq d \leq a$, we find that

$$L = (n+1)(\frac{d^r - c^r}{a^r - c^r}).$$

Case of $r \geq 1$ (continuing)

Here again m_k are decreasing in k.

Case (B): $n + c \geq a$.

Of interest are the points $P_0 := (0,0)$, $Q_k := ((k+c)^r, k); k = 0, 1, \ldots, n-1$ and $Q_* := (a^r, n); n := [a]$. Here denote by $Q_{n-1} := ((n-1+c)^r, n-1), Q_0 := (c^r, 0)$.

Subcase (i) of case (B) Assume that slope $(\overline{Q_{n-1}Q_*}) \geq$ slope $(\overline{Q_0Q_{n-1}})$ and $n \geq 1$. i.e. assume

$$\frac{1}{a^r - (n-1+c)^r} \geq \frac{n-1}{(n-1+c)^r - c^r}.$$

The line through (Q_0, Q_{n-1}) has equation

$$y = \frac{(n-1)(x - c^r)}{((n-1+c)^r - c^r)}.$$

When $c \leq d \leq n-1+c$ we get that

$$L = \frac{(n-1)(d^r - c^r)}{((n-1+c)^r - c^r)}.$$

(This is true also if $a \in \mathbb{N}$.)

The line through (Q_{n-1}, Q_*) has equation

$$y = (n-1) + \frac{(x - (n-1+c)^r)}{(a^r - (n-1+c)^r)}.$$

When $n - 1 + c \leq d \leq a$ we get that

$$L = (n-1) + \frac{(d^r - (n-1+c)^r)}{(a^r - (n-1+c)^r)}.$$

(The last is true for $a \in \mathbb{N}$.)

Subcase (ii) of case (B)

Here assume that slope $(\overline{Q_{n-1}Q_*}) \leq$ slope $(\overline{Q_0Q_{n-1}})$ and $n \geq 2$. The line through (Q_0, Q_*) has equation

$$y = \frac{n(x - c^r)}{(a^r - c^r)}.$$

When $c \leq d \leq a$ we get that

$$L = \frac{n(d^r - c^r)}{(a^r - c^r)}.$$

(The last is also true for $a \in \mathbb{N} - \{1\}$). Finally, we have to meet the special case of $c < a \leq 1 + c, r > 0$.

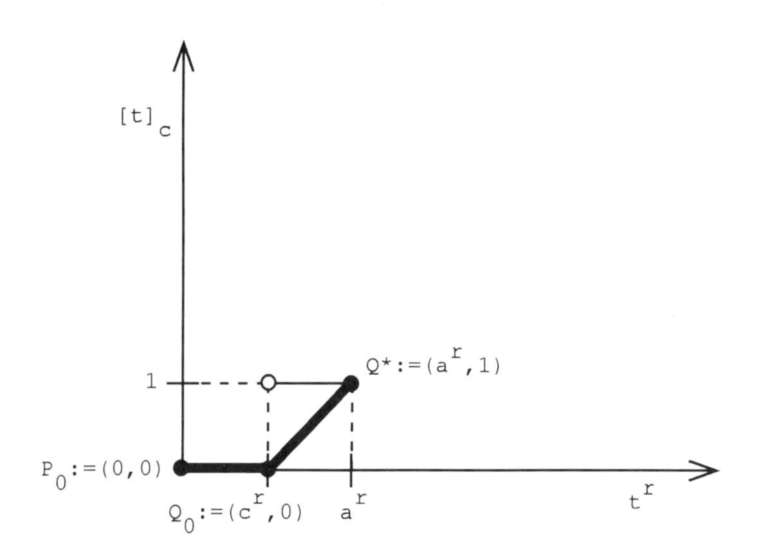

Figure 4.1.2, when $c < a \leq 1 + c$

The line through (Q_0, Q^*) has equation

$$y = (\frac{x - c^r}{a^r - c^r}).$$

When $c \leq d \leq a$ we obtain that

$$L = \frac{d^r - c^r}{a^r - c^r}.$$

□

The case of $a = +\infty$ is treated as follows.

Theorem 4.1.4. *Let $c \in (0,1), r > 0, d > 0$ and*

$$L = L_{[\cdot]_c}(r,d) := \inf\{E[X]_c : X \geq 0 \text{ a.s.}, (EX^r)^{1/r} = d\}. \tag{4.1.9}$$

(I) *If $r > 1$, then $L = 0$.*

(II) *If $0 < r \leq 1, 0 < d \leq c$, then $L = 0$.*

(III) *If $r = 1, c \leq d < +\infty$, then $L = d - c$.*

(IV) *If $0 < r \leq 1$ define $k \in \mathbb{N} \cup \{0\}$ by $k + c \leq d < k + 1 + c$. Then*

$$L = k + \frac{d^r - (k+c)^r}{(k+1+c)^r - (k+c)^r}.$$

PROOF: As similar and simpler of the proof of Theorem 4.1.3 is omitted. □

4.2. Moment Problems Related To c–Rounding Proportions Subject To Two Moment Conditions.

Let $0 \leq c \leq 1$ and $x \geq 0, [x]_c$ was defined in (4.1.1).

Moment Problem 4.2.1. Find

$$\sup_\mu \int_A [t]_c \mu(dt), \tag{4.2.1}$$

where μ is a probability measure on

$$A := [0,a] \quad \text{or} \quad [0,+\infty)$$

and

$$\inf_\mu \int_A [t]_c \mu(dt), \tag{4.2.2}$$

subject to the given two moment conditions

$$\int_A t\mu(dt) = d_1, d_1 > 0 \tag{4.2.3}$$

and

$$\int_A t^r \mu(dt) = d_r, r > 0, d_r > 0. \tag{4.2.4}$$

The next theorems deal with the solution of Moment Problem 4.2.1. Again in the sequel, the underlined probability space is assumed to be nonatomic and thus the space of laws of nonnegative random variables coincides with the space of all Borel probability measures on $\mathbb{R}_+$.

Theorem 4.2.1. *Let $c \in (0,1), 0 < r \neq 1, 0 < a < +\infty$, $d_1 > 0, d_r > 0$ and*

$$U := U_{[\cdot]_c}(a, r, d_1, d_r) := \sup\{E[X]_c : 0 \le X \le a \ \ a.s.,$$
$$EX = d_1, EX^r = d_r\}. \tag{4.2.5}$$

(I) *Suppose there exists $\lambda \in [0,1]$ such that $d_1 = \lambda c, d_r = \lambda c^r$. Then $U = d_1/c$.*

(II) *Suppose that $0 < a \le c$. Then $U = 0$.*

(III) *Suppose $c < a \le 1 + c$.*

(i) *If there exist $\lambda_1, \lambda_2 \ge 0$ such that*

$$\lambda_1 + \lambda_2 \le 1, \quad \text{and} \quad d_1 = \lambda_1 c + \lambda_2 a, d_r = \lambda_1 c^r + \lambda_2 a^r,$$

then

$$U = \frac{(c-a)d_r - (c^r - a^r)d_1}{ca^r - c^r a}.$$

(ii) *Suppose $c \le d_1 \le a$ and assume that either $r > 1$ and*

$$d_1^r \le d_r \le \Delta_r := a^r + (\frac{a^r - c^r}{a - c})(d_1 - a)$$

or

$$0 < r < 1$$

and

$$\Delta_r \le d_r \le d_1^r,$$

then $U = 1$.

From now on suppose that $a > 1 + c$ and set $n = [a]$; the integral part of a,

$$m := \begin{cases} n+1, & \text{if } n + c < a, \\ n, & \text{if } n + c \ge a. \end{cases}$$

(IV) *Suppose $m-1+c \le d_1 \le a$ and assume that either $r>1$ and*

$$d_1^r \le d_r \le d_{r,m,a} := a^r + \left(\frac{a^r-(m-1+c)^r}{a-(m-1+c)}\right)(d_1-a),$$

or $r \in (0,1)$ and $d_{r,m,a} \le d_r \le d_1^r$. Then $U=m$.

(V) *Suppose one of the following holds:*

(i) *$r>1$ and there exists $k \in \{0,1,\ldots,m-2\}$ such that $k+c \le d_1 < k+1+c$ and $\tilde{d}_k := (k+c)^r + [(k+1+c)^r-(k+c)^r]\cdot(d_1-k-c) \le d_r \le d_{r,m,c} := c^r + [\frac{(m-1+c)^r-c^r}{m-1}]\cdot(d_1-c)$;*

(ii) *$0<r<1$ and there exists $k \in \{0,1,\ldots,m-2\}$ such that $k+c \le d_1 < k+1+c$ and $d_{r,m,c} \le d_r \le \tilde{d}_k$. Then*

$$U = d_1+1-c.$$

Call

$$\Delta := \frac{m(c^r a - ca^r) - (m-1)(c^r(m-1+c) - c(m-1+c)^r)}{(a-c)(m-1+c)^r - (m-1+c)(a^r-c^r)}.$$

(VI) *Assume that $\Delta \ge 1$.*

(i) *Suppose there exists $\lambda_1, \lambda_2 \ge 0$ with $\lambda_1+\lambda_2 \le 1$ such that $d_1 = \lambda_1 c + \lambda_2(m-1+c)$, $d_r = \lambda_1 c^r + \lambda_2(m-1+c)^r$. Then*

$$U = \frac{d_r(m-1+c-mc) - d_1((m-1+c)^r - mc^r)}{(m-1+c)c^r - (m-1+c)^r c}.$$

(ii) *Suppose there exist $\lambda_1, \lambda_2 \ge 0$ with $\lambda_1+\lambda_2 \le 1$ and such that*

$$d_1 = \lambda_1(m-1+c) + \lambda_2 a$$

and

$$d_r = \lambda_1(m-1+c)^r + \lambda_2 a^r.$$

Then

$$U = \frac{m(a-(m-1+c))d_r - m(a^r-(m-1+c)^r)d_1}{a(m-1+c)^r - a^r(m-1+c)}.$$

(VII) *Assume $\Delta \le 1$.*

(i) *Suppose there exists $\lambda_1, \lambda_2 \ge 0$ with $\lambda_1+\lambda_2 \le 1$ such that $d_1 = \lambda_1 c + \lambda_2 a$, $d_r = \lambda_1 c^r + \lambda_2 a^r$. Then*

$$U = \frac{d_1(mc^r - a^r) + d_r(a-cm)}{ac^r - ca^r}.$$

(ii) *Suppose there exists $\lambda_1, \lambda_2, \lambda_3 \ge 0$ with $\lambda_1+\lambda_2+\lambda_3 = 1$ and such that*

$$\begin{aligned} d_1 &= \lambda_1 c + \lambda_2(m-1+c) + \lambda_3 a \\ d_r &= \lambda_1 c^r + \lambda_2(m-1+c)^r + \lambda_3 a^r. \end{aligned}$$

Then

$$U = 1 + \left\{ \frac{\begin{array}{c}(m-1)(a-c-m+1)(d_r - c^r) - (m-1)\cdot \\ \cdot(a^r - (m-1+c)^r)(d_1 - c)\end{array}}{(a-c)((m-1+c)^r - c^r) - (m-1)(a^r - c^r)} \right\}.$$

PROOF: Observe that the points $v_k = (k+c, (k+c)^r, k+1), k \in \mathbb{N} \cup \{0\}$ belong to the plane $Q : z = x + 1 - c$. Set $n = [a]$, and assume that $a \notin \mathbb{N}$. We meet the following cases:

1) $\underline{n + c < a}$
Here the points of the graph $(t, t^r, [t]_c), 0 \le t \le a, 0 < r \neq 1$ that determine the upper envelope of its convex hull are: $0 := (0,0,0), v_k; k = 0, 1, \ldots, n$ and $B := (a, a^r, n+1)$.

2) $\underline{n + c \ge a}$
Same as in #1, however here $k = 0, 1, \ldots, n-1$ and $B := (a, a^r, n)$. Therefore the last point of the graph will be $B = (a, a^r, m)$, where

$$m := \begin{cases} n+1, & \text{if } n + c < a \\ n, & \text{if } n + c \ge a. \end{cases}$$

Hence $k = 0, 1, \ldots, m-1$ in either case of $n + c < a$ or $n + c \ge a$.

Thus in either case, the upper envelope of the convex hull of $(t, t^r, [t]_c), 0 \le t \le a$, $0 < r \neq 1$ is determined by the points $0, v_k; k = 0, 1, \ldots, m-1$ and B. Set $A := v_{m-1} = (m-1+c, (m-1+c)^r, m)$ and $D := v_0 = (c, c^r, 1)$. Next we observe that the part S we are interested in, of the upper-envelope of the associated convex hull is a concave surface formed by the following planes:

$\underline{\text{Case } \Delta \le 1:}$ $Q : z = x + 1 - c, P : z = m, (D\overset{\triangle}{A}B), (O\overset{\triangle}{D}B)$.

$\underline{\text{Case } \Delta > 1:}$ $Q, P, (O\overset{\triangle}{D}A), (O\overset{\triangle}{A}B)$.

The values of U in the different arising cases are now clearly read off by applying the geometric moment theory method of optimal distance. All the above are also true when $a \in \mathbb{N}$.

For this proof see the following figures.

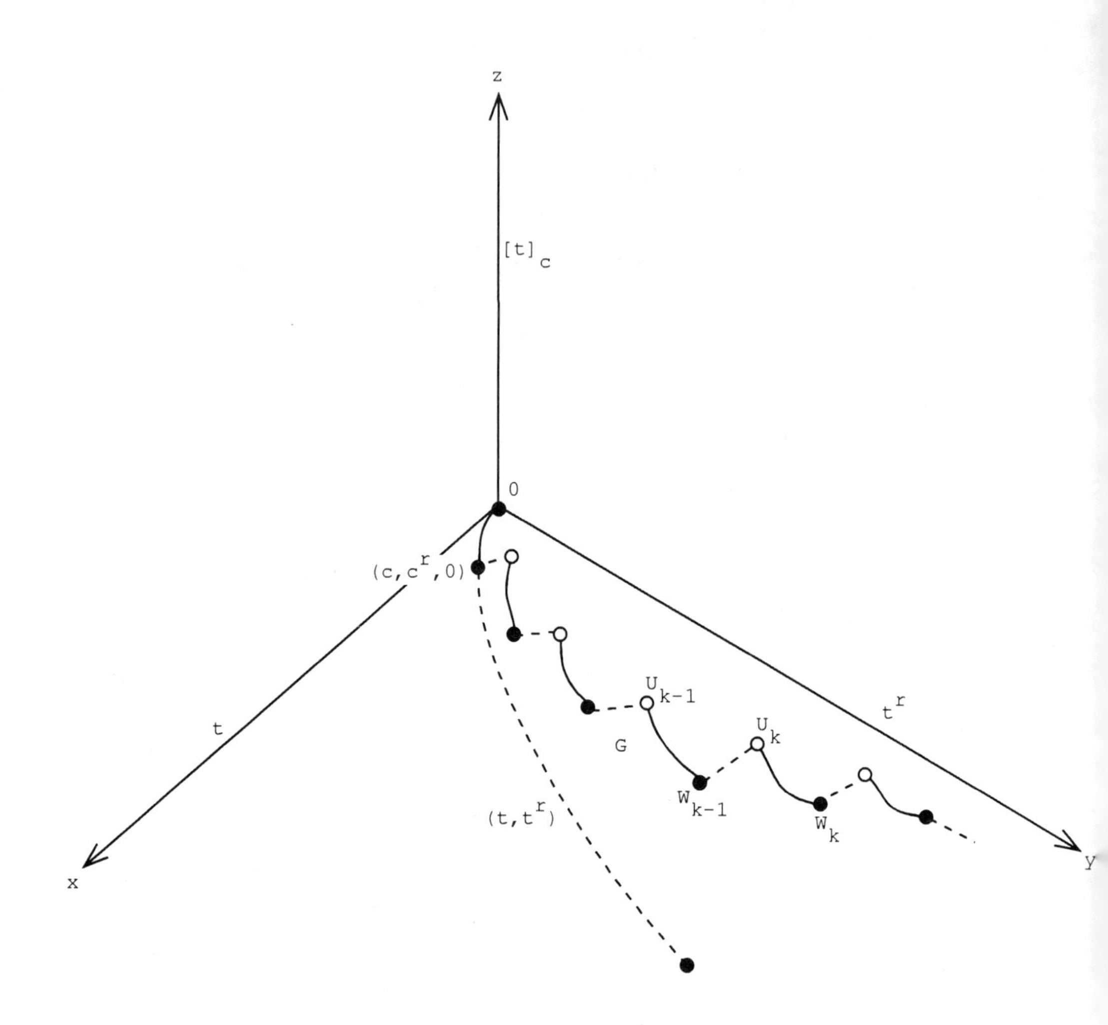

Figure 4.2.1: graph $G = (t, t^r, [t]_c), c \in (0,1), r > 1, t \geq 0$.

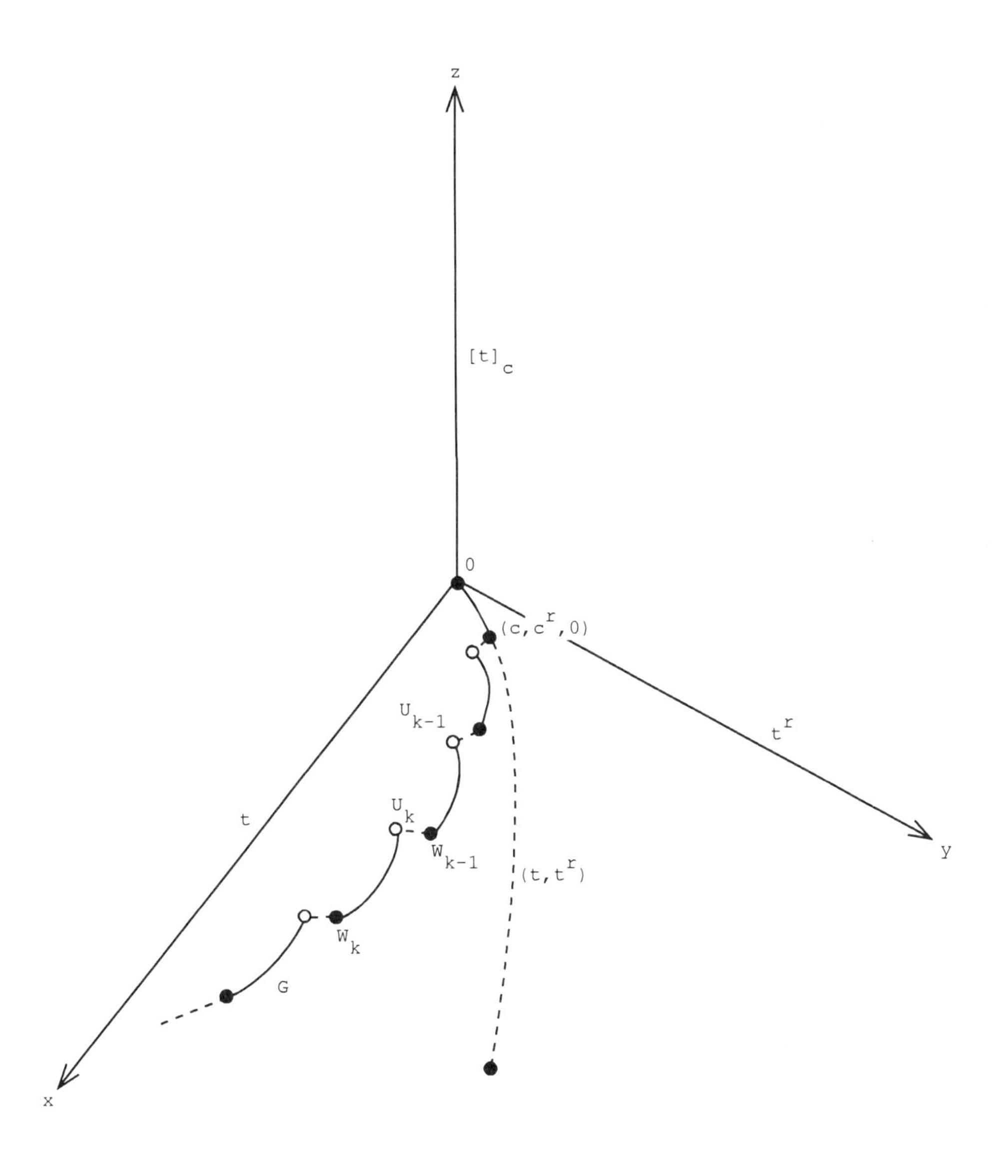

Figure 4.2.2: graph $G = (t, t^r, [t]_c), c \in (0, 1), 0 < r < 1, t \geq 0$.

The case of $a = +\infty$ is similar and simpler and we shall state the related results without proof.

Theorem 4.2.2. *Let $0 < c < 1, 0 < r \neq 1, d_1 > 0, d_r > 0$ and*

$$\begin{aligned} U = U_{[\cdot]_c}(r, d_1, d_r) : &= \sup\{E[X]_c : X \geq 0 \\ a.s., EX &= d_1, EX^r = d_r\} \end{aligned} \tag{4.2.6}$$

(I) *Suppose $r > 1$.*

(i) *Let $0 < d_1 \leq c$ and $c^r \leq d_r < +\infty$. Then*

$$U = d_1/c.$$

(ii) *Suppose that there exist $\lambda_1, \lambda_2 \geq 0$ with $\lambda_1 + \lambda_2 \leq 1$ and such that $d_1 = \lambda_1 c$, $d_r = (\lambda_1 + \lambda_2)c^r$. Then $U = d_1/c$.*

(iii) *Suppose there exists integer $k \geq 0$ such that $k + c \leq d_1 < k + 1 + c$ and $(k+c)^r + [(k+1+c)^r - (k+c)^r](d_1 - k - c) \leq d_r < +\infty$. Then*

$$U = d_1 + 1 - c.$$

(II) *Suppose $0 < r < 1$.*

(i) *Let $c \leq d_1 < +\infty, 0 < d_r \leq c^r$. Then*

$$U = d_r c^{-r}.$$

(ii) *Suppose there exist $\lambda_1 \geq 0$, $\lambda_2 \geq 0$, such that $\lambda_1 + \lambda_2 \leq 1$ and $d_1 = (\lambda_1 + \lambda_2)c, d_r = \lambda_2 c^r$. Then*

$$U = d_r c^{-r}.$$

(iii) *Suppose that there exists an integer $k \geq 0$ with $k + c \leq d_1 < k + 1 + c$ and such that*

$$c^r < d_r \leq (k+c)^r + [(k+1+c)^r - (k+c)^r](d_1 - k - c).$$

Then

$$U = d_1 + 1 - c.$$

Open Moment Problem 4.2.2

Find $U = \sup_\mu \int_A [t]_c \mu(dt)$, subject to the two moment conditions (4.2.3) and (4.2.4), when the moment point (d_1, d_r) is located in the region:

1) Case of $r > 1$. Let $k \in \mathbb{Z}_+$ be such that $k + c \le d_1 < k + 1 + c$ and

$$d_1^r \le d_r \le [(k+c)^r + ((k+1+c)^r - (k+c)^r) \cdot (d_1 - k - c)].$$

2) Case of $0 < r < 1$. Let $k \in \mathbb{Z}_+$ be such that $k + c \le d_1 < k + 1 + c$ and

$$[(k+c)^r + ((k+1+c)^r - (k+c)^r) \cdot (d_1 - k - c)] \le d_r \le d_1^r.$$

The next two theorems deal with the lower bound in Moment Problem (4.2.1), see (4.2.2), (4.2.3), (4.2.4).

Theorem 4.2.3. *Let $a > 0, 0 < c < 1, d_1 > 0, d_r > 0, 0 < r \ne 1$. Denote*

$$\begin{aligned} L \ &: = L_{[\cdot]_c}(a, r, d_1, d_r) \\ &: = \inf\{E[X]_c : 0 \le X \le a \quad \text{a.s. and} \quad EX = d_1, EX^r = d_r\}. \end{aligned} \tag{4.2.7}$$

(I) *Suppose there exist t_1, t_2, λ such that $0 \le t_1 \le t_2 \le \min(c, a), 0 \le \lambda \le 1$ and*

$$d_1 = (1-\lambda)t_1 + \lambda t_2, d_r = (1-\lambda)t_1^r + \lambda t_2^r.$$

Then $L = 0$.

(II) *Suppose $c < a \le 1 + c$ and there exist $\lambda_1, \lambda_2 \ge 0$ such that $\lambda_1 + \lambda_2 \le 1$ and*

$$d_1 = \lambda_1 c + \lambda_2 a, d_r = \lambda_1 c^r + \lambda_2 a^r.$$

Then

$$L = (d_r - c^{r-1} d_1)/(a^r - a c^{r-1}).$$

From now on assume $a > 1 + c$.

Set $n = [a]$ and

$$m = \begin{cases} n+1, & \text{if } n + c < a \\ n, & \text{if } n + c \ge a. \end{cases}$$

(III) *Suppose that one of the following two conditions holds:*

(i) $r > 1$ and there exists $k \in \{0, 1, \ldots, m-2\}$ such that $k + c \le d_1 < k + 1 + c$ and $\tilde{d}_k := (k+c)^r + ((k+1+c)^r - (k+c)^r)(d_1 - k - c) \le d_r \le d_{r,m,c} := c^r + [\frac{(m-1+c)^r - c^r}{m-1}] \cdot (d_1 - c)$;

(ii) $0 < r < 1$ and there exists $k \in \{0, 1, \ldots, m-2\}$ such that $k + c \le d_1 < k + 1 + c$ and $d_{r,m,c} \le d_r \le \tilde{d}_k$. Then $L = d_1 - c$.

(IV) *Assume that*

$$\Delta_{r,m,c} := \frac{(m-1)(a^r - c^{r-1}a)}{(m-1+c)^r - c^{r-1}(m-1+c)} \leq m.$$

(i) *Suppose there exist* $\lambda_1, \lambda_2 \geq 0, \lambda_1 + \lambda_2 \leq 1$ *such that* $d_1 = \lambda_1 c + \lambda_2(m-1+c)$ *and* $d_r = \lambda_1 c^r + \lambda_2(m-1+c)^r$. *Then*

$$L = \frac{(m-1)(d_r - c^{r-1}d_1)}{(m-1+c)^r - c^{r-1}(m-1+c)}.$$

(ii) *Suppose there exist* $\lambda_1, \lambda_2 \geq 0, \lambda_1 + \lambda_2 \leq 1$ *such that* $d_1 = \lambda_1(m-1+c) + \lambda_2 a, d_r = \lambda_1(m-1+c)^r + \lambda_2 a^r$. *Then*

$$L = \frac{((m-1+c)m - a(m-1))d_r - ((m-1+c)^r m - a^r(m-1))d_1}{(m-1+c)a^r - a(m-1+c)^r}.$$

(V) *Assume that* $\Delta_{r,m,c} > m$.

(i) *Suppose that there exist* $\lambda_1, \lambda_2 \geq 0$, *with* $\lambda_1 + \lambda_2 \leq 1$ *and such that* $d_1 = \lambda_1 c + \lambda_2 a, d_r = \lambda_1 d^r + \lambda_2 a^r$. *Then*

$$L = \frac{m(d_r - c^{r-1}d_1)}{a^r - ac^{r-1}}.$$

(ii) *Suppose that there exist* $\lambda_1, \lambda_2, \lambda_3 \geq 0$ *with* $\lambda_1 + \lambda_2 + \lambda_3 = 1$ *and such that* $d_1 = \lambda_1 c + \lambda_2(m-1+c) + \lambda_3 a, \quad d_r = \lambda_1 c^r + \lambda_2(m-1+c)^r + \lambda_3 a^r$. *Then*

$$L = \frac{(m-1)(m-a+c)(d_r - c^r) - (m(m-1+c)^r - (m-1)a^r - c^r)(d_1 - c)}{(m-1)(a^r - c^r) - (a-c)((m-1+c)^r - c^r)}.$$

PROOF: See figures (4.2.1), (4.2.2) and observe that the points $w_k := (k+c, (k+c)^r, k), k \in N \cup \{0\}$ belong to the same plane $Q : z = x - c$. Set $n := [a]$, and assume $a \notin \mathbb{N}$. We meet the following cases:

(1) $n + c < a$. Here the points of the graph $(t, t^r, [t]_c), 0 \leq t \leq a, 0 < r \neq 1$ that determine the lower envelope of its convex hull are:

$$0 = (0,0,0), w_k; k = 0, 1, \ldots, n$$

and

$$B := (a, a^r, n+1).$$

(2) $n+c \geq a$. Same as in (1), however $k = 0, 1, \ldots, n-1$ and $B := (a, a^r, n)$. Therefore, the last point of the graph will be $B = (a, a^r, m)$, where

$$m := \begin{cases} n+1, & \text{if } n+c < a \\ n, & \text{if } n+c \geq a. \end{cases}$$

Hence, $k = 0, 1, \ldots, m-1$ in either case of $n+c < a$ or $n+c \geq a$.

Thus, in either case, the lower envelope of convex hull of $(t, t^r, [t]_c), 0 \leq t \leq a, 0 < r \neq 1$ is determined by the points $0, w_k; k = 0, 1, \ldots, m-1$ and B. Set

$$A := w_{m-1} = (m-1+c, (m-1+c)^r, m-1)$$

and

$$W := w_0 = (c, c^r, 0).$$

Next we observe that the part S we are interested in, of the lower envelope of the associated convex hull is a convex surface generated by the following convex hulls:

Case $\Delta_{r,m,c} \leq m$: conv. hull $(\overset{\frown}{OW})$, conv. hull $\{w_k\}_{k=0}^{m-1}$, conv. hull $(O \overset{\Delta}{W} A)$ and conv. hull $(O \overset{\Delta}{A} B)$.

Case $\Delta_{r,m,c} \geq m$: conv. hull $(\overset{\frown}{OW})$, conv. hull $\{w_k\}_{k=0}^{m-1}$, conv. hull $(W \overset{\Delta}{O} B)$ and conv. hull $(W \overset{\Delta}{A} B)$.

The values of L in the different arising cases are now clearly given by the geometric moment method of optimal distance.

The same proof goes through when $a \in \mathbb{N}$, in fact is much simpler. □

Theorem 4.2.3 for $a = +\infty$ reads as follows. The proof as similar to that is omitted.

Theorem 4.2.4 *Let* $0 < c < 1, d_1 > 0, d_r > 0, 0 < r \neq 1$. *Denote*

$$L := L_{[\cdot]_c}(r, d_1, d_r) := \inf\{E[X]_c : X \geq 0 \ \ a.s., EX = d_1, EX^r = d_r\}. \qquad (4.2.8)$$

(I) *Suppose one of the following three conditions holds:*

(i) *there exist* t_1, t_2, λ *such that* $0 \leq t_1 \leq t_2 \leq c, 0 \leq \lambda \leq 1$ *such that*

$$d_1 = (1-\lambda)t_1 + \lambda t_2, d_r = (1-\lambda)t_1^r + \lambda t_2^r;$$

(ii) $r > 1$ *and there exist* $\lambda_1, \lambda_2 \geq 0, \lambda_1 + \lambda_2 \leq 1$ *such that* $d_1 = \lambda_1 c, d_r = (\lambda_1 + \lambda_2)c^r$;

(iii) $r > 1$ *and* $0 < d_1 \leq c$ *and* $d_r \geq c^r$.

Then $L = 0$.

(II) *Suppose there exists integer* k *such that* $k + c \leq d_1 < k + 1 + c$ *and either when* $r > 1$ *we have* $d_r \geq d_{r,k} := (k+c)^r + ((k+1+c)^r - (k+c)^r)(d_1 - k - c)$; *or when* $0 < r < 1$ *we have* $d_{r,k} \geq d_r \geq c^r$. *Then* $L = d_1 - c$.

(III) *Let* $0 < r < 1$, *and either there exist* $\lambda_1 \geq 0, \lambda_2 \geq 0, \lambda_1 + \lambda_2 \leq 1$ *such that* $d_1 = (\lambda_1 + \lambda_2)c, d_r = \lambda_2 c^r$, *or* $d_1 \geq c$ *and* $0 < d_r \leq c^r$. *Then*

$$L = (c^{r-1}d_1 - d_r)c^{1-r}.$$

Open Moment Problem 4.2.3.

Find $L = \inf_{\mu} \int_A [t]_c \mu(dt)$, subject to the two moment conditions (4.2.3) and (4.2.4), when the moment point (d_1, d_r) is located in the region described in the open moment problem 4.2.2.

4.3. Moment Problems Related To Jefferson-Rounding Proportions Subject To Two Moment Conditions.

In case of conventional (Jefferson) rounding or MYZ–rounding $[x] = [x]_1, x \geq 0$; the integral part of x, the last four theorems take a simpler form.

Moment Problem 4.3.1 Find

$$\sup_{\mu} \int_A [t]\mu(dt), \tag{4.3.1}$$

where μ is a probability measure on $A := [0, a]$ or $[0, +\infty)$ and

$$\inf_{\mu} \int_A [t]\mu(dt), \tag{4.3.2}$$

subject to the given two moment conditions (4.2.3) and (4.2.4):

$$\int_A t\mu(dt) = d_1, d_1 > 0, \tag{4.2.3}$$

$$\int_A t^r \mu(dt) = d_r, r > 0, d_r > 0. \tag{4.2.4}$$

The next theorems deal with the solution of Moment Problem (4.3.1).

In the sequel, the underlined probability space is assumed to be nonatomic and thus the space of laws of nonnegative random variables coincides with the space of all Borel probability measures on $\mathbb{R}_+$.

Theorem 4.3.1. *Let $a > 0, 0 < r \neq 1, d_1 > 0, d_r > 0$ and*

$$U = U_{[\cdot]}(a, r, d_1, d_r) := \sup\{E[X] : 0 \leq X \leq a \text{ a.s. and } EX = d_1, EX^r = d_r\}. \tag{4.3.3}$$

Call $\theta := [a]$.

(I) *Denote*

$$\Delta_{r,a} := a^r + \left(\frac{a^r - \theta^r}{a - \theta}\right)(d_1 - \theta).$$

Suppose that $a \neq \theta, \theta \leq d_1 \leq a$ and either $r > 1, d_1^r \leq d_r \leq \Delta_{r,a}$, or $0 < r < 1$ and $\Delta_{r,a} \leq d_r \leq d_1^r$. Then $U = \theta$.

(II) *Suppose $0 < \theta \neq a$ and there are $\lambda_1, \lambda_2 \geq 0$ with $\lambda_1 + \lambda_2 \leq 1$ and such that $d_1 = \lambda_1\theta + \lambda_2 a$ and $d_r = \lambda_1\theta^r + \lambda_2 a^r$. Then*

$$U = \frac{(\theta^r - a^r)d_1 + (a - \theta)d_r}{a(\theta^{r-1} - a^{r-1})}.$$

(III) *Let $\theta \geq 1$ and there exists $k \in \{0, 1, \ldots, \theta - 1\}$ such that $k \leq d_1 < k + 1$ and either $r > 1$ and $d_{r,k} := k^r + [(k+1)^r - k^r](d_1 - k) \leq d_r \leq \theta^{r-1}d_1$ or $0 < r < 1$ and $\theta^{r-1}d_1 \leq d_r \leq d_{r,k}$. Then $U = d_1$.*

PROOF: Observe that the points $v_k := (k, k^r, k), k \in \mathbb{N} \cup \{0\}$ belong to the plane $Q : z = x$. Set $\theta = [a]$. Call $0 := (0,0,0), A := (\theta, \theta^r, \theta)$ and $B := (a, a^r, \theta)$. Consider also the plane $P : z = \theta$. Observe that the plane Q is above ($P \cap$ conv. hull $(\overset{\frown}{AB})$), also plane Q is above the convex hull of $(0 \overset{\Delta}{A} B)$. Furthermore the plane P is above the convex hull $(0 \overset{\Delta}{A} B)$.

Note that ($P \cap$ conv. hull $(\overset{\frown}{AB})$) = conv. hull $(\overset{\frown}{AB})$. Therefore the part of the upper envelope of convex hull of $(t, t^r, [t]), 0 \leq t \leq a$ we are interested in, is formed by {convex hull$\{v_k\}_{k=0}^{\theta}$, convex hull $(\overset{\frown}{AB})$, convex hull $(O \overset{\Delta}{A} B)$}. This is a concave surface in $\mathbb{R}^3$.

Now the values of U in the three different arising cases are clearly read off by applying the geometric moment theory method of optimal distance. See figures (4.3.1) and (4.3.2).

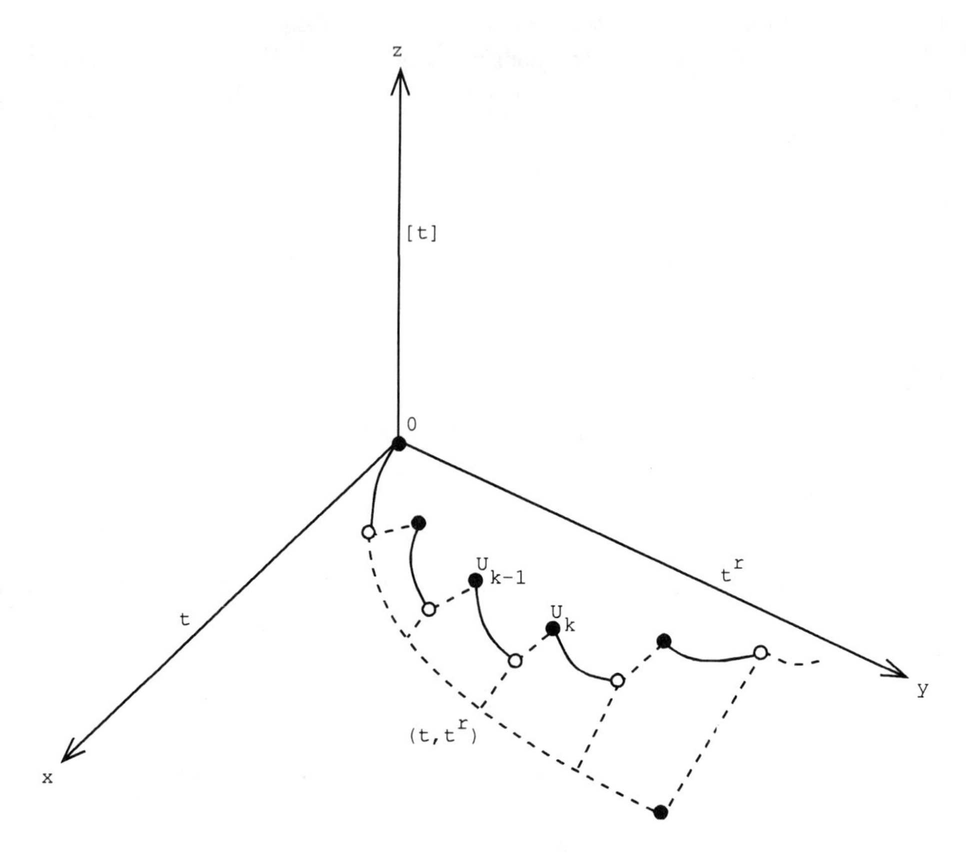

Figure (4.3.1): Graph of $(t, t^r, [t]), t \geq 0; r > 1$.

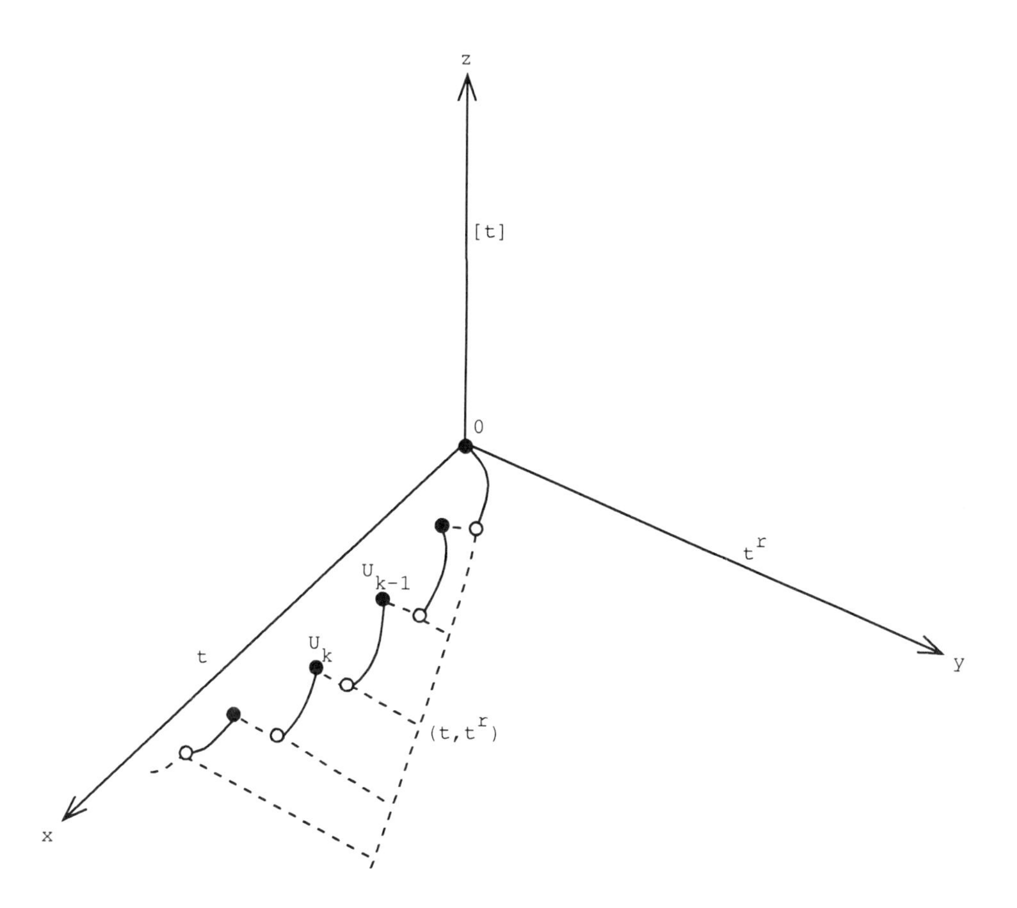

Figure (4.3.2): Graph of $(t, t^r, [t]), t \geq 0; 0 < r < 1$.

□

For the case $a = +\infty$ we have the following version of the last theorem.

Theorem 4.3.2 *Let $0 < r \neq 1, d_1 > 0, d_r > 0$ and*

$$U := U_{[\cdot]}(r, d_1, d_r) := \sup\{E[X] : X \geq 0 \ a.s$$

and

$$EX = d_1, EX^r = d_r\}. \tag{4.3.4}$$

Suppose there exists $k \geq 0$ integer such that $k \leq d_1 < k+1$ and either $r > 1$ and $d_r \geq d_{r,k} := k^r + ((k+1)^r - k^r)(d_1 - k)$ or $0 < r < 1$ and $0 < d_r \leq d_{r,k}$. Then $U = d_1$.

PROOF: Similar to theorem 4.3.1. □

Open Moment Problem 4.3.2

Find

$$U = \sup_{\mu} \int_A [t]\mu(dt)$$

subject to (4.2.3) and (4.2.4), given that the moment point (d_1, d_r) belongs to the following region:

1) Case of $r > 1$; $k \leq d_1 < k+1, k \in \mathbb{Z}_+$ and $d_1^r \leq d_r \leq (k^r + ((k+1)^r - k^r)(d_1 - k))$.

2) Case of $0 < r < 1$; $k \leq d_1 < k+1, k \in \mathbb{Z}_+$ and $(k^r + ((k+1)^r - k^r)(d_1 - k)) \leq d_r \leq d_1^r$.

If we change in Theorems 4.3.1 and 4.3.2 the upper bound U with the corresponding lower bound (see (4.2. (2-4)) for $c = 1$) we obtain the following two theorems.

Theorem 4.3.3. *Let $a > 0, 0 < r \neq 1, d_1 > 0, d_r > 0$ and*

$$L := L_{[\cdot]}(a, r, d_1, d_r) := \inf\{E[X] : 0 \leq X \leq a \text{ a.s.}, EX = d_1, EX^r = d_r\} \tag{4.3.5}$$

(I) *Suppose there exist t_1, t_2, λ such that $0 \leq t_1 \leq t_2 \leq 1, 0 \leq \lambda \leq 1$, and $d_1 = (1-\lambda)t_1 + \lambda t_2$ and $d_r = (1-\lambda)t_1^r + \lambda t_2^r$. Then $L = 0$.*

(II) *If $0 < a \leq 1$ then $L = 0$.*

(III) *If $1 < a < 2$ and there exist $\lambda_1, \lambda_2 \geq 0$ with $\lambda_1 + \lambda_2 \leq 1$ and such that $d_1 = \lambda_1 + \lambda_2 a, d_r = \lambda_1 + \lambda_2 a^r$. Then $L = (d_r - d_1)/(a^r - a)$.*

From now on assume $a \geq 2$, and let $\theta := [a]$, the integral part of a.

(IV) *Suppose* $\Delta_\theta := (\theta - 1)(a^r - a)/(\theta^r - \theta) \le \theta$.

(i) *Suppose* $d_1 = \lambda_1 + \lambda_2\theta, d_r = \lambda_1 + \lambda_2\theta^r$ *for some* $\lambda_1, \lambda_2 \ge 0, \lambda_1 + \lambda_2 \le 1$. *Then* $L = \Delta_\theta$.

(ii) *Suppose* $d_1 = \lambda_1\theta + \lambda_2 a$ *and* $d_r = \lambda_1\theta^r + \lambda_2 a^r$ *for some* $\lambda_1, \lambda_2 \ge 0, \lambda_1 + \lambda_2 \le 1$. *Then*

$$L = \frac{(\theta^2 - a(\theta - 1))d_r - (\theta^{r+1} - a^r(\theta - 1))d_1}{\theta a^r - a\theta^r}.$$

(V) *Suppose* $\Delta_\theta > \theta$.

(i) *Suppose* $d_1 = \lambda_1 + \lambda_2\theta + \lambda_3 a$ *and* $d_r = \lambda_1 + \lambda_2\theta^r + \lambda_3 a^r$ *for some* $\lambda_1, \lambda_2, \lambda_3 \ge 0, \lambda_1 + \lambda_2 + \lambda_3 = 1$. *Then*

$$L = \frac{(\theta - 1)(\theta - a + 1)(d_r - 1) - ((\theta^r - 1)\theta - (a^r - 1)(\theta - 1))(d_1 - 1)}{(\theta - 1)(a^r - 1) - (a - 1)(\theta^r - 1)}.$$

(ii) *Suppose* $d_1 = \lambda_1 + \lambda_2 a, d_r = \lambda_1 + \lambda_2 a^r$ *for some* $\lambda_1, \lambda_2 \ge 0, \lambda_1 + \lambda_2 \le 1$. *Then* $L = \theta(d_r - d_1)/(a^r - a)$.

(VI) *Suppose* $\theta > 1$ *and one of the following holds.*

(i) $r > 1$ *and there exists* $k \in \{1, \dots, \theta - 1\}$ *such that* $k \le d_1 < k + 1$ *and*

$$d_{r,k} := k^r + ((k+1)^r - k^r)(d_1 - k) \le d_r \le \Delta_{r,\theta} := 1 + \frac{(\theta^r - 1)(d_1 - 1)}{(\theta - 1)};$$

(ii) $0 < r < 1$ *and there exists* $k \in \{1, \dots, \theta - 1\}$ *such that* $k \le d_1 < k + 1$ *and* $\Delta_{r,\theta} \le d_r \le d_{r,k}$. *Then* $L = d_1 - 1$.

PROOF: See figures (4.3.1) and (4.3.2) and observe that the points

$$v_k := (k + 1, (k + 1)^r, k), k = 0, 1, \dots$$

belong to the plane $Q : z = x - 1$. Set $\theta := [a]$. Call $O := (0, 0, 0), W := (1, 1, 0)$, $A := (\theta, \theta^r, \theta - 1)$ and $B := (a, a^r, \theta)$. Obviously $W, A \in Q$ and $O \notin Q$. The last point of the graph $(t, t^r, [t]); r > 0, 0 \le t \le a$ will be B and the last lower point determining the lower envelope of conv. hull $(t, t^r, [t])_{t=0}^a$ is A. Note that B is above the plane Q, also $v_k \in (Q \cap$ conv. hull $(t, t^r, [t])_{t=0}^a)$, for $k = 0, 1, \dots, \theta - 1$. Here assume that $\theta > 1$ and observe that Q is below conv. hull $(O \overset{\Delta}{W} A)$.

1) Case of $\Delta_\theta \le \theta$. Then the part of the lower envelope of the conv. hull $(t, t^r, [t])_{t=0}^a$ we are interested in, is formed by

$$\left\{ \begin{array}{l} \text{conv. hull } (\overset{\frown}{OW}), \text{ conv. hull } (\{v_k\}_{k=0}^{\theta-1}), \\ \text{conv. hull } (W \overset{\Delta}{A} O), \text{ conv. hull } (A \overset{\Delta}{O} B) \end{array} \right\}.$$

2) <u>Case of $\Delta_\theta > \theta$.</u> Then the part of the lower envelope of the conv. hull $(t, t^r, [t])_{t=0}^{a}$, we are interested in, is formed by

$$\left\{ \begin{array}{l} \text{conv. hull } (\overset{\frown}{OW}), \text{ conv. hull } (\{v_k\}_{k=0}^{\theta-1}), \\ \text{conv. hull } (W \overset{\Delta}{A} B), \text{ conv. hull } (W \overset{\Delta}{O} B) \end{array} \right\}.$$

Obviously the above are convex surfaces in $\mathbb{R}^3$.

Now the values of L in the different cases are given by the geometric moment method of optimal distance. □

The special case $a = +\infty$ is treated as follows.

Theorem 4.3.4. *Let $0 < r \neq 1, d_1 > 0, d_r > 0$ and*

$$L := L_{[\cdot]}(r, d_1, d_r) := \inf\{E[X] : X \geq 0 \text{ a.s.}, EX = d_1, EX^r = d_r\}. \tag{4.3.6}$$

(I) *Suppose there exist $0 \leq t_1 \leq t_2 \leq 1, 0 \leq \lambda \leq 1$ such that $d_1 = (1-\lambda)t_1 + \lambda t_2$ and $d_r = (1-\lambda)t_1^r + \lambda t_2^r$. Then $L = 0$.*

(II) *Suppose $0 < r < 1$ and either $d_1 = \lambda_1 + \lambda_2, d_r = \lambda_1, \lambda_i \geq 0, i = 1, 2, \lambda_1 + \lambda_2 \leq 1$ or $d_1 \geq 1, 0 < d_r \leq 1$. Then $L = d_1 - d_r$.*

(III) *Suppose $0 < r < 1$ and for some positive integer $k, k \leq d_1 < k+1$, and $1 \leq d_r \leq (k^r + ((k+1)^r - k^r))(d_1 - k)$. Then $L = d_1 - 1$.*

(IV) *Suppose $r > 1$ and either $d_1 = \lambda_1, d_r = \lambda_1 + \lambda_2, \lambda_i \geq 0, i = 1, 2, \lambda_1 + \lambda_2 \leq 1$, or $0 < d_1 \leq 1$ and $1 \leq d_r$. Then $L = 0$.*

(V) *Suppose $r > 1$ and for some positive integer $k, k \leq d_1 < k+1$, and $(k^r + ((k+1)^r - k^r))(d_1 - k) \leq d_r$. Then $L = d_1 - 1$.*

PROOF: As similar to Theorem 4.3.3 is omitted. □

Open Moment Problem 4.3.3.

Find $L = \inf_{\mu} \int_A [t] \mu(dt); A := [0, a]$ or $[0, t\infty)$, where μ is a probability measure on A subject to (4.2.3) and (4.2.4), given that the moment point (d_1, d_r) belongs to the following region:

1) <u>Case of $r > 1$</u>; $k \leq d_1 < k+1; k \in \mathbb{N}$ and $d_1^r \leq d_r \leq \{k^r + ((k+1)^r - k^r)(d_1 - k)\}$.

2) <u>Case of $0 < r < 1$</u>; $k \leq d_1 < k+1; k \in \mathbb{N}$ and $\{k^r + ((k+1)^r - k^r)(d_1 - k)\} \leq d_r \leq d_1^r$.

4.4 Moment Problems Related To Adams Rule of Rounding Subject To Two Moment Conditions.

In the case of Adams rule of rounding we have $[x]_0 = \lceil x \rceil; x \geq 0 (c = 0)$, where $\lceil \cdot \rceil$ is the *ceiling* of the number:

Definition 4.4.1. Let $x \geq 0$, then $\lceil x \rceil$ is defined to be the smallest integer greater equal the number. e.g. $\lceil 0 \rceil = 0, \lceil 3.7 \rceil = 4, \lceil 5 \rceil = 5$.

Moment Problem 4.4.1. Find

$$\sup_{\mu} \int_A \lceil t \rceil \mu(dt) \tag{4.4.1}$$

where μ is a probability measure on

$$A = [0, a] \quad \text{or} \quad [0, +\infty)$$

and

$$\inf_{\mu} \int_A \lceil t \rceil \mu(dt) \tag{4.4.2}$$

subject to the given two moment conditions

$$\int_A t\mu(dt) = d_1, d_1 > 0, \tag{4.2.3}$$

and

$$\int_A t^r \mu(dt) = d_r, r > 0, d_r > 0. \tag{4.2.4}$$

The next theorems deal with the solution of Moment Problem (4.4.1).

Here again the underlined probability space is assumed to be nonatomic.

Theorem 4.4.1. *Let $a > 0, 0 < r \neq 1, d_1 > 0, d_r > 0$. Denote*

$$\begin{aligned} U &:= U_{\lceil \cdot \rceil}(a, r, d_1, d_r) \\ &= \sup\{E\lceil X \rceil : 0 \leq X \leq a \text{ a.s.}, EX = d_1, EX^r = d_r\}. \end{aligned} \tag{4.4.3}$$

Set $\theta := \lceil a \rceil$, the ceiling of a.

(I) *Suppose that $\theta - 1 \leq d_1 \leq a$ and either*

(II) $r > 1$ *and* $d_1^r \leq d_r \leq \Delta_r := a^r + [\frac{a^r - (\theta-1)^r}{a-\theta+1}](d_1 - \theta)$,

or

(I2) $0 < r < 1$ *and* $\Delta_r \leq d_r \leq d_1^r$. *Then* $U = \theta$.

(II) *Suppose $\theta > 1$ and there are $\lambda_1, \lambda_2 \geq 0$ with $\lambda_1 + \lambda_2 = 1$ such that $d_1 = \lambda_1(\theta - 1) + \lambda_2 a$, $d_r = \lambda_1(\theta-1)^r + \lambda_2 a^r$. Then*

$$U = 1 + \frac{((\theta-1)^r - a^r)d_1 + (a - \theta + 1)d_r}{a(\theta-1)^{r-1} - a^r}.$$

(III) *Suppose $\theta > 1$, and let $k := [d_1]$, the integral part of d_1 and*

$$\Delta_r^* := k^r + ((k+1)^r - k^r)(d_1 - k), \widetilde{\Delta}_r := (\theta - 1)^{r-1} d_1.$$

Suppose that either

(III-1) *$r > 1$ and $\Delta_r^* \leq d_r \leq \widetilde{\Delta}_r$ or*

(III-2) *$0 < r < 1$ and $\widetilde{\Delta}_r \leq d_r \leq \Delta_r^*$.*

Then $U = d_1 + 1$.

PROOF: Observe that the points $v_k := (k-1, (k-1)^r, k), k \in \mathbb{N}$ belong to the plane $Q: z = x + 1$. Set $\theta := \lceil a \rceil$, the ceiling of a. Call $0 := (0,0,1), A := (\theta - 1, (\theta-1)^r, \theta)$ and $B := (a, a^r, \theta)$. Consider also the plane $P: z = \theta$. Observe that plane Q is above the convex hull of $(\overset{\frown}{AB}, \overline{AB})$ and above the convex hull of $(O\overset{\Delta}{A}B)$. Furthermore the plane P (through the convex hull $(\overset{\frown}{AB}, \overline{AB})$) is higher than the convex hull $(O\overset{\Delta}{A}B)$. Therefore the part of the upper envelope of the convex hull of $(t, t^r, \lceil t \rceil)_{t=0}^{a}, r > 0$ we are interested in is formed by

$$\{\text{conv. hull } (\{v_k\}_{k=1}^{\theta}), \text{conv. hull } (\overset{\frown}{AB}, \overline{AB}), \text{conv. hull } (O\overset{\Delta}{A}B)\}.$$

That is a concave surface in $\mathbb{R}^3$. Now the values of U in the different cases are read off by applying the geometric moment theory method of optimal distance. See figures (4.4.1) and (4.4.2).

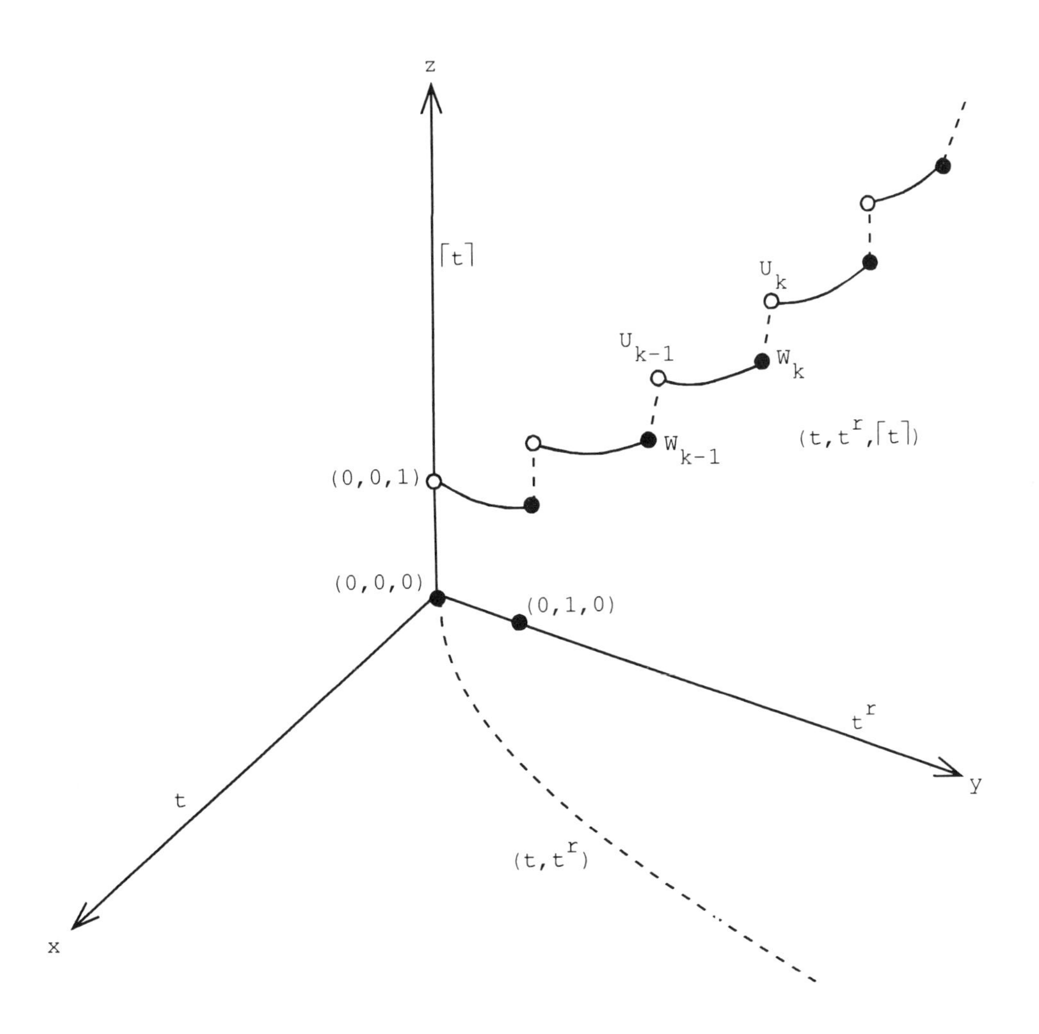

Figure (4.4.1): Graph of $(t, t^r, \lceil t \rceil), t \geq 0; r > 1$.

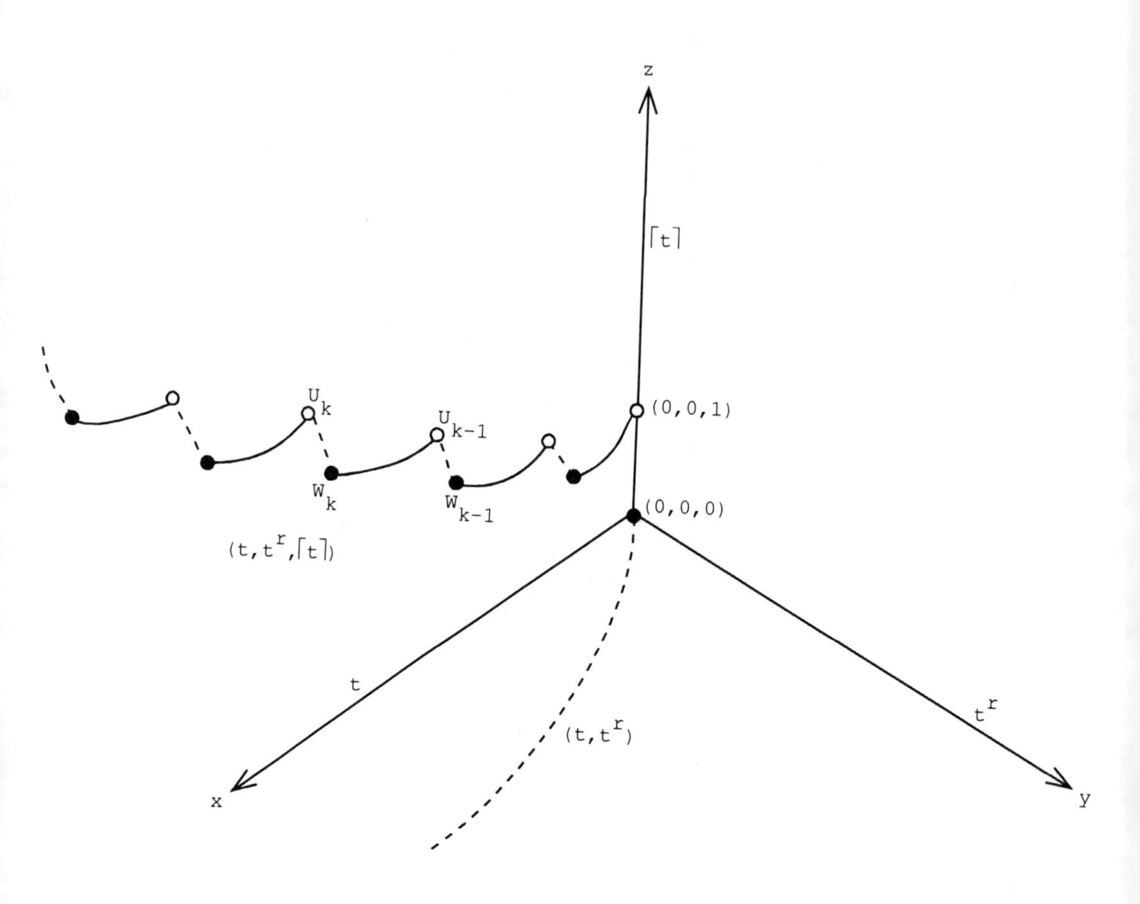

Figure (4.4.2): Graph of $(t, t^r, \lceil t \rceil), t \geq 0; 0 < r < 1$.

□

The case $a = +\infty$ is considerably simpler.

Theorem 4.4.2. *Let $0 < r \neq 1, d_1, d_r > 0$. Denote*

$$U := U_{\lceil \cdot \rceil}(r, d_1, d_r) = \sup\{E\lceil X \rceil : X \geq 0 \text{ a.s.}, EX = d_1, EX^r = d_r\}. \tag{4.4.4}$$

Denote $k := [d_1]$ and

$$\Delta_r := k^r + [(k+1)^r - k^r](d_1 - k).$$

Suppose that either

(I) $r > 1, d_r \geq \Delta_r$, *or*

(II) $0 < r < 1, d_r \leq \Delta_r$.

Then $U = d_1 + 1$.

PROOF: Similar to Theorem 4.4.1. □

Open Moment Problem 4.4.2.

Find

$$U = \sup_\mu \int_A \lceil t \rceil \mu(dt)$$

subject to (4.2.3) and (4.2.4), given that the moment point (d_1, d_r) belongs to the region described in open moment Problem 4.3.2.

Next we find (4.4.2) subject to (4.2.3) and (4.2.4).

Theorem 4.4.3. *Let $0 < r \neq 1, d_1, d_r > 0, a > 0$. Denote*

$$L := L_{\lceil \cdot \rceil}(a, r, d_1, d_r) = \inf\{E\lceil X \rceil : 0 \leq X \leq a \, a.s. \text{ and } EX = d_1, EX^r = d_r\}. \tag{4.4.5}$$

Call $\theta := \lceil a \rceil$, the ceiling of a.

(I) *Suppose $0 < a \leq 1$ and $d_1 = \lambda a, d_r = \lambda a^r$ for some $\lambda \in [0, 1]$. Then $L = d_1/a$.*

(II) *Let $a > 1$ and suppose $d_1 = \lambda_1(\theta - 1) + \lambda_2 a, d_r = \lambda_1(\theta - 1)^r + \lambda_2 a^r$ for some $\lambda_1, \lambda_2 \geq 0, \lambda_1 + \lambda_2 \leq 1$. Then $L = \frac{(a^r - \theta(\theta-1)^{r-1})d_1 + (\theta - a)d_r}{a^r - a(\theta - a)^{r-1}}$.*

(III) *Suppose one of the following holds:*

(i) *$r > 1$ and for some $k \in \{0, 1, \ldots, \theta - 2\}, k \leq d_1 < k + 1, \Delta_r := k^r + [(k+1)^r - k^r](d_1 - k) \leq d_r \leq (\theta - 1)^{r-1} d_1$;*

(ii) $0 < r < 1$ *and for some* $k \in \{0, 1, \ldots, \theta - 2\}, k \le d_1 < k + 1, (\theta - 1)^{r-1} d_1 \le d_r \le \Delta_r$.

Then $L = d_1$.

PROOF: See figures (4.4.1) and (4.4.2) and observe that the points $w_k := (k, k^r, k), k \in \mathbb{Z}_+$ belong to the same plane $Q : z = x$. Set $\theta := \lceil a \rceil$. Call $0 := (0,0,0) \in Q, A := (\theta - 1, (\theta - 1)^r, \theta - 1) \in Q$ and $B := (a, a^r, \theta)$. Note that B is higher than $(a, a^r, a) \in Q$. Here $w_k \in$ conv. hull $(t, t^r, \lceil t \rceil)_{t=0}^{a}; r > 0$ for $k = 0, 1, \ldots, \theta - 1$. Also Q is below conv. hull $(0 \overset{\Delta}{A} B)$. We observe that $\{$conv. hull $(O \overset{\Delta}{A} B)\} \cup \{$ conv. hull $(\{w_k\}_{k=0}^{\theta-1})\}$ generates a convex surface in $\mathbb{R}^3$. This is the part of the lower envelope of the conv. hull $(t, t^r, \lceil t \rceil)_{t=0}^{a}; r > 0$ that we are interested. Thus the values of L in the different cases are read off by applying the geometric moment method of optimal distance. □

The case $a = +\infty$ is treated as follows.

Theorem 4.4.4. *Let* $0 < r \ne 1, d_1 > 0, d_r > 0$. *Denote*

$$L := L_{\lceil \cdot \rceil}(r, d_1, d_r) := \inf\{E\lceil X \rceil : X \ge 0 \, a.s. \text{ and } EX = d_1, DX^r = d_r\}. \tag{4.4.6}$$

Then case (III) of Theorem 4.4.3 takes place.

PROOF: Similar to Theorem 4.4.3. □

Open Moment Problem 4.4.3.

Find

$$L = \inf_{\mu} \int_A \lceil t \rceil \mu(dt)$$

subject to (4.2.3) and (4.2.4), given that the moment point (d_1, d_r) belongs to the region described in the open moment problem 4.3.2.

4.5. Moment Problems Related To Jefferson And Adams Rules Of Rounding Subject To One Moment Condition.

Moment Problem 4.5.1

Find

$$\sup_{\mu} \int_A [t] \mu(dt) \tag{4.5.1}$$

and

$$\inf_{\mu} \int_A [t]\mu(dt), \tag{4.5.2}$$

where μ is a probability measure on $A = [0, a]$ or $[0, +\infty)$ such that

$$\int_A t^r \mu(dt) = d_r, r > 0, d_r > 0. \tag{4.5.3}$$

The calculation of the above leads to Jefferson rule of rounding. Also, find

$$\sup_{\mu} \int_A \lceil t \rceil \mu(dt) \tag{4.5.4}$$

and

$$\inf_{\mu} \int_A \lceil t \rceil \mu(dt) \tag{4.5.5}$$

subject to (4.5.3). The calculation of (4.5.4) and (4.5.5) leads to Adams rule of rounding.

The following theorems solve the Moment Problem 4.5.1. Here the underlined probability space is assumed to be nonatomic.

Theorem 4.5.1. *Let $d > 0, r > 0$ and $0 < a < +\infty$. Consider*

$$U := U_{[\cdot]}(a, r, d) = \sup\{E[X] : 0 \leq X \leq a \ \ a.s., \ \ EX^r = d^r\}. \tag{4.5.6}$$

Here E is the expectation operator, $[\cdot]$ the integral part of the number and X is a random variable. Denote $n := [a]$ and $k := [d]$. Since $d \leq a$ then $0 \leq k \leq n$. Then

(i) *$U = n$, when $k = n$, that is, when $d \geq n$.*

(ii) *$U = d^r n^{1-r}$, when $r \leq 1$ and $k < n$.*

(iii) *$U = (k + \theta_k) \leq d$, when $r \geq 1$ and $k < n$.*

Here

$$\theta_k := \frac{d^r - k^r}{(k+1)^r - k^r}, \quad \text{where } k \in \{0, 1, \ldots, n-1\}.$$

PROOF: From the geometric moment method of optimal distance, it follows that $U = \psi(d^r)$, where $A_1 = \{(u, \psi(u)) : 0 \leq u \leq a^r\}$ describes the upper boundary of the convex hull conv A_0 of the curve

$$A_0 := \{(t^r, [t]) : 0 \leq t \leq a\}.$$

Note that A_0 consists of the following parts:

(i) For $k = 0, \ldots, n-1$, the half open horizontal line segments $[P_k, Q_k)$ where $P_k = (k^r, k)$ and $Q_k = ((k+1)^r, k)$.

(ii) The closed non-empty horizontal line segment $[P_n, Q_*]$, where $Q_* = (a^r, n)$. See figure 4.5.1.

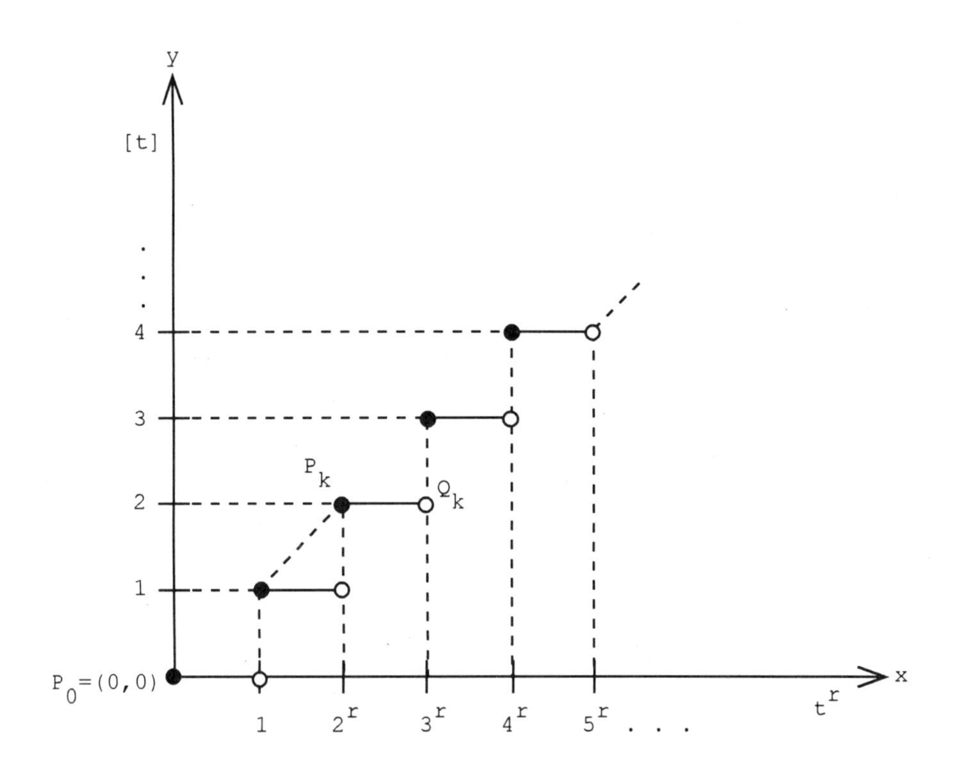

Figure 4.5.1: graph of $(t^r, [t]); r > 0, t \geq 0$.

$[P_n, Q_*]$ is always a part of the upper boundary A_1 of conv A_0, yielding assertion (i) of the theorem. The line segment from P_k to P_{k+1} has a slope $\frac{1}{(k+1)^r - k^r}, k = 0, 1, \ldots, n-1$ which is *decreasing* in k when $r \geq 1$ and *increasing* in k when $0 < r \leq 1$ (since the last denominator has derivative $r((k+1)^{r-1} - k^{r-1})$). Consequently, if $0 < r \leq 1$ then A_1 consists of the two line segments $[P_0, P_n]$ and $[P_n, Q_*]$. Thus $\psi(u) = un^{-r+1}$ when $0 \leq u \leq n^r$, while $\psi(u) = n$ when $n^r \leq u \leq a^r$.

This yields assertion (ii). On the other hand, if $r \geq 1$ then A_1 is composed of the line segments $[P_k, P_{k+1}], k = 0, 1, \ldots, n-1$ together with the horizontal line segment $[P_n, Q_*]$. This yields assertion (iii). The assertions of the theorem follow from the geometry of conv A_0. □

The case of $a = +\infty$ is treated next.

Theorem 4.5.2. *Let $d > 0, r > 0$. Consider*

$$U := U_{[\cdot]}(r, d) = \sup\{E[X] : X \geq 0 \text{ a.s., } EX^r = d^r\} \tag{4.5.7}$$

Set $k := [d]$. Then

(i) $U = (k + \theta_k) \leq d$, *when* $r \geq 1$.
Here $\theta_k := \frac{d^r - k^r}{(k+1)^r - k^r}$, *where* $k \in \mathbb{Z}_+$.

(ii) $U = +\infty$, *if* $r < 1$.

PROOF: As similar and simpler of the proof of Theorem 4.5.1 is omitted.

Theorem 4.5.3. *Let $d > 0, r > 0$ and $0 < a < +\infty$. Consider*

$$U := U_{\lceil\cdot\rceil}(a, r, d) = \sup\{E\lceil X\rceil : 0 \leq X \leq a \text{ a.s., } EX^r = d^r\}. \tag{4.5.8}$$

Here $\lceil\cdot\rceil$ is the ceiling of the number. Denote $n := \lceil a\rceil, k := \lceil d\rceil$. Since $d \leq a$ then $1 \leq k \leq n$. Then

(i) $U = n$, *when* $k = n$, *that is, when* $d > n - 1$.

(ii) $U = (1 + d^r(n-1)^{1-r})$, *when* $r \leq 1$ *and* $k < n$.

(iii) $U = (k + \theta_k) \leq (1 + d)$, *when* $r \geq 1$ *and* $k < n$.

Here $\theta_k := \frac{d^r - (k-1)^r}{k^r - (k-1)^r}$.

PROOF: From the geometric moment method of optimal distance, it follows that $U = \psi(d^r)$, where $A_1 = \{(u, \psi(u)) : 0 \leq u \leq a^r\}$ describes the upper boundary of the convex hull conv A_0 of the curve

$$A_0 = \{(t^r, \lceil t \rceil) : 0 \leq t \leq a\}.$$

Note that A_0 consists of the following parts:

(i) The origin $(0, 0)$ corresponding to $t = 0$.

(ii) For $k = 1, \ldots, n-1$, the half open horizontal line segments $(P_k, Q_k]$, where $P_k := ((k-1)^r, k)$ and $Q_k := (k^r, k)$.

(iii) The half open non-empty horizontal line segment $(P_n, Q_*]$, where $Q_* = (a^r, n)$. It is always a part of the upper boundary A_1 of conv A_0, yielding assertion (i) of the theorem. The line segment from P_k to P_{k+1} has a slope $\frac{1}{k^r-(k-1)^r}$ which is *decreasing* in k when $r \geq 1$ and *increasing* in k when $0 < r \leq 1$ (since the last denominator has derivative $r(k^{r-1} - (k-1)^{r-1})$). Consequently, if $0 < r \leq 1$ then A_1 consists of the two line segments $[P_1, P_n]$ and $[P_n, Q_*]$. Thus $\psi(u) = (1 + u(n-1)^{-r+1})$ when $0 \leq u \leq (n-1)^r$, while $\psi(u) = n$ when $(n-1)^r \leq u \leq a^r$. This yields assertion (ii). On the other hand, if $r \geq 1$ then A_1 is composed of the line segments $[P_k, P_{k+1}], (k = 1, \ldots, n-1)$ together with the horizontal line segment $[P_n, Q_*]$. This yields assertion (iii). The assertions of the theorem follow from the geometry of conv A_0. See figure 4.5.2.

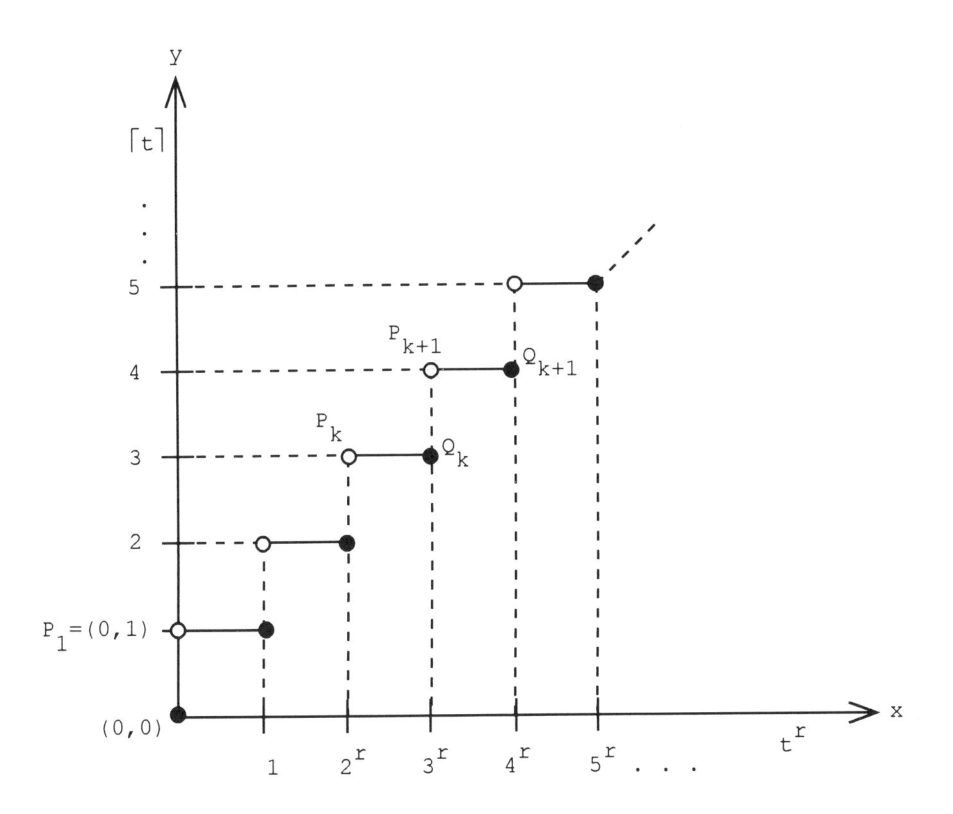

Figure 4.5.2: graph of $(t^r, \lceil t \rceil); r > 0, t \geq 0$.

□

The case $a = +\infty$ is treated as follows:

Theorem 4.5.4 *Let $d > 0, r > 0$. Consider*

$$U := U_{\lceil \cdot \rceil}(r, d) = \sup\{E\lceil X \rceil : X \geq 0 \ a.s., \ \ EX^r = d^r\}. \tag{4.5.9}$$

Denote $k := \lceil d \rceil$. Then

(i) $U = (k + \theta_k) \leq (1 + d)$, *when* $r \geq 1$. *Here* $\theta_k := \frac{d^r - (k-1)^r}{k^r - (k-1)^r}, k \in \mathbb{N}$.

(ii) $U = +\infty$, *if* $r < 1$.

PROOF: As similar to the proof of Theorem 4.5.3 is omitted. □

Next we find the corresponding best lower bounds.

Theorem 4.5.5. *Let $d > 0, r > 0$ and $0 < a < +\infty$. Consider*

$$L := L_{[\cdot]}(a, r, d) = \inf\{E[X] : 0 \le X \le a \ \ a.s., EX^r = d^r\}. \tag{4.5.10}$$

Denote $n := [a]$. When $0 < a \le 1$, then $L = 0$. When $0 < d \le 1$, then $L = 0$.

Case $r \ge 1$

Subcase (i) Assume that $(\frac{n-1}{n^r-1}) \ge \frac{1}{a^r-n^r}$. If $1 \le d \le a$, then

$$L = \left(\frac{d^r - 1}{a^r - 1}\right) n.$$

However, if $1 \le d \le a$ and $a \in \mathbb{N} - \{1\}$, then

$$L = (\frac{d^r - 1}{a^r - 1})(a - 1).$$

Subcase (ii) Here assume that $\frac{n-1}{n^r-1} \le \frac{1}{a^r-n^r}$. If $1 \le d \le n$, then

$$L = (\frac{d^r - 1}{n^r - 1})(n - 1).$$

If $n \le d \le a$, then

$$L = (n - 1) + (\frac{d^r - n^r}{a^r - n^r}).$$

Case $0 < r \le 1$

When $k + 1 \le d < k + 2, k \in \{0, 1, \dots, n - 2\}$, then

$$L = k + (\frac{d^r - (k+1)^r}{(k+2)^r - (k+1)^r})$$

(the last is true also when $a \in \mathbb{N} - \{1\}$).

When $n \le d \le a$, then

$$L = n + (\frac{d^r - a^r}{a^r - n^r}).$$

PROOF: See figure 4.5.1 and apply the geometric moment method of optimal distance. Call $0 := (0,0), W := (1,0), Q_k := ((k+1)^r, k), (Q_{k+1} = ((k+2)^r, k+1)), A := (n^r, n-1)$ and $B := (a^r, n)$.

Call $m_k :=$ slope $(\overline{Q_k Q_{k+1}}) = \frac{1}{(k+2)^r - (k+1)^r}$, where $k = 0, 1, \dots, n-2$.

Denote $f(k) := (k+2)^r - (k+1)^r, k \geq 0$. Thus $f'(k) = r((k+2)^{r-1} - (k+1)^{r-1})$. If $r \geq 1$ then $f'(k) \geq 0$ and f is increasing, hence m_k is decreasing in k. If $0 < r \leq 1$ then $f'(k) \leq 0$ and f is decreasing, hence m_k is increasing in k.

Case $r \geq 1$

Here m_k are decreasing in k. Therefore the lower envelope of the convex hull of $(t^r, [t])$ is to be determined by $\overline{OW}, \overline{WA}$ and $\overline{AB}$. Here slope $(\overline{WA}) = \frac{n-1}{n^r-1}$ and slope $(\overline{AB}) = \frac{1}{a^r-n^r}$.

Subcase (i) Assume slope $(\overline{WA}) \geq$ slope $(\overline{AB})$ i.e. $\frac{n-1}{n^r-1} \geq \frac{1}{a^r-n^r}$.

Then the lower envelope of the convex hull of $(t^r, [t])$ will be the line segments $\overline{OW}$ and $\overline{WB}$. If $0 < d \leq 1$, then $L = 0$. If $1 \leq d \leq a$, then $\frac{L}{n} = \frac{d^r-1}{a^r-1}$ and $L = (\frac{d^r-1}{a^r-1})n$. If $a \in \mathbb{N} - \{1\}$ then $L = (\frac{d^r-1}{a^r-1})(a-1)$. If $0 < a \leq 1$ then $L = 0$.

Subcase (ii) Assume slope $(\overline{WA}) \leq$ slope $(\overline{AB})$ i.e. assume that $\frac{n-1}{n^r-1} \leq \frac{1}{a^r-n^r}$. Thus the lower envelope of the convex hull of $(t^r, [t])$ will be made by the line segments $\overline{OW}, \overline{WA}$ and $\overline{AB}$. If $0 < d \leq 1$ and $L = 0$. If $1 \leq d \leq n$ then $\frac{L}{n-1} = \frac{d^r-1}{n^r-1}$, that is $L = (\frac{d^r-1}{n^r-1})(n-1)$. If $n \leq d \leq a$, then

$$\frac{L-(n-1)}{1} = \frac{d^r - n^r}{a^r - n^r}$$

and

$$L = (n-1) + \frac{d^r - n^r}{a^r - n^r}.$$

Case $0 < r \leq 1$

Here m_k are increasing in k. Therefore the convex hull of $(t^r, [t])$ is formed by $\overline{OW}, \{\overline{Q_kQ_{k+1}}\}_{k=0}^{n-2}$ and $\overline{AB}$. Obviously slope $(\overline{AB}) \geq$ slope $(\overline{A, ((n+1)^r, n)})$. The line (l) through Q_k and Q_{k+1} has equation

$$y = k + \frac{(x-(k+1)^r)}{((k+2)^r - (k+1)^r)}.$$

If $0 < d \leq 1$ then $L = 0$ and if $0 < a \leq 1$ again $L = 0$. When $(k+1) \leq d < (k+2)$ for some $k \in \{0, 1, \ldots, n-2\}$, at $x = d^r$, we get

$$L = k + (\frac{d^r - (k+1)^r}{(k+2)^r - (k+1)^r}).$$

(The last is true also when $a \in \mathbb{N} - \{1\}$.)

If $n \leq d \leq a$, then the line (l^*) through A, B has equation

$$y = n + \frac{(x - a^r)}{(a^r - n^r)}.$$

At $x = d^r$ we get

$$L = n + (\frac{d^r - a^r}{a^r - n^r}).$$

□

The case of $a = +\infty$ has as follows.

Theorem 4.5.6. *Let $d > 0, r > 0$. Consider*

$$L := L_{[\cdot]}(r,d) = \inf\{E[X] : X \geq 0 \ \ a.s., EX^r = d^r\} \tag{4.5.11}$$

When $0 < a \leq 1$ the $L = 0$. Also, when $0 < d \leq 1$ then $L = 0$. Assume that $a \geq 1$ and $d \geq 1$.

Case $r > 1$

We obtain $L = 0$.

Case $r = 1$

We obtain $L = d - 1$.

Case $0 < r < 1$

Assume that $k + 1 \leq d < k + 2; k \in \mathbb{Z}_+$, then

$$L = k + \left(\frac{d^r - (k+1)^r}{(k+2)^r - (k+1)^r}\right).$$

PROOF: As similar to Theorem 4.5.5 is omitted. □

The corresponding results for $\lceil\cdot\rceil$ come next.

Theorem 4.5.7. *Let $d > 0, r > 0$ and $0 < a < +\infty$. Consider*

$$L := L_{\lceil\cdot\rceil}(a,r,d) = \inf\{E\lceil X\rceil : 0 \leq X \leq a \ \ a.s., EX^r = d^r\}. \tag{4.5.12}$$

Call $n := \lceil a\rceil$.

Case $r \geq 1$

Subcase (i) Assume that

$$(n-1)^{1-r} \geq \frac{1}{a^r - (n-1)^r}.$$

Then

$$L = n\left(\frac{d}{a}\right)^r.$$

(If $a \in \mathbb{N}$ then $L = d^r a^{1-r}$.)

Subcase (ii) Assume that

$$(n-1)^{1-r} \leq \frac{1}{a^r - (n-1)^r}.$$

When $0 < d \leq n-1$, then

$$L = d^r (n-1)^{1-r}.$$

When $n - 1 \leq d \leq a$, then

$$L = (n-1) + \frac{d^r - (n-1)^r}{a^r - (n-1)^r}.$$

(The last is true also when $0 < a \leq 1$.)

<u>Case $0 < r \leq 1$</u>

Assume that $k \leq d < (k+1)$ for some $k \in \{0, 1, \ldots, n-2\}$. Then

$$L = k + (\frac{d^r - k^r}{(k+1)^r - k^r}).$$

If $n - 1 \leq d \leq a$, then

$$L = n + (\frac{d^r - a^r}{a^r - (n-1)^r}).$$

(The last is true in the cases of $0 < a \leq 1$ or $a \in \mathbb{N}$.)

PROOF: See figure 4.5.2 and apply the geometric moment method of optimal distance. Call $O := (0,0), Q_k := (k^r, k); k = 0, 1, \ldots, n-1, A := ((n-1)^r, n-1)$ and $B := (a^r, n)$. Call $m_k :=$ slope $(\overline{Q_k Q_{k+1}}) = \frac{1}{(k+1)^r - k^r}$. Consider $f(k) = (k+1)^r - k^r, k \geq 0$ then $f'(k) = r((k+1)^{r-1} - k^{r-1})$. If $r \geq 1$, then $f'(k) \geq 0$ and f is increasing, thus m_k is decreasing in k.

If $0 < r \leq 1$, then $f'(k) \leq 0$ and f is decreasing, thus m_k is increasing in k.

<u>Case $r \geq 1$</u>

Here m_k are decreasing in k. Therefore the lower envelope of the convex hull of the graph of $(t^r, \lceil t \rceil)$ is to be determined by $\overline{OA}$ and $\overline{AB}$. Here slope $(\overline{OA}) = (n-1)^{1-r}$ and slope $(\overline{AB}) = \frac{1}{a^r - (n-1)^r}$.

<u>Subcase (i)</u> Assume that slope $(\overline{OA}) \geq$ slope $(\overline{AB})$ i.e.

$$(n-1)^{1-r} \geq \frac{1}{a^r - (n-1)^r}.$$

Then the lower envelope of the convex hull of $(t^r, \lceil t \rceil)$ will be just the line segment $\overline{OB}$. Hence $\frac{L}{n} = \frac{d^r}{a^r}$, i.e. $L = \frac{d^r n}{a^r}$. (If $a \in \mathbb{N}$ then $L = d^r a^{1-r}$.)

<u>Subcase (ii)</u> Assume that slope $(\overline{OA}) \leq$ slope $(\overline{AB})$ i.e. $(n-1)^{1-r} \leq \frac{1}{a^r - (n-1)^r}$.

Then the lower envelope of the convex hull of $(t^r, \lceil t \rceil)$ will be the two line segments $\overline{OA}$ and $\overline{AB}$. If $0 < d \leq n-1$, then

$$\frac{L}{n-1} = \frac{d^r}{(n-1)^r}$$

and

$$L = d^r (n-1)^{1-r}.$$

If $n - 1 \leq d \leq a$, then

$$\frac{d^r - (n-1)^r}{a^r - (n-1)^r} = \frac{L - (n-1)}{n - (n-1)} = \frac{L - (n-1)}{1}.$$

Thus

$$L = (n-1) + \frac{d^r - (n-1)^r}{a^r - (n-1)^r}.$$

(The last covers also the case of $0 < a \leq 1$.)

Case $0 < r \leq 1$

Here m_k are increasing in k. Therefore the convex hull of $(t^r, \lceil t \rceil)$ is formed by $\{\overline{Q_k Q_{k+1}}\}_{k=0}^{n-2}$ and $\overline{AB}$. Obviously, slope $(\overline{AB}) \geq$ slope $(\overline{A, (n^r, n)})$. The line (l) through Q_k, Q_{k+1} has equation $y = k + (\frac{x-k^r}{(k+1)^r - k^r})$. Assume $k \leq d < k+1$ for some $k \in \{0, 1, \ldots, n-2\}$. At $x = d^r$ we obtain

$$L = k + (\frac{d^r - k^r}{(k+1)^r - k^r}).$$

If $n - 1 \leq d \leq a$, we consider the line (l^*) through A and B. This has equation

$$y = n + (\frac{x - a^r}{a^r - (n-1)^r}).$$

At $x = d^r$ we obtain

$$L = n + (\frac{d^r - a^r}{a^r - (n-1)^r}).$$

(The last covers also the cases of $a \in \mathbb{N}$ and $0 < a \leq 1$. □

The case $a = +\infty$ is given by

Theorem 4.5.8. *Let $d > 0, r > 0$. Consider*

$$L := L_{\lceil \cdot \rceil}(r, d) = \inf\{E\lceil X \rceil : X \geq 0 \ a.s., EX^r = d^r\}. \tag{4.5.13}$$

Case $r > 1$

$$L = 0.$$

Case $r = 1$

$$L = d.$$

Case $0 < r \leq 1$ Assume $k \leq d < k+1$ for some $k \in \mathbb{Z}_+$. Then

$$L = k + (\frac{d^r - k^r}{(k+1)^r - k^r}).$$

PROOF: As similar to Theorem 4.5.7 is omitted. □

CHAPTER FIVE

THE LEVY RADIUS

5.1 The Levy Radius Of A Set Of Probability Measures Satisfying Moment Conditions Involving $\{t, t^2\}$

5.5.1. Introduction

Here we consider probability measures μ on $\mathbb{R}$ such that both $\int |t| d\mu, \int t^2 d\mu$ are finite. We consider

$$M(\epsilon_1, \epsilon_2) := \{\mu : |\int t^j d\mu - s_0^j| \leq \epsilon_j, j = 1, 2\}, \tag{5.1.1}$$

where s_0 is a given point in $\mathbb{R}$, also $0 < \epsilon_j \leq 1; j = 1, 2$. We would like to measure the *"size"* of $M(\epsilon_1, \epsilon_2)$. Since weak convergence of probability measures is of great importance and their standard weak topology is well described by the *Levy distance* d_L (Chow-Teicher (1978), p. 255), it is natural to define the *Levy radius* for $M(\epsilon_1, \epsilon_2)$ as

$$D := \sup_{\mu \in M(\epsilon_1, \epsilon_2)} d_L(\mu, \delta_{s_0}), \tag{5.1.2}$$

where δ_{s_0} is the unit measure at s_0.

Using a moment theoretical result due to Selberg (1940) we present an algorithm for the exact calculation of D (see Anastassiou (1) (1987)).

The above problem was motivated by Korovkin's theorem (Korovkin (1960), p. 6): If the conditions $l_n(1) \longrightarrow 1, l_n(t) \longrightarrow s_0$, and $l_n(t^2) \longrightarrow s_0^2$ are satisfied for the sequence of positive linear functionals $l_n(f)$, then $l_n(f) \longrightarrow f(s_0)$ for any function $f(t)$ bounded on $\mathbb{R}$ and continuous at s_0.

It is convenient to mention:

Definition 5.1.1. Let μ, ν be probability measures on $\mathbb{R}$ with distribution functions F and G, respectively. The *Levy distance between* μ, ν is defined as follows:

$$d_L(\mu, \nu) := d_L(F, G) := \inf\{h > 0 : F(x - h) - h \leq G(x) \leq F(x + h) + h, \forall x \in \mathbb{R}\}.$$

Remark 5.1.1. Observe that $d_L(F, G) \leq r$ if $G(x) \leq F(x + r + 0) + r, \forall x \in \mathbb{R}$ and $G(x) \geq F(x - r - 0) - r, \forall x \in \mathbb{R}$. Now if $\nu = \delta_{s_0}$, then the distribution function G of ν is the unit step function at s_0 and

$$d_L(F,G) \le r \quad \text{if} \quad F(x) \begin{cases} \le r & \text{for} \quad x < s_0 - r; \\ \ge 1 - r & \text{for} \quad x > s_0 + r, \end{cases}$$

which is the same as $F(s_0 - r - 0) \le r$ together with $F(s_0 + r + 0) \ge 1 - r$.

This in turn is the same as

$$\mu((-\infty, s_0 - r)) \le r; \quad \mu((s_0 + r, +\infty)) \le r. \tag{5.1.3}$$

Therefore

$$d_L(\mu, \delta_{s_0}) = \inf\{r > 0 : \mu((-\infty, s_0 - r)) \le r; \mu((s_0 + r, +\infty)) \le r\}. \tag{5.1.4}$$

Thus we get

$$D = \sup_{\mu \in M(\epsilon_1, \epsilon_2)} (\inf\{r > 0; \mu((-\infty, s_0 - r)) \le r; \mu((s_0 + r, +\infty)) \le r\}). \tag{5.1.5}$$

We have $r \ge D$ if $\mu((-\infty, s_0 - r)) \le r$ and $\mu((s_0 + r, +\infty)) \le r$ all $\mu \in M(\epsilon_1, \epsilon_2)$. Equivalently, the moment conditions in $M(\epsilon_1, \epsilon_2)$ imply $\mu((-\infty, s_0 - r)) \le r$ and $\mu((s_0 + r, +\infty)) \le r$. Only $r \in (0, 1]$ will be of our interest.

5.1.2. Results

The main result we present here has as follows:

Theorem 5.1.1. *Set $s_0 \ge 0, \epsilon_1, \epsilon_2 > 0$ and $\epsilon_1, \epsilon_2 \le 1$. Let $M(\epsilon_1, \epsilon_2)$ denote the set of all probability measures μ on $\mathbb{R}$ such that*

$$|y_1 - s_0| \le \epsilon_1, \quad |y_2 - s_0^2| \le \epsilon_2, \tag{5.1.6}$$

where $y_j := \int t^j \mu(dt), j = 1, 2$.

The quantity $D := \sup_{\mu \in M(\epsilon_1, \epsilon_2)} d_L(\mu, \delta_{s_0})$ is calculated as follows:

Define r_1 as follows: If $s_0 \le (1 - \epsilon_2)/2$, then $r_1 \in (0, 1]$ is the unique root of g in $(0, 1]$ where

$$g(x) := x^3 - 2s_0 x^2 + s_0^2 x - (s_0^2 + \epsilon_2). \tag{5.1.7}$$

If $s_0 > (1 - \epsilon_2)/2$, then set $r_1 := 1$.

Case (i). If $\epsilon_1 > s_0 + (s_0^2 + \epsilon_2)^{1/2}$ then $D = r_1$.

Case (ii). If $\epsilon_1 \leq s_0 + (s_0^2 + \epsilon_2)^{1/2}$, *let* $r_2 \in (0,1]$ *be the smallest root of the cubic polynomial*

$$F(x) := x^3 - 2\epsilon_1 x^2 + (2\epsilon_1 s_0 + \epsilon_2)x - (2\epsilon_1 s_0 + \epsilon_2 - \epsilon_1^2) \tag{5.1.8}$$

satisfying both

$$0 \leq x \leq r_1 \quad \text{and} \quad x^2 - s_0 x + (s_0 - \epsilon_1) \geq 0. \tag{5.1.9}$$

If r_2 *exists, then* $D = r_2$. *If not, then* $D = r_1$.

Next a special case is easily proved:

Corollary 5.1.1. *Let* $s_0 = 0$ *in the above. If* $\epsilon_1^{1/2} \geq \epsilon_2^{1/3}$ *then* $D = \epsilon_2^{1/3}$. *If* $\epsilon_1^{1/2} < \epsilon_2^{1/3}$ *then* D *equals the unique root of*

$$x^3 - 2\epsilon_1 x^2 + \epsilon_2 x - \epsilon_2 + \epsilon_1^2 = 0 \tag{5.1.10}$$

in $[\epsilon_1^{1/2}, \epsilon_2^{1/3}]$.

In the proof of Theorem 5.1.1. is used:

Lemma 5.1.1. (Selberg 1940) *Let* μ *be a probability measure on* $\mathbb{R}$ *and* $y_j := \int t^j \mu(dt), j = 1, 2$, *its first two moments. Then the best upper bound on* $\mu((-\infty, c))$ *is given by*

$$\mu((-\infty, c)) \leq \begin{cases} 1 & , \text{ if } y_1 \leq c. \\ \frac{\sigma^2}{[(y_1 - c)^2 + \sigma^2]} & , \text{ if } y_1 \geq c. \end{cases} \tag{5.1.11}$$

where $\sigma^2 := y_2 - y_1^2$.

Similarly, we have the best upper bound

$$\mu((c, +\infty)) \leq \begin{cases} 1 & , \text{ if } y_1 \geq c, \\ \frac{\sigma^2}{[(y_1 - c)^2 + \sigma^2]} & , \text{ if } y_1 \leq c. \end{cases} \tag{5.1.12}$$

Consequently, from Lemma 5.1.1 and relation (5.1.5) we obtain

Corollary 5.1.2. *Let* $0 < r < 1$. *Then* $r \geq D$ *if conditions (5.1.6) imply that* $s_0 - r \leq y_1 \leq s_0 + r$ *and, further,*

$$\sigma^2/[(y_1 - s_0 + r)^2 + \sigma^2] \leq r; \quad \sigma^2/[(y_1 - s_0 - r)^2 + \sigma^2] \leq r. \tag{5.1.13}$$

Proof of Theorem 5.1.1.

Part I

Condition (5.1.13) depends on y_2 and is more difficult the larger $\sigma^2 = y_2 - y_1^2$, that is, the larger the moment y_2. In view of (5.1.6), the worst case is that where

$$y_2 = s_0^2 + \epsilon_2. \tag{5.1.14}$$

Thus, we may assume (5.1.14) true without loss of generality.

As to y_1, it is subject to the inequalities (5.1.6) and Jensen's inequality $y_1^2 \leq y_2$. That is,

$$\max(s_0 - \epsilon_1, -(s_0^2 + \epsilon_2)^{1/2}) \leq y_1 \leq \min(s_0 + \epsilon_1, (s_0^2 + \epsilon_2)^{1/2}). \tag{5.1.15}$$

It will be convenient to introduce $u := y_1 - s_0$. Then (5.1.15) says that $-m \leq u \leq M$, where

$$m := \min(\epsilon_1, \tilde{m}) \quad \text{and} \quad M := \min(\epsilon_1, \tilde{M}) \tag{5.1.16}$$

with $\tilde{m} := s_0 + (s_0^2 + \epsilon_2)^{1/2}$ and $\tilde{M} := -s_0 + (s_0^2 + \epsilon_2)^{1/2}$.

Since $s_0 \geq 0$ we have $0 \leq M \leq m$. Let $0 < r < 1$. We see that

"$r \geq D$ if $-m \leq u \leq M$ implies that $|u| \leq r$ and further (5.1.13)".

Also, we easily find "$r \geq D$ if $m \leq r$ and further (5.1.13) holds whenever $-m \leq u \leq M$." The first condition of (5.1.13) requires that $f_1(u) \geq 0$ for $-m \leq u \leq M$, where

$$f_1(u) := u^2 + 2[(1-r)s_0 + r^2]u + [r^3 - (1-r)\epsilon_2]. \tag{5.1.17}$$

The second condition of (5.1.13) requires that, for $-m \leq u \leq M$, we have $f_2(u) \geq 0$, where

$$f_2(u) := u^2 + 2[(1-r)s_0 - r^2]u + [r^3 - (1-r)\epsilon_2]. \tag{5.1.18}$$

Note that $f_1(u) - f_2(u) = 4r^2u$ changes sign only at $u = 0$. In fact, $f_1(u) \leq f_2(u)$ for $u \leq 0$, while $f_1(u) \geq f_2(u)$ for $u \geq 0$. Thus, it suffices that $f_1(u) \geq 0$ for $-m \leq u \leq 0$ and, further, $f_2(u) \geq 0$ for $0 \leq u \leq M$. However, here the first condition implies the second one. Namely,

$$f_2(u) - f_1(-u) = 4(1-r)s_0u \geq 0 \quad \text{for} \quad u \geq 0.$$

Therefore, if $f_1(u) \geq 0$ for $-m \leq u \leq 0$ then for each value u with $0 \leq u \leq M$ we have $0 \geq -u \geq -M \geq -m$, hence $f_2(u) \geq f_1(-u) \geq 0$. Consequently, it suffices to require the condition $f_1(u) \geq 0$ for $-m \leq u \leq 0$. Replacing u by $-u$, this becomes $\varphi(u) \geq 0$ for $0 \leq u \leq m$, where

$$\varphi(u) := u^2 - 2[(1-r)s_0 + r^2]u + [r^3 - (1-r)\epsilon_2]. \tag{5.1.19}$$

So far we have found: let $0 < r < 1$

$$\text{“}r \geq D \quad \text{if } r \geq m \text{ and } \varphi(u) \geq 0 \quad \text{for } 0 \leq u \leq m.\text{”}$$

Here we see that $\varphi(u) \geq 0$ for $0 \leq u \leq m$ implies $m \leq r$, since if $0 < r < m$ then $r < s_0 + (s_0^2 + \epsilon_2)^{1/2}$ and thus $\varphi(r) = (1-r)[r^2 - 2s_0 r - \epsilon_2] < 0$.

As a result

$$\text{“}r \geq D \quad \text{if} \quad \varphi(u) \geq 0 \quad \text{for} \quad 0 \leq u \leq m.\text{”}$$

Note that $m \leq \tilde{m}$ and observe that $\varphi(u)$ is strictly increasing in r as long as $u \leq \tilde{m}$ and $u \leq r$ (which is true if $u \leq m \leq r$).

Namely, one may write

$$\varphi(u) = B(r-1) + r(r-u)^2, \tag{5.1.20}$$

where

$$B := \epsilon_2 + 2s_0 u - u^2 = (\tilde{m} - u)(\tilde{M} + u) \geq 0, \quad 0 \leq u \leq \tilde{m}. \tag{5.1.21}$$

We also see from (5.1.20) that $\varphi(u)$ is negative when $r = u$ and positive when $r = 1$.

Hence, for each value $0 \leq u \leq m$ there is a unique value $r(u) > u$ (depending on s_0 and ϵ_2) such that $\varphi(u) \geq 0$ for $r \geq r(u)$ and $\varphi(u) \leq 0$ for $u \leq r \leq r(u)$. Naturally, $r(u)$ is simply equal to the unique root $r \geq u$ of

$$r(r-u)^2 = B(1-r), \tag{5.1.22}$$

where B is as in (5.1.21). In order that $\varphi(u) \geq 0$ it is necessary and sufficient that $r \geq r(u)$.

Therefore we have established

$$D = \sup\{r(u) : 0 \leq u \leq m\}. \tag{5.1.23}$$

PART II

Here, based on formula (5.1.23) we develop an algorithm for the exact calculation of D. Let the polynomial

$$g(x) := x^3 - 2s_0 x^2 + s_0^2 x - (s_0^2 + \epsilon_2). \tag{5.1.24}$$

Then $g'(x) = 3x^2 - 4s_0 x + s_0^2 = (3x - s_0)(x - s_0)$, so that g has a local maximum at $s_0/3$ and a local minimum at s_0. Namely, $g(s_0) = -s_0^2 - \epsilon_2 < 0$ and

$$g(s_0/3) = s_0^2(4s_0/27 - 1) - \epsilon_2.$$

Also, note that $g(1) = 1 - 2s_0 - \epsilon_2; g(0) = -(s_0^2 + \epsilon_2) < 0$. Consider first the possibility that $g(s_0/3) > 0$. This requires that $s_0 > 27/4$ in which case $g(1) < 0$. In fact, $1 < 9/4 < s_0/3$ so that $g(x) < 0$ for all $0 \leq x \leq 1$ as g is strictly increasing theoren. Hence a value $0 \leq x \leq 1$ with $g(x) \geq 0$ can only exist when $g(s_0/3) \leq 0$ and $g(1) \geq 0$, thus, $s_0 \leq (1 - \epsilon_2)/2$.

In that case, there is a unique positive root r_1 of g which satisfies

$$0 \leq s_0 \leq r_1 \leq 1 \quad (s_0 \leq (1 - \epsilon_2)/2). \tag{5.1.25}$$

In the case $s_0 > (1 - \epsilon_2)/2$, we define $r_1 := 1$. Consequently, the definition of r_1 in the statement of the theorem makes sense.
The discriminant of (5.1.19) is given by

$$Q := [(1 - r)s_0 + r^2]^2 - [r^3 - (1 - r)\epsilon_2] = (r - 1)g(r). \tag{5.1.26}$$

Thus, if $s_0 > (1 - \epsilon_2)/2$ then $Q > 0$ for all $r \in (0, 1)$. If $s_0 \leq (1 - \epsilon_2)/2$ then there exists a unique root $r_1 \in (0, 1]$ of g and then we have for all values $r_1 \leq r \leq 1$ that $Q \leq 0$, so that $\varphi(u) \geq 0$ for all $0 \leq u \leq m$. In this same case $Q > 0$ when $r \in (0, 1)$ satisfies $r < r_1$. Naturally, the quantity r_1 is an upper bound on every value $r(u)$, $0 \leq u \leq m$, giving us $D \leq r_1$ in both cases.

Let us study the requirement $\varphi(u) \geq 0$ for $0 \leq u \leq m$ in the case when $Q > 0$. We have $\varphi(u) \geq 0$ for $u \leq u^* - \sqrt{Q}$ and $u \geq u^* + \sqrt{Q}$, where

$$u^* := (1 - r)s_0 + r^2. \tag{5.1.27}$$

In order that $\varphi(u) \geq 0$ for $0 \leq u \leq m$ it is clearly necessary and sufficient that

$$m \leq u^* - \sqrt{Q}. \tag{5.1.28}$$

We conclude $r \geq D$ if $Q \geq 0$ in such a way that (5.1.28) holds. Also note that $Q \geq 0$ (for a value $0 \leq r \leq 1$) is the same as $r \leq r_1$. Inequality (5.1.28) is equivalent to $Q \leq (u^* - m)^2$, provided that $m \leq u^*$. Thus

$$(1 - r)s_0 + r^2 \geq m = \min(\epsilon_1, s_0 + (s_0^2 + \epsilon_2)^{1/2}). \tag{5.1.29}$$

Therefore, given (5.1.29) condition (5.1.28) is equivalent to

$$2(1 - r)ms_0 \leq r^3 - 2r^2m - (1 - r)\epsilon_2 + m^2. \tag{5.1.30}$$

Thus, a value r with $0 < r \leq r_1 \leq 1$ (i.e., $Q \geq 0$) satisfies $r \geq D$ if both (5.1.29) and (5.1.30) hold if $\varphi(u) \geq 0$ for all $0 \leq u \leq m$. Since $\varphi(u)$ is increasing in r in the case when $0 \leq u \leq m \leq r$, we are led to determine the smallest root $r_2 \in (0, 1]$ of the equation

$$2(1 - r)ms_0 = r^3 - 2r^2m - (1 - r)\epsilon_2 + m^2 \tag{5.1.31}$$

which also satisfies (5.1.29).

The case $m = \tilde{m}$ happens if

$$s_0 + (s_0^2 + \epsilon_2)^{1/2} \leq \epsilon_1; \quad \text{thus,} \quad m = s_0 + (s_0^2 + \epsilon_2)^{1/2}, \tag{5.1.32}$$

and in that case $r = 0$ is a root of (5.1.31), however does not satisfy (5.1.29).

Obviously, equation (5.1.31) implies

$$r^2 - 2rm + \epsilon_2 + 2ms_0 = 0 \quad (m = \tilde{m}). \tag{5.1.33}$$

Its discriminant equals $m^2 - \epsilon_2 - 2ms_0 = 0$, thus it has a double root $r = m$. Substituting this in (5.1.29) we have

$$(1 - m)s_0 + m^2 \geq m, \quad \text{that is,} \quad (m - 1)(m - s_0) \geq 0. \tag{5.1.34}$$

However, that condition is not satisfied because $s_0 \leq m \leq 1$.

Consequently, the above case of (5.1.31) counts only when $m = \epsilon_1$. In that case, if such a root r_2 does not exist or $r_2 > r_1$, then clearly $D = r_1$. And if $r_2 \leq r_1$, then it is also clear that $D = r_2$. In the alternative case of $\epsilon_1 > s_0 + (s_0^2 + \epsilon_2)^{1/2}$, we have $D = r_1$ always. Now the proof of the theorem has finished. □

Remark 5.1.2

When $s_0 = 0$, one can easily find from (5.1.7) and $\varphi(0) \geq 0$ that $(1 - \epsilon_2^{1/3})\epsilon_2 \leq D^3 \leq \epsilon_2$ and when ϵ_2 is small the estimate $D \cong \epsilon_2^{1/3}$ is quite accurate.

5.2. Levy Radius Of A Set Of Probability Measures Satisfying Two Moment Conditions Involving A Tchebycheff System

This is a generalization to §5.1.

5.2.1. Introduction

The next definition comes from Karlin and Studden (1966), p.1.

Definition 5.2.1. *Let $g_0, g_1, \ldots, g_n$ denote continuous real-valued functions defined on $[a, b] \subset \mathbb{R}$. These functions are called a Tchebycheff System over $[a, b]$, shortly a T-system,* *provided the $n+1$ st order determinants*

$$\begin{vmatrix} g_0(t_0) & g_0(t_1) & \ldots & g_0(t_n) \\ g_1(t_0) & g_1(t_1) & \ldots & g_1(t_n) \\ \vdots & & & \\ g_n(t_0) & g_n(t_1) & \ldots & g_n(t_n) \end{vmatrix} > 0,$$

whenever $a \le t_0 < t_1 < \ldots < t_n \le b$.
e.g. $g_i(t) = t^i, i = 0, 1, \ldots, n$ defined on any $[a, b] \subset \mathbb{R}$. Also $\{1, \cos t, \sin t, \ldots, \cos mt, \sin mt\}$, $0 \le t < 2\pi$ is a T- system over any $[a, b] \subset \mathbb{R}$ of length less than 2π. From the same reference, see pp. 21, 22; Theorem 4.1, basically we see that $\{g_i\}_0^n$ is a T-system on $[a, b]$ iff any generalized polynomial $\sum_{i=0}^{n} a_i g_i(t), (a_i \in \mathbb{R})$ has at most n distinct zeros on $[a, b]$, unless all the coefficients a_i are equal to zero.

Here we consider probability measures μ on R, defined on a $\sigma-$ algebra containing all the singletons, such that both $\int |g_1| d\mu, \int |g_2| d\mu$ are finite, where $\{1, g_1, g_2\}$ is a Tchebycheff system of continuous functions from R into R.

We consider

$$M(\epsilon_1, \epsilon_2) = \{\mu : |\int g_j d\mu - g_j(s_0)| \le \epsilon_j, j = 1, 2\} \tag{5.2.1}$$

where s_0 is a given point in R, also $0 < \epsilon_j \le 1; j = 1, 2$.

We would like again to measure the *"size"* of $M(\epsilon_1, \epsilon_2)$. Since weak convergence of probability measures is of central importance and their standard weak topology is well described by the Levy distance d_L, it is natural to define again the Levy radius for $M(\epsilon_1, \epsilon_2)$ as:

$$D = \sup d_L(\mu, \delta_{s_0}), \mu \in M(\epsilon_1, \epsilon_2), \tag{5.2.2}$$

where δ_{s_0} is the unit measure at s_0. The introduced quantity indicates to us how big $M(\epsilon_1, \epsilon_2)$ is. Note $D = D(\epsilon_1, \epsilon_2)$ and as $\epsilon_1, \epsilon_2 \longrightarrow 0$ get that $D \longrightarrow 0$, thus $d_L(\mu, \delta_{s_0}) \longrightarrow 0$, consequently, $\mu \overrightarrow{\longrightarrow} \delta_{s_0}$ weakly. Given that we know D then we know the rate of weak convergence of μ to δ_{s_0}.

Under mild assumptions on $\{1, g_1, g_2\}$, by the use of geometric moment method of optimal ratio, we find theoretically the magnitude of D and then by the use of Newton-Puiseux diagram (Hille (1973)), we give asymptotic expansions for D when ϵ, ϵ_2 are small (see Anastassiou (2) (1987)).

Remark 5.2.1. The next is naturally the same as (5.1.4)

$$d_L(\mu, \delta_{s_0}) = \inf \left\{ r > 0 : \begin{array}{r} \mu((-\infty, s_0 - r]) \le r; \\ \\ \mu([s_0 + r, +\infty)) \le r \end{array} \right\}. \tag{5.2.3}$$

Thus we get

$$D = \sup_{\mu \in M(\epsilon_1, \epsilon_2)} \left(\inf \left\{ r > 0 : \begin{array}{r} \mu((-\infty, s_0 - r]) \le r; \\ \\ \mu([s_0 + r, +\infty)) \le r \end{array} \right\} \right). \tag{5.2.4}$$

One can easily see that $d = \max(D', D'')$, where

$$D' := \sup_{\mu \in M(\epsilon_1, \epsilon_2)} (\inf\{r > 0 : \mu((-\infty, s_0 - r]) \le r\}) \tag{5.2.5}$$

and

$$D'' := \sup_{\mu \in M(\epsilon_1, \epsilon_2)} (\inf\{r > 0 : \mu([s_0 + r, +\infty)) \le r\}). \tag{5.2.6}$$

We have $r \ge D'$ if $\mu((-\infty, s_0 - r]) \le r$ all $\mu \in M(\epsilon_1, \epsilon_2)$, equivalently, the moment conditions in $M(\epsilon_1, \epsilon_2)$ imply $\mu((-\infty, s_0 - r]) \le r$, if there exists no probability measure $\mu \in M(\epsilon_1, \epsilon_2)$ and $\mu((-\infty, s_0 - r]) > r$. The same arguments hold for D''. In the case of $\{1, t, t^2\}$ and $s_0 = 0$ we have $D' = D''$. Of interest will be only $r \in (0, 1]$.

Being more specific we state:

Problem 5.2.1. Let the continuous functions $g_1, g_2 : \mathbb{R} \longrightarrow \mathbb{R}$ are such that $\{1, g_1, g_2\}$ is a Tchebycheff system with $g_2(t) \longrightarrow +\infty$ as $t \longrightarrow \pm\infty$ and $g_1(t)/g_2(t) \downarrow 0 (\uparrow 0)$ as $t \uparrow \infty (\downarrow -\infty)$ strictly monotone. (*Then g_1 is strictly increasing and thus it is in one-to-one correspondence.*) Also assume that the graph $g := (g_1, g_2)$ is strictly convex. (*Therefore every boundary point is an extreme point.*) Be given $s_0 \in \mathbb{R}$ and $0 < \epsilon_1, \epsilon_2 \le 1$, find

$$D = \sup_{\mu \in M(\epsilon_1, \epsilon_2)} d_L(\mu, \delta_{s_0}),$$

where $M(\epsilon_1, \epsilon_2)$ is defined as in (5.2.1).

Sketch of the Solution. We define $U'(y) := \sup_\mu\{\mu((-\infty, s_0 - r])\}$ and $U''(y) := \sup_\mu\{\mu([s_0 + r, +\infty))\}$, where $r \in (0, 1)$ is given, while μ ranges over all probability measures on $\mathbb{R}$ with $\int g_j d\mu := y_j, j = 1, 2; y := (y_1, y_2) \in \mathbb{R}^2$. We want the smallest r, i.e., the optimal D, such that $U'(y) \le D$ and $U''(y) \le D$ for all $y = (y_1, y_2)$ satisfying

$$|y_j - g_j(s_0)| \le \epsilon_j, j = 1, 2, \tag{5.2.7}$$

where

$$y_j := \mu(g_j) := \int g_j d\mu.$$

If we cannot find such a minimal r, we set $D := 1$.

In fact $D = \max(D', D'')$, where D' is the smallest r, such that $U'(y) \leq r$ and D'' the smallest r such that $U''(y) \leq r$, for all $y = (y_1, y_2)$ satisfying (5.2.7).

The solution is carried out through several remarks, comments and lemmas. The values of D are given in four lemmas, corresponding to the number of the main cases we have to meet. These in turn break into several subcases, for each of which we find a different value of D.

Remark 5.2.2. Let $E :=$ conv $(g(\mathbb{R}))$. By Lemma 2 p.96 of Kemperman (1968) we have that $y \in E$ if there exists a probability measure μ on $\mathbb{R}$ with $\mu(|g_j|) < +\infty, j = 1, 2$, such that $\mu(g) = y$. Thus, we can find such μ only for those y that belong to E. Let $V :=$int (E). Since E is a convex set it is $E = V \cup Bd(E)$ and $V \cap Bd(E) = \phi$. In case of $y \in Bd(E)$ we define as V_y, the union of all closed line segments $[y', y'']$ that are entirely contained in E and have y as an interior point. Since g is strictly convex, we have $V_y = \{y\}$. Let $T^y := g^{-1}(V_y)$, because g_1 is one-to-one we get $T^y = \{t_y\}$ just a singleton.

Theorem 9, p.103 of Kemperman (1968) says: if μ is a probability measure such that $\mu(|g_i|) < +\infty, j = 1, 2$ and $\mu(g) \in V_y$, then μ is concentrated on the measurable subset T^y. That is, $\mu(T^y) = 1$, i.e., $\mu(\{t_y\}) = 1$. Therefore, if $t_y \in$ (resp. $\notin$)$(-\infty, s_0 - r]$ then $\mu((-\infty, s_0 - r]) = 1$ (resp. 0) and also if $t_y \in$ (resp. $\notin$)$[s_0 + r, +\infty)$ then $\mu([s_0 + r, +\infty)) = 1$ (resp. 0), where $r \in (0, 1]$.

Next call: Y the set of all $y's$ satisfying (5.2.7), $A := Y \cap V, B := Y \cap Bd(E)$ and $\Gamma := Y \cap E$. Then $D = \max(D', D'')$, where

$$D' \quad \text{smallest } r : U'(y) \leq r, \forall y \in \Gamma,$$

$$D'' \quad \text{smallest } r : U''(y) \leq r, \forall y \in \Gamma.$$

so for $y \in B$ we consider only $\mu = \delta_{t_y}$, i.e., $\mu(g) = y = (y_1, y_2)$ where $y_j = g_j(t_y), j = 1, 2$.

From (3.14), p.101 of Kemperman (1968) and the previous comments, we can easily conclude for every $y = (y_1, y_2) \in \Gamma$ with $y_1 \notin (g_1(s_0 - r), g_1(s_0 + r))$, where $r \in (0, 1]$, that we have $U'(y)$ or $U''(y) = 1$, thus $D = 1$.

Let $y \in B$ such that $g_1(s_0 - r) < y_1 < g_1(s_0 + r)$ i.e., $t_y \in (s_0 - r, s_0 + r)$. That is $\delta_{t_y}(s_0 - r, s_0 + r) = 1$ and $\delta_{t_y}((-\infty, s_0 - r]) = \delta_{t_y}([s_0 + r, +\infty)) = 0$. Therefore $U'(y) = U''(y) = 0$ and these boundary $y's$ do not contribute anything towards the calculation of $D \in (0, 1]$. To find $D < 1$ we should treat only the interior points $y = (y_1, y_2) \in A$ such that

$$g_1(s_0 - r) < y_1 < g_1(s_0 + r), r \in (0, 1].$$

Remark 5.2.3. We introduce the following sets:

$$R^{(1)} := \{r \in (0, 1) : \mu((-\infty, s_0 - r]) \leq r, \quad \text{all } \mu \in M(\epsilon_1, \epsilon_2)\}$$

and

$$R^{(2)} := \{r \in (0,1) : \mu([s_0 + r, +\infty)) \leq r, \quad \text{all } \mu \in M(\epsilon_1, \epsilon_2)\}.$$

The pairs (ϵ_1, ϵ_2) fall in four subsets relative to the graph $g = (g_1, g_2)$.

Namely, let $E := \{(\epsilon_1, \epsilon_2) : 0 < \epsilon_1, \epsilon_2 \leq 1\}$, then $E = \bigcup_{i=1}^{4} E_i$ with $E_i \cap E_j = \phi$ for all $i \neq j : i, j = 1, 2, 3, 4$ where

$$E_1 := \{(\epsilon_1, \epsilon_2) : g_2(s_0) + \epsilon_2 \geq T, L\},$$

$$E_2 := \{(\epsilon_1, \epsilon_2) : g_2(s_0) + \epsilon_2 < T, L\},$$

$$E_3 := \{(\epsilon_1, \epsilon_2) : L \leq g_2(s_0) + \epsilon_2 < T\}$$

and

$$E_4 := \{(\epsilon_1, \epsilon_2) : T \leq g_2(s_0) + \epsilon_2 < L\}.$$

Here $T := g_2(\ell'), L := g_2(\ell'')$, where $g_1(\ell') = g_1(s_0) - \epsilon_1$ and $g_1(\ell'') = g_1(s_0) + \epsilon_1$. It can be $E_3 = \phi$ or $E_4 = \phi$, but they are never both empty.

For each main case $E_i, i = 1, 2, 3, 4$ we find a different set of values for D, as the structure of any E_i is totally different from another.

Accordingly, $R^{(1)}, R^{(2)}$ break into unions of proper subsets for each E_i.

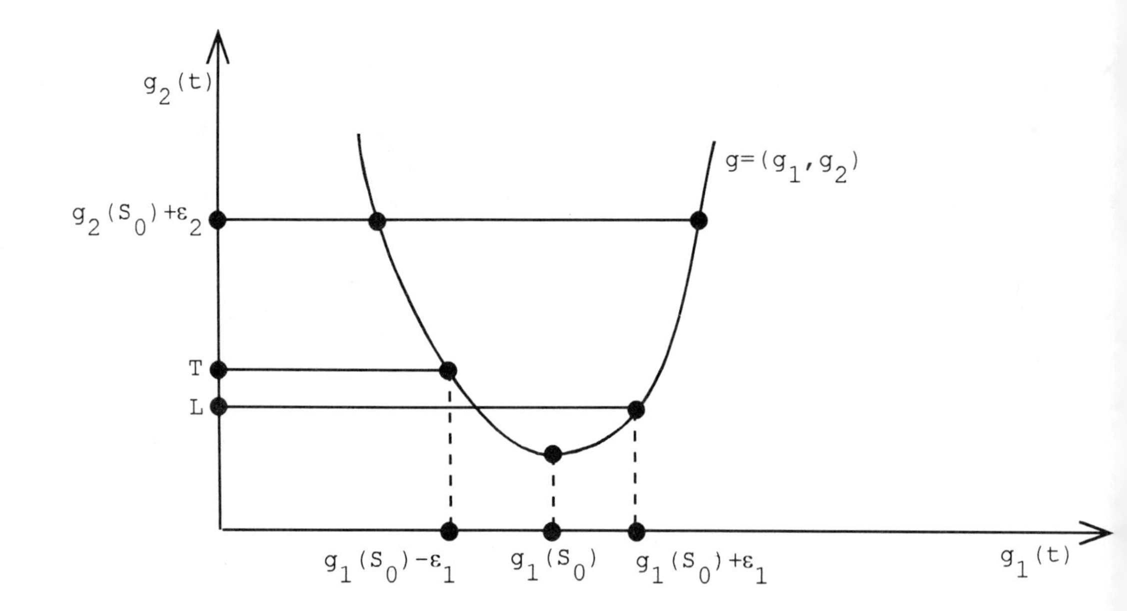

Figure 5.2.1: Graph (g_1, g_2)

Remark 5.2.4 From Kemperman (1968) it follows that $U'(y)$ (resp. $U''(y)$) is attained by a two- points supported measure at $(s_0 - r)$ (resp. $(s_0 + r)$) and t with $t > s_0 - r$ (resp. $t < s_0 + r$) where the points

$$(g_1(s_0 - r), g_2(s_0 - r))[\text{resp. } (g_1(s_0 + r), g_2(s_0 + r))],$$

(y_1, y_2) and $(z_1, z_2) := (g_1(t), g_2(t))$ are on a line.

Thus, from (6.4), p. 112 of Kemperman (1968) we get that

$$U'(y) = (z_1 - y_1)/(z_1 - g_1(s_0 - r)) \tag{5.2.8}$$

and

$$U''(y) = (z_1 - y_1)/(z_1 - g_1(s_0 + r)).$$

Observe that for fixed y_1 and increasing y_2, the point (z_1, z_2) moves to the right of the graph of g, that is z_1 increases.

Therefore we may assume that y_2 is maximal (given y_1), as a result

$$y_2 = g_2(s_0) + \epsilon_2. \tag{5.2.9}$$

Comments 5.2.1.

(i) If there exist $r \in R^{[1]}$ such that $g_2(s_0 - r) \leq g_2(s_0) + \epsilon_2$ and $r \in R^{[2]}$ such that $g_2(s_0 + r) \leq g_2(s_0) + \epsilon_2$, then D', D'' come from these r's.

(ii) In the following graph consider $\overline{A_3A_4}$ parallel to $\overline{A_1A_2}$, where the point O is fixed. Obviously,

$$\frac{OA_5}{A_2A_5} > \frac{OA_4}{A_2A_4} = \frac{OA_3}{A_1A_3}.$$

Thus, in our problem a lower point A_2 is preferable over a higher point A_1.

(iii) From (i), (ii), Remark 5.2.4 and by comparison of ratios, there exist exactly one $r_0 \in (0,1)$ and exactly one interior point $y_0 \in A$ such that $U'(y) \leq U'_{r_0}(y_0)$ for all interior points $y \in A$ and all $r \in (0,1)$. Therefore $U'_{r_0}(y_0) = D'$ i.e., $D' = r_0$, similarly for D''.

(iv) Each case E_i breaks into other subcases according to where $r \in R^{[1]}$ and $r \in R^{[2]}$ are located.

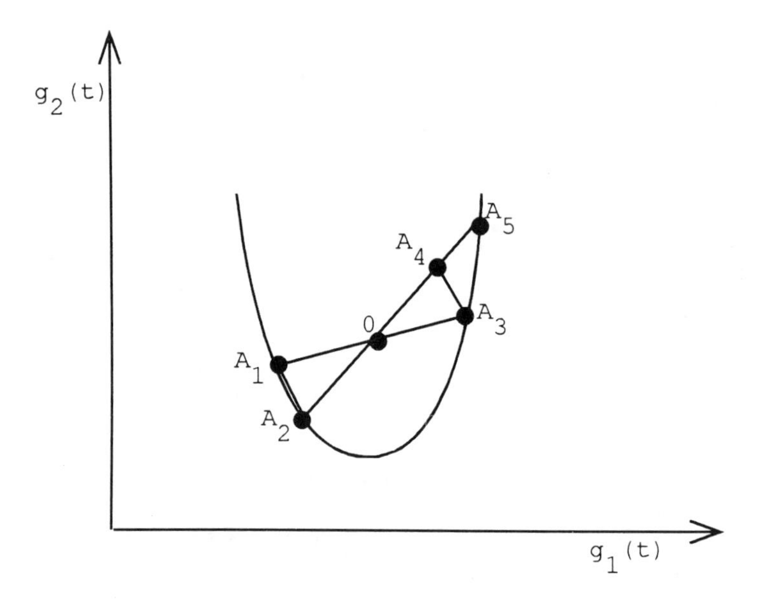

Figure 5.2.2: Comparison of ratios

Now we are ready to present the various cases of evaluation of D in the next Lemmas.

Lemma 5.2.1. Here we are working on the set

$$E_1 = \{(\epsilon_1, \epsilon_2) : g_2(s_0) + \epsilon_2 \geq T, L\}.$$

The graph of this case has as follows:

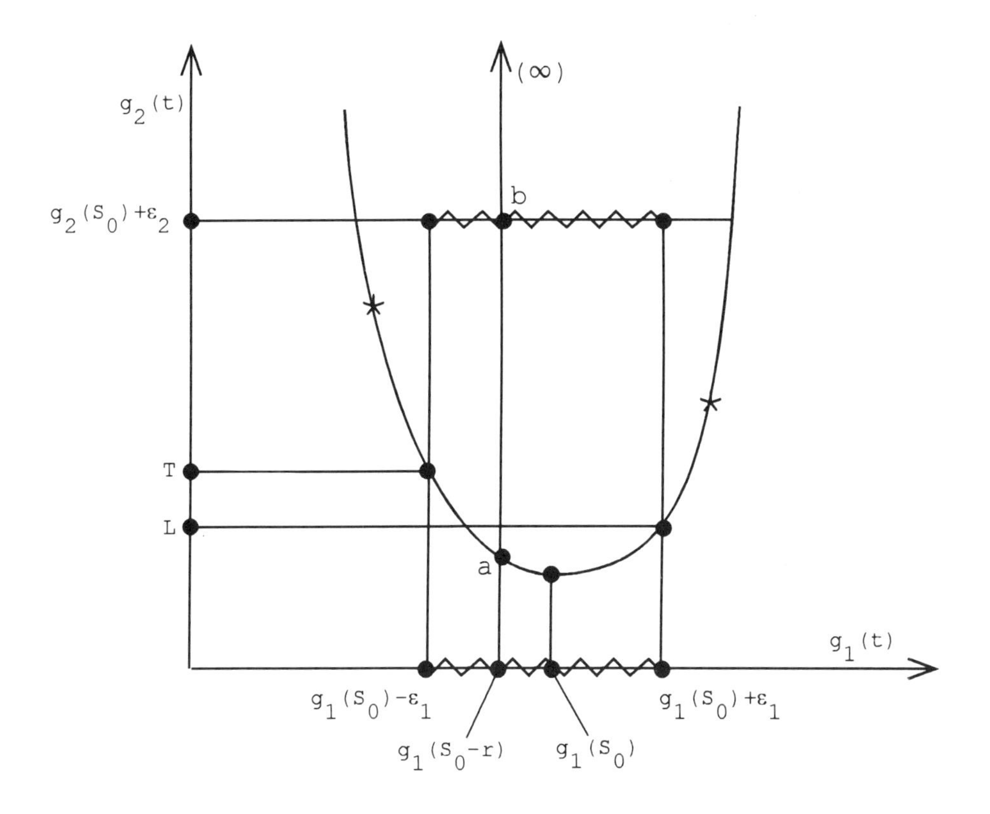

Figure 5.2.3: Set E_1

Subcases:

(i) Assume that $\exists r \in R^{[1]}$ or $\exists r \in R^{[2]}$ or both, so that $g_1(s_0) - \epsilon_1 \leq g_1(s_0 - r) \leq g_1(s_0)$ or $g_1(s_0) \leq g_1(s_0 + r) \leq g_1(s_0) + \epsilon_1$ or both (respectively). Then either

$$U'(y) = (\frac{\overline{b\infty}}{\overline{a\infty}}) = \frac{\infty}{\infty} \quad \text{or} \quad U''(y) = \frac{\infty}{\infty}$$

or both, so we set $D := 1$. (This comes for the fact that we are in a vertical strip on $g_1(t)$–axis).

(ii) Assume that

[all $r \in R^{[1]}$ are s.t. $g_1(s_0 - r) < g_1(s_0) - \epsilon_1$, with

$$\text{some of these } r : g_2(s_0 - r) \le g_2(s_0) + \epsilon_2]. \tag{5.2.10}$$

Then the optimal D' is assumed by a measure with moments $y_1 = g_1(s_0) - \epsilon_1, y_2 = g_2(s_0) + \epsilon_2$ and is given by the ratio property (see Comments 5.2.1).

$$D' = [g_1(t(D')) - g_1(s_0) + \epsilon_1]/[g_1(t(D')) - g_1(s_0 - D')]. \tag{5.2.11}$$

Assume that

$$\begin{aligned}&[\text{all } r \in R^{[2]} \text{ are s.t. } g_1(s_0) + \epsilon_1 < g_1(s_0 + r),\\ &\text{with some of these } r : g_2(s_0 + r) \le g_2(s_0) + \epsilon_2].\end{aligned} \tag{5.2.12}$$

Then the optimal D'' is assumed by a measure with moments $y_1 = g_1(s_0) + \epsilon_1, y_2 = g_2(s_0) + \epsilon_2$ and is given by the ratio property

$$D'' = [g_1(t(D'')) - g_1(s_0) - \epsilon_1]/[g_1(t(D'')) - g_1(s_0 + D'')]. \tag{5.2.13}$$

CONCLUSION. If (5.2.10), (5.2.12) hold, then $D = \max(D', D'')$, where D', D'' are given from (5.2.11) and (5.2.13) respectively. (If we cannot find such D' or D'' we set $D := 1$.)

(iii) Assume that $r \in R^{[1]}$ are as in (5.2.10), thus D' comes from (5.2.11), and assume that

$$\begin{aligned}&[\text{all } r \in R^{[2]} \text{ are s.t. } g_1(s_0) + \epsilon_1 < g_1(s_0 + r) \text{ and}\\ &\qquad g_2(s_0 + r) > g_2(s_0) + \epsilon_2].\end{aligned} \tag{5.2.14}$$

Let $s\#$ be the unique point where g_2 is minimal, since g is strictly convex.

The optimal D'' is obtained, through the ratio property, at the moment point $y^* := (y_1^*, y_2^*)$, which is the intersection of the straight line carrying the segment $\overline{g(s_0 + D'')g(s\#)}$ and the line carrying the segment

$$\{(y_1, g_2(s_0) + \epsilon_2) : g_1(s_0) - \epsilon_1 \le y_1 \le g_1(s_0) + \epsilon_1\}.$$

We have

$$y_1^* = \frac{[(g_2(s_0) + \epsilon_2 - g_2(s\#))g_1(s_0 + D'') + (g_2(s_0 + D'') - g_2(s_0) - \epsilon_2)(g_1(s\#)]}{[g_2(s_0 + D'') - g_2(s\#)]},$$

[If $g_1(s_0) - \epsilon_1 < y_1^* < g_1(s_0) + \epsilon_1$, then the optimal moment point is $(y_1^*, g_2(s_0) + \epsilon_2)$. If $y_1^* \le g_1(s_0) - \epsilon_1$ or $y_1^* \ge g_1(s_0) + \epsilon_1$, then the optimal moment points are respectively

$$(g_1(s_0) - \epsilon_1, g_2(s_0) + \epsilon_2) \quad \text{and} \quad (g_1(s_0) + \epsilon_1, g_2(s_0) + \epsilon_2)] \tag{5.2.15}$$

Therefore in each of these three sub-subcases we get a different D''.

CONCLUSION. If (5.2.10) and (5.2.14) are true, then $D = \max(D', D'')$, where D', D'' come from (5.2.11) and (5.2.15) respectively. (5.2.16)

(If we cannot find such D' or D'' we set $D := 1$.)

(iv) Assume that:

$$[\text{all } r \in R^{[2]} \text{ are s.t. } g_1(s_0) + \epsilon_1 < g_1(s_0 + r), \text{ with some of these } r : g_2(s_0 + r) \leq g_2(s_0) + \epsilon_2] \quad (5.2.17)$$

also

$$[\text{all } r \in R^{[1]} \text{ are s.t. } g_1(s_0 - r) < g_1(s_0) - \epsilon_1 \text{ and } g_2(s_0 - r) > g_2(s_0) + \epsilon_2]. \quad (5.2.18)$$

CONCLUSION. $D = \max(D', D'')$, where D'' comes from (5.2.13) and D' comes as we found D'' in (5.2.16): here the role of $g(s_0 + D'')$ is taken over by $g(s_0 - D')$, the rest works the same way. (If we cannot find such D' or D'' we set $D := 1$.) (5.2.19)

(v) Assume that [all $r \in R^{[1]}$ are s.t. $g_1(s_0 - r) < g_1(s_0) - \epsilon_1$ and $g_2(s_0 - r) > g_2(s_0) + \epsilon_2$, furthermore all $r \in R^{[2]}$ are s.t. $g_1(s_0 + r) > g_1(s_0) + \epsilon_1$ and $g_2(s_0 + r) > g_2(s_0) + \epsilon_2$].

Then $D = \max(D', D'')$, where D' is found as in (5.2.19) and D'' as in (5.2.16). (If we cannot find such D' or D'' we set $D := 1$.)

Lemma 5.2.2. Here we are working on the set

$$E_2 = \{(\epsilon_1, \epsilon_2) : g_2(s_0) + \epsilon_2 < T, L\}.$$

The graph of this case as follows:

Let t', t'' such that $s_0 \in (t', t'')$ and $g_2(t') = g_2(t'') = g_2(s_0) + \epsilon_2$. Note $g_1(t') := M$ and $g_1(t'') := N$.

Subcases:

(i) Assume that $\exists r \in R^{[1]}$ so that $M \leq g_1(s_0 - r) \leq g_1(s_0)$ or $\exists r \in R^{[2]}$ so that $g_1(s_0) \leq g_1(s_0 + r) \leq N$ or both. Again we are in a vertical strip case, therefore $U'(y) = \frac{\infty}{\infty}$ or $U''(y) = \frac{\infty}{\infty}$ or both (respectively) and we set $D := 1$.

(ii) Assume that: all $r \in R^{[1]}$ and all $r \in R^{[2]}$ are such that $g_1(s_0 - r) < M$ and $g_1(s_0 + r) > N$. This implies $g_2(s_0 - r), g_2(s_0 + r) > g_2(s_0) + \epsilon_2$. The potential optimal moments y_1 are such that $g_1(t') < y_1 < g_1(t'')$. Let again $s\#$ be the unique point where g_2 is minimal.

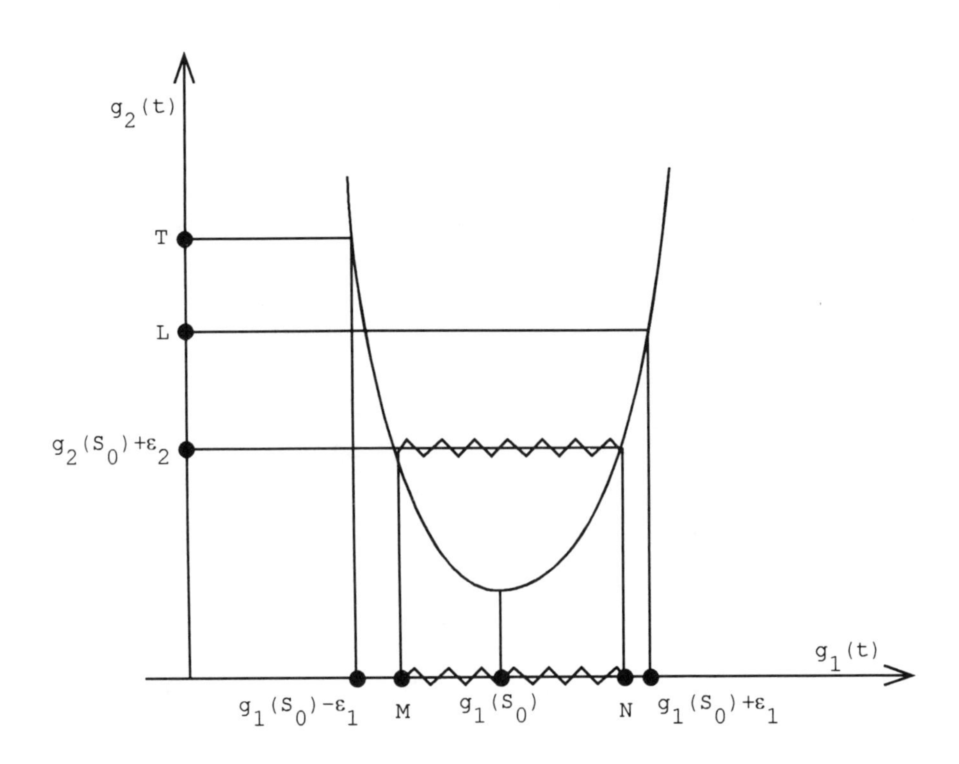

Figure 5.2.4: Set E_2

Through $g(s\#)$ draw lines to both branches of g: namely connect $g(s\#)$ with $g(s_0 - r) \quad \forall r \in R^{[1]}$ and $g(s\#)$ with $g(s_0 + r) \quad \forall r \in R^{[2]}$. Then for the optimal D', D'' call:

$$y^* := \overline{g(s\#)g(s_0 - D')} \cap \overline{g(t')g(t'')}$$

and

$$y^{**} := \overline{g(s\#)g(s_0 + D'')} \cap \overline{g(t')g(t'')}.$$

We find

$$(D' = \frac{g_1(s\#) - y_1^*}{g_1(s\#) - g_1(s_0 - D')}, D'' = \frac{y_1^{**} - g_1(s\#)}{g_1(s_0 + D'') - g_1(s\#)}). \tag{5.2.20}$$

As a result $D = \max(D', D'')$. (If we cannot find such D' or D'' we set $D := 1$.)

Lemma 5.2.3. Here we are working on the set

$$E_3 = \{(\epsilon_1, \epsilon_2) : L \leq g_2(s_0) + \epsilon_2 < T\}.$$

The graph of this case has as follows:

Let $t_0 < s_0$ so that $g_2(t_0) = g_2(s_0) + \epsilon_2$ and denote $g_1(t_0) := W$. The potential optimal y_1's are such that $W < y_1 \leq g_1(s_0) + \epsilon_1$.

Subcases:

(i) Assume that $\exists r \in R^{[1]}$ or $\exists r \in R^{[2]}$ or both, such that $W \leq g_1(s_0 - r) \leq g_1(s_0)$ or $g_1(s_0) \leq g_1(s_0 + r) \leq g_1(s_0) + \epsilon_1$ or both (respectively). Then $U'(y) = \frac{\infty}{\infty}$ or $U''(y) = \frac{\infty}{\infty}$ or both, thus we set $D := 1$.

(ii) Assume that: all $r \in R^{[1]}$ are s.t. $g_1(s_0 - r) < W$ and all $r \in R^{[2]}$ are s.t. $g_1(s_0+r) > g_1(s_0)+\epsilon_1$ and $\exists r \in R^{[2]} : g_2(s_0+r) < g_2(s_0)+\epsilon_2$. Then $D = \max(D', D'')$, where D' is found as D' in (5.2.19): Lemma 5.2.1, and D'' is found as D'' in (5.2.13), same Lemma 5.2.1.

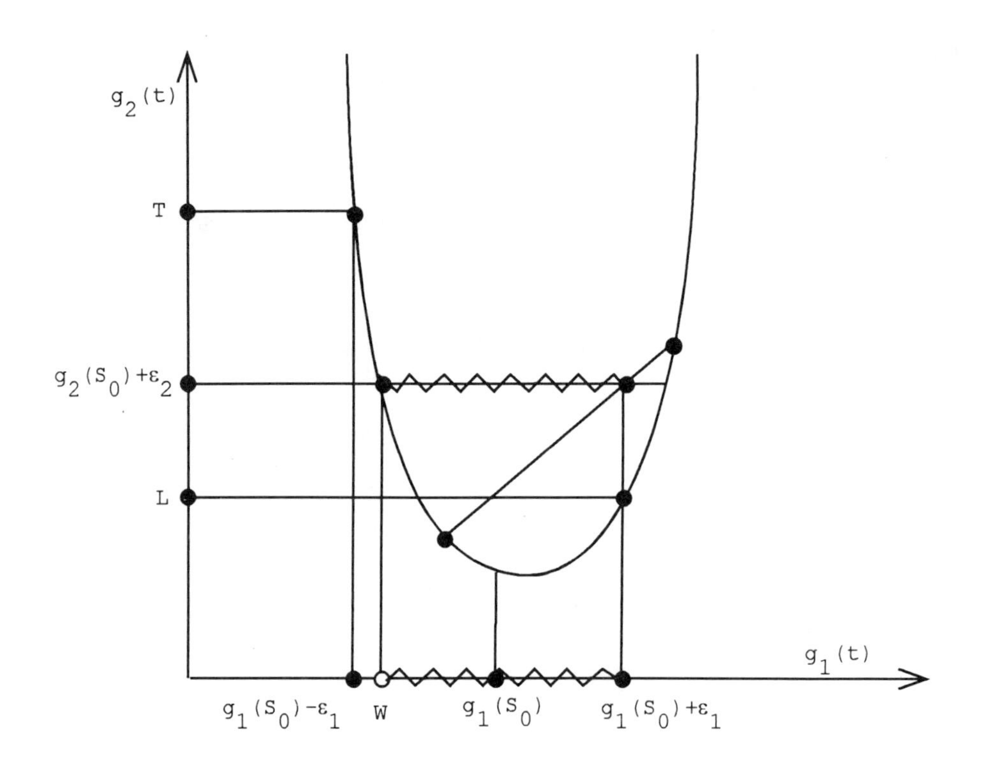

Figure 5.2.5: Set E_3

Again the optimal moment point for D'' is $(g_1(s_0)+\epsilon_1, g_1(s_0)+\epsilon_2)$. (If we cannot find such D' or D'' we set $D := 1$.)

(iii) Assume that all $r \in R^{[1]}$ are such that $g_1(s_0-r) < W$ and all $r \in R^{[2]}$ fulfill $g_1(s_0+r) > g_1(s_0)+\epsilon_1$ and $g_2(s_0+r) \geq g_2(s_0)+\epsilon_2$. Then D' is found as D' in subcase (ii) of this lemma and D'' is given by

$$D'' = \frac{g_1(s_0)+\epsilon_1 - g_1(t(D''))}{g_1(s_0+D'') - g_1(t(D''))}, \text{ if } y_1^* \geq g_1(s_0)+\epsilon_1.$$

In this last case of D'' the optimal moment point is $(g_1(s_0)+\epsilon_1, g_2(s_0)+\epsilon_2)$. If $y_1^* < g_1(s_0)+\epsilon_1$ then the optimal moment point is $(y_1^*, g_2(s_0)+\epsilon_2)$ and D'' is found by the ratio property applied on $s\#$ (here $y_1^*, s\#$ are as in Lemma 5.2.1 (iii)).

Therefore $D = \max(D', D'')$. (If we cannot find such D' or D'' we set $D := 1$.)

Finally we have

Lemma 5.2.4. Here we are working on the set

$$E_4 = \{(\epsilon_1, \epsilon_2) : T \le g_2(s_0)+\epsilon_2 < L\}.$$

The graph of this case has as follows:

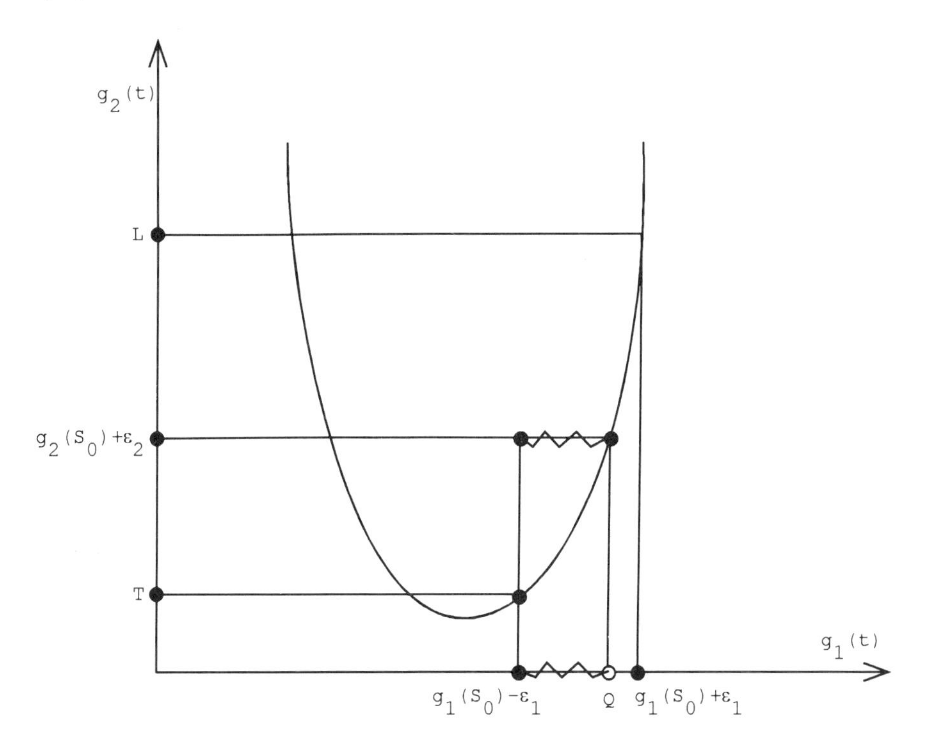

Figure 5.2.6: Set E_4

This setting is symmetric to E_3: Lemma 5.2.3. Therefore its treatment should be analogous and symmetric. The roles of W,L,T are taken over by Q,T,L respectively. Again we meet three subcases with similar solutions. This is the end of the solution of Problem 5.2.1.

Prerequisites 5.2.1. Now we deal with asymptotic expansions for the Levy radius D. In the setting of Problem 5.2.1 for small ϵ_1, ϵ_2 we find asymptotic expansions for D', D'' separately and then we pick $D = \max(D', D'')$. Basic tool here will be the Newton-Puiseux diagram, see Hille (1973), pp 105–107.

In particular we treat only the subcase (ii) of Lemma 5.2.1 as the most typical, several other subcases in these lemmas can be worked out similarly. Namely, here we have $g_2(s_0) + \epsilon_2 \geq T, L$ and all the admissible r's, giving us D', D'', fulfill (5.2.10) and (5.2.12).

Since g_1 is strictly increasing, acts as a variable and for convenience from now on we let $g_1(t) = t$ and denote $g_2 := g$. Also we assume the existence of the continuous derivatives g', g'', g'''.

Locally, we would assume that the error of the approximation

$$g(x) \cong g(s_0) + g'(s_0)(x - s_0) + \frac{g''(s_0)}{2}(x - s_0)^2 \tag{5.2.21}$$

is small. For instance this is true when $\| g''' \|_\infty < \infty$ e.t.c.

Let again $g_1(t) = t$ and $g_2 = g$. The first parts of the conditions (5.2.10) and (5.2.12) are equivalent to

$$\begin{cases} \exists \mu_1 \in M(\epsilon_1, \epsilon_2) : \mu_1((-\infty, s_0 - \epsilon_1]) > \epsilon_1 \\ \text{and} \\ \exists \mu_2 \in M(\epsilon_1, \epsilon_2) : \mu_2([s_0 + \epsilon_1, +\infty)) > \epsilon_1. \end{cases} \tag{5.2.22}$$

For let $t' < t''$ such that $g(t') := g(t'') := g(s_0) + \epsilon_2$, with $0 < s_0 - t' < 1$ and $0 < t'' - s_0 < 1$, the second parts of these conditions are equivalent to

$$[\text{all } \mu \in M(\epsilon_1, \epsilon_2) \text{ fulfill } \mu((-\infty, t']) \leq s_0 - t' \ \text{ and } \ \mu([t'', +\infty)) \leq t'' - s_0]. \tag{5.2.23}$$

As an example: let $g(t) := t^2, s_0 := 0, \epsilon_1 := 0.1, \epsilon_2 := 0.8$ (i.e., $\epsilon_1^2 < \epsilon_2$). Observe that $\mu_1 = \delta_{-\epsilon_1}, \mu_2 = \delta_{\epsilon_1} \in M(\epsilon_1, \epsilon_2)$. Furthermore (5.2.22) is fulfilled: $\mu_1((-\infty, \epsilon_1]) = 1 > \epsilon_1$ and $\mu_2([\epsilon_1, +\infty)) = 1 > \epsilon_1$. From Corollary 5.1.1 of Anastassiou (1) (1987) we find $D = 0.6924074924 < \epsilon_2^{1/2} = 0.895$, and since $-t' = t'' = \epsilon_2^{1/2} < 1$ we have $\mu((-\infty, -\epsilon_2^{1/2}]) \leq \epsilon_2^{1/2}$ and $\mu([\epsilon_2^{1/2}, +\infty)) \leq \epsilon_2^{1/2}$ for all $\mu \in M(\epsilon_1, \epsilon_2)$. That is (5.2.23) is satisfied. Next from (5.2.11) and (5.2.13) we get

$$(D' = \frac{t(D') - s_0 + \epsilon_1}{t(D') - s_0 + D'}; D'' = \frac{t(D'') - s_0 - \epsilon_1}{t(D'') - s_0 - D''}) \tag{5.2.24}$$

and projecting onto $g(t)$ axis we have

$$(D' = \frac{g(t(D')) - g(s_0) - \epsilon_2}{g(t(D')) - g(s_0 - D')}; \quad D'' = \frac{g(t(D'')) - g(s_0) - \epsilon_2}{g(t(D'')) - g(s_0 + D'')}). \tag{5.2.25}$$

As a result we have to deal with two systems: each of two equations in two unknowns, in order to find the asymptotic expansions of D', D''. Thus we have

$$\left(D' = \frac{t(D') - s_0 + \epsilon_1}{t(D') - s_0 + D'}; \quad D' = \frac{g(t(D')) - g(s_0) - \epsilon_2}{g(t(D')) - g(s_0 - D')}\right) \tag{5.2.26}$$

and

$$\left(D'' = \frac{t(D'') - \epsilon_0 - \epsilon_1}{t(D'') - s_0 - D''}; \quad D'' = \frac{g(t(D'')) - g(s_0) - \epsilon_2}{g(t(D'')) - g(s_0 + D'')}\right). \tag{5.2.27}$$

Solving for $t(D')$ the first equation of (5.2.26) and substituting into the second we get

$$(1 - D')g(s_0 + \frac{D'^2 - \epsilon_1}{1 - D'}) = g(s_0) + \epsilon_2 - D'g(s_0 - D'). \tag{5.2.28}$$

We can do the same for the system (5.2.27).

Note 5.2.1. Let $f(x) = x^3 + ax^2 + \beta x + \gamma$, then $f'(s) = 3x^2 + 2ax + \beta$. The resultant of f, f' is given by

$$R(f, f') = 4\beta^3 + 4a^3\gamma + 27\gamma^2 - \alpha^2\beta^2 - 18\alpha\beta\gamma. \tag{5.2.29}$$

It is well known that if $R(f, f') > 0$, then $f(x)$ has two complex roots.

Now we are ready to give the following theorems:

Theorem 5.2.1. *Let $\epsilon_2 := c\epsilon_1$ with $0 < \epsilon_1 < \epsilon_2 \ll 1$ (ϵ_2 much smaller than 1). We assume that $s_0 \geq s\#$ the point where $g(t)$ is minimal. Let also $\epsilon := \epsilon_1, \theta := 2/g''(s_0), \rho := 2\frac{g'(s_0)}{g''(s_0)}$ and $\lambda := \rho + \theta c$. Then we obtain the following asymptotic expansion for D' :*

$$D'(\epsilon) = \lambda^{1/3}\epsilon^{1/3} - \frac{\lambda^{2/3}}{3}\epsilon^{2/3} + \frac{2}{3}\epsilon + (\frac{\lambda^2 - 18\lambda - 27}{81\lambda^{2/3}})\epsilon^{4/3} + O(\epsilon^{5/3}). \tag{5.2.30}$$

Note: (i) It is always $\lambda > 0$ and $c > 1$. (ii) When $g(t) = t^2$, it is $\lambda = 2s_0 + c$.

PROOF: Since ϵ_1, ϵ_2 are small we can apply (5.2.21) in both sides of (5.2.28) to get

$$g''(s_0)D'^3 - 2g''(s_0)\epsilon_1 D'^2 + 2g'(s_0)\epsilon_1 D' + 2\epsilon_2 D'$$

$$+ g''(s_0)\epsilon_1^2 - 2g'(s_0)\epsilon_1 - 2\epsilon_2 = 0. \tag{5.2.31}$$

Because $s_0 \geq s\#$ it is $g'(s_0) \geq 0$, also by the strict convexity of g we have $g''(s_0) > 0$, thus $\rho \geq 0$ and $\theta > 0$. Dividing (5.2.31) by $g''(s_0)$ and setting $\epsilon_2 := c\epsilon_1$ one obtains

$$D'^3 - 2\epsilon_1 D'^2 + (\rho + \theta c)\epsilon_1 D' + \epsilon_1^2 - (\rho + \theta c)\epsilon_1 = 0. \tag{5.2.32}$$

For convenience we set $w := D', z := \epsilon := \epsilon_1$ and $\lambda := \rho + \theta c$ and finally we find

$$w^3 - 2zw^2 + \lambda zw + z^2 - \lambda z = 0. \tag{5.2.33}$$

The last is of the form $f(w) = w^3 + \alpha w^2 + \beta w + \gamma$, where $a := -2z < 0, \beta := \lambda z > 0, \gamma := z^2 - \lambda z < 0$ for $z \ll 1$. Since $z \ll 1$ one obtains $4\beta^3 - \alpha^2\beta^2 > 0$ and $4\alpha^3\gamma + 27\gamma^2 > 18\alpha\beta\gamma$. So from (5.2.29) the resultant $R(f, f')$ is positive, therefore (5.2.33) has two complex roots. Because the product of its roots is $-\gamma > 0$, we conclude that there exists a positive one.

Of this real root we find the asymptotic expansion using the Newton-Puiseux diagram. For $F(z,w) \equiv \sum \alpha_{jk} z^j w^k = w^3 - 2zw^2 + \lambda zw + z^2 - \lambda z = 0$ we have $\alpha_{00} = 0$, $\alpha_{03} = 1, \alpha_{12} = -2, \alpha_{11} = \lambda$, $\alpha_{20} = 1, \alpha_{10} = -\lambda$ and the following diagram.

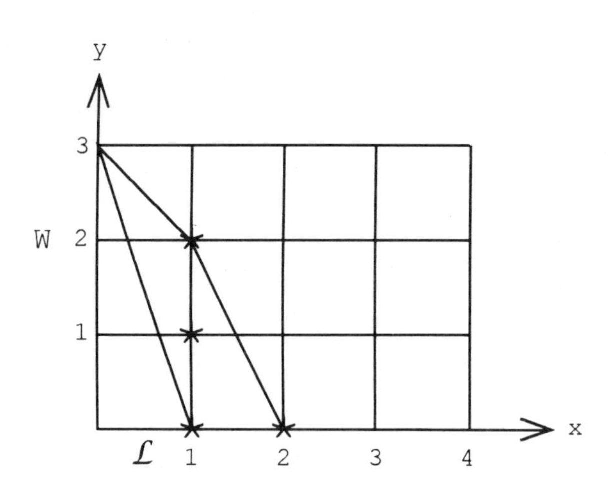

Figure 5.2.7. Newton-Puiseux diagram.

In the Cartesian plane we plot the points $(j,k) : \alpha_{jk} \neq 0$. We then form the least convex polygon which contains these points. This polygon always has at least one point on each of the axes since $F(z,w)$ is irreducible. Let $L := (\tilde{\rho}, 0), W := (0, \sigma)$ be the points on the axes nearest to the origin. Thus σ is the number of the roots which tend to zero with $z \longrightarrow 0; (\sigma = 3)$. The polygon has a broken line joining W with L and this line consists of at most σ segments. Here we have exactly one segment. This is $\overline{WL}$. Consider the line carrying it, it has equation $y + 3x = 3$ with slope -3. Setting $z := t^3, w := tu$ into (5.2.33) and dividing by t^3 we get

$$G(t,u) := (u^3 - \lambda) + (-2t^2u^2 + \lambda tu + t^3) = 0. \tag{5.2.34}$$

For $p_1(u^3) := u^3 - \lambda$, we have $p_1(0) = -\lambda \neq 0$ and $p_1(x) = x - \lambda$ has one root at $x^* = \lambda$. Rewrite (5.2.34) as

$$G(t,u) = (u^3 - \lambda) + tG_1(t,u) = 0 \tag{5.2.35}$$

where

$$G_1(t,u) := -2tu^2 + \lambda u + t^2.$$

Then $G(t,u)$ vanishes for $t = 0; u = \lambda^{1/3}, \lambda^{1/3} \cdot e^{2\pi i/3}, \lambda^{1/3} \cdot e^{4\pi i/3}$. Thus we obtain the three branches of $w(z)$. Since we are interested for the real one, we consider only $u = \lambda^{1/3}$. Because

$$\frac{\partial G}{\partial u}|_{(0,\lambda^{1/3})} = 3(\lambda^{2/3}) \neq 0,$$

by the implicit function theorem there exists exactly one $u(t)$ so that

$$u(t) = \lambda^{1/3} + \sum_{k=1}^{\infty} c_k t^k,$$

then

$$w(z) = z^{1/3}(\lambda^{1/3} + \sum_{k=1}^{\infty} c_k z^{k/3}).$$

Going back to the original notation, we finally have

$$\begin{cases} D'(\epsilon) = \epsilon^{1/3}(\lambda^{1/2} + \sum_{k=1}^{\infty} c_k \epsilon^{k/3}), \\ \text{where } c_k := \frac{d^k u}{k! dt^k}|_{(0,\lambda^{1/3})}. \end{cases} \tag{5.2.36}$$

We find:

$$c_1 = -\frac{1}{3}\lambda^{2/3}, c_2 = 2/3, c_3 = 3^{-4}\lambda^{-2/3}(\lambda^2 - 18\lambda - 27).$$

The counterpart of the last theorem has as follows:

Theorem 5.2.2. *Let $\epsilon_2 := c\epsilon_1$ with $c > g'(s_0)$ and $0 < \epsilon_1 < \epsilon_2 \ll 1$. We assume that $s_0 \geq s\#$. Let also $\epsilon := \epsilon_1, \theta := \frac{2}{g''(s_0)}, \rho := 2\frac{g'(s_0)}{g''(s_0)}$ and $l := \theta c - \rho$. Then we obtain the asymptotic expansion of D'' :*

$$D''(\epsilon) = l^{1/3}\epsilon^{1/3} - \frac{l^{2/3}}{3}\epsilon^{2/3} + \frac{2}{3}\epsilon$$

$$+(\frac{l^2-18l-27}{81l^{2/3}})\epsilon^{4/3}+O(\epsilon^{5/3}). \tag{5.2.37}$$

PROOF: As similar to Theorem 5.2.1 is omitted.

Note.

(i) It is always $l<0$ and $c>1$.

(ii) When $g(t)=t^2$ it is $l=c-2s_0$.

(iii) Since $\lambda>l$ obtain $\lambda^{1/3}\epsilon^{1/3}>l^{1/3}\epsilon^{1/3}$ and for $\epsilon\ll 1$ we get $D'(\epsilon)\geq D''(\epsilon)$. Thus $D\cong D'(\epsilon)$ for $\epsilon\ll 1$.

Examples. Let $g(t)=t^2, s_0=0$. The following numerical cases fall to the subcase (ii) of Lemma 5.2.1.

From Corollary 5.1.1 of Anastassiou (1) (1987) we find the theoretical value of D using the Newton method. And from Theorems 5.2.1-2 using the first four terms of the corresponding asymptotic expansions for D', D'' we find the approximate value of D, which is not far from the theoretical one.

ϵ_1	ϵ^2	$\epsilon_2^{1/2}$	D	D approx.
0.1	0.4	0.64	0.6048	0.6033
0.1	0.8	0.895	0.6924	0.692
0.2	0.5	0.7	0.6728	0.666

Using again the Newton-Puiseux diagram and going through the same procedure as in Theorems 5.2.1-2 we obtain the following related result.

Theorem 5.2.3. *Let $\epsilon_1:=c\epsilon_2^q$ with $q>1$ positive integer, $0<c\leq 1$ constant and $0<\epsilon_1<\epsilon_2\ll 1$. We assume that $s_0\geq s\#$. Let also $\epsilon:=\epsilon_2$ and $\theta:=\frac{2}{g''(s_0)}$. Then we obtain the asymptotic expansions for $D'(\epsilon), D''(\epsilon)$ which are of the form*

$$[\theta^{1/3}\epsilon^{1/3}-\frac{\theta^{2/3}}{3}\epsilon^{2/3}+\frac{\theta^{4/3}}{3^4}\epsilon^{4/3}+O(\epsilon^{5/3})]. \tag{5.2.38}$$

Note.

(i) It is always $\theta>0$ and when $g(t)=t^2$ it is $\theta=1$.

(ii) For $\epsilon\ll 1$ we have $D\cong D'(\epsilon)\cong D''(\epsilon)\cong \theta^{1/3}\epsilon^{1/3}-(\theta^{2/3}/3)\epsilon^{2/3}+(\theta^{4/3}/3^4)\epsilon^{4/3}$.

Example. Let $g(t) := t^2, s_0 := 0.006, \epsilon_1 = 0.001, \epsilon_2 = 0.36$. From Theorem 5.1.1 of Anastassiou (1) (1987) using the Newton method we find $D = 0.547$. Furthermore, we easily see that (5.2.22) and (5.2.23) are fulfilled, i.e., this numerical case corresponds to subcase (ii) of Lemma 5.2.1. Since $0.001 \cong (0.36)^7$ we have $q = 7$ and $c = 1$, also $\theta = 1$ and $\epsilon = 0.36$. Thus, we can apply Theorem 5.2.3 to get the approximate value of $D \cong 0.5458$, which is not far from the theoretical one.

CHAPTER SIX

THE PROKHOROV RADIUS

6.1. The Prokhorov Radius of a Set of Probability Measures Satisfying Moment Conditions Involving $\{t, t^2\}$.

6.1.1. Introduction

Here we consider probability measures μ on $\mathbb{R}$ such that both $\int |t| d\mu$, $\int t^2 d\mu$ are finite. We consider

$$M(\epsilon_1, \epsilon_2) = \{\mu : \left| \int t^j d\mu - \alpha^j \right| \leq \epsilon_j, j = 1, 2\} \tag{6.1.1}$$

where α is a given point in $\mathbb{R}$, also $0 < \epsilon_j < 1, j = 1, 2$ and $0 < \epsilon_2 + 2|\alpha|\epsilon_1 < 1$. We would like to measure the *"size"* of $M(\epsilon_1, \epsilon_2)$ to be given by a simple formula involving only $\epsilon_1, \epsilon_2, \alpha$.

Since weak convergence of probability measures is of great importance and their standard weak topology is well described by the Prokhorov distance p, it is natural to define the *Prokhorov radius* for $M(\epsilon_1, \epsilon_2)$ as

$$D := \sup_{\mu \in M(\epsilon_1, \epsilon_2)} p(\mu, \delta_\alpha) \tag{6.1.2}$$

where δ_α is the Dirac measure at α. Using the geometric moment method of optimal ratio due to Kemperman (1968) we are able to calculate the exact value of D, see Anastassiou (1992).

It will be helpful to mention:

Definition 6.1.1. (see Rachev & Shortt (2)(1989)). Let U be a Polish space with a metric d and C is the set of all nonempty closed subsets of U. Let $A \in C$, then for $\epsilon > 0$

$$A^\epsilon := \{x : d(x, A) < \epsilon\}.$$

Consider μ, ν probability measures on U. Prokhorov in 1956 introduced his famous metric

$$p(\mu, \nu) := \inf\{\epsilon > 0 : \mu(A) \leq \nu(A^\epsilon) + \epsilon, \nu(A) \leq \mu(A^\epsilon) + \epsilon, \forall A \in C\}.$$

When μ is a probability measure on $\mathbb{R}$, then

$$p(\mu, \delta_\alpha) = \inf\{r > 0 : \mu([\alpha - r, \alpha + r]) \geq 1 - r\}, \alpha \in \mathbb{R}. \tag{6.1.3}$$

Remark 6.1.1. One can restate that

$$D = \sup_{\mu \in M(\epsilon_1, \epsilon_2)} (\inf\{r > 0 : \mu([\alpha - r, \alpha + r]) \geq 1 - r\}). \tag{6.1.4}$$

Thus

$$r \geq D \text{ iff } \mu([\alpha - r, \alpha + r]) \geq 1 - r, \text{ all } \mu \in M(\epsilon_1, \epsilon_2) \tag{6.1.5}$$

iff

$$\lambda_r := \inf_{\mu \in M(\epsilon_1, \epsilon_2)} \mu([\alpha - r, \alpha + r]) \geq 1 - r.$$

Obviously $0 < D \leq 1$, therefore we are interested only for $r \in (0, 1]$. One can easily see that

$$D = \min\{r \in (0, 1] : \inf_{\mu \in M(\epsilon_1, \epsilon_2)} \mu([\alpha - r, \alpha + r]) \geq 1 - r\}. \tag{6.1.6}$$

Remark 6.1.2. (i) When $\epsilon_1, \epsilon_2 \longrightarrow 0$, then $\int t d\mu \longrightarrow \alpha$ and $\int t^2 d\mu \longrightarrow \alpha^2$. Thus $\int (t - \alpha)^2 d\mu \longrightarrow 0$, implying that $\int |t - \alpha| d\mu \longrightarrow 0$. Hence for arbitrarily small $\epsilon > 0$ we have

$$\epsilon \mu(\{t : |t - \alpha| > \epsilon\}) \leq \int_{\{t : |t - \alpha| > \epsilon\}} |t - \alpha| d\mu \leq \int |t - \alpha| d\mu \leq \epsilon^2.$$

That is, $\mu(\{t : |t - \alpha| > \epsilon\}) \leq \epsilon$. Therefore $p(\mu, \delta_\alpha) \leq \epsilon$ and $D = D(\epsilon_1, \epsilon_2, \alpha) \leq \epsilon$. Clearly $p(\mu, \delta_\alpha) \longrightarrow 0$ and $D \longrightarrow 0$ as $\epsilon_1, \epsilon_2 \longrightarrow 0$, giving us that $\mu \overrightarrow{\longrightarrow} \delta_\alpha$ weakly. The knowledge of D gives the rate of weak convergence of μ to δ_α.

(ii) Let $|\alpha| < 1$. If $D \geq |\alpha| > 0$, then we cannot have $D \longrightarrow 0$ as $\epsilon_1, \epsilon_2 \longrightarrow 0$. Thus we are interested in $D < |\alpha|$, in that case there are r such that $D \leq r < |\alpha|$. More precisely: If $0 < \alpha < 1$, then $D \leq r < \alpha$, giving us $\alpha - r < \alpha$ and $(\alpha - r)^2 < \alpha^2$, always true for sufficiently small ϵ_1, ϵ_2. If $-1 < \alpha < 0$, then $D \leq r < -\alpha$, giving us $\alpha < \alpha + r$ and $(\alpha + r)^2 < \alpha^2$, always true for sufficiently small ϵ_1, ϵ_2.

(iii) If $|\alpha| \geq 1$, then $D \leq r \leq |\alpha|$ true for any $0 < \epsilon_j < 1, j = 1, 2$. More precisely: If $\alpha \geq 1$, then $D \leq r \leq \alpha$, giving us $\alpha - r < \alpha$ and $(\alpha - r)^2 < \alpha^2$. If $\alpha \leq -1$, then $D \leq r \leq -\alpha$, giving us $\alpha < \alpha + r$ and $(\alpha + r)^2 < \alpha^2$.

Comments (ii) and (iii) will be used in the proof of Theorem 6.1.1.

6.1.2. Results

The main theorem of this section follows:

Theorem 6.1.1.

i) *If* $\alpha = 0$ *we get* $D = \epsilon_2^{1/3}$.

ii) *If* $|\alpha| \geq 1$ *we get* $D = (\epsilon_2 + 2|\alpha|\epsilon_1)^{1/3}$.

iii) *For* $0 < \alpha| < 1$ *and sufficiently small* ϵ_1, ϵ_2 *we obtain again*

$$D = (\epsilon_2 + 2|\alpha|\epsilon_1)^{1/3}.$$

Comment 6.1.1. Our basic tools for the proof of Theorem 6.1.1 will come from Kemperman (1968), namely see the pages: 93, 94, 95, 96, 101, 102, 103, 104, 107, 111, 120, 121, see also Chapter 2.

PROOF OF THEOREM 6.1.1 (I).
Case of $\alpha = 0$.

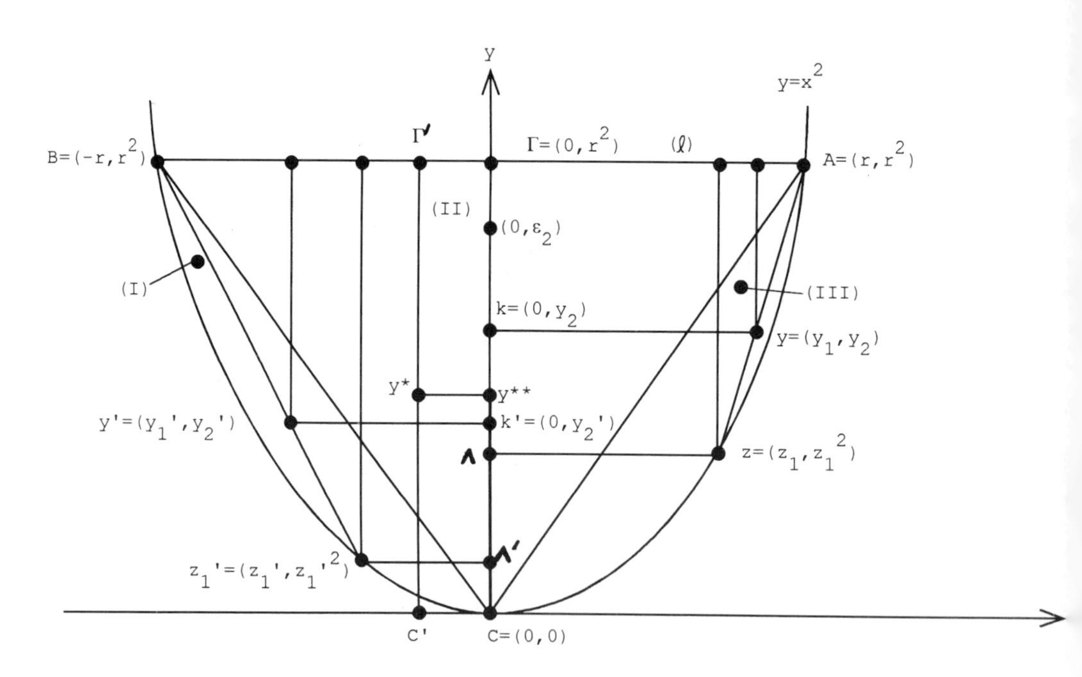

Figure 6.1.1: $\alpha = 0$

Here we have: (ℓ) the line through $(\overline{BA}), Y^* := (y_1^*, y_2^*), Y^{**} := (0, y_2^*)$, Area (I):= $\text{conv}(\widehat{BC}, \overline{BC})$, Area (II):=$\text{conv}(B \overset{\triangle}{C} A)$, Area (III):=$\text{conv}(\widehat{AC}, \overline{AC}), g(x) := (x, x^2)$, $V :=$ $\text{convg}(R), V^0$ interior of V and

$$\mathcal{R} := \{(y_1, y_2) : |y_1 - \alpha| \leq \epsilon_1, |y_2 - \alpha^2| \leq \epsilon_2\}.$$

Let $\epsilon_2 > r^2$, then the moment point $(0, \epsilon_2) \in \mathcal{R} \cap V^0 \subset M(\epsilon_1, \epsilon_2)$ is above the line (ℓ) of Figure 6.1.1.

Hence

$$L := L_{[\alpha - r, \alpha + r]}(0, \epsilon_2) := \inf_{\mu}(\mu([\alpha - r, \alpha + r])) = 0,$$

where μ is such that

$$\int t d\mu = 0 \quad \text{and} \quad \int t^2 d\mu = \epsilon_2$$

(see Kemperman (1968), pp. 96, 101, 120, also see Chapter 2). Therefore

$$\lambda_r := \inf_{\mu \in M(\epsilon_1, \epsilon_2)} \mu([\alpha - r, \alpha + r]) = 0 \geq 1 - r,$$

giving us $r \geq 1$ and $D = r = 1$. Consequently, $\epsilon_2 > 1$, which is a contradiction to the assumption that $\epsilon_2 < 1$.

We have proved that $\epsilon_2 < r^2$ for all r such that $\lambda_r \geq 1 - r$, that is $(0, \epsilon_2)$ is below the line (ℓ). Furthermore any other moment point $(0, y_2)$ with $0 < y_2 \leq \epsilon_2$ belongs to $\mathcal{R} \cap V^0$ and it is below the line (ℓ).

From Kemperman (1968), pp. 111, 120, 121, also from Chapter 2

$$L := L_{[\alpha - r, \alpha + r]}(y_1, y_2) := \inf_{\mu}(\mu([\alpha - r, \alpha + r])),$$

where μ is such that $\int t d\mu := y_1, \int t^2 d\mu := y_2$, is given by the ratios

$$L = \frac{YA}{ZA} = \frac{K\Gamma}{\Lambda\Gamma} > \frac{K\Gamma}{C\Gamma},$$

where $Y := (y_1, y_2) \in$ Area (III), and

$$L = \frac{Y'B}{Z'B} = \frac{K'\Gamma}{\Lambda'\Gamma} > \frac{K'\Gamma}{C\Gamma},$$

where $Y' := (y_1', y_2')$ belongs to Area (I), see Figure 6.1.1.

Thus Areas (I), (III) in terms of ratios are transferred into Area (II). In fact as we can see from the above inequalities, in Area (II) we get smaller L's. Therefore λ_r can be found from Area (II).

Also observe that for any $Y^* \in \text{conv}(BCA)$ we have

$$\frac{Y^*\Gamma'}{C'\Gamma'} = \frac{Y^{**}\Gamma}{C\Gamma},$$

see again Figure 6.1.1. So all it matters is the segment $C\Gamma$. Hence a typical $L = L(0, y_2)$ is given by the formula

$$L = \frac{K\Gamma}{C\Gamma} = \frac{r^2 - y_2}{r^2}, \quad \text{all } 0 < y_2 \leq \epsilon_2.$$

We would like to find the minimal $r \in (0,1]$ such that $\lambda_r \geq 1-r$. That is we would like $L \geq 1-r$, all $0 < y_2 \leq \epsilon_2$

$$\begin{array}{lll} \text{iff} & 1-\frac{y_2}{r^2} \geq 1-r, & \text{all } 0 < y_2 \leq \epsilon_2 \\ \text{iff} & y_2 \leq r^3, & \text{all } 0 < y_2 \leq \epsilon_2 \\ \text{iff} & \epsilon_2 \leq r^3 & \text{all } \epsilon_2^{1/3} \leq r. \end{array}$$

Because the boundary points of $g(\mathbb{R})$ of the arc $\widehat{BCA}$, see Figure 6.1.1., play no role towards the calculation of D we get that

$$D = \min r = \epsilon_2^{1/3}.$$

For the exclusion of the above boundary points see Kemperman (1968), pp. 102, 103, 104; Section 4 of non-interior points, also see Chapter 2 and the proofs of parts (ii) and (iii) of this theorem. The reasoning is exactly the same. □

To prove parts (ii) and (iii) of Theorem 6.1.1 we need

Theorem 6.1.2. *Let μ be probability measures on $\mathbb{R}$ such that $\int t d\mu := y_1, \int t^2 d\mu := y_2$. Let $\alpha \in \mathbb{R}, 0 < r \leq 1$. Set*

$$L := L_{[\alpha-r,\alpha+r]}(y_1, y_2) = \inf_{\mu}(\mu([\alpha - r, \alpha + r])).$$

Then we find

1) $L = 0$, *if* $y_2 + \alpha^2 - 2\alpha y_1 > r^2$;

2) $L = 1$, *if* $y_2 = y_1^2$ *with* $y_1 \in [\alpha - r, \alpha + r]$;

3) *If* $y_2 \neq y_1^2$ *and* $r|y_1 - \alpha| \leq y_2 + \alpha^2 - 2\alpha y_1 \leq r^2$, *then*

$$L = 1 - \frac{[y_2 + \alpha^2 - 2\alpha y_1]}{r^2};$$

4) *If* $y_2 \neq y_1^2$ *and* $y_2 + \alpha^2 - 2\alpha y_1 \leq r(y_1 - \alpha)$, *then*

$$L = \frac{((\alpha + r) - y_1)^2}{(\alpha + r)^2 - 2(\alpha + r)y_1 + y_2};$$

5) *If* $y_2 \neq y_1^2$ *and* $y_2 + \alpha^2 - 2\alpha y_1 \leq r(\alpha - y_1)$, *then*

$$L = (y_1 - (\alpha - r))^2/((\alpha - r)^2 - 2(\alpha - r)y_1 + y_2).$$

Comment 6.1.2. Here $y_1^2 \leq y_2$ and $(y_1, y_2) \in \text{convg}(\mathbb{R})$, where $g(x) := (x, x^2), x \in \mathbb{R}$.

Proof of Theorem 6.1.2.

Case (2). Set again $V :=$convg$((\mathbb{R})$. We use the terminology and the results of Kemperman (1968), pp. 102, 103, 104; Section 4 of non-interior points, see also Chapter 2. Let y be a boundary point of V, then $V_y = \{y\}$. And $\Gamma^y := g(\mathbb{R}) \cap V_y = \{y\}$, i.e. $\Gamma^y = \{y\}$. Also $T^y := g^{-1}(\{y\}) = t_y$, by g being one to one mapping. That is, $T^y = \{t_y\}$. From Theorem 9 of Kemperman (1968), p. 103, see also Chapter 2 we have: Let y be a fixed boundary point of V and let μ be a probability measure on $\mathbb{R}$ such that $\int |t| d\mu < \infty, \int t^2 d\mu < \infty$ with $\int t d\mu := y_1, \int t^2 d\mu := y_2; y := (y_1, y_2)$. Then μ is concentrated at $\{t_y\}$. Conclusion $L = 1$.

Case (1). Let $y_1 \in S' := \mathbb{R} - [\alpha - r, \alpha + r]$, then $y = (y_1, y_1^2)$ is a boundary point of $g(S')$. Therefore any measure μ as in case (2) will be concentrated at $\{y_1\}$. Consequently,

$$L = L_{[\alpha-r,\alpha+r]}(y - \text{boundary} \quad \in g(S')) = 0.$$

Furthermore

$$L = L_{[\alpha-r,\alpha+r]}(y \in (\text{convg}(S'))^0) = 0.$$

By calling $W_{S'} := \overline{\text{convg}(S')}$, we have got that

$$L = L_{[\alpha-r,\alpha+r]}(y) = 0, \quad \forall y \in W_{S'}.$$

See also Kemperman (1968), pp. 101, 120, 121 and Chapter 2.

Case (3). Let $\alpha \geq 0$. When $\alpha < 0$, the proof is exactly the same, as such is omitted.

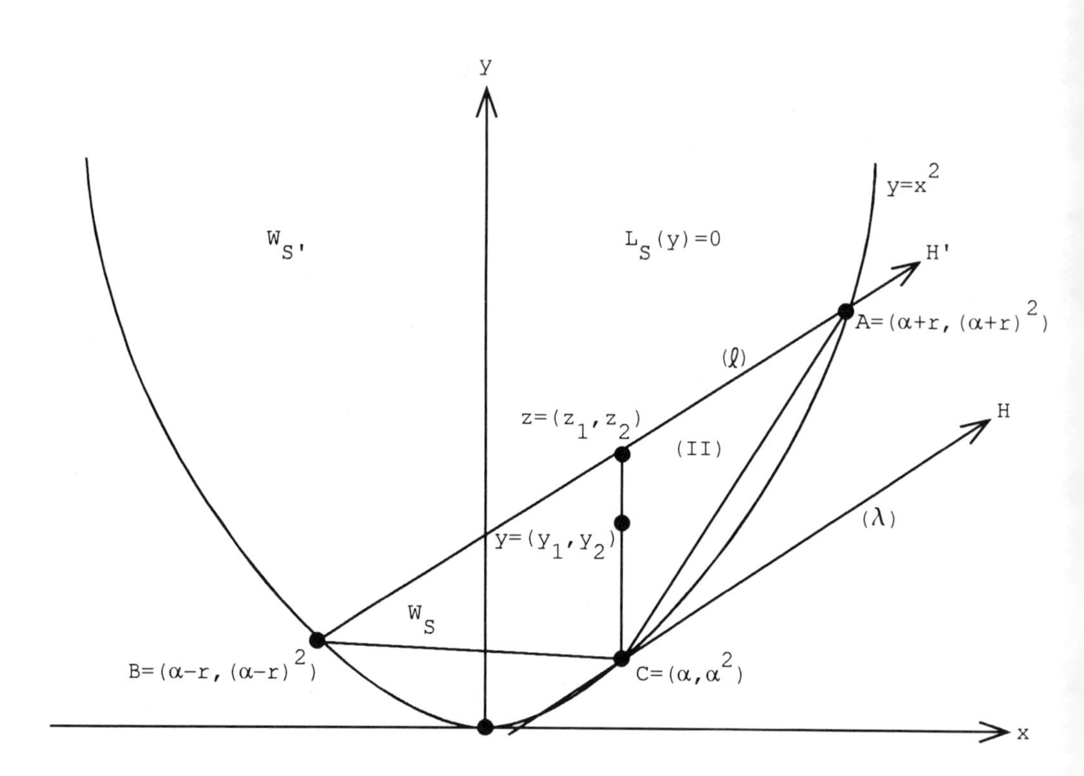

Figure 6.1.2: Case of $y \in \text{conv}(BCA)$

Let $S := [\alpha - r, \alpha + r], S' := \mathcal{R} - S, 0 < r \leq 1$. Set Area (II):= $\text{conv}(BCA), W_S = \overline{convg(S)}$. Consider (ℓ) the line through $(\overline{BA})$; call it also (H') and (λ) the line tangent to $y = x^2$ and parallel to (ℓ); call it also (H), it goes through the point $C := (\alpha, \alpha^2)$. Here see Kemperman (1968), pp. 101, 102, 111, 120, 121 and Chapter 2. One can easily prove that $y := (y_1, y_2) \in \text{conv}(BCA)$ iff

$$r|y_1 - \alpha| \leq y_2 + \alpha^2 - 2\alpha y_1 \leq r^2.$$

And line (ℓ) is given by

$$(-2\alpha)x + 1y + (\alpha^2 - r^2) = 0,$$

i.e., $\tilde{A} = -2\alpha, \tilde{B} = 1, \tilde{C} = \alpha^2 - r^2$ for the line $(\xi) : \tilde{A}x + \tilde{B}y + \tilde{C} = 0$. We know that the distance

$$d((x_0, y_0), \xi) = \frac{|\tilde{A}x_0 + \tilde{B}y_0 + \tilde{C}|}{\sqrt{\tilde{A}^2 + \tilde{B}^2}}.$$

Here

$$d((y_1, y_2), \ell) = \frac{|-2\alpha y_1 + y_2 + (\alpha^2 - r^2)|}{\sqrt{4\alpha^2 + 1}}$$

and

$$d((\alpha, \alpha^2), \ell) = \frac{r^2}{\sqrt{4\alpha^2 + 1}}.$$

Therefore from the geometric moment method of optimal ratio

$$L = L_S(y) = \frac{d((y_1, y_2), \ell)}{d((\alpha, \alpha^2), \ell)}$$

i.e.

$$L = \frac{|-2\alpha y_1 + y_2 + (\alpha^2 - r^2)|}{r^2}, \quad \text{all } y := (y_1, y_2) \in \text{Area (II)}.$$

Hence

$$L = L_S(y) = 1 - \frac{[y_2 + \alpha^2 - 2\alpha y_1]}{r^2};$$

for all $y := (y_1, y_2)$ such that

$$r|y_1 - \alpha| \leq y_2 + \alpha^2 - 2\alpha y_1 \leq r^2.$$

Cases (4,5). Here we consider $\alpha \geq 0$, the case $\alpha < 0$ as similar is omitted.

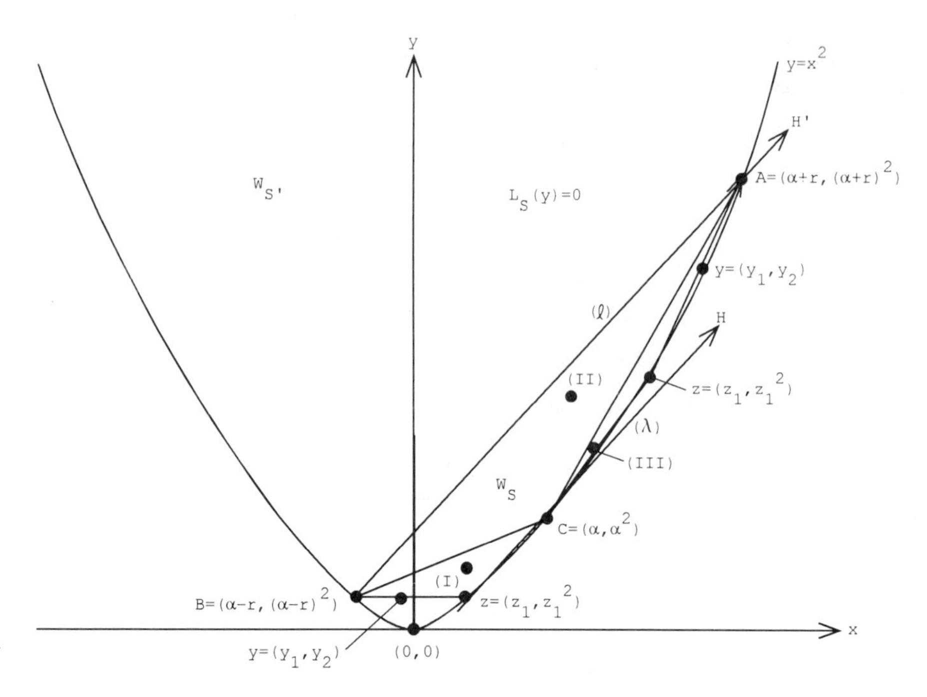

Figure 6.1.3 α: Case of Y below area (II)

Here $S = [\alpha - r, \alpha + r], 0 < r \leq 1$, Area (I)=conv($\widehat{BC}$, $\overline{BC}$) and Area (III)=conv($\widehat{AC}$, $\overline{AC}$). Consider $W_S = \overline{convg(S)}$, (ℓ) the line through $(\overline{BA})$; call it also (H'), and (λ) the line tangent to $y = x^2$ and parallel to (ℓ); call it also (H), it goes through the point $C = (\alpha, \alpha^2)$.

One can easily see that (y_1, y_2) is on or below (CA) iff $y_2 + \alpha^2 - 2\alpha y_1 \leq r(y_1 - \alpha), (y_1 \geq \alpha)$. And (y_1, y_2) is on or below (BC) iff $y_2 + \alpha^2 - 2\alpha y_1 \leq r(\alpha - y_1), (y_1 \leq \alpha)$. Furthermore, (y_1, y_2) is on or above (BA) iff $y_2 + \alpha^2 - 2\alpha y_1 \geq r^2$. Again we use the techniques of Kemperman (1968), see pp. 93, 94, 95, 96, 101, 102, 111, 120, 121; see also Chapter 2.

Area (III). From

$$L = L_S(y) = \frac{YA}{ZA} = \frac{\alpha + r - y_1}{\alpha + r - z_1} = \frac{(\alpha + r)^2 - y_2}{(\alpha + r)^2 - z_1^2}$$

we get

$$z_1 = \frac{y_1(\alpha + r) - y_2}{(\alpha + r) - y_1}$$

and

$$L = \frac{((\alpha + r) - y_1)^2}{(\alpha + r)^2 - 2(\alpha + r)y_1 + y_2}, \tag{6.1.7}$$

for all $Y := (y_1, y_2)$ such that

$$y_2 + \alpha^2 - 2\alpha y_1 \leq r(y_1 - \alpha), \quad (y_1 \geq \alpha).$$

Area (I). The subcase $y_1 = \alpha - r = z_1$ is trivial. For $z_1 \neq -(\alpha - r)$ we get

$$L = L_S(y) = \frac{BY}{BZ} = \frac{y_1 - (\alpha - r)}{z_1 - (\alpha - r)} = \frac{y_2 - (\alpha - r)^2}{z_1^2 - (\alpha - r)^2}.$$

Thus

$$z_1 = \frac{y_2 - y_1(\alpha - r)}{y_1 - (\alpha - r)}$$

and

$$L = \frac{(y_1 - (\alpha - r))^2}{(\alpha - r)^2 - 2(\alpha - r)y_1 + y_2}, \tag{6.1.8}$$

for all $Y := (y_1, y_2)$ such that

$$y_2 + \alpha^2 - 2\alpha y_1 \leq r(\alpha - y_1), \quad (y_1 \leq \alpha).$$

Remark 6.1.3. Assume $y_2 = (\alpha - r)^2$, then $z_1^2 = (\alpha - r)^2$ and $z_1 = \pm(\alpha - r)$. If $z_1 = \alpha - r$, then this is a trivial subcase. If $z_1 = -(\alpha - r)$, then

$$L = L_S(y) = \frac{y_1 - (\alpha - r)}{2(r - \alpha)}. \tag{6.1.9}$$

But (6.1.9) is covered by (6.1.8) for this particular case of $y_2 = (\alpha - r)^2$ and $z_1 = r - \alpha$.

Remark 6.1.4. Proving the case of $\alpha < 0$ we deal with the following: Assume $y_2 = (\alpha + r)^2$, then $z_1^2 = (\alpha + r)^2$ and $z_1 = \pm(\alpha + r)$. If $z_1 = \alpha + r$, then this is a trivial subcase. If $z_1 = -(\alpha + r)$, then

$$L = L_S(y) = \frac{(\alpha + r) - y_1}{2(\alpha + r)}. \tag{6.1.10}$$

Substituting $y_2 = (\alpha + r)^2, z_1 = -(\alpha + r)$ into (6.1.7) we get (6.1.10). That is, this subcase is covered by formula (6.1.7).

The above end the proof of Theorem 6.1.2. □

Proof of Theorem 6.1.1 is continued:

Remark 6.1.5.

1) $L = 1$ means nothing towards the calculation of D. Consider $\alpha \geq 0$, the case of $\alpha < 0$ as similarly treated is omitted. We know from Theorem 6.1.2 (2) that the points (y_1, y_2) such that $y_2 = y_1^2, y_1 \in [\alpha - r, \alpha + r]$ produce $L = 1$.

Consider the rectangle $\mathcal{R}$ of points (y_1, y_2) such that $|y_j - \alpha^j| \le \epsilon_j, j = 1, 2$. That is, $\alpha - \epsilon_1 \le y_1 \le \alpha + \epsilon_1$ and $\alpha^2 - \epsilon_2 \le y_2 \le \alpha^2 + \epsilon_2$. Note that $\alpha + \epsilon_1 > \alpha$ and $\alpha^2 + \epsilon_2 > \alpha^2$. Call again $V :=$conv$g(\mathbb{R}); g(x) := (x, x^2)$. Obviously $\mathcal{R} \cap V \neq \emptyset$, containing some of the interior points of V, call them (y_1^0, y_2^0). Notice $y_1^2 \le y_2$.

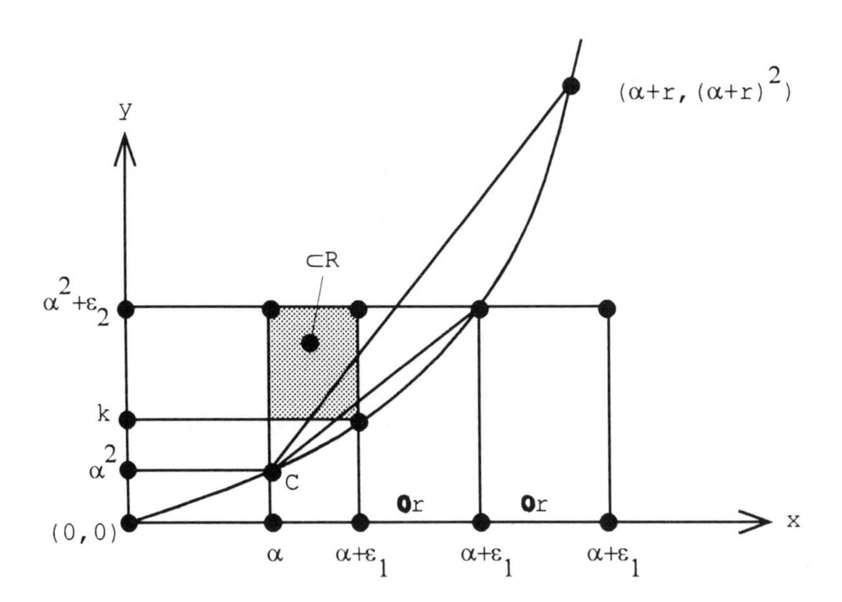

Figure 6.1.3b: Case of $L = 1$

More precisely it can happen one of the following: Either

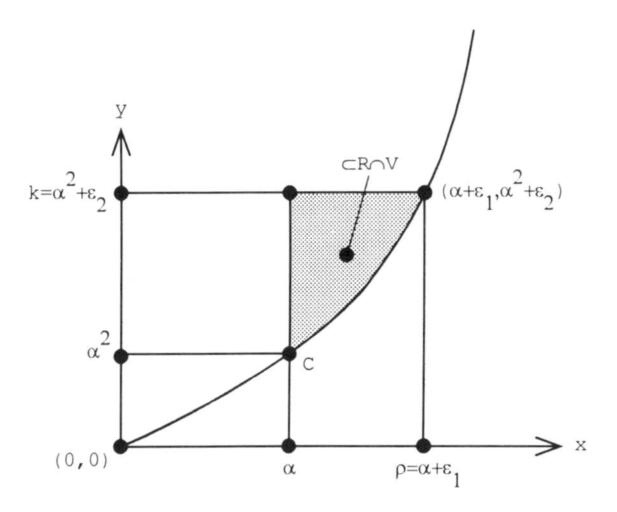

Figure 6.1.4a: $\rho^2 = k$

or

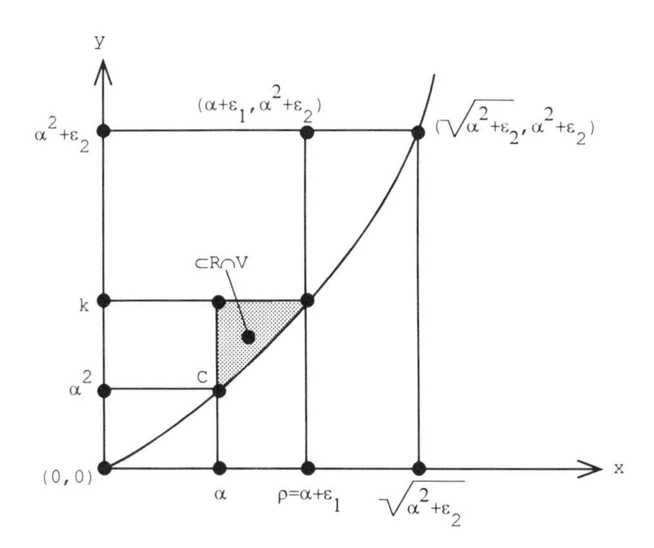

Figure 6.1.4b: $\alpha \leq \rho \leq \sqrt{\alpha^2 + \epsilon_2}$

or

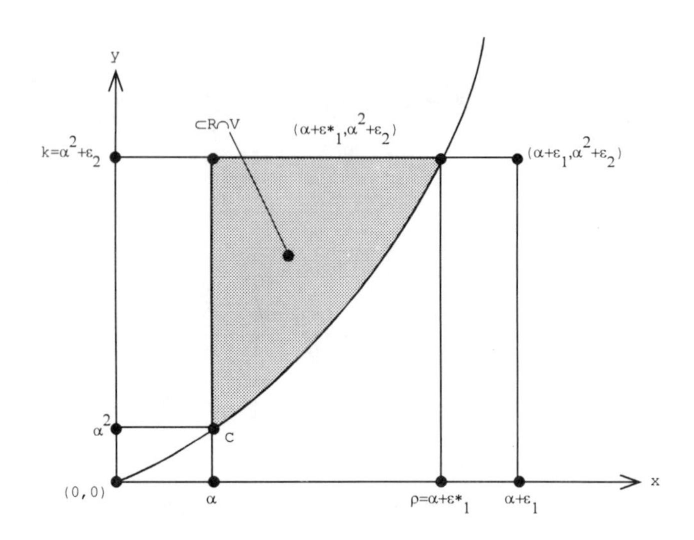

Figure 6.1.4c: $\alpha \leq \rho \leq \alpha + \epsilon_1, \epsilon_1^* \leq \epsilon_1$

Conclusion. Always we have

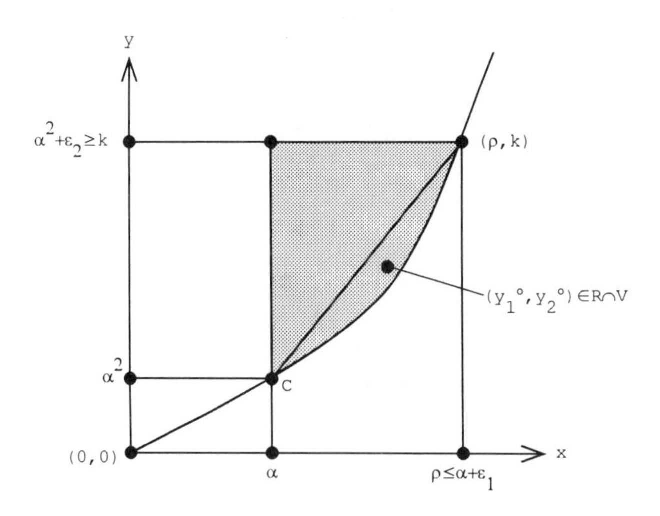

Figure 6.1.4d: $\rho \leq \alpha + \epsilon_1$

The point $\rho = \alpha + \epsilon_1^*, \epsilon_1^* \leq \epsilon_1$. Thus for any $r \in (0,1]$ we have $\Gamma := [\alpha, \alpha+r] \cap [\alpha, \alpha+\epsilon_1^*] \neq \emptyset$. If $r \leq \epsilon_1^*, \Gamma = [\alpha, \alpha + r]$, and if $r \geq \epsilon_1^*, \Gamma = [\alpha, \alpha + \epsilon_1^*]$. If $r \leq \epsilon_1^*$, then

$$\text{convg}([\alpha, \alpha + r]) \subset \text{convg}([\alpha, \alpha + \epsilon_1^*])$$

If $r \geq \epsilon_1^*$, then

$$\text{convg}([\alpha, \alpha + r]) \supset \text{convg}([\alpha, \alpha + \epsilon_1^*]).$$

In any case there exist interior points of V : $(y_1^0, y_2^0) \in \mathcal{R} \cap V$ which also belong to convg($[\alpha, \alpha + r]$)$\cap$ convg($[\alpha, \alpha + \epsilon_1^*]$) $\neq \emptyset$. Fix such (y_1^0, y_2^0). By Lemma 2, p. 96 of Kemperman (1968), see also Chapter 2, there exists a probability measure μ on $\mathcal{R}$ such that

$$\int t d\mu^0 = y_1^0, \quad \int t^2 d\mu^0 = y_2^0. \tag{6.1.11}$$

Obviously

$$\text{convg}([\alpha, \alpha + r]) \cap \text{convg}([\alpha, \alpha + \epsilon_1^*]) \subset \text{convg}([\alpha, \alpha + r]) \subset \text{convg}([\alpha - r, \alpha + r]).$$

I.e. (y_1^0, y_2^0) is an interior point of convg($[\alpha - r, \alpha + r]$) for which we find from Theorem 6.1.2 that $\inf_\mu \mu([\alpha - r, \alpha + r]) < 1, \mu$ as in (6.1.11)), for any $r \in (0, 1]$.

Therefore

$$\lambda_r := \inf_{\mu \in M(\epsilon_1, \epsilon_2)} \mu([\alpha - r, \alpha + r]) < 1 = L.$$

Therefore

$$\lambda_r := \inf_{\mu \in M(\epsilon_1, \epsilon_2)} \mu([\alpha - r, \alpha + r]) < 1 = L.$$

$L = 1$ comes from the associated boundary points in that neighborhood. *Consequently, the boundary points* $(y_1, y_1^2), y_1 \in [\alpha - r, \alpha + r]$, *do not contribute anything towards the calculation of* $D \in (0, 1]$. As such, are excluded from consideration.

2) $L = 0$ **means nothing towards the calculation of** D. Consider $\alpha \geq 0$, the case of $\alpha < 0$ as similarly treated is omitted. Let $L = 0$, this can happen iff $y_2 + \alpha^2 - 2\alpha y_1 \geq r^2$, with $(y_1, y_2) \neq ((\alpha \pm r), (\alpha \pm r)^2)$. Here $\alpha - \epsilon_1 \leq y_1 \leq \alpha + \epsilon_1, \alpha^2 - \epsilon_2 \leq y_2 \leq \alpha^2 + \epsilon_2, r \in (0, 1]$. Therefore $\lambda_r = 0 \geq 1 - r$ can only be true for $r = 1$, that is, $D = 1$.

Note that

$$y_2 + \alpha^2 \leq 2\alpha^2 + \epsilon_2 \tag{6.1.12}$$

and

$$-2\alpha y_1 \leq 2\alpha\epsilon_1 - 2\alpha^2. \tag{6.1.13}$$

Adding (6.1.12), (6.1.13) we obtain

$$1 = r^2 \leq y_2 + \alpha^2 - 2\alpha y_1 \leq \epsilon_2 + 2\alpha\epsilon_1.$$

Hence $\epsilon_2 + 2\alpha\epsilon_1 \geq 1$, which is a contradiction since we assumed $\epsilon_2 + 2\alpha\epsilon_1 < 1$.

The last assumption is justified by the following: Let $\epsilon_2 + 2\alpha\epsilon_1 \geq 1$ and $\epsilon := \max(\epsilon_1, \epsilon_2)$. Then $1 \leq \epsilon_2 + 2\alpha\epsilon_1 \leq \epsilon(1 + 2\alpha)$, giving us $\epsilon \geq (1 + 2\alpha)^{-1}$, that is ϵ is bounded away from zero. But we would like to have $\epsilon_1, \epsilon_2 \to 0$, that is, it should be $\epsilon \to 0$. Consequently the case of $\epsilon_2 + 2\alpha\epsilon_1 \geq 1$ is absurd.

We have established that $L = 0$ plays no role in the calculation of D. I.e. we have proved that *the points* $(y_1, y_2) \in \mathcal{R} \cap V - \{((\alpha \underline{+} r), (\alpha \underline{+} r)^2)\}$, which are on or above the line through $\{((\alpha \underline{+} r), (\alpha \underline{+} r)^2)\}$ play no role in the calculation of D. As such, are excluded from consideration.

Proof of Theorem 6.1.1 is continued:

Remark 6.1.6. Case of $\alpha > 0$. Note that any $(y_1, y_2) \in \mathcal{R} \cap V$ fulfills

$$y_1 \geq \max(\alpha - \epsilon_1, -\sqrt{\alpha^2 + \epsilon_2}).$$

Take $\bar{y}_1 := \max(\alpha - \epsilon_1, -\sqrt{\alpha^2 + \epsilon_2}), \bar{y}_2 := \alpha^2 + \epsilon_2$. Obviously $(\bar{y}_1, \bar{y}_2) \in \mathcal{R} \cap V$ and there exists a probability measure μ on $\mathbb{R}$ such that

$$\int t d\mu = \bar{y}_1, \quad \int t^2 d\mu = \bar{y}_2$$

(see Kemperman (1968), Lemma 2, p. 96 and Chapter 2).

Consider any *admissible* $0 < r \leq 1$ i.e., $\lambda_r \geq 1 - r$. Assume $\bar{y}_2 + \alpha^2 - 2\alpha\bar{y}_1 > r^2$, then

$$2\alpha^2 + \epsilon_2 - 2\alpha\bar{y}_1 > r^2.$$

If $\bar{y} = \alpha - \epsilon_1$, then $\epsilon_2 + 2\alpha\epsilon_1 > r^2$. If $\bar{y}_1 = -\sqrt{a^2 + \epsilon_2}$, then $(\alpha + \sqrt{\alpha^2 + \epsilon_2})^2 > r^2$, then $r < \alpha + \sqrt{\alpha^2 + \epsilon_2} \leq \alpha + \epsilon_1 - \alpha$. Hence $\epsilon_1 > r$.

Furthermore from Theorem 6.1.2 we obtain

$$\lambda_r = L = 0 \geq 1 - r, \quad \text{then} \quad r \geq 1.$$

Hence $r = 1$. Consequently we get either $\epsilon_2 + 2\alpha\epsilon_1 > 1$ or $\epsilon_1 > 1$. Both contradict the assumption of the theorem.

Therefore $(\bar{y}_1, \bar{y}_2)$ *is on or below the line* (ℓ) *through the points* $\{((\alpha \underline{+} r), (\alpha \underline{+} r)^2)\}$. *I.e.* $\bar{y}_2 + \alpha^2 - 2\alpha\bar{y}_1 \leq r^2$, *for any admissible* r.

Remark 6.1.6′. Case of $\alpha < 0$. Note that any $(y_1, y_2) \in \mathcal{R} \cap V$ fulfills

$$y_1 \leq \min(\alpha + \epsilon_1, \sqrt{\alpha^2 + \epsilon_2}).$$

Take $\bar{y}_1 := \min(\alpha + \epsilon_1, \sqrt{\alpha^2 + \epsilon_2}), \bar{y}_2 := \alpha^2 + \epsilon_2$. Obviously $(\bar{y}_1, \bar{y}_2) \in \mathcal{R} \cap V$ and there exists a probability measure μ on $\mathbb{R}$ such that

$$\int t d\mu = \bar{y}_1 \quad \text{and} \quad \int t^2 d\mu = \bar{y}_2$$

(see Kemperman (1968), Lemma 2, p. 96 and Chapter 2).

Consider any *admissible* $0 < r \leq 1$ i.e. $\lambda_r \geq 1 - r$. Assume $\bar{y}_2 + \alpha^2 - 2\alpha\bar{y}_1 > r^2$, then $2\alpha^2 + \epsilon_2 - 2\alpha\bar{y}_1 > r^2$. If $\bar{y}_1 = (\alpha + \epsilon_1)$, then $\epsilon_2 - 2\alpha\epsilon_1 > r^2$. If $\bar{y}_1 = \sqrt{\alpha^2 + \epsilon_2}$, then $(\alpha - \sqrt{\alpha^2 + \epsilon_2})^2 > r^2$, then $\sqrt{\alpha^2 + \epsilon_2} - \alpha > r$, then $\alpha + \epsilon_1 \geq \sqrt{\alpha^2 + \epsilon_2} > r + \alpha$. Hence $\epsilon_1 > r$.

Furthermore from Theorem 6.1.2 we have $\lambda_r = L = 0 \geq 1 - r$, then $r \geq 1$, then $r = 1$. Thus, we get either $\epsilon_2 - 2\alpha\epsilon_1 > 1$ or $\epsilon_1 > 1$. Both contradict the assumptions of the theorem.

Therefore $(\bar{y}_1, \bar{y}_2)$ *is on or below the line* (ℓ) *through the points* $\{((\alpha \pm r), (\alpha \pm r)^2)\}$. *I.e.* $\bar{y}_2 + \alpha^2 - 2\alpha\bar{y}_1 \leq r^2$, *for any admissible* r.

Proof of Theorem 6.1.1 is continued:

Remark 6.1.7. Case of $\alpha > 0$.

1) From Remark 6.1.2 (ii) and (iii) we have: when $0 < \alpha < 1$ it holds that $(\alpha - r)^2 < \alpha^2$, for sufficiently small ϵ_1, ϵ_2; and when $\alpha \geq 1$ it holds again that $(\alpha - r)^2 < \alpha^2$ for any $0 < \epsilon_j < 1, j = 1, 2$. Let

$$\bar{y}_1 := \max(\alpha - \epsilon_1, -\sqrt{\alpha^2 + \epsilon_2}), \quad \bar{y}_2 := \alpha^2 + \epsilon_2.$$

Then $(\bar{y}_1, \bar{y}_2)$ is above line (BC), in fact even strictly above line (ΦC), see Figure 6.1.5.

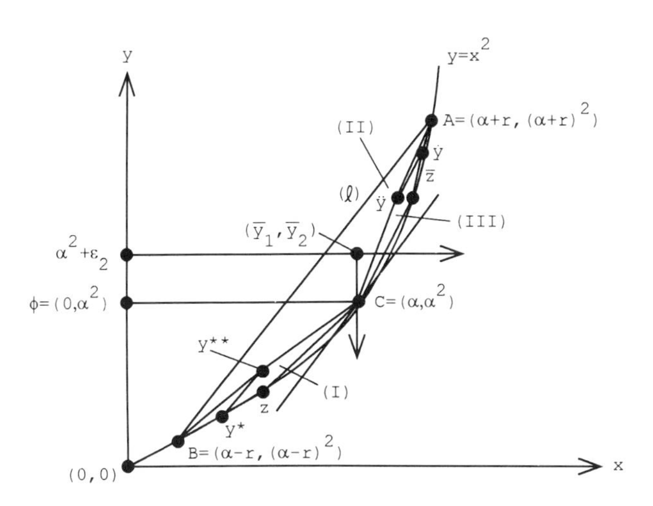

Figure 6.1.5: $(\bar{y}_1, \bar{y}_2)$ is above line (ΦC)

2) Let $Y^* := (y_1^*, y_2^*) \in \mathcal{R} \cap$ interior (Area (I))$\cap V^0$. That is $|y_1^* - \alpha| \leq \epsilon_1, |y_2^* - \alpha^2| \leq \epsilon_2$. As before, from geometric moment method of optimal ratio we get that

$$L = \frac{BY^*}{BZ}.$$

Consider ZC and Y^{**} on BC such that Y^*Y^{**} is parallel to ZC. Then

$$\frac{BY^*}{BZ} = \frac{BY^{**}}{BC} \quad \text{and} \quad L = \frac{BY^{**}}{BC}.$$

Here $Y^{**} := (y_1^{**}, y_2^{**}) \in V^0$. One can easily see that $y_1^* < y_1^{**} < \alpha, y_2^* < y_2^{**} < \alpha^2$. Therefore $|y_1^{**} - \alpha| \leq \epsilon_1, |y_2^{**} - \alpha^2| \leq \epsilon_2$ i.e.$(y_1^{**}, y_2^{**}) \in \mathcal{R}$.

Conclusion. Area (I) is transferred into Area (II) and then all is taken care of by the formula of Area (II) for L: see Theorem 6.1.2(3).

3) Let $\dot{Y} := (\dot{y}_1, \dot{y}_2) \in \mathcal{R} \cap$ interior (Area (III)) $\cap V^0$. That is, $|\dot{y}_1 - \alpha| \leq \epsilon_1, |\dot{y}_2 - \alpha^2| \leq \epsilon_2$. As before, from optimal ratio method we get that

$$L = \frac{A\dot{Y}}{A\bar{Z}}.$$

Consider $\bar{Z}C$ and $\ddot{Y}$ on AC such that $\ddot{Y}\dot{Y}$ is parallel to $C\bar{Z}$. Then

$$\frac{A\ddot{Y}}{AC} = \frac{A\dot{Y}}{A\bar{Z}} \quad \text{and} \quad L = \frac{A\ddot{Y}}{AC}.$$

Here $\ddot{Y} := (\ddot{y}_1, \ddot{y}_2) \in V^0$. One can easily see that $\alpha < \ddot{y}_1 < \dot{y}_1$ and $\alpha^2 < \ddot{y}_2 < \dot{y}_2$. Therefore $|\ddot{y}_1 - \alpha| \leq \epsilon_1$ and $|\ddot{y}_2 - \alpha^2| \leq \epsilon_2$, i.e. $(\ddot{y}_1, \ddot{y}_2) \in \mathcal{R}$.

Conclusion. Area (III) is transferred into Area (II) and then all is taken care of by the formula of (II) for L. See Theorem 6.1.2 (3).

Final Conclusion. ($\alpha > 0$). D can be calculated only from Area (II), by the use of Theorem 6.1.2 (3).

Remark 6.1.8. Case of $\alpha > 0$. Calculation of D (Theorem 6.1.1, cases (ii), (iii)). Here, all $y_1 \geq \bar{y}_1 = \max(\alpha - \epsilon_1, -\sqrt{\alpha^2 + \epsilon_2}) = \alpha - \epsilon_1$ and all $y_2 \leq \bar{y}_2 = \alpha^2 + \epsilon_2$. Thus, all

$$\Delta := y_2 + \alpha^2 - 2\alpha y_1 \leq (\alpha^2 + \epsilon_2) + \alpha^2 - 2\alpha(\alpha - \epsilon_1) = \epsilon_2 + 2\alpha\epsilon_1 =: K,$$

i.e. $1 > K \geq \Delta$. Here we find λ_r over conv(BCA). We would like that any

$$L = 1 - \frac{[y_2 + \alpha^2 - 2\alpha y_1]}{r^2} \geq 1 - r$$

iff $\Delta = [y_2 + \alpha^2 - 2\alpha y_1] \leq r^3$, for all $(y_1, y_2) \in (\text{conv}(BCA) \cap \mathcal{R} \cap V^0)$ iff $K \leq r^3$ iff $K^{1/3} \leq r$. Thus $D = \min r = K^{1/3}$.

Conclusion. $D = (\epsilon_2 + 2\alpha\epsilon_1)^{1/3}$ is true for $\alpha \geq 1$ with any $0 < \epsilon_j < 1, j = 1, 2$; and true for $0 < \alpha < 1$ when ϵ_1, ϵ_2 are sufficiently small.

Proof of Theorem 6.1.1 is continued:

Remark 6.1.7′. Case of $\alpha < 0$.

1) From Remark 6.1.2 (ii), (iii) we have that when $-1 < \alpha < 0$ it holds $(\alpha + r)^2 < \alpha^2$, always true for sufficiently small ϵ_1, ϵ_2; and when $\alpha \leq -1$ it holds again $(\alpha + r)^2 < \alpha^2$, for

any $0 < \epsilon_j < 1, j = 1,2$. Let $\bar{y}_1 := \min(\alpha + \epsilon_1, \sqrt{\alpha^2 + \epsilon_2}), \bar{y}_2 := \alpha^2 + \epsilon_2$. Then $(\bar{y}_1, \bar{y}_2)$ is above the line (AC), in fact is even strictly above the line (ΦC), see Figure 6.1.6.

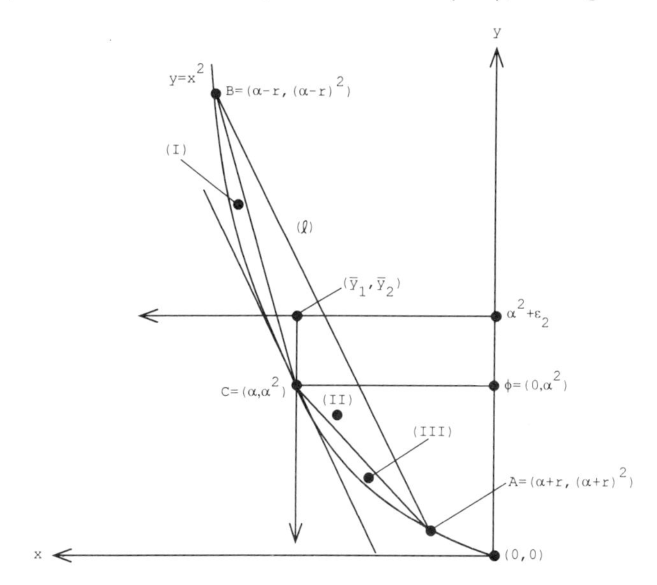

Figure 6.1.6: $(\bar{y}_1, \bar{y}_2)$ is above the line (AC)

2) Let $Y^* := (y_1^*, y_2^*) \in \mathcal{R} \cap$ interior (Area (I))$\cap V^0$. That is, $|y_1^* - \alpha| \leq \epsilon_1, |y_2^* - \alpha^2| \leq \epsilon_2$. As before, from optimal ratio method we get that $L = \frac{BY^*}{BZ}$. Consider ZC and $Y^{**} := (y_1^{**}, y_2^{**})$ on BC such that Y^*Y^{**} is parallel to ZC. Then

$$\frac{BY^*}{BZ} = \frac{BY^{**}}{BC}, \quad \text{i.e.} \quad L = \frac{BY^{**}}{BC}.$$

Here $Y^{**} \in V^0$.

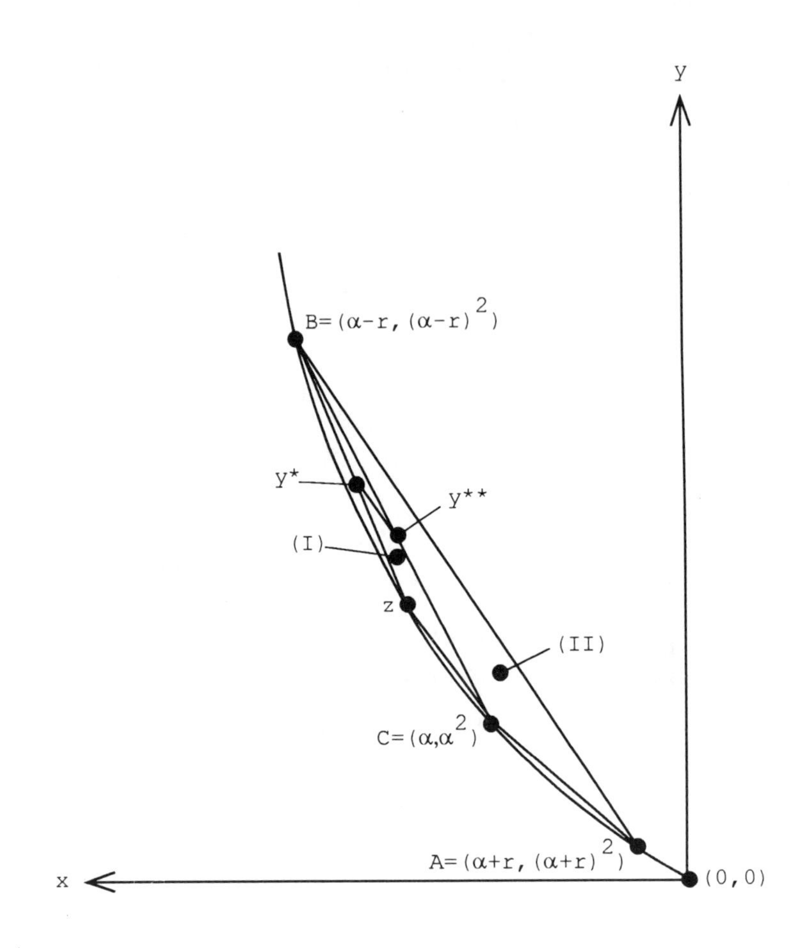

Figure 6.1.7: Area (I) is transferred in Area (II)

Also $y_1^* < y_1^{**} < \alpha$ and $\alpha^2 < y_2^{**} < y_2^*$, i.e. $|y_1^{**} - \alpha| \leq \epsilon_1$ and $|y_2^{**} - \alpha^2| \leq \epsilon_2$, giving us $(y_1^{**}, y_2^{**}) \in \mathcal{R}$.

Conclusion. Area (I) is transferred into Area (II) and then all is taken care of by the formula of (II) for L; see Theorem 6.1.2 (3).

3) Let $\dot{Y} := (\dot{y}_1, \dot{y}_2) \in \mathcal{R} \cap$ interior (Area (III))$\cap V^0$. That is, $|\dot{y}_1 - \alpha| \leq \epsilon_1$ and $|\dot{y}_2 - \alpha^2| \leq \epsilon_2$. As before, from optimal ratio method we get that

$$L = \frac{A\dot{Y}}{A\bar{Z}}.$$

Consider $\bar{Z}C$ and $\ddot{Y} := (\ddot{y}_1, \ddot{y}_2)$ on AC such that $\ddot{Y}\dot{Y}$ is parallel to $C\bar{Z}$. Then

$$\frac{A\ddot{Y}}{AC} = \frac{A\dot{Y}}{A\bar{Z}}, \quad \text{i.e.} \quad L = \frac{A\ddot{Y}}{AC}.$$

Here $\ddot{Y} \in V^0$.

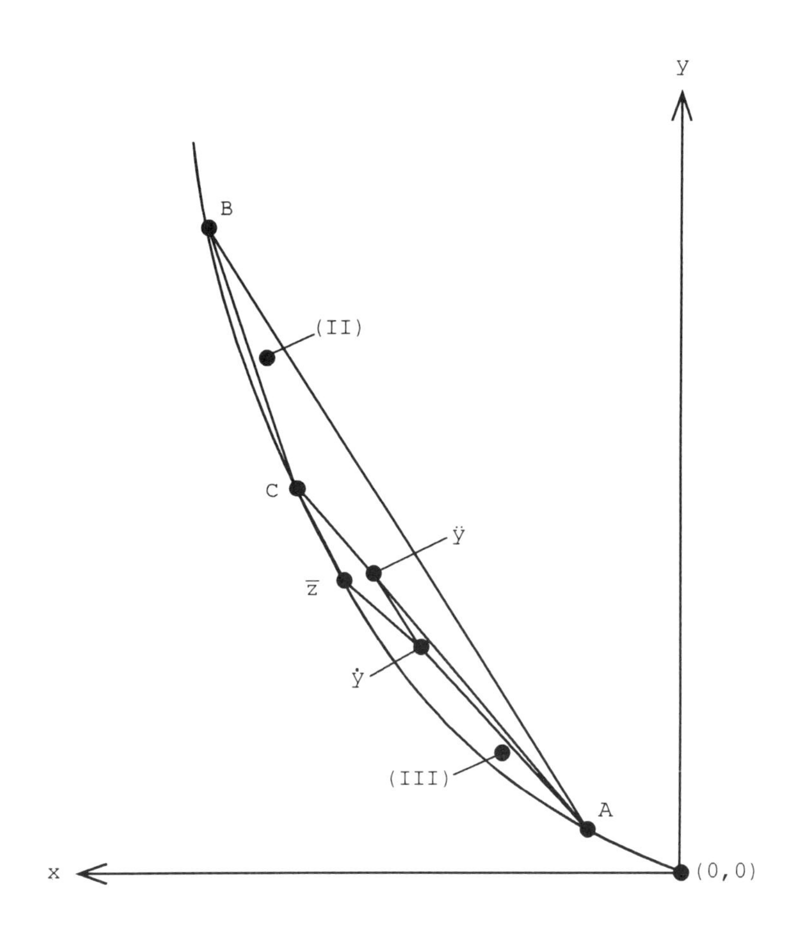

Figure 6.1.8: Area (III) transferred into Area (II)

Note that $\alpha < \ddot{y}_1 < \dot{y}_1$ and $\dot{y}_2 < \ddot{y}_2 < \alpha^2$, that is, $|\ddot{y}_1 - \alpha| \leq \epsilon_1$ and $|\ddot{y}_2 - \alpha^2| \leq \epsilon_2$, i.e. $(\ddot{y}_1, \ddot{y}_2) \in \mathcal{R}$.

Conclusion. Area (III) is transferred into Area (II) and then all is taken care of by the formula of (II) for L. See Theorem 6.1.2 (3).
Final Conclusion. ($\alpha < 0$). D can be calculated only from Area (II), by the use of Theorem 6.1.2(3).

Remark 6.1.8′. Case of $\alpha < 0$. Calculation of D (Theorem 6.1.1, cases (ii), (iii)). Here, all $y_1 \leq \bar{y}_1 = \min(\alpha + \epsilon_1, \sqrt{\alpha^2 + \epsilon_2}) = \alpha + \epsilon_1$ and all $y_2 \leq \bar{y}_2 = \alpha^2 + \epsilon_2$. Thus, all

$$\Delta := y_2 + \alpha^2 - 2\alpha y_1 \leq (\alpha^2 + \epsilon_2) + \alpha^2 - 2\alpha(\alpha + \epsilon_1) = \epsilon_2 - 2\alpha\epsilon_1 =: K,$$

i.e. $1 > K \geq \Delta$. Here we find λ_r over conv(BCA). We would like that any

$$L = 1 - \frac{[y_2 + \alpha^2 - 2\alpha y_1]}{r^2} \geq 1 - r$$

iff $\Delta = [y_2 + \alpha^2 - 2\alpha y_1] \leq r^3$, for all $(y_1, y_2) \in (\text{conv}(BCA) \cap \mathcal{R} \cap V^0)$ iff $K \leq r^3$ iff $K^{1/3} \leq r$. Thus $D = \min r = K^{1/3}$.
Conclusion. $D = (\epsilon_2 - 2\alpha\epsilon_1)^{1/3}$ is true for $\alpha \leq -1$ with any $0 < \epsilon_j < 1, j = 1, 2$; and true for $-1 < \alpha < 0$ when ϵ_1, ϵ_2 are sufficiently small. The last finishes the proof of Theorem 6.1.1. □

6.2 The Trigonometric Prokhorov Radius

6.2.1. Introduction

Here we consider probability measures μ on $[0, 2\pi]$.
We consider

$$M(\epsilon) := \left\{ \begin{array}{ll} \mu : & |\int cos\, t d\mu - cos\alpha| \leq \epsilon, \\ & |\int sin\, t\, d\mu - sin\alpha| \leq \epsilon \end{array} \right\}, \tag{6.2.1}$$

where $\alpha \in [1, 2\pi - 1]$ be given and $0 < \epsilon < \frac{1}{\sqrt{2}}(1 - cos1)$.

We would like to measure the “size” of $M(\epsilon)$ to be given by a simple equation involving only ϵ, α.

Since weak convergence of probability measures is of central importance and their standard weak topology is well described by the Prokhorov distance p it is natural to define the Trigonometric Prokhorov radius for $M(\epsilon)$ as

$$D := \sup_{\mu \in M(\epsilon)} p(\mu, \delta_\alpha) \tag{6.2.2}$$

where δ_α is the Dirac measure at α.

Using the geometric moment method of optimal ratio due to Kemperman (1968), (see also Chapter 2) we are able to calculate the exact value of D; see also Anastassiou–Rachev, Memphis State University Report series #1988–27. "How precise is the approximation of a random queue by means of deterministic queueing models".

It will be helpful to state:

Fact 6.2.1 Let μ be a probability measure on $[0, 2\pi]$ and δ_α the Dirac measure at $\alpha, \alpha \in [1, 2\pi - 1]$. Then

$$p(\mu, \delta_\alpha) = \inf\{r > 0 : \mu([\alpha - r, \alpha + r]) \geq 1 - r\} \tag{6.2.3}$$

is the Prokhorov distance of μ from δ_α (see also (6.1.3)). Obviously, we are concerned only for $r \in [0, 1]$.

Remark 6.2.1. One can restate that

$$D = \sup_{\mu \in M(\epsilon)} (inf\{r > 0 : \mu([\alpha - r, \alpha + r]) \geq 1 - r\}) \tag{6.2.4}$$

Thus

$$\begin{aligned} r \geq D \text{ iff } \mu([\alpha - r, \alpha + r]) \geq 1 - r, \text{ all } \mu \in M(\epsilon) \\ \text{iff } \lambda := \inf_{\mu \in M(\epsilon)} \mu([\alpha - r, \alpha + r]) \geq 1 - r. \end{aligned} \tag{6.2.5}$$

Obviously $0 < D \leq 1$, therefore we are interested only for $r \in (0, 1]$.

One can easily see that

$$D = min\{r \in (0, 1] | \lambda := \inf_{\mu \in M(\epsilon)} \mu([\alpha - r, \alpha + r]) \geq 1 - r\}. \tag{6.2.6}$$

6.2.2. Results

The main result here has as follows:

Theorem 6.2.1. *Let $\alpha \in [1, 2\pi - 1]$ be fixed and $0 < \epsilon < \frac{1}{\sqrt{2}}(1 - cos1)$. Then $D = r_0$ is the unique solution of*

$$r_0 - r_0 cos\, r_0 = \epsilon(|cos\, \alpha| + |sin\, \alpha|). \tag{6.2.7}$$

Here $0 < r_0 < 1$.

Comment 6.2.1.

From (6.2.7) we have:

Let $\epsilon \longrightarrow 0$, then $r_0 - r_0 \cdot cos\, r_0 \longrightarrow 0$, that is $D \longrightarrow 0$. The last implies that $p(\mu, \delta_\alpha) \longrightarrow 0$, which in turn gives us that $\mu \overset{\rightarrow}{\longrightarrow} \delta_\alpha$ weakly.

In fact D converges to zero quantitavely through (6.2.7). The knowledge of D gives the rate of weak convergence of μ to δ_α.

Comment 6.2.2. Our basic tools for the proof of Theorem 6.2 will come from Kemperman (1968), namely see the pages 93, 94, 95, 96, 101, 102, 103, 104, 107, 111, 116; see also Chapter 2.

Comment 6.2.3

Using that cos $r_0 \approx (1 - \frac{r_0^2}{2})$ into (6.2.7) we get that

$$D \approx (2\epsilon(|cos\,\alpha| + |sin\,\alpha|))^{1/3}, \alpha \in [1, 2\pi - 1], \quad 0 < \epsilon < \frac{1}{\sqrt{2}}(1 - cos\,1).$$

Proof of Theorem 6.2.1. It is going to be carried out by several parts and remarks:

Remark 6.2.2

From Comment 6.2.2, we have the following scheme for our related moment problem

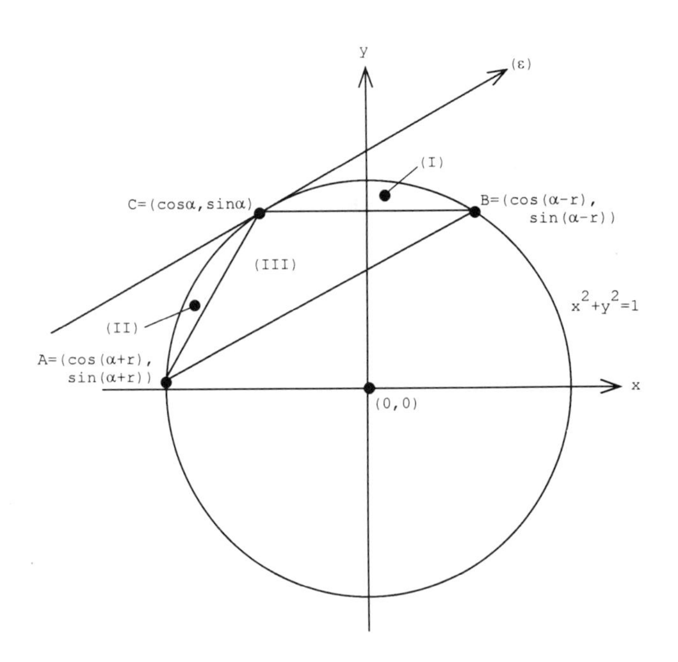

Figure 6.2.1: Moments in unit circle

Here $0 \le t \le 2\pi, x =$cos $t, y = sint, \alpha \in [1, 2\pi - 1], 0 < r \le 1$ i.e. $0 <$ cos $r < 1$. Since $\overset{\frown}{BC} = \overset{\frown}{CA}$, that implies $\overline{BC} = \overline{CA}$. The tangent (ϵ) at C is parallel to (AB) with slope $\frac{dy}{dx} = \frac{dy/dt}{dx/dt} = \frac{cos\,t}{-sin\,t}|_{t=\alpha} = -\frac{cos\,\alpha}{sin\,\alpha}(sin\,\alpha \neq 0)$. Thus slope $(AB) = -\frac{cos\,\alpha}{sin\alpha}(sin\,\alpha \neq 0)$.

The line (AB) has equation

$$\frac{cos\,\alpha}{sin\,\alpha}x + 1y - \frac{cos\,r}{sin\,\alpha} = 0, (sin\,\alpha \neq 0).$$

Call $A^* := \frac{cos\,\alpha}{sin\,\alpha}, B^* := 1, C^* := -\frac{cos\,r}{sin\,\alpha}$. We have that the distance

$$d((x_0, y_0), (AB)) = \frac{|A^* x_0 + B^* y_0 + C^*|}{\sqrt{A^{*2} + B^{*2}}}.$$

The quantity $L := L_{(y_1, y_2)} := \inf_{\mu}([\alpha - r, \alpha + r])$, μ such that

$$\int_{[0,2\pi]} cos\,t d\mu = y_1, \int_{[0,2\pi]} sin\,t d\mu = y_2 \tag{6.2.8}$$

where $(y_1, y_2) \in conv(A \overset{\triangle}{B} C) - \{A, B, C\}$ is fixed, is given by (see Kemperman (1968), pp. 111, 116 and Chapter 2)

$$L = \frac{d((y_1, y_2), (AB))}{d(C, (AB))} = \frac{|A^* y_1 + B^* y_2 + C^*|}{|A^* cos\,\alpha + B^* sin\,\alpha + C^*|}.$$

We would like that $L \geq 1 - r$. Hence, any such L should fulfill

$$L = \frac{|(cos\,\alpha)y_1 + (sin\,\alpha)y_2 - cos\,r|}{|1 - cos\,r|} \geq 1 - r, (sin\,\alpha \neq 0). \tag{6.2.9}$$

Here notice that $1 - cos\,r > 0$. Also $\overset{\frown}{BA} = 2r \leq 2 < \pi$, i.e. $(0,0)$ is out of the conv $(A \overset{\triangle}{C} B)$ and here we deal with $(y_1, y_2) \in conv(A \overset{\triangle}{C} B) - \{A, B, C\}$. Since $\cos r > 0$ we get $(cos\,\alpha)y_1 + (sin\,\alpha)y_2 - cos\,r \geq 0$, all $(y_1, y_2) \in conv(A \overset{\triangle}{C} B) - \{A, B, C\}, sin\,\alpha \neq 0$.

Equivalently, from (6.2.9) we have

$$\begin{bmatrix} (cos\,\alpha)y_1 + (sin\,\alpha)y_2 \geq r cos\,r - r + 1, \\ \text{all}\,(y_1, y_2) \in conv(A \overset{\triangle}{C} B) - \{A, B, C\}, sin\,\alpha \neq 0. \end{bmatrix} \tag{6.2.10}$$

Notice that $r cos\,r - r + 1$ is strictly decreasing in $r, 0 < r \leq 1$. More specifically, here $(y_1, y_2) \in [conv(A \overset{\triangle}{C} B) - \{A, B, C\}] \cap \mathcal{R}$, where $\mathcal{R}$ is the square

$$\mathcal{R} := \{(y_1, y_2) : |y_1 - cos\,\alpha| \leq \epsilon, |y_2 - sin\,\alpha| \leq \epsilon\}.$$

We would like to find $\Pi := min\,r$ such that (6.2.10) is true for all $(y_1, y_2) \in [conv(A \overset{\triangle}{C} B) - \{A, B, C\}] \cap \mathcal{R}$.

For optimal such y_1^*, y_2^* we get this

$$\Pi := min\,r := r_0$$

as the unique solution of

$$[(\cos\alpha)y_1^* + (\sin\alpha)y_2^* = r_0 \cdot \cos r_0 - r_0 + 1, \text{ where } \sin\alpha \neq 0, 0 < r_0 \leq 1] \quad (6.2.11)$$

The above is found in several different cases. Similarly is met the case of $\sin\alpha = 0$.

Remark 6.2.3.

Points located on the open line segment (AB), or on the side of (AB) where $(0,0)$ is located, play no role towards the calculation of L.

In particular the square

$$\mathcal{R} := \{(y_1, y_2) : |y_1 - \cos\alpha| \leq \epsilon, |y_2 - \sin\alpha| \leq \epsilon\}$$

is not located in the side of (AB) where the origin is located and does not intersect (AB).

Here we assume $\epsilon < \frac{1}{\sqrt{2}} \cdot (1 - \cos 1), \alpha \in [1, 2\pi - 1]$, $0 < r \leq 1$.

Proof: Let $(y_1, y_2) \in \mathcal{R} \cap$ unit disk be in the side of (AB) where $(0,0)$ is located or on the line (AB), within the disk, $\neq \{A, B\}$. Then

$$(\cos\alpha)y_1 + (\sin\alpha)y_2 \leq \cos r, (\cos r > 0).$$

Furthermore we have

$$\cos\alpha - \epsilon \leq y_1 \leq \cos\alpha + \epsilon$$

and

$$\sin\alpha - \epsilon \leq y_2 \leq \sin\alpha + \epsilon.$$

Case (1) *Assume that $\sin\alpha \geq 0$, $\cos\alpha \geq 0$.*

Then $(\cos\alpha - \epsilon)\cos\alpha \leq (\cos\alpha)y_1$ and $(\sin\alpha - \epsilon)\sin\alpha \leq (\sin\alpha)y_2$. Adding the last two inequalities we get that $\cos^2\alpha + \sin^2\alpha - \epsilon(\cos\alpha + \sin\alpha) \leq (\cos\alpha)y_1 + (\sin\alpha)y_2 \leq \cos r$. That is

$$1 - \epsilon(\cos\alpha + \sin\alpha) \leq \cos r.$$

In that case from optimal ratio method we find $L_{(y_1,y_2)} = 0$, implying $\lambda = 0$ i.e. $0 \geq 1 - r$, thus $r = 1$. Hence $1 - \cos 1 \leq \epsilon(\cos\alpha + \sin\alpha) \leq \sqrt{2} \cdot \epsilon$, giving us $\epsilon \geq \frac{1}{\sqrt{2}} \cdot (1 - \cos 1)$, therefore a contradiction by the assumption

$$\epsilon < \frac{1}{\sqrt{2}}(1 - \cos 1).$$

Therefore the square $\mathcal{R}$ is above the line (AB).

Case (2) *Assume that $\sin\alpha \geq 0, \cos\alpha \leq 0$.*

Here we have

$$(\sin\alpha - \epsilon)\sin\alpha \leq (\sin\alpha)y_2$$

and

$$(\cos\alpha + \epsilon)\cos\alpha \leq (\cos\alpha)y_1.$$

Adding these inequalities we find

$$\sin^2\alpha + \cos^2\alpha + \epsilon(\cos\alpha - \sin\alpha) \leq (\cos\alpha)y_1 + (\sin\alpha)y_2 \leq \cos r.$$

Hence

$$1 - \epsilon(\sin\alpha - \cos\alpha) \leq \cos r,$$

and similarly as before $L = 0$ and $\lambda = 0 \geq 1 - r$, thus $r = 1$. Therefore $1 - \cos 1 \leq \epsilon(\sin\alpha - \cos\alpha) \leq \sqrt{2}\cdot\epsilon$, getting again a contradiction.

Therefore the square $\mathcal{R}$ is to the left of (AB) and neither in the side where the origin is located nor does intersect (AB).

Case (3) *Assume that* $\sin\alpha \leq 0, \cos\alpha \leq 0$.

We have

$$(\cos\alpha + \epsilon)\cos\alpha \leq (\cos\alpha)y_1$$

and

$$(\sin\alpha + \epsilon)\sin\alpha \leq (\sin\alpha)y_2.$$

Adding the last two we get

$$1 + \epsilon(\cos\alpha + \sin\alpha) \leq (\cos\alpha)y_1 + (\sin\alpha)y_2 \leq \cos r.$$

Again, as before, we get

$$1 - \cos 1 \leq \epsilon((-\cos\alpha) + (-\sin\alpha)) \leq \sqrt{2}\cdot\epsilon$$

i.e. $\epsilon \geq \frac{1}{\sqrt{2}}(1 - \cos 1)$, a contradiction by $\epsilon < \frac{1}{\sqrt{2}}(1 - \cos 1)$.

Therefore the square $\mathcal{R}$ is not in the side of (AB) where $(0,0)$ is located and does not intersect (AB).

Case (4) Assume that $\sin\alpha \leq 0$, $\cos\alpha \geq 0$.

We have

$$(\cos\alpha - \epsilon)\cos\alpha \leq (\cos\alpha)y_1$$

and

$$(\sin\alpha + \epsilon)\sin\alpha \leq (\sin\alpha)y_2.$$

Adding the last two inequalities we get $1 + \epsilon(\sin\alpha - \cos\alpha) \leq (\cos\alpha)y_1 + (\sin\alpha)y_2 \leq \cos r$. Then $1 - \cos r \leq \epsilon(\cos\alpha - \sin\alpha) \leq \sqrt{2}$. Again $L = 0$ and $\lambda = 0 \geq 1 - r$, thus $r = 1$, that is

$$1 - \cos 1 \leq \sqrt{2}\cdot\epsilon \quad \text{and} \quad \epsilon \geq \frac{1}{\sqrt{2}}(1 - \cos 1),$$

which is a contradiction by $\epsilon < \frac{1}{\sqrt{2}}(1 - \cos 1)$.

Therefore the square $\mathcal{R}$ is not in the side of (AB) where $(0,0)$ is located and does not intersect (AB). □

Remark 6.2.4

The boundary points of the unit disk play no role towards the calculation of D.

Proof: We start with some observations.

(i) When $\alpha = 1, r = 1$:

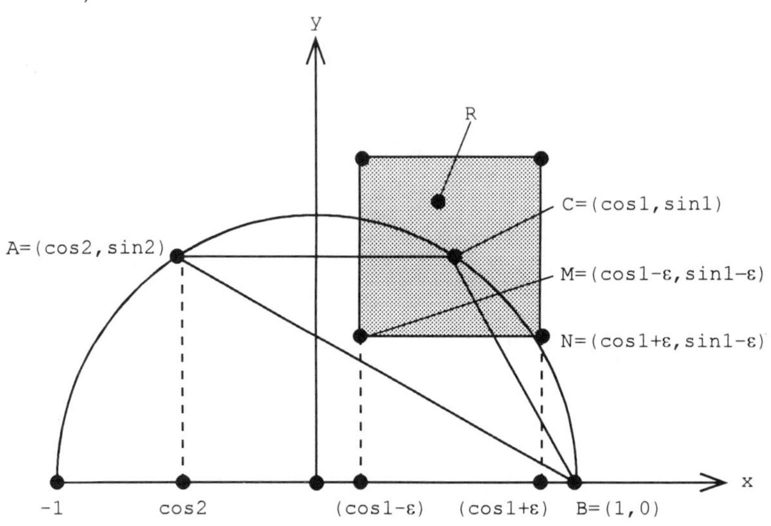

Figure 6.2.2: $\alpha = 1, r = 1$

By assumption $\epsilon < 1/\sqrt{2}(1 - cos\,1) < (1 - cos\,1)$ i.e. $\epsilon < 1 - cos\,1$ and $\epsilon + cos\,1 < 1$, that is the point $(1,0)$ does not belong to the square $\mathcal{R}$.

(ii) When $\alpha = 2\pi - 1, r = 1$.

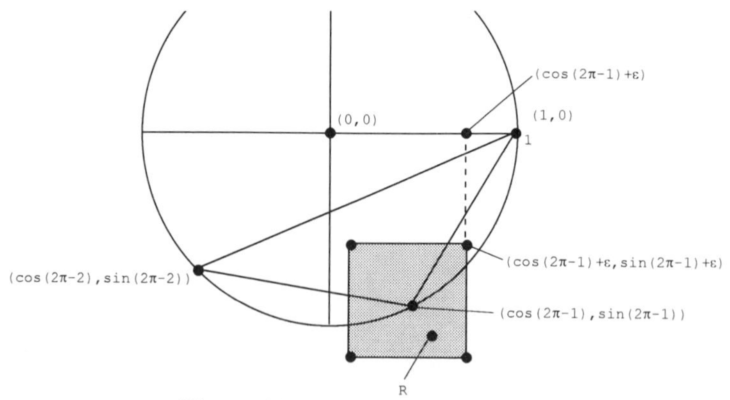

Figure 6.2.3: $\alpha = 2\pi - 1, r = 1$

We would like that $\cos(2\pi - 1) + \epsilon = \cos 1 + \epsilon < 1$, equivalently $\epsilon < 1 - \cos 1$, which is true by the assumption that $\epsilon < \frac{1}{\sqrt{2}}(1 - \cos 1)$. Therefore the point $(1,0)$ does not belong to the square $\mathcal{R}$.

Conclusion

The point $(1,0)$ cannot belong to any such square $\mathcal{R}$; $\alpha \in [1, 2\pi - 1]$, $\epsilon < \frac{1}{\sqrt{2}}(1 - \cos 1)$.

Let now $g(t) := (\cos t, \sin t)$, $0 \le t \le 2\pi$. Call $V := conv\, g([0, 2\pi])$ which is the unit disk. We use the terminology and the results of Kemperman (1968), pp. 102, 103, 104; §4 of non-interior points, see also Chapter 2. Let y be a boundary point of V, then $V_y = \{y\}$. And

$$\Gamma^y := g([0, 2\pi]) \cap V_y = \{y\}, \text{ i.e. } \Gamma^y = \{y\}.$$

Also $T^y := g^{-1}(\{y\}) = t_y$, by g being an one-to-one and onto mapping for all $t \neq \{0, 2\pi\}$, where $y \neq (1,0)$. That is $T^y = \{t_y\}$, all $y \neq (1,0)$. From Theorem 9 of Kemperman (1968), p. 103, see also Chapter 2, we get: Let $y \neq (1,0)$ be a fixed boundary point of V and let μ be a probability measure on $[0, 2\pi]$ with

$$\int \cos t\, d\mu = y_1, \quad \int \sin t\, d\mu = y_2; \quad y := (y_1, y_2).$$

Then μ is concentrated at $\{t_y\}$. So if $(1,0) \neq y \in g([\alpha - r, \alpha + r])$, where $\alpha \in [1, 2\pi - 1]$, any $0 < r \le 1$, that is a boundary point $\neq (1,0)$, then $L := L_y = 1$.

Conclusion. The points $y := (y_1, y_2) \in (\text{conv}(A \overset{\Delta}{C} B) - \{A, B, C\})$ for any $0 < r \le 1$, give always $L < 1$ (any non-boundary point $y \in \mathcal{R} \cap$ unit disk, not in the side of (AB) where $(0,0)$ is located, give $L < 1$, for any $0 < r \le 1$), see Chapter 2, furthermore we want

$$\lambda := \inf_{\mu \in M(\varepsilon)} \mu([\alpha - r, \alpha + r]), \quad \textit{for any admissible } r \in (0, 1],$$

where

$$M(\varepsilon) := \{\mu : |y_1 - \cos\alpha| \le \varepsilon, |y_2 - \sin\alpha| \le \varepsilon\},$$

with

$$y_1 := \int_{[0,2\pi]} \cos t\, d\mu, \quad y_2 := \int_{[0,2\pi]} \sin t\, d\mu.$$

Obviously $\lambda < 1 = L$ (for a boundary point $\neq (1,0)$). Therefore λ is achieved by non-boundary points of unit disk $\cap \mathcal{R}$. *Therefore the boundary points of the unit disk play no role towards the calculation of D.*

Remark 6.2.5.

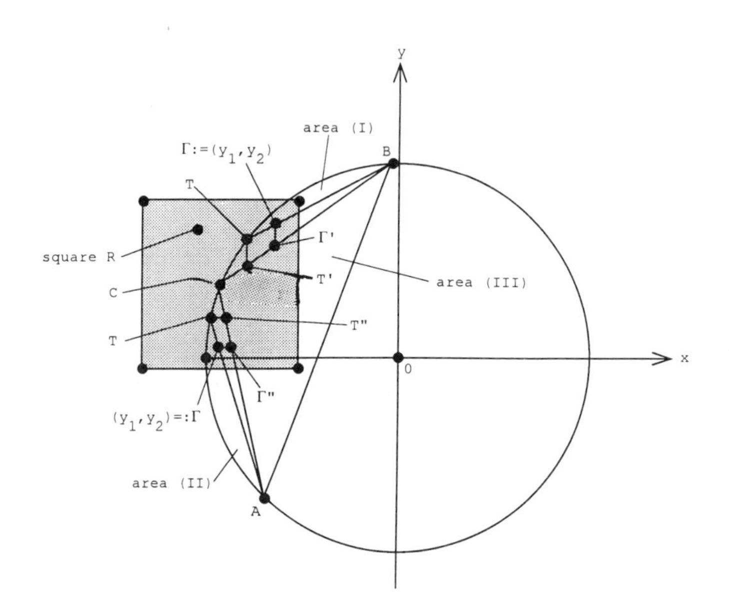

Figure 6.2.4. $\Gamma' := (y_1, y_2')$, $\Gamma'' := (y_1', y_2)$

We will prove that: *for every point $\Gamma := (y_1, y_2)$ in area$(I) \cap \mathcal{R}$ or area$(II) \cap \mathcal{R}$, there exists a point Γ' or Γ'' in area$(III) \cap \mathcal{R}$ producing a smaller L. So all it matters for the calculation of D is area$(III) \cap \mathcal{R}$.* One can prove that T', T'', Γ', Γ'' belong to the unit disk.

PROOF: The line segments $(\overline{AC})$, $(\overline{BC})$ belong to the unit disk, $\alpha \in [1, 2\pi - 1]$. The lines (AC), (BC) are given by $(y - \sin\alpha) = m(x - \cos\alpha)$, where m is the slope of (AC) or (BC). Let $\Gamma := (y_1, y_2) \in \mathcal{R} \cap$ unit disk, located under $\widehat{AC}$ – above $\overline{AC}$ or under $\widehat{BC}$ – above $\overline{BC}$. If $m = 0$, take $\Gamma' = (y_1, \sin\alpha)$, it is on (AC) or (BC) and belongs to $\mathcal{R} \cap$

unit disk. If $m = \infty$, take $\Gamma'' = (\cos\alpha, y_2)$, it is on (AC) or (BC) and belongs to $\mathcal{R}\cap$ unit disk. Let us now assume that $m \neq \{0,\infty\}$: If $|m| \leq 1$, consider $\Gamma' := (y_1, y_2')$ such that $y_2' - \sin\alpha = m(y_1 - \cos\alpha)$, thus $|y_2' - \sin\alpha| \leq \varepsilon$. Therefore $\Gamma' = (y_1, y_2') \in \mathcal{R}\cap$ unit disk and it is on (AC) or (BC).

If $|m| > 1$, consider $\Gamma'' := (y_1', y_2)$ such that $y_2 - \sin\alpha = m(y_1' - \cos\alpha)$, thus

$$|y_1' - \cos\alpha| = \frac{|y_2 - \sin\alpha|}{|m|} < \varepsilon.$$

Therefore $\Gamma'' \in \mathcal{R}\cap$ unit disk and it is on (AC) or (BC). Here $\Gamma\Gamma'//TT'$ and $\Gamma\Gamma''//TT''$. Also $\Gamma\Gamma'//y$-axis and $\Gamma\Gamma''//x$-axis. In general we deal with the following scheme:

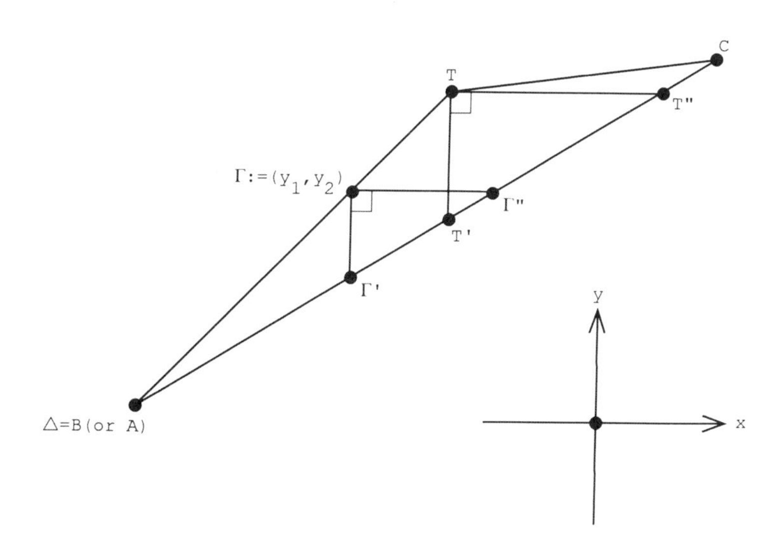

Figure 6.2.5. Optimal triangle directions.

Notice that

$$L_\Gamma = \frac{\Delta\Gamma}{\Delta T} = \frac{\Delta\Gamma'}{\Delta T'} > \frac{\Delta\Gamma'}{\Delta C} = L_{\Gamma'},$$

and

$$L_\Gamma = \frac{\Delta\Gamma}{\Delta T} = \frac{\Delta\Gamma''}{\Delta T''} > \frac{\Delta\Gamma''}{\Delta C} = L_{\Gamma''}.$$

Hence for each point Γ of areas [(I) or (II)] $\cap\ \mathcal{R}$ there exists a point $\Gamma'^{('')}$ of area$(III) \cap \mathcal{R}$ producing a smaller L. *Since we are interested for smaller L's, (from the above) the areas*

$(I) \cap \mathcal{R}$, $(II) \cap \mathcal{R}$ play no role towards the calculation of D. Therefore all depends on area$(III) \cap \mathcal{R}$.

Remark 6.2.6. From Remarks 6.2.3, 6.2.4, and 6.2.5 we get

$$D = \Pi = \min r.$$

Remark 6.2.7. Case of $\alpha = \pi/2$, i.e., $C = (0,1)$.

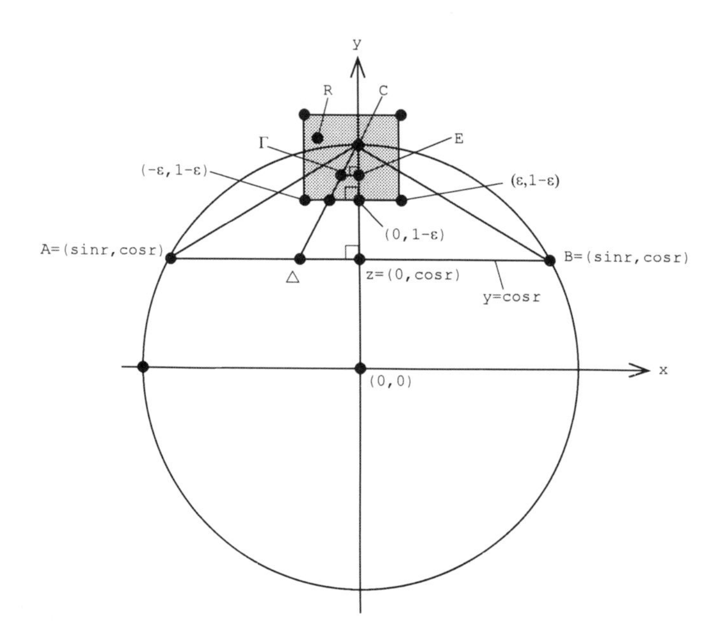

Figure 6.2.6. $\alpha = \pi/2$, $E := (0, y_2)$, $\Gamma := (y_1, y_2)$.

Here $|y_1| \leq \varepsilon$, $|y_2 - 1| \leq \varepsilon$, i.e., $-\varepsilon \leq y_1 \leq \varepsilon$, $1 - \varepsilon \leq y_2 \leq 1 + \varepsilon$. From Remark 6.2.3 we have $1 - \varepsilon > \cos r$, that is $(0, 1-\varepsilon)$ is above $y = \cos r$.

Let $\Gamma \in \mathcal{R} \cap$ unit disk, then $E \in \mathcal{R} \cap$ unit disk. We have

$$L_\Gamma = \frac{\Delta\Gamma}{\Delta C} = \frac{ZE}{ZC} = L_E,$$

so all it matters from $\operatorname{conv}(A \overset{\Delta}{B} C) - \{A, B, C\}$ is the line segment $\overline{CZ} - \{C\}$. Here we would like to have

$$L = L_E = \frac{y_2 - \cos r}{1 - \cos r} \geq 1 - r \Leftrightarrow y_2 \geq r\cos r - r + 1.$$

But $y_2 \geq 1-\varepsilon$ and $(0,1-\varepsilon) \in \mathcal{R} \cap$ unit disk. Thus, equivalently

$$1-\varepsilon \geq r\cos r - r + 1.$$

Hence, optimality happens when

$$r_0 - r_0 \cos r_0 = \varepsilon.$$

Therefore $D = r_0$, $0 < r_0 < 1$ ($\varepsilon < \frac{1}{\sqrt{2}}(1 - cos1)$).

Remark 6.2.8. Case of $\alpha = \pi$, i.e., $C = (-1,0)$.

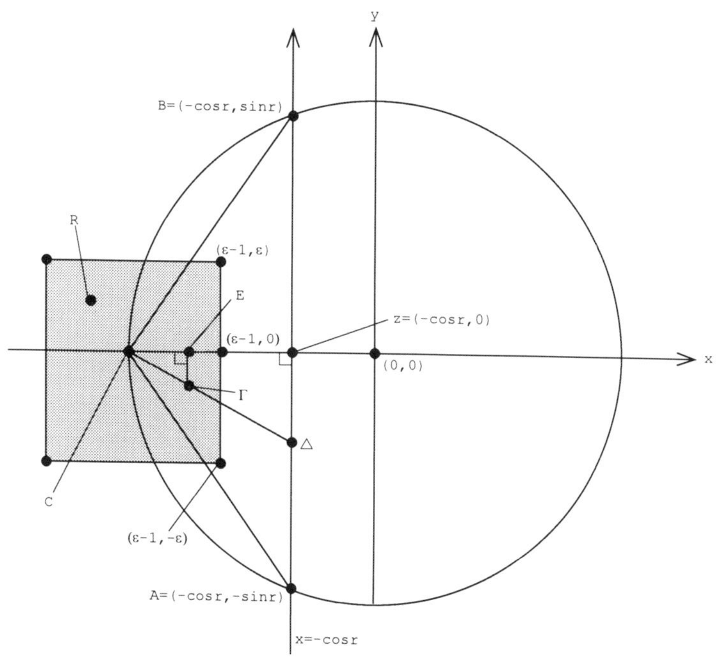

Figure 6.2.7. $\alpha = \pi$, $E := (y_1, 0)$, $\Gamma := (y_1, y_2)$.

Here $|y_1 + 1| \leq \varepsilon$, $|y_2| \leq \varepsilon$, i.e., $-\varepsilon - 1 \leq y_1 \leq \varepsilon - 1$, $-\varepsilon \leq y_2 \leq \varepsilon$. From Remark 6.2.3 we have $\varepsilon - 1 < -\cos r$, that is, $(\varepsilon - 1, 0)$ is to the left of $x = -\cos r$.

Let $\Gamma \in \mathcal{R} \cap$ unit disk, then $E \in \mathcal{R} \cap$ unit disk. We have

$$L_\Gamma = \frac{\Delta\Gamma}{\Delta C} = \frac{ZE}{ZC} = L_E,$$

so all it matters from conv$(A \overset{\Delta}{B} C) - \{A, B, C\}$ is the line segment $\overline{CZ} - \{C\}$. Here we would like to have

$$L = L_E = \frac{-\cos r - y_1}{1 - \cos r} \geq 1 - r$$

$$\Leftrightarrow$$

$$-y_1 \geq 1 - r + r\cos r$$

$$\Leftrightarrow$$

$$y_1 \leq r - r\cos r - 1.$$

But $y_1 \leq \varepsilon - 1$ and $(\varepsilon - 1, 0) \in \mathcal{R} \cap$ unit disk.

Thus, equivalently

$$\varepsilon - 1 \leq r - r\cos r - 1.$$

Hence, optimality happens when

$$r_0 - r_0 \cos r_0 = \varepsilon.$$

Therefore $D = r_0$, $0 < r_0 < 1$

$$\left(\varepsilon < \frac{1}{\sqrt{2}} \cdot (1 - \cos 1\right).$$

Remark 6.2.9. Case of $\alpha = 3\pi/2$, i.e., $C = (0, -1)$.

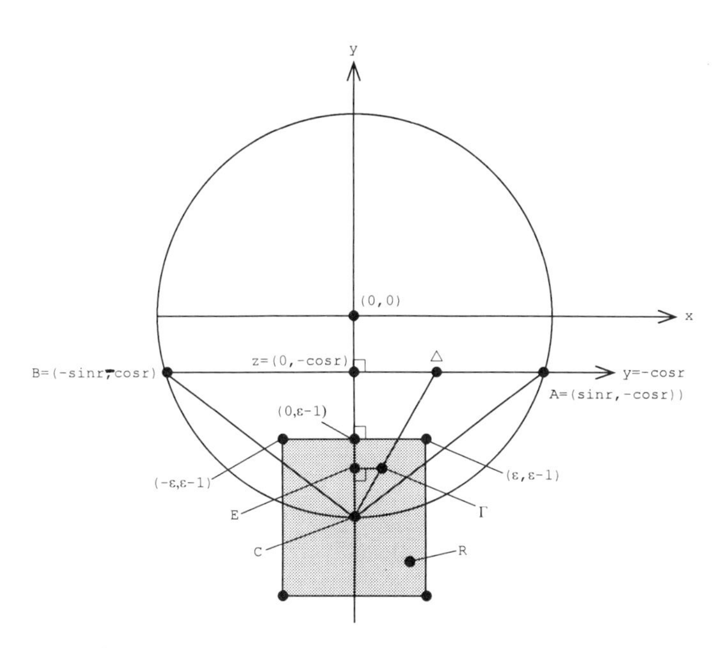

Figure 6.2.8. $\alpha = 3\pi/2$, $E := (0, y_2)$, $\Gamma := (y_1, y_2)$.

Here $|y_1| \leq \varepsilon$, $|y_2 + 1| \leq \varepsilon$, i.e., $-\varepsilon \leq y_1 \leq \varepsilon$, $-1 - \varepsilon \leq y_2 \leq -1 + \varepsilon$. From Remark 6.2.3 we have $(\varepsilon - 1) < -\cos r$, that is $(0, \varepsilon - 1)$ is below $y = -\cos r$.

Let $\Gamma \in \mathcal{R} \cap$ unit disk, then $E \in \mathcal{R} \cap$ unit disk. We have

$$L_\Gamma = \frac{\Delta\Gamma}{\Delta C} = \frac{ZE}{ZC} = L_E,$$

so all it matters from $\mathrm{conv}(A \overset{\Delta}{B} C) - \{A, B, C\}$ is the line segment $\overline{CZ} - \{C\}$. Here we would like to have

$$L = L_E = \frac{-\cos r - y_2}{1 - \cos r} \geq 1 - r$$

$$\Leftrightarrow$$

$$-y_2 \geq 1 - r + r\cos r$$

$$\Leftrightarrow$$

$$y_2 \leq r - r\cos r - 1.$$

But $y_2 \leq \varepsilon - 1$ and $(0, \varepsilon - 1) \in \mathcal{R} \cap$ unit disk.

Thus, equivalently

$$\varepsilon - 1 \leq r - r\cos r - 1.$$

Hence, optimality happens when

$$r_0 - r_0 \cos r_0 = \varepsilon.$$

Therefore $D = r_0$, $0 < r_0 < 1$

$$\left(\varepsilon < \frac{1}{\sqrt{2}} \cdot (1 - \cos 1)\right).$$

Remark 6.2.10. Case of $\alpha \in [1, \pi/2)$, i.e., $\sin\alpha > 0$, $\cos\alpha > 0$.

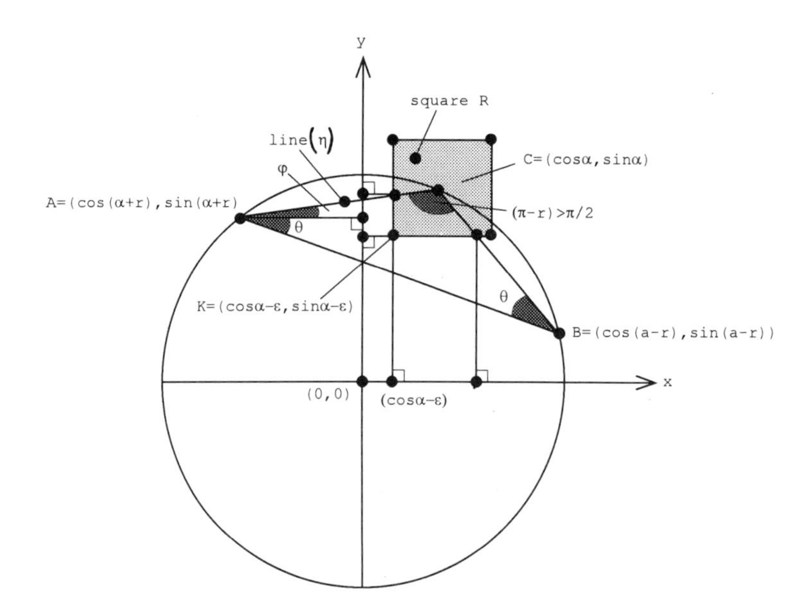

Figure 6.2.9. $\alpha \in [1, \pi/2)$.

We always have $\theta < \pi/4$ and $\pi - r > \pi/2$. From Remark 6.2.3, the square $\mathcal{R}$ is above the line (AB), that is, the point $K = (\cos\alpha - \varepsilon, \sin\alpha - \varepsilon)$ is above line (AB). Furthermore for any $(y_1, y_2) \in \mathcal{R} \cap$ unit disk we have

$$(\cos\alpha)y_1 + (\sin\alpha)y_2 > \cos r,$$

any $0 < r \leq 1$. Pick as optimal $y_1^* = \cos\alpha - \varepsilon$, $y_2^* = \sin\alpha - \varepsilon$.

The line $(AC) := \text{line}(\eta)$ has equation:

$$y - \sin\alpha = \left(\frac{\sin(\alpha + r) - \sin\alpha}{\cos(\alpha + r) - \cos\alpha}\right) \cdot (x - \cos\alpha).$$

At $x = \cos\alpha - \varepsilon$ we get

$$y = \sin\alpha - \left(\frac{\sin(\alpha + r) - \sin\alpha}{\cos(\alpha + r) - \cos\alpha}\right) \cdot \varepsilon.$$

We would like that K be below AC. That is, we need

$$\sin\alpha - \varepsilon < \sin\alpha - \left(\frac{\sin(\alpha + r) - \sin\alpha}{\cos(\alpha + r) - \cos\alpha}\right) \cdot \varepsilon$$

$$\Leftrightarrow$$

$$1 > \left(\frac{\sin(\alpha + r) - \sin\alpha}{\cos(\alpha + r) - \cos\alpha}\right) = \text{slope}(\eta) = \tan\varphi.$$

(If $\sin\alpha = \sin(\alpha + r)$, then $y = \sin\alpha > \sin\alpha - \varepsilon$ and K is below (AC).)

Call $\mathcal{N} := \sin(\alpha + r) - \sin\alpha$ and $\mathcal{P} := \cos(\alpha + r) - \cos\alpha$. Since $\alpha \in [1, \pi/2)$ we have that $\mathcal{P} < 0$. If $\text{slope}(\eta) < 0$, then $\tan\varphi < 0$, i.e., $\tan\varphi < 1$. If $\text{slope}(\eta) \geq 0$ it means that $\mathcal{N} \leq 0$, that is, $\sin(\alpha + r) \leq \sin\alpha$, i.e., $\alpha + r \in (\pi/2, \pi)$. Since $\text{slope}(AB) < 0$ we have $0 \leq \varphi < \theta < \pi/4$, thus $\tan\varphi < 1$. Hence K is below AC. (Note that $\theta = r/2$.)

The line (BC) has equation:

$$y - \sin\alpha = \left(\frac{\sin\alpha - \sin(\alpha - r)}{\cos\alpha - \cos(\alpha - r)}\right) \cdot (x - \cos\alpha),$$

here it is always true that $\cos\alpha \neq \cos(\alpha - r)$. At $y = \sin\alpha - \varepsilon$ we get

$$x = \cos\alpha - \varepsilon \cdot \left(\frac{\cos\alpha - \cos(\alpha - r)}{\sin\alpha - \sin(\alpha - r)}\right).$$

We would like that K be to the left of (BC). That is, we need

$$\cos\alpha - \varepsilon < \cos\alpha - \varepsilon \cdot \left(\frac{\cos\alpha - \cos(\alpha - r)}{\sin\alpha - \sin(\alpha - r)}\right)$$

$$\Leftrightarrow$$

$$1 > \left(\frac{\cos\alpha - \cos(\alpha - r)}{\sin\alpha - \sin(\alpha - r)}\right) =: \cot\tilde{\varphi}, \quad \alpha \in [1, \pi/2).$$

The last is true because $\cot\tilde{\varphi} < 0$. Therefore K is to the left of (BC).

Conclusion. The moment point

$$K = (\cos\alpha - \varepsilon, \sin\alpha - \varepsilon) \in \operatorname{conv}(A \overset{\Delta}{C} B) - \{A, B, C\},$$

in fact, K belongs to the interior of $\operatorname{conv}(A \overset{\Delta}{C} B)$.

Finally: for all $(y_1, y_2) \in \mathcal{R} \cap$ unit disk we have $y_1 \geq \cos\alpha - \varepsilon$, $y_2 \geq \sin\alpha - \varepsilon$; $\cos\alpha > 0$, $\sin\alpha > 0$. Therefore

$$(\cos\alpha)y_1 + (\sin\alpha)y_2 \geq (\cos\alpha)(\cos\alpha - \varepsilon) + (\sin\alpha)(\sin\alpha - \varepsilon) = 1 - \varepsilon(\cos\alpha + \sin\alpha),$$

for all $(y_1, y_2) \in \mathcal{R} \cap$ unit disk. It should be

$$1 - \varepsilon(\cos\alpha + \sin\alpha) \geq r\cos r - r + 1.$$

Optimality can happen only when

$$r_0 - r_0 \cos r_0 = \varepsilon(\cos\alpha + \sin\alpha).$$

That is $D = r_0$, where r_0 is unique and $0 < r_0 < 1$.

Remark 6.2.11. Case of $\alpha \in (\pi/2, \pi)$, i.e., $\sin\alpha > 0$, $\cos\alpha < 0$.

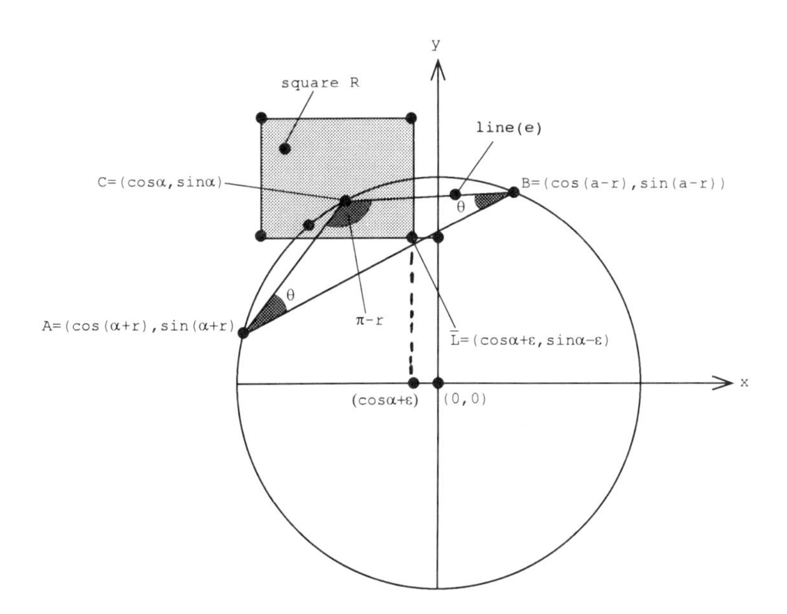

Figure 6.2.10. $\alpha \in (\pi/2, \pi)$.

It is always true that $\pi - r > \pi/2$, $\theta = r/2 < \pi/4$. From Remark 6.2.3 we have that $\mathcal{R}$ is above AB, so that all $(y_1, y_2) \in \mathcal{R} \cap$ unit disk fulfill $(\cos\alpha)y_1 + (\sin\alpha)y_2 > \cos r$, for any $0 < r \leq 1$. Our optimal y_1, y_2 are $y_1^* = \cos\alpha + \varepsilon$, $y_2^* = \sin\alpha - \varepsilon$, that is, the optimal moment point is $\overline{L} = (\cos\alpha + \varepsilon, \sin\alpha - \varepsilon)$.

The line $(BC) :=$ line(e) has equation

$$y - \sin\alpha = \left(\frac{\sin\alpha - \sin(\alpha - r)}{\cos\alpha - \cos(\alpha - r)}\right) \cdot (x - \cos\alpha)$$

and at $x = \cos\alpha + \varepsilon$ we get

$$y = \sin\alpha + \varepsilon\left(\frac{\sin\alpha - \sin(\alpha - r)}{\cos\alpha - \cos(\alpha - r)}\right).$$

We want $\overline{L}$ to be below (BC). That is, we need

$$\sin\alpha - \varepsilon < \sin\alpha + \varepsilon\left(\frac{\sin\alpha - \sin(\alpha - r)}{\cos\alpha - \cos(\alpha - r)}\right)$$

$$\Leftrightarrow$$

$$-1 < \left(\frac{\sin\alpha - \sin(\alpha - r)}{\cos\alpha - \cos(\alpha - r)}\right) =: \tan\varphi.$$

(If $\sin\alpha = \sin(\alpha - r)$, then $y = \sin\alpha > \sin\alpha - \varepsilon$, therefore $\overline{L}$ is below (BC).)

Call $\mathcal{N} := \sin\alpha - \sin(\alpha - r)$ and $\mathcal{P} := \cos\alpha - \cos(\alpha - r)$. Here $\tan\varphi = \text{slope}(e)$, furthermore $\mathcal{P} < 0$ always by $\alpha \in (\pi/2, \pi)$. If $\text{slope}(e) \geq 0$, then $\tan\varphi \geq 0$ and $\tan\varphi > -1$. If $\text{slope}(e) < 0$, it means that $\mathcal{N} > 0$, i.e., $\sin\alpha > \sin(\alpha - r)$, hence $\alpha - r \in [1, \pi/2)$. And $\text{slope}(AB) > 0$. Again $0 < \tilde{\varphi} < \theta < \pi/4$. Here $\text{slope}(e) = \tan(\pi - \tilde{\varphi})$, but $\pi > \varphi := \pi - \tilde{\varphi} > 3\pi/4$. Therefore $\text{slope}(e) = \tan(\pi - \tilde{\varphi}) > -1$, thus $\overline{L}$ is below (BC).

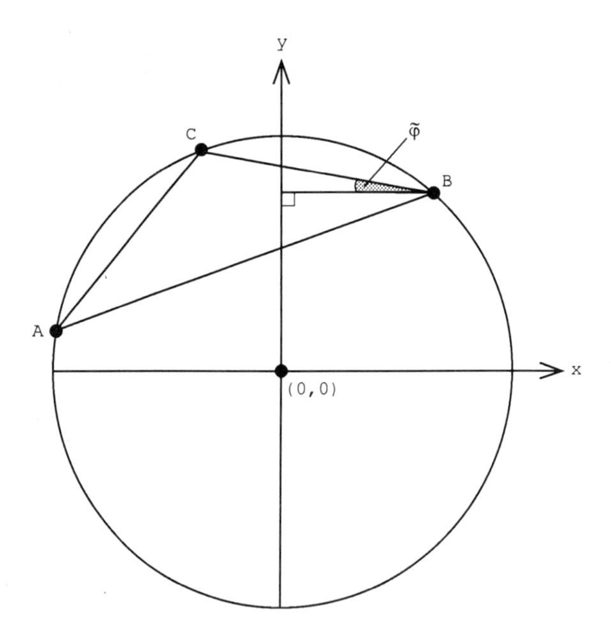

Figure 6.2.11. $\overline{L}$ below (BC).

The line (AC) has equation

$$y - \sin\alpha = \left(\frac{\sin(\alpha + r) - \sin\alpha}{\cos(\alpha + r) - \cos\alpha}\right) \cdot (x - \cos\alpha),$$

assuming that $\cos(\alpha + r) - \cos\alpha \neq 0$. At $y = \sin\alpha - \varepsilon$ we get

$$x = \cos\alpha - \varepsilon \left(\frac{\cos(\alpha + r) - \cos\alpha}{\sin(\alpha + r) - \sin\alpha} \right).$$

We want that $\overline{L}$ be in the side of (AC) where the origin is located, i.e., we want that

$$\cos\alpha + \varepsilon > \cos\alpha - \varepsilon \left(\frac{\cos(\alpha + r) - \cos\alpha}{\sin(\alpha + r) - \sin\alpha} \right)$$

$$\Leftrightarrow$$

$$-1 < \left(\frac{\cos(\alpha + r) - \cos\alpha}{\sin(\alpha + r) - \sin\alpha} \right) =: \cot\varphi.$$

Call $\mathcal{D} := \cos(\alpha + r) - \cos\alpha$ and $\mathcal{N} := \sin(\alpha + r) - \sin\alpha$. Here always we have that slope$(AC) \neq 0$. Also always holds $\mathcal{N} < 0$, by $\alpha \in (\pi/2, \pi)$.

If slope$(AC) > 0$ then $\tan\varphi > 0$ and $\cot\varphi > 0$, hence $\cot\varphi > -1$. If slope$(AC) < 0$ then $D > 0$ and $\cos(\alpha + r) > \cos\alpha$, hence $(\alpha + r) \in (\pi, 3\pi/2)$. Also slope$(AB) > 0$.

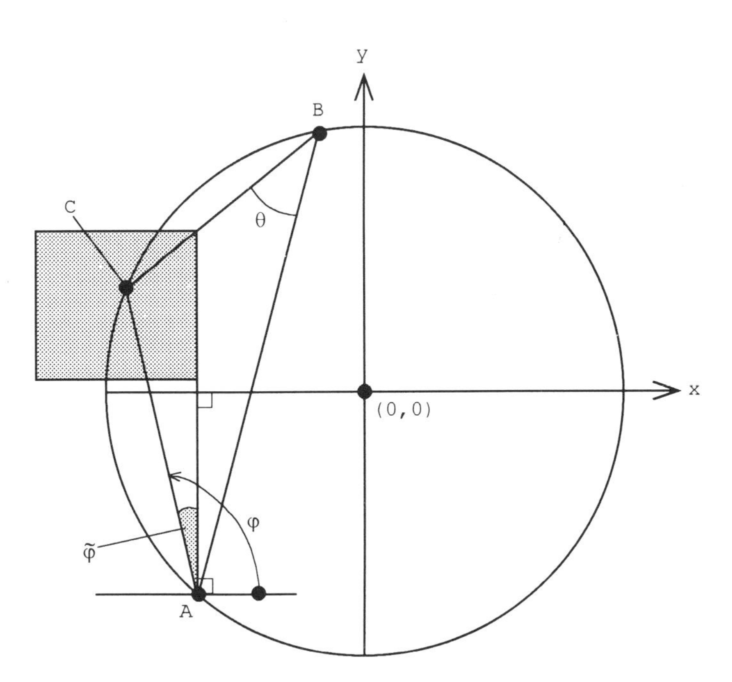

Figure 6.2.12. $\overline{L}$ is right of (AC).

Here

$$\text{slope}(CA) = \tan\varphi = \tan\left(\frac{\pi}{2} + \tilde{\varphi}\right) = \tan\left(\frac{\pi}{2} - (-\tilde{\varphi})\right) = \cot(-\tilde{\varphi}) = -\cot\tilde{\varphi},$$

but $0 < \tilde{\varphi} < \theta < \pi/4$. Thus $0 < \tan\tilde{\varphi} < 1$ and $\cot\tilde{\varphi} > 1$, i.e., $-\cot\tilde{\varphi} < -1$, therefore, $\text{slope}(CA) < -1$ and $\tan\varphi < -1$, i.e., $\cot\varphi > -1$. (If $\cos(\alpha + r) = \cos\alpha$, then line (AC) is given by $x = \cos\alpha$, and $\cos\alpha < \cos\alpha + \varepsilon$, hence $\overline{L}$ is in the side of (AC) where $(0,0)$ is located.)

We have proved that $\overline{L}$ is located in the side of (AC) where the origin is, in all possible cases.

Conclusion. The moment point $\overline{L} = (\cos\alpha + \varepsilon, \sin\alpha - \varepsilon)$ is an interior point of $\text{conv}(A \overset{\Delta}{C} B)$.

Here we have that $\sin\alpha > 0$, $\cos\alpha < 0$ and for all $(y_1, y_2) \in \mathcal{R} \cap$ unit disk holds that $y_1 \le \cos\alpha + \varepsilon$ and $y_2 \ge \sin\alpha - \varepsilon$. Therefore

$$(\cos\alpha)y_1 \ge (\cos\alpha)(\cos\alpha + \varepsilon)$$

and

$$(\sin\alpha)y_2 \ge (\sin\alpha)(\sin\alpha - \varepsilon).$$

Hence

$$(\cos\alpha)y_1 + (\sin\alpha)y_2 \ge 1 - \varepsilon(-\cos\alpha + \sin\alpha),$$

for all $(y_1, y_2) \in \mathcal{R} \cap$ unit disk. It should be

$$1 - \varepsilon(-\cos\alpha + \sin\alpha) \ge r\cos r - r + 1.$$

Optimality happens only when

$$r_0 - r_0\cos r_0 = \varepsilon(-\cos\alpha + \sin\alpha),$$

where r_0 is unique.

Therefore $D = r_0$, $0 < r_0 < 1$.

Remark 6.2.12. Case of $\alpha \in (\pi, 3\pi/2)$, i.e., $\sin\alpha < 0$, $\cos\alpha < 0$.

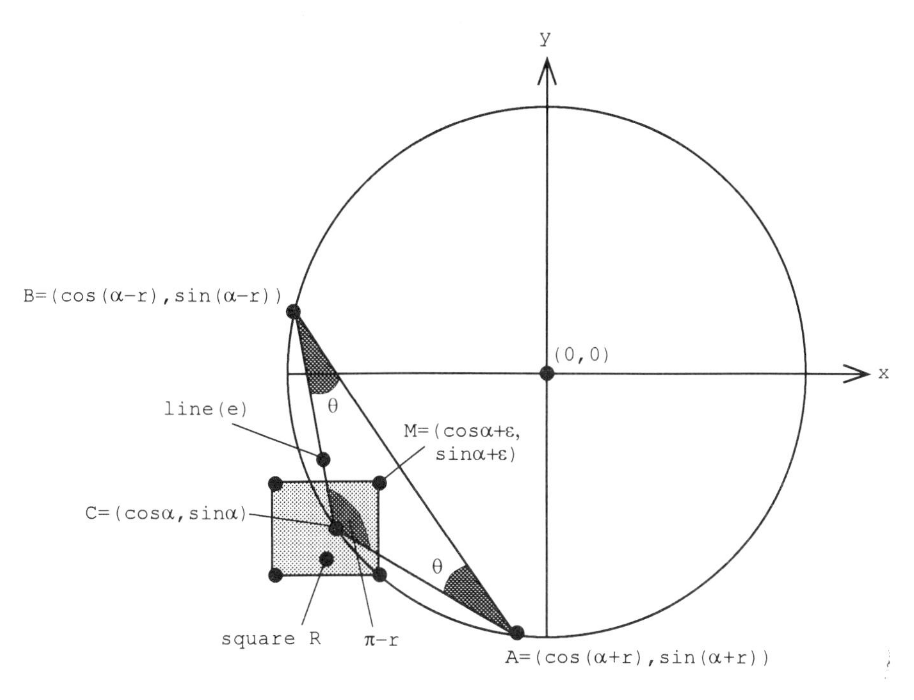

Figure 6.2.13. $\alpha \in (\pi, 3\pi/2)$.

It is always true that $\pi - r > \pi/2$, $\theta = r/2 < \pi/4$. From Remark 6.2.3 we get that $\mathcal{R}$ is below (AB), i.e., for all $(y_1, y_2) \in \mathcal{R} \cap$ unit disk we get $(\cos\alpha)y_1 + (\sin\alpha)y_2 > \cos r$, for any $0 < r \leq 1$. Here slope$(AB) < 0$. Also our optimal (y_1, y_2) is $(y_1^* = \cos\alpha + \varepsilon,\ y_2^* = \sin\alpha + \varepsilon)$, that is, the optimal moment point is $M = (\cos\alpha + \varepsilon,\ \sin\alpha + \varepsilon)$.

The line (BC) has equation

$$y - \sin\alpha = \left(\frac{\sin\alpha - \sin(\alpha - r)}{\cos\alpha - \cos(\alpha - r)}\right) \cdot (x - \cos\alpha),$$

assuming that $\cos\alpha - \cos(\alpha - r) \neq 0$. At $y = \sin\alpha + \varepsilon$ we get

$$x = \cos\alpha + \varepsilon \cdot \left(\frac{\cos\alpha - \cos(\alpha - r)}{\sin\alpha - \sin(\alpha - r)}\right).$$

We would like M to be to the right of (BC), where the origin is located. That is, we need:

$$\cos\alpha+\varepsilon > \cos\alpha+\varepsilon\cdot\left(\frac{\cos\alpha-\cos(\alpha-r)}{\sin\alpha-\sin(\alpha-r)}\right)$$

$$\Leftrightarrow$$

$$1 > \left(\frac{\cos\alpha-\cos(\alpha-r)}{\sin\alpha-\sin(\alpha-r)}\right) =: \cot\varphi.$$

Call $\mathcal{P} := \cos\alpha - \cos(\alpha-r)$ and $\mathcal{N} := \sin\alpha - \sin(\alpha-r)$. Always holds that slope$(AB) < 0$ and slope$(BC) \neq 0$.

Here always $\mathcal{N} < 0$, since $\alpha \in (\pi, 3\pi/2)$. If slope$(BC) < 0$, then $\cot\varphi < 0 < 1$. If slope$(BC) > 0$, it means that $\mathcal{P} < 0$ (the case of $\mathcal{P} = 0$ means that $x = \cos\alpha < \cos\alpha + \varepsilon$, therefore, M is to the right of (BC) in the side of the origin), then $\cos\alpha < \cos(\alpha - r)$, i.e., $(\alpha - r) \in (\pi/2, \pi)$. Therefore, here

$$\text{slope}(BC) = \tan\varphi = \tan\left(\frac{\pi}{2} - \tilde{\varphi}\right) = \cot\tilde{\varphi} > 1,$$

by $0 < \tilde{\varphi} < \theta < \pi/4$. Thus $\cot\varphi < 1$.

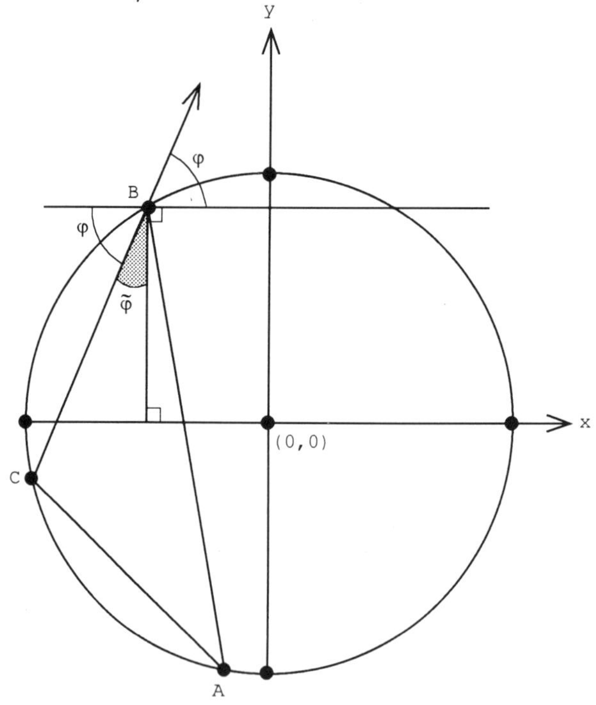

Figure 6.2.14. M is to the right (BC).

Hence M is in the side of (BC) where the origin is located, to the right of (BC).

The line (CA) has equation:

$$y - \sin\alpha = \left(\frac{\sin(\alpha + r) - \sin\alpha}{\cos(\alpha + r) - \cos\alpha}\right) \cdot (x - \cos\alpha)$$

and at $x = \cos\alpha + \varepsilon$ we get

$$y = \sin\alpha + \varepsilon \cdot \left(\frac{\sin(\alpha + r) - \sin\alpha}{\cos(\alpha + r) - \cos\alpha}\right).$$

We would like M to be above (CA), in the side of the origin. That is we need

$$\sin\alpha + \varepsilon > \sin\alpha + \varepsilon \cdot \left(\frac{\sin(\alpha + r) - \sin\alpha}{\cos(\alpha + r) - \cos\alpha}\right)$$

$$\Leftrightarrow$$

$$1 > \left(\frac{\sin(\alpha + r) - \sin\alpha}{\cos(\alpha + r) - \cos\alpha}\right) =: \tan\varphi.$$

Call $\mathcal{N} := \sin(\alpha + r) - \sin\alpha$ and $\mathcal{P} := \cos(\alpha + r) - \cos\alpha$. Always slope$(AB) < 0$ and always $\mathcal{P} > 0$. If slope$(CA) < 0$, then $\tan\varphi < 0 < 1$.

If slope$(CA) > 0$ then $\mathcal{N} > 0$, (if slope$(CA) = 0$ then $\mathcal{N} = 0$ and $\sin\alpha = \sin(\alpha + \varepsilon)$ thus $y = \sin\alpha < \sin\alpha + \varepsilon$, therefore we are above the line (CA)), then $\sin(\alpha + r) > \sin\alpha$, i.e., $(\alpha + r) \in (3\pi/2, 2\pi)$. Thus slope$(CA) = \tan\varphi < 1$, by $0 < \varphi < \theta < \pi/4$. Hence the point M is above the line (CA).

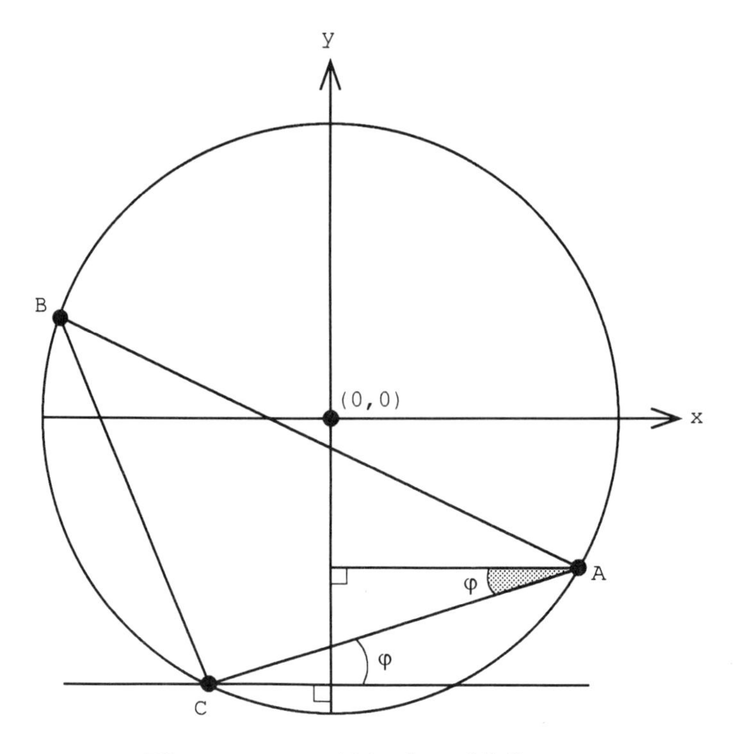

Figure 6.2.15. M is above (CA).

Conclusion. *We have proved that* $M = (\cos\alpha+\varepsilon, \sin\alpha+\varepsilon)$ *is an interior point of* $\operatorname{conv}(A \overset{\Delta}{C} B)$.

Here we have that $\sin\alpha < 0$, $\cos\alpha < 0$ and $y_1 \leq \cos\alpha + \varepsilon$, $y_2 \leq \sin\alpha + \varepsilon$, for all $(y_1, y_2) \in \mathcal{R} \cap$ unit disk. Thus

$$(\cos\alpha)y_1 \geq (\cos\alpha)(\cos\alpha + \varepsilon)$$

and

$$(\sin\alpha)y_2 \geq (\sin\alpha)(\sin\alpha + \varepsilon).$$

Hence

$$(\cos\alpha)y_1 + (\sin\alpha)y_2 \geq 1 + \varepsilon(\cos\alpha + \sin\alpha),$$

for all $(y_1, y_2) \in \mathcal{R} \cap$ unit disk. It should be

$$1 - \varepsilon((-\cos\alpha) + (-\sin\alpha)) \geq r\cos r - r + 1.$$

Optimality can happen only when $r_0 - r_0 \cos r_0 = \varepsilon((-\cos\alpha) + (-\sin\alpha))$, r_0 is unique. Therefore $D = r_0$, $0 < r_0 < 1$.

Remark 6.2.13. Case of $\alpha \in (3\pi/2, 2\pi - 1]$, i.e., $\sin\alpha < 0$, $\cos\alpha > 0$.

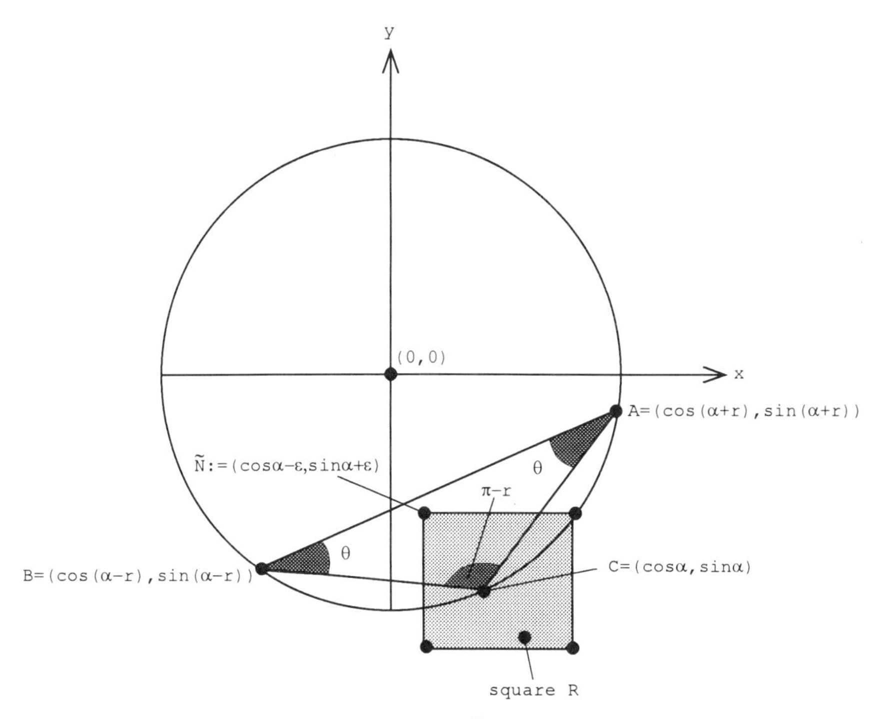

Figure 6.2.16. $\alpha \in (3\pi/2, 2\pi - 1]$, $\tilde{\mathcal{N}} := (\cos\alpha - \varepsilon, \sin\alpha + \varepsilon)$.

It is always true that $\pi - r > \pi/2$, $\theta = r/2 < \pi/4$. By Remark 6.2.3 we get that $\mathcal{R}$ is below (AB), that is, for all $(y_1, y_2) \in \mathcal{R} \cap$ unit disk we have $(\cos\alpha)y_1 + (\sin\alpha)y_2 > \cos r$, for any $0 < r \le 1$. Here slope$(AB) > 0$. Our optimal (y_1, y_2) is $(y_1^* = \cos\alpha - \varepsilon,\ y_2^* = \sin\alpha + \varepsilon)$. That is the optimal moment point is $\tilde{\mathcal{N}} := (\cos\alpha - \varepsilon, \sin\alpha + \varepsilon)$.

The line (BC) has equation:

$$y - \sin\alpha = \left(\frac{\sin\alpha - \sin(\alpha - r)}{\cos\alpha - \cos(\alpha - r)}\right) \cdot (x - \cos\alpha)$$

and at $x = \cos\alpha - \varepsilon$ we get

$$y = \sin\alpha - \varepsilon \cdot \left(\frac{\sin\alpha - \sin(\alpha - r)}{\cos\alpha - \cos(\alpha - r)}\right).$$

We would like that $\tilde{\mathcal{N}}$ is above (BC). That is, we need

$$\sin\alpha + \varepsilon > \sin\alpha - \varepsilon \cdot \left(\frac{\sin\alpha - \sin(\alpha - r)}{\cos\alpha - \cos(\alpha - r)}\right)$$

$$\Leftrightarrow$$

$$-1 < \left(\frac{\sin\alpha - \sin(\alpha - r)}{\cos\alpha - \cos(\alpha - r)}\right) =: \tan\varphi.$$

Call $\mathcal{N} := \sin\alpha - \sin(\alpha - r)$ and $\mathcal{D} := \cos\alpha - \cos(\alpha - r)$. Always $\text{slope}(AB) > 0$ and $\mathcal{D} > 0$ by $\alpha \in (3\pi/2, 2\pi - 1]$. If $\text{slope}(BC) \geq 0$, then $\tan\varphi \geq 0 > -1$.

If $\text{slope}(BC) < 0$, then $\mathcal{N} < 0$ and $\sin\alpha < \sin(\alpha - r)$, hence $(\alpha - r) \in (\pi, 3\pi/2)$. Here $0 < \tilde{\varphi} < \theta < \pi/4$. Thus

$$\text{slope}(BC) = \tan\varphi = \tan(\pi - \tilde{\varphi}) = -\tan\tilde{\varphi} > -1$$

(by $0 < \tan\tilde{\varphi} < 1$).

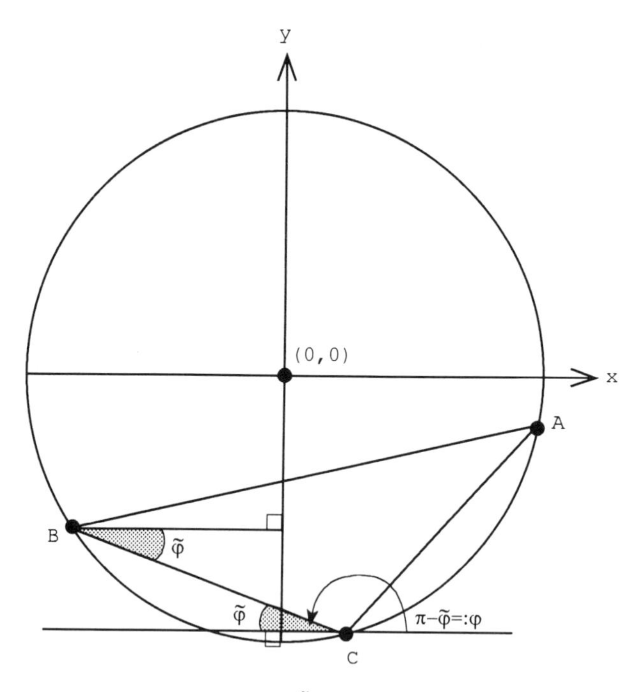

Figure 6.2.17. $\tilde{\mathcal{N}}$ is above (BC).

Hence $\tilde{\mathcal{N}}$ is above the line (BC), where the origin is located.

The line (CA) has equation:

$$y - \sin\alpha = \left(\frac{\sin(\alpha + r) - \sin\alpha}{\cos(\alpha + r) - \cos\alpha}\right) \cdot (x - \cos\alpha).$$

Note that $\sin(\alpha + r) - \sin\alpha > 0$ and $\cos(\alpha + r) - \cos\alpha > 0$. And at $y = \sin\alpha + \varepsilon$ we get

$$x = \cos\alpha + \varepsilon \cdot \left(\frac{\cos(\alpha + r) - \cos\alpha}{\sin(\alpha + r) - \sin\alpha}\right).$$

We want $\tilde{\mathcal{N}}$ to be above (CA), in the side of the origin. That is we need

$$\cos\alpha - \varepsilon < \cos\alpha + \varepsilon \left(\frac{\cos(\alpha + r) - \cos\alpha}{\sin(\alpha + r) - \sin\alpha}\right)$$

$$\Leftrightarrow$$

$$-1 < \left(\frac{\cos(\alpha + r) - \cos\alpha}{\sin(\alpha + r) - \sin\alpha}\right) =: \cot\varphi.$$

The last is true by $\cot\varphi > 0$. Therefore $\tilde{\mathcal{N}}$ is above (CA).

Conclusion. $\tilde{\mathcal{N}} = (\cos\alpha - \varepsilon, \sin\alpha + \varepsilon)$ is an interior point of $\mathrm{conv}(A \overset{\Delta}{C} B)$.

Here we have $\sin\alpha < 0$, $\cos\alpha > 0$ and $y_1 \geq \cos\alpha - \varepsilon$, $y_2 \leq \sin\alpha + \varepsilon$. Thus

$$(\cos\alpha)y_1 \geq (\cos\alpha)(\cos\alpha - \varepsilon)$$

and

$$(\sin\alpha)y_2 \geq (\sin\alpha)(\sin\alpha + \varepsilon).$$

Hence

$$(\cos\alpha)y_1 + (\sin\alpha)y_2 \geq 1 - \varepsilon \cdot (\cos\alpha + (-\sin\alpha)).$$

It should be

$$1 - \varepsilon \cdot (\cos\alpha + (-\sin\alpha)) \geq r\cos r - r + 1.$$

Optimality happens only when

$$r_0 - r_0 \cos r_0 = \varepsilon \cdot (\cos\alpha + (-\sin\alpha)),$$

for unique r_0. Therefore again $D = r_0$, $0 < r_0 < 1$.

The proof of Theorem 6.2.1 is now completed. □

CHAPTER SEVEN

PROBABILITY MEASURES, POSITIVE LINEAR OPERATORS AND KOROVKIN TYPE INEQUALITIES

7.1. Introduction

Denote by $C([a,b])$, the space of continuous real valued functions on $[a,b] \subset \mathbb{R}$.

Definition 7.1.1. Let L be a linear operator mapping $C([a,b])$ into itself. L is called *positive* iff whenever $f \geq g$; $f, g \in C([a,b])$, we have that $L(f) \geq L(g)$.

Remark 7.1.1. Let L be a positive linear operator mapping $C([a,b])$ into itself. It follows from the *Riesz representation theorem*, for every $x \in [a,b]$ there is a finite non-negative measure μ_x such that

$$L(f;x) := (Lf)(x) = \int_{[a,b]} f(t)\mu_x(dt), \quad \forall\, f \in C([a,b]). \tag{7.1.1}$$

Naturally, the converse is not true. That is, only special kernels $\mu_x(\cdot)$ will transform continuous functions into continuous functions. Call $\mu_x([a,b]) := m_x$, then μ_x/m_x is a probability measure on $[a,b]$, and this way we can be dealing with probability theory instead of functional analysis, in certain respects, and be using moment methods when needed, but this comes a bit later in a natural way. At this point we need to create some more background for that. We present the following amazing well known result.

Korovkin's First Theorem 7.1.1. (Korovkin (1960), p. 14). *Let $[a,b]$ be a compact interval in $\mathbb{R}$ and $(L_n)_{n\in\mathbb{N}}$ be a sequence of positive linear operators L_n mapping $C([a,b])$ into itself. Suppose that $(L_n f)$ converges uniformly to f for the three test functions $f = 1, x, x^2$. Then $(L_n f)$ converges uniformly to f on $[a,b]$ for all functions $f \in C([a,b])$.*

PROOF: (Bauer (1978)) Denote $f(x) = x$ by id and $f(x) = x^2$ by id^2. Every $f \in C([a,b])$ is bounded:

$$|f(x)| \leq \gamma, \quad \forall\, x \in [a,b]$$

where $\gamma > 0$. Also, f is uniformly continuous on $[a,b]$: given $\varepsilon > 0$ there exists a number $\delta > 0$ such that, for all $x, y \in [a,b]$,

$$|x - y| \leq \sqrt{\delta} \text{ then } |f(x) - f(y)| \leq \varepsilon,$$

$$\Leftrightarrow$$

$$(x - y^2) \leq \delta \text{ then } |f(x) - f(y)| \leq \varepsilon.$$

The technique consists of putting $\sqrt{\delta}$ where one normally puts δ. It leads for arbitrary points $x, y \in [a,b]$ to the inequality

$$|f(x) - f(y)| \leq \varepsilon + \alpha(x - y)^2,$$

with $\alpha := 2\gamma\delta^{-1}$, which one immediately derives by considering the two cases $(x - y)^2 \leq \delta$ and $(x - y)^2 > \delta$. Thus we have for all $y \in [a,b]$ the following inequality between functions

$$|f - f(y)| \leq \varepsilon + \alpha(id - y)^2.$$

Linearity and positivity of the operators L_n then imply

$$|L_n f - f(y) \cdot L_n(1)| \leq \varepsilon L_n(1) + \alpha(L_n(id^2) - 2yL_n(id) + y^2 L_n(1)).$$

Evaluating the above inequality at $x = y$, we get

$$|L_n f - f \cdot L_n 1| \leq \varepsilon L_n 1 + \alpha(L_n(id^2) - 2(id) \cdot L_n(id) + (id^2) \cdot L_n(1)).$$

From the assumption on $(L_n f)$ for the three functions $f = 1, id, id^2$ and from the triangle inequality it follows that $L_n f$ converges to f uniformly, true for all $f \in C([a,b])$. □

Here we denote $L_n f(x) := (L_n f)(x)$, observe that $L_n(f(x)) = f(x) \cdot L_n(1)$ by linearity of L_n.

The trigonometric counterpart of the above has as follows:

Korovkin's Second Theorem 7.1.2. (Korovkin (1960), p. 17). *Let $(L_n)_{n \in \mathbb{N}}$ be a sequence of positive linear operators mapping the space of 2π-periodic continuous functions $C([0,2\pi])$ into itself. Suppose that $(L_n f)$ converges uniformly to f for the three test functions $f = 1$, $\sin t$, $\cos t$. Then $(L_n f)$ converges uniformly to f for all functions $f \in C([0,2\pi])$.*

Also we would like to mention Korovkin's most general related result.

Theorem 7.1.3. (Korovkin (1960), p. 49). *Let $(L_n)_{n\in\mathbb{N}}$ be a sequence of positive linear operators mapping $C([a,b])$ into itself. Suppose that $(L_n f_i)$ converges uniformly to f_i, $i = 1,2,3$; where $\{f_1, f_2, f_3\}$ is a Tschebyshev system of continuous functions on $[a,b]$. Then $(L_n f)$ converges uniformly to f for all functions $f \in C([a,b])$.*

Definition 7.1.2. A linear functional is called *positive* if it preserves order when applied to a space of functions.

Again, when $\ell: C([a,b]) \to \mathbb{R}$ is a positive linear functional, from the *Riesz representation theorem* there is a finite non-negative measure μ_ℓ such that

$$\ell(f) = \int_{[a,b]} f(t)\mu_\ell(dt), \quad \forall f \in C([a,b]). \tag{7.1.2}$$

Korovkin's related basic result has as follows:

Theorem 7.1.4. (Korovkin (1960), p. 6). *If the conditions $\ell_n(1) \to 1$, $\ell_n(t) \to s$ and $\ell_n(t^2) \to s^2$, for a fixed $s \in \mathbb{R}$ are satisfied for the sequence of positive linear functionals $\{\ell_n\}_{n\in\mathbb{N}}$, then $\ell_n(f) \to f(s)$ for any real valued function f which is bounded on $\mathbb{R}$ and continuous at s.*

Next we would like to mention some well-known explicit upper bounds for the supremum norm

$$\|L_n f - f\| := \|L_n f - f\|_\infty := \sup_{x\in[a,b]} |L_n f(x) - f(x)|,$$

where $f \in C([a,b])$, and $\{L_n\}_{n\in\mathbb{N}}$ is a sequence of positive linear operators. These have motivated deeply our work and the results we will present later.

For this we need the following basic definition (from Lorentz (1966), pp. 43–44) and properties.

Definition 7.1.3. Let $f \in C([a,b])$ and $0 \le h \le b-a$. Call

$$\omega_1(h) := \omega(h) := \omega_1(f,h) := \omega(f,h) := \sup_{\substack{\text{all } x,y \\ |x-y|\le h}} |f(x) - f(y)|$$

the (first) modulus of continuity of f at h.

Remark 7.1.2. $\omega(h)$ has the following properties:

(i) $\omega(h) \to 0$ for $h \to 0$;

(ii) $\omega(h)$ is positive and increasing;

(iii) ω is subadditive in h, and thus

(iv) ω is continuous in h.

Conditions (i), (ii), and (iii) are also *sufficient* for a given function $\omega(h)$, $0 \leq h \leq b-a$, to be the modulus of continuity of some function $f \in C([a,b])$.

Also it holds

(v) $\omega(nh) \leq n\omega(h)$, all $n \in \mathbb{N}$;

(vi) $\omega(\lambda h) \leq \lceil \lambda \rceil \omega(h)$, all $\lambda > 0$, where $\lceil \lambda \rceil := \min\{i \in \mathbb{N}\colon\ i \geq \lambda\}$ is the ceiling of λ;

(vii) $\omega(\alpha f, h) = |\alpha| \omega(f,h)$; and

(viii) $\omega(f_1 + f_2, h) \leq \omega(f_1, h) + \omega(f_2, h)$.

As a related important result we mention

Theorem 7.1.5. (Lorentz (1966), p. 45). *For each modulus of continuity $\omega(x)$, $0 \leq x \leq \ell$, there is a concave modulus of continuity $\omega^*(x)$ with the property*

$$\omega(x) \leq \omega^*(x) \leq 2\omega(x), \quad \textit{all } 0 \leq x \leq \ell. \tag{7.1.3}$$

The first quantitative estimate for the convergence in Korovkin's first Theorem 7.1.1 was proved by Mamedov (1959) and is presented next.

Theorem 7.1.6. *Let $\{L_n\}_{n \in \mathbb{N}}$ be a sequence of positive linear operators in the space $C([a,b])$, for which $L_n 1 = 1$, $L_n(t;x) = x + \alpha_n(x)$, $L_n(t^2;x) = x^2 + \beta_n(x)$. Then*

$$\|L_n(f;x) - f(x)\|_\infty \leq 3\omega_1(f;\sqrt{d_n}), \tag{7.1.4}$$

where $d_n := \|\beta_n(x) - 2x\alpha_n(x)\|_\infty$.

Inequality (7.1.4) was greatly improved by Shisha and Mond (1968), whose result comes next. Their celebrated inequality (7.1.5) started essentially the area of Korovkin type inequalities and it was a great motivation for a major part of the author's so far research, using as proving tools methods from geometric moment theory.

Theorem 7.1.7. (Shisha and Mond (1) (1968)). *Let* $[a,b] \subset \mathbb{R}$ *be a compact interval. Let* $\{L_n\}_{n\in\mathbb{N}}$ *be a sequence of positive linear operators acting on* $C([a,b])$ *into itself. For* $n = 1,2,\ldots$, *suppose* $L_n(1)$ *is bounded. Let* $f \in C([a,b])$. *Then for* $n = 1,2,\ldots$, *we have*

$$\|L_n f - f\|_\infty \le \|f\|_\infty \cdot \|L_n 1 - 1\|_\infty + \|L_n(1) + 1\|_\infty \cdot \omega_1(f,\mu_n), \tag{7.1.5}$$

where

$$\mu_n := \|(L_n((t-x)^2))(x)\|_\infty^{1/2}, \tag{7.1.6}$$

and $\|\cdot\|_\infty$ *stands for the sup-norm over* $[a,b]$. *In particular, if* $L_n(1) = 1$ *then* (7.1.5) *reduces to*

$$\|L_n(f) - f\|_\infty \le 2\omega_1(f,\mu_n). \tag{7.1.7}$$

Note: (i) In forming μ_n^2, x is kept fixed, however t forms the functions t, t^2 on which L_n acts.

(ii) One can easily see, for $n = 1,2,\ldots$

$$\mu_n^2 \le \|L_n(t^2;x) - x^2\|_\infty + 2c \cdot \|L_n(t;x) - x\|_\infty + c^2 \cdot \|L_n(1;x) - 1\|_\infty,$$

where $c := \max(|a|,|b|)$.

So if the assumptions of first Korovkin's Theorem 7.1.1 are fulfilled then $\mu_n \to 0$, therefore $\omega_1(f,\mu_n) \to 0$, as $n \to +\infty$, and we obtain from (7.1.5) that $\|L_n f - f\|_\infty \to 0$, as $n \to +\infty$, which is Korovkin's conclusion!!! I.e., Korovkin's result has been recast in a quantitative form.

Proof of Theorem 7.1.7: (Shisha–Mond (1) (1968)). Let $x \in [a,b]$ and $\delta > 0$. Let $t \in [a,b]$. If $|t-x| > \delta$, then

$$|f(t) - f(x)| \le \omega_1(f,|t-x|) = \omega_1(f,|t-x|\delta^{-1}\delta)$$

$$\leq (1 + |t - x| \cdot \delta^{-1}) \cdot \omega_1(f, \delta) \leq (1 + (t - x)^2\delta^{-2}) \cdot \omega_1(f, \delta).$$

The estimate

$$|f(t) - f(x)| \leq (1 + (t - x)^2\delta^{-2}) \cdot \omega_1(f, \delta)$$

holds, easily, also if $|t - x| \leq \delta$. Let $n \in \mathbb{N}$, then by positivity and linearity of L_n we have

$$\begin{aligned} |(L_n f - f(x) \cdot L_n 1)(x)| &\leq \omega_1(f, \delta) \cdot [(L_n 1 + \delta^{-2} \cdot L_n((t - x)^2))(x)] \\ &\leq \omega_1(f, \delta) \cdot \left[(L_n 1)(x) + \left(\frac{\mu_n}{\delta}\right)^2\right]. \end{aligned}$$

If $\mu_n > 0$, choose as $\delta := \mu_n$. Hence

$$|[L_n f - f(x) \cdot L_n 1](x)| \leq \omega_1(f, \mu_n) \cdot \|L_n 1 + 1\|_\infty,$$

$$|-f(x) + f(x) \cdot (L_n 1)(x)| \leq \|f\|_\infty \cdot \|L_n 1 - 1\|_\infty. \tag{7.1.8}$$

Adding the above, we get (7.1.5).

If $\mu_n = 0$, we have $\forall \delta > 0$,

$$|(L_n f - f(x) \cdot (L_n 1))(x)| \leq \omega_1(f, \delta) \cdot (L_n 1)(x).$$

Taking $\delta \to 0$ we obtain $(L_n f)(x) = f(x) \cdot (L_n 1)(x)$. Therefore from (7.1.8) we have

$$|(L_n f)(x) - f(x)| \leq \|f\| \cdot \|L_n 1 - 1\|_\infty,$$

which implies (7.1.5). □

Example 7.1.1. Let $B_n: C([0,1]) \to C([0,1])$, $n \in \mathbb{N}$ be the well-known Bernstein polynomial operators (these are positive linear operators), defined by

$$(B_n f)(x) := \sum_{k=0}^{n} f\left(\frac{k}{n}\right) \cdot \binom{n}{k} x^k (1 - x)^{n-k}, \quad \forall x \in [0, 1],$$

where $f \in C([0,1])$. Here $(B_n 1)(x) = 1$, $(B_n t)(x) = x$ and

$$(B_n t^2)(x) = (n - 1)n^{-1}x^2 + n^{-1}x, \quad \forall x \in [0, 1].$$

One can easily check that

$$(B_n((t - x)^2))(x) = n^{-1}(x - x^2), \quad \forall x \in [0, 1].$$

Thus, from inequality (7.1.5) we get that

$$\|B_n f - f\|_\infty \le 2\omega_1\left(f, \frac{1}{2\sqrt{n}}\right), \quad \forall\, n \in \mathbb{N} \tag{7.1.9}$$

which is a well-known result.

The convergence of a sequence of positive linear operators $\{L_n\}_{n\in\mathbb{N}}$ to the unit operator I is closely related to the weak convergence of a sequence of finite measures μ_n to the unit (Dirac) measure δ_{x_0}, $x_0 \in [a,b]$, see (7.1.1). Later we will describe estimates for the remainder

$$\left|\int_{[a,b]} f\, d\mu_n - f(x_0)\right|; \quad f \in C^k([a,b]),\ k \in \mathbb{Z}_+.$$

Using geometric moment methods, Shisha–Mond type best or nearly best upper bounds will be established for various choices of $[a,b]$, k and given moments of μ_n. Some of them lead to attainable inequalities. The optimal functions/measures are typically spline functions and finitely supported measures. The corresponding inequalities involve the first modulus of continuity of $f^{(k)}$ (the kth derivative of f) or a modification of it. Several applications of these results will be given.

It will be useful to mention the following on the weak convergence of probability measures as a related background; initially see Billingsley (1979, p. 329) and then Dudley (1989).

If F_n and F are distribution functions on $\mathbb{R}^k$; $k \ge 1$, F_n *converges weakly* to F, written $F_n \Rightarrow F$, iff $\lim_{n\to+\infty} F_n(x) = F(x)$ for all continuity points x of F. The corresponding probability measures μ_n and μ are in this case also said to *converge weakly*: $\mu_n \Rightarrow \mu$. Let $A \subset \mathbb{R}^k$, the closure $\overline{A}$ of A is the set of limits of sequences in A and the interior $A^0 := \mathbb{R}^k - \overline{(\mathbb{R}^k - A)}$. We denote by $Bd(A) = \overline{A} - A^0$, the boundary of A. A Borel set A is a *μ-continuity set* iff $\mu(Bd(A)) = 0$. Let $\mathcal{R}^k$ denote the k-dimensional Borel sets of $\mathbb{R}^k$.

Theorem 7.1.8. *For probability measures μ_n and μ on $(\mathbb{R}^k, \mathcal{R}^k)$, each of the following conditions is equivalent to the weak convergence of μ_n to μ.*

(1) $\lim_{n\to+\infty} \int f\, d\mu_n = \int f\, d\mu$, *for bounded continuous* $f\colon \mathbb{R}^k \to \mathbb{R}$;

(2) $\limsup_{n\to+\infty} \mu_n(C) \le \mu(C)$, *for closed C in* $\mathbb{R}^k$;

(3) $\liminf_{n\to+\infty} \mu_n(G) \geq \mu(G)$, *for open* G *in* $\mathbb{R}^k$;

(4) $\lim_{n\to+\infty} \mu_n(A) = \mu(A)$ *for* μ*-continuity sets* A *in* $\mathbb{R}^k$.

One can consider weak convergence in a much more general setting, see Dudley (1989), pp. 228–229, 302–303, 306–307, 309–310: Let $(S, \mathcal{T})$ be a topological space and P_n, P be probability measures on the Borel σ-algebra $\mathcal{B}$ generated by $\mathcal{T}$.

Definition 7.1.4. Let $C_b(S)$ be the set of all bounded, continuous, real-valued functions on S. We say that P_n *converges weakly* to P ($P_n \Rightarrow P$) iff for every $f \in C_b(S)$, $\int f\, dP_n \to \int f\, dP$, as $n \to +\infty$.

Theorem 7.1.9. *Let* P_n, P *be probability measures on* (S, d), *which is a metric space. Then the following are equivalent:*

(1) $P_n \Rightarrow P$;

(2) *for all open* U, $\liminf_{n\to+\infty} P_n(U) \geq P(U)$;

(3) *for all closed* F, $\limsup_{n\to+\infty} P_n(F) \leq P(F)$;

(4) *for all continuity sets* A *of* P, $\lim_{n\to+\infty} P_n(A) = P(A)$.

On a general metric space, the most natural way to express smoothness is through Lipschitz conditions, as follows. Let (S, d) be a metric space and $f: S \to \mathbb{R}$ a function. The Lipschitz seminorm is defined by

$$\|f\|_L := \sup_{x \neq y} \frac{|f(x) - f(y)|}{d(x, y)}.$$

Call $\|f\|_\infty := \sup_{x\in S} |f(x)|$, the supremum norm of f. Call $\|f\|_{BL} := \|f\|_\infty + \|f\|_L$ and

$$BL(S, d) := \{f: S \to \mathbb{R} \mid \|f\|_{BL} < +\infty\},$$

which is the set of all bounded, real-valued Lipschitz functions on S. Note that $\|\cdot\|_{BL}$ is a norm with the submultiplicative property in addition. If (S, d) is any compact metric space, then $BL(S, d)$ is dense in $C(S)$ for $\|\cdot\|_\infty$.

Let $A \subset (S,d)$—a metric space and $\varepsilon > 0$. Call

$$A^{\varepsilon} := \{y \in S : d(x,y) < \varepsilon \text{ for some } x \in A\}.$$

Definition 7.1.5. For any probability measures P, Q on (S,d) call

$$\rho(P,Q) := \inf\{\varepsilon > 0 : P(A) \leq Q(A^{\varepsilon}) + \varepsilon, \text{ for all Borel sets } A \subset S\}.$$

It is proved that ρ is a metric on the set of all probability measures on S, called the *Lévy–Prokhorov metric.* Also define

$$\beta(P,Q) := \sup\left\{\left|\int f\,d(P-Q)\right| : \|f\|_{BL} \leq 1\right\}.$$

It is proved that this is also a metric on the set of probability measures of the metric space (S,d).

Theorem 7.1.10. *For any separable metric space (S,d) and the probability measures P_n and P on S, the following are equivalent:*

(i) $P_n \Rightarrow P$, *as* $n \to +\infty$;

(ii) $\int f\,dP_n \to \int f\,dP, \forall f \in BL(S,d)$;

(iii) $\rho(P_n, P) \to 0$;

(iv) $\beta(P_n, P) \to 0$.

After all this presented background on the weak convergence of probability measures, we want to emphasize on the fact that: over compact Hausdorff spaces the pointwise convergence of a sequence of positive linear operators to the unit operator, is equivalent (by Riesz representation theorem) to the weak convergence of a sequence of finite measures to the unit (Dirac) measure at the given point. Here is assumed that the associated kernels transform continuous functions into continuous functions. In proving related results soon, we use methods from geometric moment theory. However, still we have to develop some necessary machinery.

Definition 7.1.6. (Anastassiou (3) (1968)). Let $(V_1, \|\cdot\|)$, $(V_2, \|\cdot\|)$ be real normed vector spaces and Q be a compact subset of V_1. For a continuous function $f: Q \to V_2$ we define its (first) modulus of continuity

$$\omega_1(f, h) := \sup\{\|f(\mathbf{x}) - f(\mathbf{y})\| : \text{all } \mathbf{x}, \mathbf{y} \in Q,\ \|\mathbf{x} - \mathbf{y}\| \le h\}. \tag{7.1.10}$$

The last enjoys most of the basic properties of the basic, on the real line, first modulus of continuity, see Definition 7.1.3 and Remark 7.1.2.

To remind, let $x \ge 0$, then the ceiling of x: $\lceil x \rceil$ is defined to be the least integer greater equal x. We will use the following:

Lemma 7.1.1. (Anastassiou (3) (1968)). *Let $(V_1, \|\cdot\|)$, $(V_2, \|\cdot\|)$ be real normed vector spaces and Q a subset of V_1 which is star-shaped relative to the fixed point $\mathbf{x}_0$. Consider $f: Q \to V_2$ with the properties:*

$$f(\mathbf{x}_0) = 0 \ \text{ and } \ \|\mathbf{s} - \mathbf{t}\| \le h \Rightarrow \|f(\mathbf{s}) - f(\mathbf{t})\| \le w;\ w, h > 0. \tag{7.1.11}$$

Then there is a maximal such function ϕ, namely

$$\phi(t) := \lceil \|\mathbf{t} - \mathbf{x}_0\| / h \rceil \cdot w \cdot \mathbf{i},$$

where $\mathbf{i}$ is any unit vector in V_2.

That is, $\|f(\mathbf{t})\| \le \|\phi(\mathbf{t})\|$ all $\mathbf{t} \in Q$.

PROOF: For $\mathbf{s}, \mathbf{t} \in Q$ such that $\|\mathbf{s} - \mathbf{t}\| \le h$ we easily have

$$\left| w \cdot \left\lceil \frac{\|\mathbf{s} - \mathbf{x}_0\|}{h} \right\rceil - w \cdot \left\lceil \frac{\|\mathbf{t} - \mathbf{x}_0\|}{h} \right\rceil \right| \le w,$$

therefore ϕ satisfies (7.1.11). Further, when $\|\mathbf{t} - \mathbf{x}_0\| \le h$ we have $\|f(\mathbf{t})\| \le w$. Note that for any other $\mathbf{t} \in Q$ there is $n \in \mathbb{N}$: $(n-1)h < \|\mathbf{t} - \mathbf{x}_0\| \le nh$. Since

$$\mathbf{t} - \mathbf{x}_0 = \sum_{k=0}^{n-1} \Delta_k,$$

where

$$\Delta_k := \frac{(n-k)\mathbf{t} + k\mathbf{x}_0}{n} - \frac{(n-(k+1))\mathbf{t} + (k+1)\mathbf{x}_0}{n} = \frac{\mathbf{t} - \mathbf{x}_0}{n}$$

(that is $\|\Delta_k\| \le h$), we obtain

$$\|f(\mathbf{t})\| \le \sum_{k=0}^{n-1} \left\| f\left(\frac{(n-k)\mathbf{t} + k\mathbf{x}_0}{n}\right) - f\left(\frac{(n-(k+1))\mathbf{t} + (k+1)\mathbf{x}_0}{n}\right) \right\| \le n \cdot w. \qquad \square$$

Corollary 7.1.1. *Let $[a,b] \subset \mathbb{R}$ be a compact interval. Let $f: [a,b] \to \mathbb{R}$ with the properties*

$$f(x_0) = 0$$

and

$$|s-t| \le h \Rightarrow |f(s) - f(t)| \le w; \quad w, h > 0.$$

Then there is a maximal such function ϕ, namely,

$$\phi(t) := \lceil |t - x_0|/h \rceil \cdot w.$$

One can generalize Lemma 7.1.1 in the following direction:

Proposition 7.1.1. (See Anastassiou (1) (1985).) *Let C be a subset of the real normed vector space $V = (V, \|\cdot\|)$ which is star-shaped relative to the fixed point x_0. Let further $\{(h_i, w_i): i \in I\}$ be a given collection of numbers ($h_i > 0$, $w_i > 0$, I arbitrary), and consider the collection $\mathcal{F}$ of functions $f: C \to \mathbb{R}$ such that $f(x_0) = 0$ while, for each $i \in I$,*

$$\|s - t\| \le h_i \Rightarrow |f(s) - f(t)| \le w_i.$$

Then

$$\sup_{f \in \mathcal{F}} |f(s)| = \rho(\|s - x_0\|) \qquad (s \in C), \tag{7.1.12}$$

where

$$\rho(\|u\|) := \inf\left\{ \sum_{i \in I} k_i w_i : \|u\| \le \sum_{i \in I} k_i h_i \right\}$$

where $k_i \in \mathbb{Z}_+$, $k := \sum_{i \in I} k_i < \infty$.

Proof: Obviously, $\rho(\|u\|)$ is an even subadditive function on $\mathbb{R}$ satisfying $\rho(0) = 0$ and $\rho(h_i) \le w_i$ ($i \in I$). Moreover, $\rho(\|u\|)$ is non-decreasing on $\mathbb{R}_+$. Hence $f_0(s) := \rho(\|s - x_0\|)$ restricted to C defines a function $f_0 \in \mathcal{F}$, showing that (7.1.12) holds with the $\ge$ sign. To prove the opposite inequality, it suffices to show that $|f(s)| \le \sum_{i\in I} k_i w_i$ as soon as $f \in \mathcal{F}$ and $\|s - x_0\| \le \sum_{i\in I} k_i h_i$ ($k_i \in \mathbb{Z}_+$, $k := \sum_{i\in I} k_i < \infty$). This is easily done by an induction on k. The cases $k = 0$ and $k = 1$ are obvious. Let $k \ge 0$ satisfy the assertions and suppose $s \in C$ satisfies

$$\|s - x_0\| \le \sum_{i\in I} k_i h_i + h_r \qquad \left(k_i \in \mathbb{Z}_+,\ \sum k_i =: k; r \in I\right).$$

Choosing s' on the line segment $\overline{x_0 s}$ such that $\|s' - x_0\| \le \sum_1^n k_i h_i$ and $\|s - s'\| \le h_r$, one easily sees that $|f(s)| \le \sum_1^n k_i w_i + w_r$. □

Remark 7.1.3. (Anastassiou (3) (1986)). An auxilary function: Let $h > 0$ be fixed. We shall often use the even function defined by

$$\phi_n(x) := \int_0^{|x|} \left\lceil \frac{t}{h} \right\rceil \frac{(|x| - t)^{n-1}}{(n-1)!} dt, \qquad (x \in \mathbb{R}). \tag{7.1.13}$$

Equivalently

$$\phi_n(x) = \int_0^{|x|} \int_0^{x_1} \cdots \left(\int_0^{x_{n-1}} \left\lceil \frac{x_n}{h} \right\rceil dx_n \right) \cdots dx_1. \tag{7.1.14}$$

Since $\lceil \frac{t}{h} \rceil = \sum_{j=0}^{\infty} 1_{t<j\cdot h}$, where $1_{t<j\cdot h}$ is the indicator (characteristic) function on the set of $t{:}\, t < j \cdot h$, the latter yields that

$$\phi_n(x) = \frac{1}{n!} \left(\sum_{j=0}^{\infty} (|x| - jh)_+^n \right). \tag{7.1.15}$$

In particular, letting $k := \lceil |x|/h \rceil$,

$$\phi_1(x) = \sum_{j=0}^{\infty} (|x| - jh)_+ = \sum_{j=0}^{k-1} (|x| - jh) = k|x| - \frac{1}{2} k(k-1)h. \tag{7.1.16}$$

Maximizing over $k \ge 0$, (attained if $k = 1/2 + |x|/h$), it follows that

$$\phi_1(x) \le \phi_{\star 1}(x) := \frac{1}{2h} (|x| + h/2)^2. \tag{7.1.17}$$

In fact, $\phi_1(x_k) = \phi_{\star 1}(x_k) = \frac{1}{2} \cdot hk^2$ and $\phi_1'(x_k) = \phi_{\star 1}'(x_k) = k$ at $x_k := (k - \frac{1}{2})h$, ($k = 1, 2, 3, \ldots$). Further, one easily gets

$$\phi_n(x) \le \phi_{\star n}(x) := \left(\frac{|x|^{n+1}}{(n+1)!h} + \frac{|x|^n}{2n!} + \frac{h|x|^{n-1}}{8(n-1)!} \right) \tag{7.1.18}$$

with equality only at $x = 0$. Note from (7.1.15) that ϕ_n is a (polynomial) spline function. In fact in each interval $((j-1)h, jh]$ it is a polynomial of degree n. At the points $jh (j = 0, 1, \ldots)$ the derivatives $\phi_n^{(k)}$ $(k = 0, 1, \ldots, n-1)$ are continuous while the nth derivative makes an upward jump of size 1. Moreover, $\phi_n(x)$ is convex on $\mathbb{R}$ and strictly increasing on $\mathbb{R}^+$, $(n \geq 1)$. Finally

$$\phi_n(x) = \int_0^x \phi_{n-1}(t)dt, \qquad (x \in \mathbb{R}_+, n \geq 1)$$

provided we define $\phi_0(t) := \lceil t/h \rceil$.

After we have given motivation, history and the necessary prerequisites on convergence of positive linear operators and weak convergence of probability measures, we are ready to present some of our related results.

Our approach is parallel, we will be working for both at the same time: pointwise convergence of positive linear operators to unit operator and weak convergence of finite measures to unit measure at a point.

Type of results: quantitative, more specifically, usually sharp inequalities giving the rate of the above convergences through a modulus of smoothness and estimating the degree of related approximation.

In theorem statements and proofs usually the visible side will be the measure theoretic one. That is because our proving method is probabilistic from geometric moment theory and is unique within approximation theory so far. Our results are optimal, and general, improving all other corresponding ones in the literature. The aim of presentation of the following results is rather solely a demonstration of the power of moment methods, and definitely is not an account of the rich work done so far in Korovkin type inequalities, which corresponds to a very long list of important articles covering all possible directions of research in this topic.

The results that follow come from Anastassiou (1) (1985); (1) (1986) and (3) (1986). In there several times we will be dealing, very generally, just with Borel-measurable functions, instead of only continuous ones.

7.2 Optimal Korovkin Type Inequalities

Here we start by finding estimates to

$$\left|\int_M f\, d\mu - f(x_0)\right|,$$

where $f \in C^n(M)$ $(n \in \mathbb{Z}^+)$ in the Fréchet sense, M is a non-empty convex compact subset of real normed vector space $(V, \|\cdot\|)$, fixed $x_0 \in M$ and μ is a finite measure on M. When $n \geq 1$, $(V, \|\cdot\|)$ is assumed to be a Banach space.

Theorem 7.2.1. *Let μ be a finite measure of mass m on the non-empty convex and compact subset M of the real normed vector space $(V, \|\cdot\|)$. Consider $x_0 \in M$ and $C(x_0) > 0$ such that $0 \leq \|x - x_0\| \leq C(x_0)$ for all $x \in M$. Assume that*

$$\left(\int_M \|x - x_0\|^r \cdot \mu(dx)\right)^{1/r} := D_r(x_0), \tag{7.2.1}$$

where $r > 0$ and $D_r(x_0) > 0$ are given. For the existence of μ we suppose that $D_r^r(x_0) \leq m \cdot C^r(x_0)$. Let $f : M \to \mathbb{R}$ be such that

$$|f(x) - f(y)| \leq w \text{ if } \|x - y\| \leq h, \quad x, y \in M \tag{7.2.2}$$

where $w, h > 0$ are given. Then the best possible constant $K(x_0) = K(m,\ r,\ D_r(x_0),\ h,\ w,\ f(x_0))$ in the inequality

$$\left|\int_M f\, d\mu - f(x_0)\right| \leq |m - 1| \cdot |f(x_0)| + m \cdot K(x_0) \tag{7.2.3}$$

is given as follows (independently of $f(x_0)$). Set

$$n(x_0) := \left\lceil \frac{C(x_0)}{h} \right\rceil, \quad k(x_0) := \left\lceil \frac{D_r(x_0)}{h \cdot m^{1/r}} \right\rceil.$$

It is $1 \leq k(x_0) \leq n(x_0)$ because $D_r^r(x_0) \leq m \cdot C^r(x_0)$.

(i) *$K(x_0) = n(x_0) \cdot w$ when $k(x_0) = n(x_0)$, i.e., when $D_r(x_0)/m^{1/r} > C(x_0) - h$.*

(ii) *$K(x_0) = \left[1 + \frac{1}{m} \cdot \left(\frac{D_r(x_0)}{h}\right)^r \cdot (n(x_0) - 1)^{(1-r)}\right] \cdot w$ when $r \leq 1$ and $k(x_0) < n(x_0)$.*

(iii) $K(x_0) = (k(x_0) + \theta_{k(x_0)}) \cdot w \leq [1 + D_r(x_0)/(h \cdot m^{1/r})] \cdot w$ *when* $r \geq 1$ *and* $k(x_0) < n(x_0)$. *Here*

$$\theta_{k(x_0)} := [(D_r^r(x_0)/m) - (k(x_0) - 1)^r \cdot h^r]/[k^r(x_0) \cdot h^r - (k(x_0) - 1)^r \cdot h^r].$$

The equality sign in (7.2.3) is usually not attained, but can be approached arbitrarily closely by

$$f(x) := \left\lceil \frac{\|x - x_0\|}{h} \right\rceil \cdot w$$

and a measure μ *of mass* m *supported by at most two points.*

PROOF: Set $g(x) := f(x) - f(x_0)$. By Lemma 7.1.1 it follows that

$$|g(x)| \leq \left\lceil \frac{\|x - x_0\|}{h} \right\rceil \cdot w, \quad \text{all } x \in M.$$

Therefore

$$\begin{aligned} \left| \int_M f \, d\mu - f(x_0) \right| &= \left| \int_M g \, d\mu + (m-1) f(x_0) \right| \\ &\leq \int_M \left\lceil \frac{\|x - x_0\|}{h} \right\rceil \cdot w \cdot \mu(dx) + |m - 1| \cdot |f(x_0)|. \end{aligned}$$

The last is equality when $f(x) := \left\lceil \frac{\|x-x_0\|}{h} \right\rceil \cdot w$, where $\left\lceil \frac{\|x-x_0\|}{h} \right\rceil \cdot w$ fulfills (7.2.2). Thus, the optimal constant $K(x_0)$ in (7.2.3) is given by

$$m \cdot K(x_0) = \sup_\mu \int_M \left\lceil \frac{\|x - x_0\|}{h} \right\rceil \cdot w \cdot \mu(dx)$$

where μ ranges over the measures on M of mass m which satisfy (7.2.1). Introducing the probability measure $v := m^{-1} \cdot \mu$, we have

$$K(x_0) = \sup_v \int_M \left\lceil \frac{\|x - x_0\|}{h} \right\rceil \cdot w \cdot v(dx),$$

where v ranges through the probability measures on M satisfying

$$\int_M \|x - x_0\|^r \cdot v(dx) = D_r^r(x_0)/m,$$

with $0 \leq \|x - x_0\| \leq C(x_0)$, all $x \in M$.

Considering the probability measure ρ induced by v and the mapping $x \to \|x - x_0\|$, calling $u := \|x - x_0\|$ and $\phi(u) := w \cdot \lceil \frac{u}{h} \rceil$ we want

$$K(x_0) = \sup_{\rho} \int \phi(u) \cdot \rho(du), \quad 0 \le u \le C(x_0),$$

over all probability measures ρ such that

$$\int u^r \cdot \rho(du) = D_r^r(x_0)/m, \quad r > 0.$$

It follows from the method of optimal distance in geometric moment theory, see Chapter 2, that

$$K(x_0) = \psi(D_r^r(x_0)/m), \quad \text{where } \Gamma_1 := \{(\tau, \psi(\tau)) \colon 0 \le \tau \le C^r(x_0)\}$$

describes the upper boundary of the convex hull conv Γ_0 of the curve

$$\Gamma_0 := \{(u^r, \phi(u)) \colon 0 \le u \le C(x_0)\}.$$

The graph of Γ_0 is formed by:

(i) The origin $(0,0)$ when $u = 0$.

(ii) For $k(x_0) = 1, \dots, n(x_0) - 1$, the half open horizontal line segments $(P_{k(x_0)}, Q_{k(x_0)}]$ where

$$P_{k(x_0)} := ((k(x_0) - 1)^r \cdot h^r, k(x_0) \cdot w)$$

and

$$Q_{k(x_0)} := (k^r(x_0) \cdot h^r, k(x_0) \cdot w).$$

(iii) The half open non-empty horizontal line segment $(P_{n(x_0)}, Q_*]$, where $Q_* := (C^r(x_0), n(x_0) \cdot w)$.

This is always part of the upper boundary Γ_1 of conv Γ_0, proving claim (i) of the theorem.

The slope of the line segment $[P_{k(x_0)}, P_{k(x_0)+1}]$ is $w \cdot h^{-r}/[k(x_0)^r - (k(x_0) - 1)^r]$, which is decreasing in $k(x_0)$ when $r \ge 1$ and increasing in $k(x_0)$ when $0 < r \le 1$. As a result, when $0 < r \le 1$ the curve Γ_1 is consisted by the two line segments $[P_1, P_{n(x_0)}]$ and $[P_{n(x_0)}, Q_*]$. Therefore

$$\psi(\tau) = w \cdot [1 + \tau \cdot h^{-r} \cdot (n(x_0) - 1)^{-r+1}]$$

when $0 \le \tau \le (n(x_0)-1)^r \cdot h^r$ while $\psi(\tau) = n(x_0) \cdot w$ when $(n(x_0)-1)^r \cdot h^r \le \tau \le C^r(x_0)$. This proves claim (ii).

If $r \ge 1$ then Γ_1 is formed by the line segments $[P_{k(x_0)}, P_{k(x_0)+1}]$ ($k(x_0) = 1, \ldots, n(x_0) - 1$) and the horizontal line segment $[P_{n(x_0)}, Q_*]$. This proves claim (iii). The last claims of the theorem follow from the geometry of conv Γ_0. That the best upper bound (7.2.3) is usually not attained comes from the fact that $[P_{k(x_0)}, P_{k(x_0)+1}]$ is part of the closure of conv Γ_0 and not part of conv Γ_0. □

Corollary 7.2.1. *If $r \ge 1$ then*

$$\left|\int_M f\,d\mu - f(x_0)\right| \le |m-1| \cdot |f(x_0)| + \omega_1(f,h) \cdot \left(m + \frac{D_r(x_0)}{h} \cdot m^{1-(1/r)}\right). \qquad (7.2.4)$$

PROOF: Apply Theorem 7.2.1 when $r \ge 1$ and $w := \omega_1(f,h)$. □

Note. Let L be a positive linear operator. The derived pointwise estimates for $|L(f,x) - f(x)|$ several times imply the uniform ones.

As related results on $(\mathbb{R}, |\cdot|)$ we would like to present:

Proposition 7.2.1. *Let $[a,b]$ be a compact interval in $\mathbb{R}$ such that $0 \in [a,b]$ and put $c := \max(|a|, b)$. Call $L := b - a$. Let $\phi: [-L, L] \to \mathbb{R}_+$ be an even non-negative function, $\Phi(0) = 0$, which is subadditive on $[0, L]$ (e.g., Φ could be nondecreasing and concave). Consider*

$$\mathcal{F} := \{f: [a,b] \to \mathbb{R} \text{ and } |f(s) - f(t)| \le \Phi(s-t) \text{ for } s,t \in [a,b]\}.$$

Let $\mathcal{F}_0 := \{f \in \mathcal{F}: f(0) = 0\}$. Clearly $\Phi \in \mathcal{F}_0$ and $|f| \le \Phi$ for all $f \in \mathcal{F}_0$ thus

$$\sup_{\mathcal{F}_0} \left|\int f\,dv\right| = \int \Phi\,dv,$$

for each probability measure v on $[a,b]$.

Suppose v is further restricted by the moment condition

$$\int \Psi(t) v(dt) := \rho$$

where Ψ is a given continuous even and non-negative function on $[-c,+c]$, which is strictly increasing on $[0,c]$, $\Psi(0)=0$. Let

$$E := \sup_{\mathcal{F},v} \left| \int f \, dv \right| .$$

(i) *If $\Phi(\Psi^{-1})$ is convex on $[0,\Psi(c)]$ then $E = \rho\Phi(c)/\Psi(c)$.*

(ii) *If $\Phi(\Psi^{-1})$ is concave on $[0,\Psi(c)]$ then $E = \Phi(\Psi^{-1}(\rho))$.*

Remark 7.2.1. If $\Phi(t) = k|t|^{\alpha}$ $(0 < \alpha \leq 1,\, k > 0)$, $\Psi(t) = |t|^r$ $(r > 0)$ then (i), (ii) happen when $r \leq \alpha$ and $r \geq \alpha$, respectively.

Remark 7.2.2. We maintain the notations of Proposition 7.2.1. Let μ be a measure on $[a,b]$ with mass m and moment

$$\int \Psi(t)\mu(dt) := m\rho.$$

Then each $f \in \mathcal{F}$ satisfies

$$\left| \int f \, d\mu - f(0) \right| \leq |f(0)|\,|m-1| + mE.$$

This inequality is sharp and in fact is attained by $f(t) := f(0) + \varepsilon\Phi(t)$ and a suitably chosen measure supported by at most two points ($\varepsilon = \pm 1$ equals the sign of $(m-1)f(0)$).

The following is well known and useful.

Lemma 7.2.1. *Be given a closed finite interval $[a,b]$ and fixed $x_0 \in [a,b]$. Consider $f \in C^n([a,b])$, $n \geq 1$, and denote $\phi(x) := f^{(n)}(x) - f^{(n)}(x_0)$. Then $(a \leq x \leq b)$*

$$f(x) = \sum_{k=0}^{n} \frac{f^{(k)}(x_0)}{k!} \cdot (x - x_0)^k + \int_{x_0}^{x} \phi(t) \cdot \frac{(x-t)^{n-1}}{(n-1)!} \cdot dt, \tag{7.2.5}$$

A nearly optimal result comes next motivating the following optimal results.

Theorem 7.2.2. *Let μ be a measure of mass $m > 0$ on $[a,b]$ and $x_0 \in [a,b]$ fixed. Let n be a fixed positive integer and put*

$$h := \left[\int |t - x_0|^{n+1} \mu(dt) \right]^{1/(n+1)} . \tag{7.2.6}$$

Suppose $f \in C^n([a,b])$ satisfies

$$|f^{(n)}(s) - f^{(n)}(t)| \leq w \quad \text{if } a \leq s,t \leq b, \text{ and } |s-t| \leq h, \tag{7.2.7}$$

where w is a given positive number. Then

$$\left|\int f\,d\mu - f(x_0)\right| \leq |f(x_0)| \cdot |m-1| + \sum_{k=1}^{n} \frac{|f^{(k)}(x_0)|}{k!} \cdot \left|\int (t-x_0)^k \mu(dt)\right|$$

$$+ \frac{wh^n}{n!} \cdot (m^{1/(n+1)} + 1/(n+1)). \tag{7.2.8}$$

PROOF: Without loss of generality let $x_0 := 0$. From (7.2.7) $\phi(x) := f^{(n)}(x) - f^{(n)}(0)$ satisfies $|\phi(s) - \phi(t)| \leq w$ when $|s-t| \leq h$, therefore $|\phi(t)| \leq w\lceil |t|/h \rceil$ by Corollary 7.1.1. It follows from (7.2.5) and (7.1.13) that

$$\left| f(x) - \sum_{k=0}^{n} \frac{f^{(k)}(0)}{k!} x^k \right| \leq w\phi_n(x).$$

From $\lceil t/h \rceil \leq 1 + t/h$ and (7.1.13),

$$w\phi_n(x) \leq \frac{w|x|^n}{n!} \cdot \left(1 + \frac{|x|}{(n+1)h}\right). \tag{7.2.9}$$

Integrating relative to μ and using Hölder's inequality we obtain (7.2.8). □

The trigonometric counterpart of Theorem 7.2.2 follows.

Proposition 7.2.2. *Let $f \in C^n([-\pi,\pi])$, $n \geq 1$, and μ a measure on $[-\pi,\pi]$ of mass $m > 0$. Put*

$$\beta := \left(\int \left(\sin \frac{|t|}{2} \right)^{n+1} \cdot \mu(dt) \right)^{1/(n+1)} \tag{7.2.10}$$

and denote by $w := \omega_1(f^{(n)}, \beta)$ the modulus of continuity of $f^{(n)}$ at β. Then

$$\left|\int f\,d\mu - f(0)\right| \leq |f(0)| \cdot |m-1| + \sum_{k=1}^{n} \frac{|f^{(k)}(0)|}{k!} \cdot \left|\int t^k \mu(dt)\right|$$

$$+ w[m^{1/(n+1)} + \pi/(n+1)] \cdot \frac{\pi^n \beta^n}{n!}. \tag{7.2.11}$$

PROOF: Analogous to the proof of Theorem 7.2.2, using the fact $|t| \leq \pi \sin(|t|/2)$. □

The following attainable inequalities are obtained by using the method of optimal distance from geometric moment theory, see Chapter 2.

Theorem 7.2.3. *Let μ be a finite measure on $[a,b] \subset \mathbb{R}$, $0 \in (a,b)$ and $|a| \le b$. Put*

$$c_k := \int t^k \mu(dt), \quad k = 0,1,\ldots,n; \;\; d_n := \left(\int |t|^n \mu(dt)\right)^{1/n}. \tag{7.2.12}$$

Let $f \in C^n([a,b])$ be such that

$$|f^{(n)}(s) - f^{(n)}(t)| \le w \;\; \text{if } a \le s,t \le b, \text{ and } |s-t| \le h \tag{7.2.13}$$

where w, h are given positive numbers.

Then we have the upper bound

$$\left|\int f\,d\mu - f(0)\right| \le |f(0)| \cdot |c_0 - 1| + \sum_{k=1}^{n} \frac{|f^{(k)}_{(0)}|}{k!} \cdot |c_k| + w\phi_n(b) \cdot \left(\frac{d_n}{b}\right)^n. \tag{7.2.14}$$

The above inequality is in a certain sense attained by the measure μ with masses $[c_0 - (d_n/b)^n]$ and $(d_n/b)^n$ at 0 and b, respectively, and when, moreover, the optimal function is

$$\tilde{f} = \begin{cases} w\phi_n, & \text{on } [0,b]; \\ 0, & \text{on } [a,0]. \end{cases} \tag{7.2.15}$$

Namely, the latter is the limit of a sequence of functions f having continuous nth derivatives satisfying (7.2.13) and $f^{(k)}(0) = 0$ $(k = 0,\ldots,n)$ and such that the difference of the two sides of (7.2.14) tends to 0. In fact, $\lim_{N\to+\infty} f_{nN}(t) = \tilde{f}(t)$, where $(a \le t \le b)$

$$f_{nN}(t) := w \int_0^t \left(\int_0^{t_1} \cdots \left(\int_0^{t_{n-1}} f_{0N}(t_n)dt_n\right)\cdots\right) dt_1. \tag{7.2.16}$$

Here, for $k = 0,1,\ldots,\lceil b/h\rceil - 1$ and $N \ge 1$ f_{0N} is the continuous function defined

$$f_{0N}(t) := \begin{cases} 0, & \text{if} \quad a \le t < 0; \\ \frac{Nwt}{2h} + kw\left(1 - \frac{N}{2}\right), & \text{if } kh \le t \le \left(k + \frac{2}{N}\right)h; \\ (k+1)w, & \text{if} \quad \left(k + \frac{2}{N}\right)h < t \le (k+1)h; \\ \lceil b/h\rceil w, & \text{if} \quad \left(\lceil b/h\rceil - 1 + \frac{2}{N}\right)h < t \le b. \end{cases} \tag{7.2.17}$$

Observe that $f_{0N}(t)$ fulfills (7.2.13) and further

$$\lim_{N\to+\infty} f_{0N}(t) = \begin{cases} \lceil t/h \rceil w, & t \in [0,b]; \\ 0, & t \in [a,0]. \end{cases}$$

PROOF: From (7.2.5), integrating relative to μ get

$$\left|\int f\,d\mu - f(0)\right| \leq |f(0)|\,|c_0 - 1| + \sum_{k=1}^{n} \frac{|f^{(k)}(0)|}{k!}|c_k| + S_n$$

where

$$S_n := w \int \phi_n(t)\mu(dt).$$

We would like to maximize S_n given that μ has preassigned moments c_0 and $d_n = [\int |t|^n \mu(dt)]^{1/n}$. Since the functions on hand $|t|^n$ and $\phi_n(t)$ are both even we are essentially concerned with a measure on $[0,b]$ (using that $|a| \leq b$). As usual, consider the curve defined by $u := t^n$ and $v := \phi_n(t)$, that is, $v = \phi_n(u^{1/n})$ where $u \geq 0$. Here

$$\phi_n(u^{1/n}) = \frac{1}{n!}\left(\sum_{j=0}^{\infty}(u^{1/n} - jh)_+^n\right).$$

The function $(u^{1/n} - jh)_+^n$ has its first derivative equal to $(1 - jh/u^{1/n})_+^{n-1}$ which is obviously increasing in u. It follows that $\phi_n(u^{1/n})$ is convex. Consequently from the method of optimal distance, see Chapter 2, the integral $\int \phi_n(t)\mu(dt)$ is maximized by a measure taking values at 0 and b only. Let μ have masses p and q at 0 and b, respectively. Thus $p \geq 0$, $q \geq 0$ while $p + q = c_0$. Further, $0 + qb^n = d_n^n$, thus $q = (d_n/b)^n$. Consequently, $\max S_n = w\phi_n(b)q = w\phi_n(b)(d_n/b)^n$. □

Corollary 7.2.2. *Consider the positive linear operator*

$$L: C^n([a,b]) \to C([a,b]), \quad n \in \mathbb{N}.$$

Let

$$\begin{aligned} c_k(x) &:= L((t-x)^k, x), \quad k = 0,1,\ldots,n; \\ d_n(x) &:= [L(|t-x|^n, x)]^{1/n}; \\ c(x) &:= \max(x-a, b-x)(c(x) \geq (b-a)/2). \end{aligned} \tag{7.2.18}$$

Let $f \in C^n([a,b])$ such that $\omega_1(f^{(n)},h) \le w$, where w,h are fixed positive numbers, $0 < h < b-a$. Then we have the upper bound

$$|L(f,x) - f(x)| \le |f(x)| \cdot |c_0(x) - 1| + \sum_{k=1}^{n} \frac{|f^{(k)}(x)|}{k!} \cdot |c_k(x)| + R_n. \tag{7.2.19}$$

Here

$$R_n := w\phi_n(c(x)) \cdot \left(\frac{d_n(x)}{c(x)}\right)^n = \frac{w}{n!} \cdot \Theta_n(h/c(x)) \cdot d_n^n(x), \tag{7.2.20}$$

where

$$\Theta_n(h/u) := n!\phi_n(u)/u^n.$$

The above inequality is sharp. Analogous to Theorem 7.2.3, it is in a certain sense attained by $w\phi_n((t-x)_+)$ and a measure μ_x supported by $\{x,b\}$ when $x-a \le b-x$, also attained by $w\phi_n((x-t)_+)$ and a measure μ_x supported by $\{x,a\}$ when $x-a \ge b-x$: in each case with masses $c_0(x) - (d_n(x)/c(x))^n$ and $(d_n(x)/c(x))^n$, respectively.

PROOF: Apply Theorem 7.2.3 with 0 shifted to x. □

The next lemma will be used in Theorem 7.2.4.

Theorem 7.2.2. *Be given $[a,b] \subset \mathbb{R}$ and $x_0 \in (a,b)$ fixed, consider all measures μ with prescribed moments*

$$\mu([a,b]) := c_0 > 0; \quad \int (t-x_0)\mu(dt) := c_1(x_0), \quad \int |t-x_0|\mu(dt) := d_1(x_0) > 0. \tag{7.2.21}$$

For $w,h > 0$ $(0 < h < b-a)$ as given numbers, put $M(x_0) := \sup_\mu \int (w/c_0) \cdot \phi_1(|t-x_0|)\mu(dt)$. Then

$$M(x_0) = w\phi_1(b-x_0) \cdot \left(\frac{d_1(x_0) + c_1(x_0)}{2c_0(b-x_0)}\right) + w\phi_1(x_0-a) \cdot \left(\frac{d_1(x_0) - c_1(x_0)}{2c_0(x_0-a)}\right). \tag{7.2.22}$$

The optimal measure is carried by $\{a, x_0, b\}$.

PROOF: Easy. □

The assertion of Theorem 7.2.3 can be improved if more is known about μ. One result in this direction is the following.

Theorem 7.2.4. *Let $[a,b] \subset \mathbb{R}$, $x_0 \in (a,b)$, and consider all measures μ on $[a,b]$ such that*

$$\mu([a,b]) := c_0 > 0; \quad \int (t - x_0)\mu(dt) := c_1(x_0), \quad \int |t - x_0|\mu(dt) := d_1(x_0) > 0. \quad (7.2.23)$$

Further, consider $f \in C^1([a,b])$ with $\omega_1(f', h) \leq w$ where w, h are given positive numbers $(0 < h < b - a)$.

Then, we get the best upper bound

$$\left| \int f \, d\mu - f(x_0) \right| \leq |f(x_0)| \cdot |c_0 - 1| + |f'(x_0)| \cdot |c_1(x_0)| + c_0 M(x_0), \quad (7.2.24)$$

where $M(x_0)$ is given by (7.2.22).

PROOF: Easy. □

Next we generalize Theorem 7.2.3 into Banach spaces.

Prerequisites 7.2.1. Let $(V, \|\cdot\|)$ be a real Banach space and V^j denote the j-fold product space $V \times \cdots \times V$ endowed with the max-norm $\|x\|_{V^j} := \max_{1 \leq k \leq j} \|x_k\|$, where $x := (x_1, \ldots, x_k) \in V^j$. Then the space $L_j := L_j(V^j; \mathbb{R})$ of all real valued multilinear continuous functions $g: V^j \to \mathbb{R}$ is a Banach space with norm

$$\|g\| := \|g\|_{L_j} := \sup_{(\|x\|_{V^j} = 1)} |g(x)| = \sup \frac{|g(x)|}{\|x_1\| \cdots \|x_j\|}.$$

Let M be a non-empty convex and compact subset of V and $x_0 \in M$ be fixed. Let $f: M \to \mathbb{R}$ be a continuous function whose Fréchet derivatives

$$f^{(j)}: M \to L_j = L_j(V^j; \mathbb{R})$$

exist and are continuous for $1 \leq j \leq n$, $n \in \mathbb{N}$. Call $(x - x_0)^j := (x - x_0, \ldots, x - x_0) \in V^j = V \times \cdots \times V$.

Then by Taylor's formula, see Cartan (1971), we get:

$$f(x) = \sum_{j=0}^{n} \frac{f^{(j)}(x_0)(x - x_0)^j}{j!} + R_n(x, x_0), \quad \text{all } x \in M, \quad (7.2.25)$$

where

$$R_n(x, x_0) := \int_0^1 \frac{(1-u)^{n-1}}{(n-1)!} \cdot (f^{(n)}(x_0 + u(x - x_0)) - f^{(n)}(x_0))(x - x_0)^n \cdot du. \quad (7.2.26)$$

Considering

$$w := \omega_1(f^{(n)}, h) = \sup_{\substack{x,y \\ \|x-y\| \le h}} \|f^{(n)}(x) - f^{(n)}(y)\|$$

we obtain

$$\begin{aligned} &|(f^{(n)}(x_0 + u(x - x_0)) - f^{(n)}(x_0))(x - x_0)^n| \\ &\le \|f^{(n)}(x_0 + u(x - x_0)) - f^{(n)}(x_0)\| \cdot \|x - x_0\|^n \\ &\le w \cdot \|x - x_0\|^n \cdot \left\lceil \frac{u \cdot \|x - x_0\|}{h} \right\rceil, \end{aligned}$$

by Lemma 7.1.1.

Therefore for all $x \in M$:

$$|R_n(x, x_0)| \le w \cdot \|x - x_0\|^n \cdot \int_0^1 \left\lceil \frac{u \cdot \|x - x_0\|}{h} \right\rceil \cdot \frac{(1-u)^{n-1}}{(n-1)!} \cdot du = w \cdot \phi_n(\|x - x_0\|) \quad (7.2.27)$$

by a change of variable, where

$$\phi_n(t) := \int_0^{|t|} \left\lceil \frac{s}{h} \right\rceil \frac{(|t| - s)^{n-1}}{(n-1)!} \cdot ds = \frac{1}{n!} \cdot \left(\sum_{j=0}^{\infty} (|t| - j \cdot h)_+^n \right), \quad t \in \mathbb{R}. \quad (7.2.28)$$

For a full list of properties of the spline function ϕ_n see Remark 7.1.3.

Following Prerequisites 7.2.1 we give:

Theorem 7.2.5. *Let μ be a finite measure on $M \neq \phi$, which is a convex and compact subset of the Banach space $(V, \|\cdot\|)$. Consider $x_0 \in M$ and $C(x_0) > 0$ such that $0 \le \|x - x_0\| \le C(x_0)$ for all $x \in M$. Put $\mu(M) := c_0$ and*

$$\left(\int \|x - x_0\|^n \cdot \mu(dx) \right)^{1/n} := D_n(x_0), \quad n \in \mathbb{N}.$$

Let $f: M \to \mathbb{R}$ be a continuous function whose Fréchet derivatives $f^{(j)}$ exist and are continuous for $1 \le j \le n$. Assume that $\omega_1(f^{(n)}, h) \le w$, where $w, h > 0$ are given.

Then we get the upper bound

$$\begin{aligned} \left| \int_M f\, d\mu - f(x_0) \right| \le |f(x_0)| \cdot |c_0 - 1| + \left| \sum_{j=1}^{n} \frac{1}{j!} \int_M f^{(j)}(x_0)(x - x_0)^j \cdot \mu(dx) \right| \\ + w \cdot \phi_n(C(x_0)) \cdot \left(\frac{D_n(x_0)}{C(x_0)} \right)^n. \end{aligned} \quad (7.2.29)$$

PROOF: From (7.2.25), integrating relative to μ we get

$$\left|\int_M f\,d\mu - f(x_0)\right| \leq |f(x_0)| \cdot |c_0 - 1| + \left|\sum_{j=1}^{n} \frac{1}{j!} \int_M f^{(j)}(x_0)(x - x_0)^j \cdot \mu(dx)\right| + S_n$$

where

$$S_n := w \cdot \int \phi_n(\|x - x_0\|) \cdot \mu(dx).$$

Using the method of optimal distance from geometric moment theory we want to maximize S_n over all μ with moments $c_0 = \mu(M)$ and $D_n(x_0) = (\int_M \|x - x_0\|^n \cdot \mu(dx))^{1/n}$.

Consider the measure ρ of mass c_0 induced by μ and the mapping $x \to \|x - x_0\|$ and call $u := \|x - x_0\|$. Thus, we want

$$\sup S_n = \sup w \cdot \int_M \phi_n(u)\rho(du)$$

over all measures ρ of mass c_0 such that

$$D_n(x_0) = \left(\int u^n \cdot \rho(du)\right)^{1/n}, \quad 0 \leq u \leq C(x_0).$$

We consider the curve defined by $\tau := u^n$, $v := \phi_n(u)$, that is $v = \phi_n(\tau^{1/n})$, $\tau \geq 0$.

Here

$$\phi_n(\tau^{1/n}) = \frac{1}{n!} \cdot \left(\sum_{j=0}^{\infty} (\tau^{1/n} - j \cdot h)_+^n\right).$$

The function $(\tau^{1/n} - j \cdot h)_+^n$ has its first derivative equal to $(1 - j \cdot h/\tau^{1/n})_+^{n-1}$ which is increasing in τ.

Thus $\phi_n(\tau^{1/n})$ is convex. As a result the integral $\int \phi_n(u) \cdot \rho(du)$ is maximized by a measure taking values at 0 and $C(x_0)$. Let ρ have masses p and q at 0 and $C(x_0)$. Note $p, q \geq 0$ and $p + q = c_0$, further $0 + q \cdot C^n(x_0) = D_n^n(x_0)$, thus $q = (D_n(x_0)/C(x_0))^n$. Therefore

$$\max S_n = w \cdot \phi_n(C(x_0)) \cdot q = w \cdot \phi_n(C(x_0)) \cdot \left(\frac{D_n(x_0)}{C(x_0)}\right)^n. \qquad \square$$

7.3 Nearly Optimal Korovkin Inequalities

Here we present some good inequalities with explicit constants better than those in the literature. They involve the first modulus of continuity, or its (smaller) modification, of the

nth derivative of $f \in C^n([a,b])$, $n \geq 1$, evaluated at $h := rd_{n+1}(x_0)$, where $d_{n+1}(x_0) := \left(\int |t - x_0|^{n+1}\mu(dt)\right)^{1/(n+1)}$, $x_0 \in [a,b]$.

The following is a refinement of a result due to Mond and Vasudevan (1980).

Theorem 7.3.1. *Consider a closed interval $[a,b] \subset \mathbb{R}$, a given point $x_0 \in [a,b]$, and a measure μ on $[a,b]$ of mass $m > 0$ satisfying*

$$\int (t - x_0)\mu(dt) := 0; \quad \left(\int (t - x_0)^2\mu(dt)\right)^{1/2} := d_2(x_0) > 0.$$

Then for $f \in C^1([a,b])$ and $r > 0$,

$$\left|\int f\, d\mu - f(x_0)\right| \leq |f(x_0)| \cdot |m - 1| + \left(\sqrt{m} + \frac{1}{r}\right) \cdot \omega_1(f', rd_2(x_0)) \cdot d_2(x_0). \tag{7.3.1}$$

PROOF: By the mean value theorem, there is $\xi \in (t, x_0)$ such that

$$f(t) - f(x_0) = (t - x_0) \cdot f'(x_0) + (t - x_0)(f'(\xi) - f'(x_0)). \tag{7.3.2}$$

Then

$$\begin{aligned} |f'(\xi) - f'(x_0)| &\leq \omega_1(f', |\xi - x_0|) \leq \omega_1(f', |t - x_0|) = \omega_1(f', |t - x_0|\delta^{-1}\delta) \\ &\leq (1 + |t - x_0|\delta^{-1}) \cdot \omega_1(f', \delta) \quad (\text{for all } \delta > 0). \end{aligned}$$

Therefore $|f'(\xi) - f'(x_0)| \leq (1 + |t - x_0|\delta^{-1}) \cdot \omega_1(f', \delta)$, for all $\delta > 0$. Multiplying by $|t - x_0|$, integrating relative to μ, and applying (7.3.2) we have

$$\begin{aligned} \left|\int f\, d\mu - mf(x_0)\right| &\leq |f'(x_0)| \cdot \left|\int (t - x_0)\mu(dt)\right| \\ &\quad + \left(\int |t - x_0|\mu(dt) + \left(\int (t - x_0)^2\mu(dt)\right) \cdot \delta^{-1}\right) \cdot \omega_1(f', \delta) \\ &\leq |f'(x_0)| \cdot \left|\int (t - x_0)\mu(dt)\right| \\ &\quad + \left[\left(\int (t - x_0)^2\mu(dt)\right)^{1/2} \cdot \sqrt{m} + \delta^{-1} \cdot \int (t - x_0)^2\mu(dt)\right] \cdot \omega_1(f', \delta). \end{aligned}$$

Letting $\delta := rd_2(x_0)$, one obtains (7.3.1). □

The following is a refinement of the main theorem of a paper due to Gonska(2) (1983). Gonska's result is one of the latest improvements in this type of inequality. Here denote $\mu(f) := \int f\, d\mu$.

Theorem 7.3.2. *Let μ be a measure on $[a,b]$ of mass $m>0$. Consider $0<h<b-a$ and $x_0\in[a,b]$ and let $f\in C^1([a,b])$.*

Then

$$\left|\int f\,d\mu - f(x_0)\right| \le |f(x_0)|\cdot|m-1| + \omega_1(f',h)\cdot\left\{\mu(|t-x_0|)+\frac{1}{2h}\cdot\mu((t-x_0)^2)\right\}$$

$$+\left|\frac{1}{2h}\int_{-h}^{h} f'_*(x_0+u)du\right|\cdot|\mu(t-x_0)|, \tag{7.3.3}$$

where

$$f'_*(t) := \begin{cases} f'(a), & t<a; \\ f'(t), & t\in[a,b]; \\ f'(b), & t>b. \end{cases}$$

Remark 7.3.1. Observe that $\left|(1/2h)\cdot\int_{-h}^{h} f'_*(x_0+u)du\right| \le \|f'\|$, where $\|\cdot\|$ denotes the sup-norm over $[a,b]$.

PROOF: If $g\in C^2([a,b])$, then clearly

$$\left|\int g\,d\mu - mg(x_0)\right| \le |g'(x_0)|\cdot\left|\int(t-x_0)\mu(dt)\right| + \frac{\|g''\|}{2}\cdot\int(t-x_0)^2\mu(dt). \tag{7.3.4}$$

Further,

$$|\mu(f)-f(x_0)\cdot m| \le |\mu(f-g)-m\cdot(f-g)(x_0)| + |\mu(g)-g(x_0)\cdot m|.$$

Hence,

$$|\mu(f)-f(x_0)\cdot m| \le \|(f-g)'\|\cdot\mu(|t-x_0|)+|g'(x_0)|\cdot|\mu(t-x_0)|+\frac{\|g''\|}{2}\cdot\mu((t-x_0)^2). \tag{7.3.5}$$

Let f'_h be the so-called *first Stekloff function* of f'; that is,

$$(f'_h)(t) := \frac{1}{2h}\cdot\int_{-h}^{h} f'_*(t+u)du, \quad (a\le t\le b).$$

By a well-known theory (Timan (1963))

$$\|f'-(f')_h\| \le \omega_1(f',h) \text{ and } \|(f')'_h\| \le h^{-1}\cdot\omega_1(f',h).$$

It is easy to find $g \in C^2([a,b])$ such that $g' = (f')_h$. Applying (7.3.5) one obtains

$$|\mu(f) - f(x_0) \cdot m| \leq \omega_1(f', h) \cdot \left\{ \mu(|t - x_0|) + \frac{1}{2h} \cdot \mu((t - x_0)^2) \right\} + \left| \frac{1}{2h} \cdot \int_{-h}^{h} f'_*(x_0 + u) du \right| \cdot |\mu(t - x_0)|. \qquad \square$$

Corollary 7.3.1. *Let the closed interval $[a,b] \subset \mathbb{R}$ and $x_0 \in [a,b]$. Also consider a measure μ on $[a,b]$ of mass $m > 0$, such that*

$$\int (t - x_0)\mu(dt) := 0; \quad \left(\int (t - x_0)^2 \mu(dt) \right)^{1/2} := d_2(x_0) > 0.$$

Let $r > 0$ and $f \in C^1([a,b])$. Then

$$\left| \int f \, d\mu - f(x_0) \right| \leq |f(x_0)| \cdot |m - 1| + \left(\sqrt{m} + \frac{1}{2r} \right) \cdot \omega_1(f', r d_2(x_0)) \cdot d_2(x_0). \tag{7.3.6}$$

Observe that (7.3.6) *is sharper than* (7.3.1).

PROOF: By Schwarz's inequality $\mu(|t - x_0|) \leq (\mu((t - x_0)^2))^{1/2} \cdot \sqrt{m}$. Now apply (7.3.3) with $h := r d_2(x_0)$. $\square$

Using (7.1.17), we obtain the following result.

Theorem 7.3.3. *Let $f \in C^1([a,b])$ and μ be a measure on $[a,b]$ of mass $m > 0$ with given moments*

$$\int (t - x_0)\mu(dt) := 0; \quad \left(\int (t - x_0)^2 \mu(dt) \right)^{1/2} := d_2(x_0) > 0,$$

where $x_0 \in [a,b]$. Consider $r > 0$. Then

$$\left| \int f \, d\mu - f(x_0) \right| \leq |f(x_0)| \cdot |m - 1| + \frac{1}{8r} \cdot (2 + \sqrt{m} r)^2 \cdot \omega_1(f', r d_2(x_0)) \cdot d_2(x_0). \tag{7.3.7}$$

Note. If $x_0 = 0$ we could get a sharper inequality by using the modified modulus of continuity $\bar{\omega}_1$ instead of ω_1, where

$$\bar{\omega}_1(f', h) := \sup\{|f'(x) - f'(y)| : x \cdot y \geq 0, |x - y| \leq h\}. \tag{7.3.8}$$

Obviously, $\bar{\omega}_1 \leq \omega_1$.

Corollary 7.3.2. *In the special case of* $m = 1$, $x_0 = 0$ *we have*

$$\left|\int f\,d\mu - f(0)\right| \le \bar{\omega}_1(f', rd_2) \cdot \frac{(2+r)^2}{8r} \cdot d_2 \tag{7.3.9}$$

(*where* $d_2 := d_2(0)$).

PROOF OF THEOREM 7.3.3: Integrating (7.2.5) relative to μ we get

$$\int f\,d\mu - mf(x_0) = f'(x_0) \cdot \int (t - x_0)\mu(dt) + \int K_1(t, x_0)\mu(dt), \tag{7.3.10}$$

where

$$K_1(t, x_0) := \int_{x_0}^{t} \phi(x)dx; \quad \phi(x) := f'(x) - f'(x_0).$$

Note that $\phi(x_0) = 0$ and $|\phi(x)| \le \omega_1(f', h) \cdot \lceil |x - x_0|/h \rceil$ for all $h > 0$.

Hence,

$$|K_1(t, x_0)| \le M_1(t, x_0) := \omega_1(f', h) \cdot \int_0^{|t-x_0|} \left\lceil \frac{z}{h} \right\rceil dz = \omega_1(f', h) \cdot \phi_1(|t - x_0|),$$

for all $t, x_0 \in [a, b]$. By (7.1.17) we obtain

$$|K_1(t, x_0)| \le M_1(t, x_0) \le \omega_1(f', h) \cdot \left[\frac{(t - x_0)^2}{2h} + \frac{|t - x_0|}{2} + \frac{h}{8}\right]. \tag{7.3.11}$$

Now integrating (7.3.11) against μ, using Schwarz's inequality, and setting $h := rd_2(x_0)$, we find

$$\int |K_1(t, x_0)|\mu(dt) \le \frac{1}{8r} \cdot (2 + \sqrt{m}r)^2 \cdot \omega_1(f', rd_2(x_0)) \cdot d_2(x_0). \tag{7.3.12}$$

Finally, from (7.3.12) and (7.3.10)

$$\left|\int f\,d\mu - mf(x_0)\right| \le \frac{1}{8r} \cdot (2 + \sqrt{m}r)^2 \cdot \omega_1(f', rd_2(x_0)) \cdot d_2(x_0).$$

Since

$$\left|\int f\,d\mu - f(x_0)\right| \le |f(x_0)| \cdot |m - 1| + \left|\int f\,d\mu - f(x_0) \cdot m\right|$$

the theorem follows. □

Remark 7.3.2. When $\omega_1(f', h) = Ah^\alpha$, $0 < \alpha \le 1$ and $A > 0$ constant, the value of $r > 0$ minimizing the right-hand side of (7.3.7) is given by

$$r = 2(1 - \alpha)/(\sqrt{m}(1 + \alpha)). \tag{7.3.13}$$

If $\alpha = 1$ then letting $r \downarrow 0$ one obtains that

$$\left|\int f\,d\mu - f(x_0)\right| \le |f(x_0)|\cdot|m-1| + \frac{1}{2}A\cdot d_2^2(x_0), \tag{7.3.14}$$

where

$$A := \sup_{s\neq t}\{|f'(s) - f'(t)|/|s-t|\}.$$

When $|a| = b$ and $x_0 = 0$, the last inequality is attained by $f(x) = x^2$ and a measure μ with masses $m/2$ at $\pm b$ (both sides are then equal mb^2).

Remark 7.3.3. When $m = 1$, $|a| = b$, $x_0 = 0$, $r = 2$, and α is small, the inequality (7.3.7) is nearly attained by $f(x) = |x|^{1+\alpha}$ and μ with mass $\frac{1}{2}$ at $\pm b$.

Remark 7.3.4. With $m = 1$, $|a| = b$, $x_0 = 0$, the measure μ having mass $\frac{1}{2}$ at $\pm b$, and $f(x) = |x|^{1+\alpha}$ $(0 < \alpha \le 1)$, the left-hand side of (7.3.7) equals $a^{1+\alpha}$, while the right-hand side equals $(1+\alpha)\cdot(r^{\alpha-1}/8)\cdot(2+r)^2\cdot a^{1+\alpha}$. Minimizing over r the right-hand side becomes $C(\alpha)\cdot a^{1+\alpha}$, where $C(\alpha) := 2^\alpha\cdot(1-\alpha)^{-1+\alpha}\cdot(1+\alpha)^{-\alpha}$. The quantity $\ln C(\alpha)$ is a concave function of α taking its largest value at $\alpha = 0.580332$ and there $C(\alpha) = 1.650485$. Further

$$\begin{array}{rr}
C(0.01) = 1.016923 & C(0.6) = 1.649385 \\
C(0.05) = 1.084313 & C(0.7) = 1.607942 \\
C(0.1) = 1.167200 & C(0.8) = 1.501667 \\
C(0.2) = 1.324023 & C(0.9) = 1.318405 \\
C(0.3) = 1.46069 & C(0.95) = 1.189863 \\
C(0.4) = 1.56700 & C(0.99) = 1.052338 \\
C(0.5) = 1.63299 & C(0.999) = 1.0074349.
\end{array}$$

So we see that (7.3.7) is never far off, in that it is attained up to a factor 1.65 at most.

Note. (i) If $0 < r \le 4/\sqrt{m}$, then inequality (7.3.7) is sharper than inequality (7.3.6). (ii) If $r \ge 4/\sqrt{m}$, then (7.3.6) is sharper than (7.3.7). Because of (7.3.13) case (i) is probably more interesting.

In terms of best constants, in this type of inequality, the next result improves all the related results we are aware of.

Corollary 7.3.3. *Let $x_0 \in [a,b]$ and the measure μ on $[a,b]$ of mass $m > 0$ satisfy the moment conditions*

$$\int (t - x_0)\mu(dt) := 0 \quad \text{and} \quad d_2(x_0) := \left(\int (t - x_0)^2 \mu(dt) \right)^{1/2}.$$

Consider $r > 0$ and $f \in C^1([a,b])$. Then

$$\left| \left| \int f\, d\mu - f(x_0) \right| - |f(x_0)|\,|m-1| \right.$$

$$\leq \begin{cases} \frac{1}{8r} \cdot (2 + \sqrt{m}r)^2 \cdot \omega_1(f', r d_2(x_0)) \cdot d_2(x_0), & \text{if } r \leq 2/\sqrt{m}; \\ \sqrt{m}\omega_1(f', r d_2(x_0)) \cdot d_2(x_0), & \text{if } r > 2/\sqrt{m}. \end{cases} \tag{7.3.15}$$

When $x_0 = 0$, we get a sharper estimate by replacing ω_1 by $\bar{\omega}_1$ (see (7.3.8)).

PROOF: Note that the first part of (7.3.15) follows from (7.3.7). If $r \geq 2/\sqrt{m}$ then apply (7.3.7) with r replaced by $r_1 = 2/\sqrt{m}$ and note that $(1/8r_1)(2 + \sqrt{m}r_1)^2 = \sqrt{m}$. □

Taking $r = \frac{1}{2}$ in (7.3.7) one obtains:

Theorem 7.3.4. *Let the random variable Y have distribution μ, $E(Y) := x_0$, and $\operatorname{Var}(Y) := \sigma^2$. Consider $f \in C_B^1(\mathbb{R})$. Then*

$$|Ef(Y) - f(x_0)| = \left| \int f\, d\mu - f(x_0) \right| \leq (1.5625) \cdot \omega_1\left(f', \frac{1}{2}\sigma\right) \cdot \sigma. \tag{7.3.16}$$

The last inequality is stronger than the corresponding pointwise results following Mond and Vasudevan (1980) and King (1975).

Application 7.3.1. Consider X_j real i.i.d. random variables and put $S_n := \sum_{j=1}^{n}, X_j$, $n \geq 1$. Let $x_0 := E(X)$, $\sigma^2 := \operatorname{Var}(X)$ thus $E(S_n/n) = x_0$ and $\operatorname{Var}(S_n/n) = \sigma^2/n$. Denote F_{S_n} the d.f. of S_n. Then (7.3.16) yields

$$|Ef(S_n/n) - f(x_0)| = \left| \int_{\mathbb{R}} f(t/n) dF_{S_n}(t) - f(x_0) \right| \leq (1.5625) \cdot \omega_1\left(f', \frac{\sigma}{2\sqrt{n}}\right) \cdot \frac{\sigma}{\sqrt{n}}. \tag{7.3.17}$$

The following Corollaries 7.3.(4–7) are applications of (7.3.17) to well-known positive linear operators arising from probability theory. The corollaries about the Baskakov, Szász–Mirakjan, and Weierstrass operators are improvements of the corresponding results from Ditzian (1975) and Singh (1981).

We start with the classical Bernstein polynomials.

Corollary 7.3.4. *For any* $f \in C^1([0,1])$ *consider*

$$(B_n f)(t) := \sum_{k=0}^{n} f(k/n) \binom{n}{k} t^k (1-t)^{n-k}, \quad t \in [0,1].$$

Then

$$|(B_n f)(t) - f(t)| \le (1.5625) \cdot \omega_1 \left(f', \frac{1}{2}\sqrt{\frac{t(1-t)}{n}} \right) \cdot \sqrt{\frac{t(1-t)}{n}}$$
$$\le \left(\frac{0.78125}{\sqrt{n}} \right) \cdot \omega_1 \left(f', \frac{1}{4\sqrt{n}} \right).$$

PROOF: Consider $(X_j)_{j \in \mathbb{N}}$ Bernoulli (i.i.d.) random variables such that $Pr\,(X_j = 0) = 1-t$, $Pr\,(X_j = 1) = t$, $t \in (0,1)$, then $E(X) = t$ and $\mathrm{Var}(X) = t(1-t)$. Now apply (7.3.17) with $x_0 = t$. Further, note that $\max_{0 \le t \le 1}(t(1-t)) = \frac{1}{4}$ at $t = \frac{1}{2}$. □

For $t \ge 0$ and $f \in C_B(\mathbb{R}_+)$ the Szász–Mirakjan operator is defined as

$$(M_n f)(t) := e^{-nt} \cdot \sum_{k=0}^{\infty} f\left(\frac{k}{n}\right) \cdot \frac{(nt)^k}{k!}$$

while the Baskakov-type operator is defined as

$$(V_n f)(t) := \sum_{k=0}^{\infty} f\left(\frac{k}{n}\right) \cdot \binom{n+k-1}{k} \cdot \frac{t^k}{(1+t)^{n+k}}.$$

Both operators are of the form $E(S_n/n)$ above. Namely, X there has the distribution

$$P_X := e^{-t} \cdot \sum_{k=0}^{\infty} \frac{t^k}{k!} \delta_k \quad \text{and} \quad P_X := \sum_{k=0}^{\infty} \left(\frac{1}{1+t}\right) \cdot \left(\frac{t}{1+t}\right)^k \cdot \delta_k,$$

(Poisson and geometric), respectively. In both cases, $E(X) = t$ while $\mathrm{Var}(X) = t$ and $\mathrm{Var}(X) = (t + t^2)$, respectively.

Thus (7.3.17) implies:

Corollary 7.3.5. *With the above notations, we have*

$$|(M_n f)(t) - f(t)| \le (1.5625) \cdot \omega_1 \left(f', \frac{1}{2} \cdot \left(\frac{t}{n}\right)^{1/2} \right) \cdot \left(\frac{t}{n}\right)^{1/2}$$

and

$$|(V_nf)(t)-f(t)| \le (1.5625)\cdot\omega_1\left(f',\frac{1}{2}\cdot\left(\frac{t+t^2}{n}\right)^{1/2}\right)\cdot\left(\frac{t+t^2}{n}\right)^{1/2},$$

for all $f \in C_B^1(\mathbb{R}_+)$.

The Weierstrass operator is defined by

$$(W_nf)(t) := \sqrt{\frac{n}{\pi}}\cdot\int_{-\infty}^{\infty} f(x)\cdot e^{-n(x-t)^2}\,dx.$$

It agrees with $Ef(S_n/n)$ *when* X *has the normal distribution* $(t,\frac{1}{2})$ *with density* $(1/\sqrt{\pi})\cdot e^{-(x-t)^2}$.

Corollary 7.3.6. *For all* $f \in C_B^1(\mathbb{R})$ *we have*

$$\|W_n(f)-f\| \le (1.5625)\cdot\omega_1\left(f',\frac{1}{2\sqrt{2n}}\right)\cdot\frac{1}{\sqrt{2n}},$$

where $\|\cdot\|$ *is the* sup-*norm.*

As our last illustration, let X *have an exponential density* $e^{-x/t}$ *on* $\mathbb{R}_+$ *so that* $E(X)=t$, $\mathrm{Var}(X)=t^2$. *Then* S_n *has a gamma density with parameters* n *and* t^{-1}, *so that* S_n/n *has a gamma density with parameters* n *and* n/t.

This leads to the operator (*see:* Feller (1966), p. 219)

$$(H_nf)(t) := \frac{n^n}{(n-1)!t^n}\int_0^{\infty} f(x)\cdot x^{n-1}\cdot e^{-nx/t}dx, \quad t>0.$$

Corollary 7.3.7. *For* $f \in C_B^1(\mathbb{R}_+)$, $t>0$ *we find*

$$|(H_nf)(t)-f(t)| \le (1.5625)\cdot\omega_1\left(f',\frac{t}{2\sqrt{n}}\right)\cdot\frac{t}{\sqrt{n}}.$$

Next we present a similar inequality for higher order derivatives ($n \ge 1$).

Theorem 7.3.5. *Let* μ *be a measure of mass* $m>0$ *on the closed interval* $[a,b]\subset\mathbb{R}$, *for which we assume that* $((1/m)\cdot\int|t-x_0|^{n+1}\mu(dt))^{1/(n+1)} := d_{n+1} > 0$, *where* $x_0\in[a,b]$ *is fixed. Consider* $f\in C^n([a,b])$, $n\ge 1$, *with* $\omega_1(f^{(n)}, rd_{n+1}) \le w$, *where* r,w *are given positive numbers. Then*

$$\left|\int f\,d\mu - f(x_0)\right| \le |m-1|\cdot|f(x_0)| + \sum_{k=1}^{n}\frac{|f^{(k)}(x_0)|}{k!}\cdot\left|\int (t-x_0)^k\mu(dt)\right|$$

$$+\frac{mw}{rn!}\cdot\left[\frac{nr^2}{8}+\frac{r}{2}+\frac{1}{(n+1)}\right]\cdot d_{n+1}^n. \tag{7.3.18}$$

Note. (i) When $x_0 = 0 \in [a,b]$ then (7.3.18) is also true when ω_1 is replaced by $\bar{\omega}_1$.

(ii) In applications r is usually small.

(iii) Inequality (7.3.18) on $[-b,b]$ for $m = 1$, $r \downarrow 0$, and $x_0 = 0$ is attained by $f(x) = |x|^{n+1}$ and μ with mass $\frac{1}{2}$ at $\pm b$.

PROOF: Exactly as the proof of Theorem 7.2.2 except that we use the bound (7.1.18) for ϕ_n and take $h := rd_{n+1}$ instead. □

Nearly optimal Korovkin type inequalities are presented now in Banach spaces.

Theorem 7.3.6. *Let μ be a finite measure of mass $m > 0$ on $M \neq \phi$, which is a convex subset of the Banach space $(V, \|\cdot\|)$ and $x_0 \in M$ be given. Assume that*

$$\left(\frac{1}{m}\cdot\int_M \|x-x_0\|^{n+1}\cdot\mu(dx)\right)^{1/(n+1)} := D_{n+1}(x_0) > 0, \quad n \in \mathbb{N}.$$

Let $f: M \to \mathbb{R}$ be a bounded continuous function, whose Fréchet derivatives $f^{(j)}$, $1 \leq j \leq n$ exist, are bounded and continuous. Assume that $\omega_1(f^{(n)}, r\cdot D_{n+1}(x_0)) \leq w$ where $r, w > 0$ are given. Then we have the estimate

$$\left|\int_M f\,d\mu - f(x_0)\right| \leq |f(x_0)|\cdot|m-1| + \left|\sum_{j=1}^n \frac{1}{j!}\cdot\int_M f^{(j)}(x_0)(x-x_0)^j\cdot\mu(dx)\right|$$

$$+\frac{m\cdot w}{r\cdot n!}\cdot\left[\frac{nr^2}{8}+\frac{r}{2}+\frac{1}{(n+1)}\right]\cdot D_{n+1}^n(x_0). \tag{7.3.19}$$

PROOF: Integrate (7.2.25) and use the estimate (see (7.1.18))

$$\phi_n(\|x-x_0\|) \leq \frac{\|x-x_0\|^{n+1}}{(n+1)!h} + \frac{\|x-x_0\|^n}{2\cdot n!} + \frac{h\cdot\|x-x_0\|^{n-1}}{8(n-1)!}$$

along with Hölder's inequality, then set $h := r\cdot D_{n+1}(x_0)$. □

Corollary 7.3.8. *When $\mu(M) = 1$ and $\left(\int_M \|x-x_0\|^2\cdot\mu(dx)\right)^{1/2} := D_2(x_0) > 0$, it holds:*

$$\left|\int_M f\,d\mu - f(x_0)\right| - \left|\int_M f^{(1)}(x_0)(x-x_0)\cdot\mu(dx)\right|$$

$$\leq (1.5625) \cdot \omega_1 \left(f^{(1)}, \frac{1}{2} D_2(x_0) \right) \cdot D_2(x_0). \tag{7.3.20}$$

PROOF: Apply Theorem 7.3.6 for $m = 1$, $r = \frac{1}{2}$ and $n = 1$. □

Proposition 7.3.1. *Let μ be a finite measure of mass $m > 0$ on $M \neq \phi$, which is a convex subset of the Banach space $(V, \|\cdot\|)$ and $x_0 \in M$ be given. Assume that*

$$\left(\int_M \|x - x_0\|^2 \cdot \mu(dx) \right)^{1/2} := D_2(x_0) > 0$$

and consider $r > 0$.

Let $f: M \to \mathbb{R}$ be a bounded continuous function, whose Fréchet derivative $f^{(1)}$ exists, is bounded and continuous. Then

$$\left| \int_M f \cdot d\mu - f(x_0) \right| - |f(x_0)| \cdot |m - 1| - \left| \int_M f^{(1)}(x_0)(x - x_0) \cdot \mu(dx) \right|$$

$$\leq \frac{1}{8 \cdot r} \cdot (2 + \sqrt{m} \cdot r)^2 \cdot \omega_1(f', r \cdot D_2(x_0)) \cdot D_2(x_0). \tag{7.3.21}$$

PROOF: Similar to Theorem 7.3.6. □

Corollary 7.3.9. *Let μ be a finite measure of mass $m > 0$ on M, which is a convex subset of the Banach space $(V, \|\cdot\|)$ and $x_0 \in M$ be given. Assume that $\left(\int_M \|x - x_0\|^2 \cdot \mu(dx)\right)^{1/2} := D_2(x_0) > 0$. Consider $r > 0$. Let $f: M \to \mathbb{R}$ be a bounded and continuous function, whose Fréchet derivative $f^{(1)}$ exists, is bounded and continuous. Then*

$$\left| \int_M f \cdot d\mu - f(x_0) \right| - |f(x_0)| \cdot |m - 1| - \left| \int_M f^{(1)}(x_0)(x - x_0) \cdot \mu(dx) \right|$$

$$\leq \begin{cases} \frac{1}{8 \cdot r} \cdot (2 + \sqrt{m} \cdot r)^2 \cdot \omega_1(f^{(1)}, r \cdot D_2(x_0)) \cdot D_2(x_0), & \text{if } r \leq 2/\sqrt{m}; \\ \sqrt{m} \cdot \omega_1(f^{(1)}, r \cdot D_2(x_0)) \cdot D_2(x_0), & \text{if } r \geq 2/\sqrt{m}. \end{cases} \tag{7.3.22}$$

PROOF: If $r \geq 2/\sqrt{m}$ then apply (7.3.21) with r replaced by $r_1 = 2/\sqrt{m}$ and note that $\frac{1}{8r_1}(2 + \sqrt{m} \cdot r_1)^2 = \sqrt{m}$. □

Application 7.3.2. Let M be a non-empty convex and compact subset of the Banach space $(V, \|\cdot\|)$. Let $x_0 \in M$ and $\{x_1, \ldots, x_n\}$ ($n \in \mathbb{N}$) be a finite set of distinct points in

M. The second Shepard metric interplation operator (see Shepard (1968) and Gordon and Wixom (1978) is defined by

$$S_n^2(f,x_0) := S^2\{x_1,\ldots,x_n\}(f,x_0)$$

$$:= \begin{cases} \displaystyle\sum_{i=1}^{n} f(x_i)\cdot \frac{\prod_{j=1,\,j\neq i}^{n}\|x_0-x_j\|^2}{\sum_{\ell=1}^{n}\prod_{k=1,\,k\neq \ell}^{n}\|x_0-x_k\|^2}, & \text{if } x_0 \notin \{x_1,\ldots,x_n\}; \\ f(x_i), & \text{if otherwise.} \end{cases} \tag{7.3.23}$$

Obviously S_n^2 is a positive linear operator.

In the setting of Corollary 7.3.9 we have:

$$|S_n^2(f,x_0)-f(x_0)| - |S_n^2(f^{(1)}(x_0)(x-x_0),x_0)|$$

$$\leq \begin{cases} \frac{1}{8\cdot r}\cdot(2+r)^2\cdot\omega_1(f^{(1)}, r\cdot D_{2,n}(x_0))\cdot D_{2,n}(x_0), & \text{if } 0<r\leq 2; \\ \omega_1(f^{(1)}, r\cdot D_{2,n}(x_0))\cdot D_{2,n}(x_0), & \text{if } r\geq 2, \end{cases} \tag{7.3.24}$$

where

$$D_{2,n}(x_0) = n^{1/2}\cdot\left(\sum_{i=1}^{n}\|x_0-x_i\|^{-2}\right)^{-1/2}. \tag{7.3.25}$$

Here note that $S_n^2(f,x_i) = f(x_i)$ for all $i=1,\ldots,n$. In particular $S_n^2(1,x)=1$ for all $x\in M$. Thus to $S_n^2(\cdot,x_0)$ corresponds a probability measure $\mu_n^{(x_0)}$ so that $S_n^2(f,x_0)=\int_M f\,d\mu_n^{(x_0)}$. Thus $S_n^2(\|x-x_0\|^2,x_0)$ is equal to

$$D_{2,n}^2(x_0) := \int_M \|x-x_0\|^2\cdot\mu_n^{(x_0)}(dx) = n\cdot\left(\sum_{i=1}^{n}\|x_0-x_i\|^{-2}\right)^{-1},$$

which is the harmonic mean of $\|x_0-x_i\|^2$, $i=1,\ldots,n$.

7.4 Multivariate Korovkin Type Inequalities

Introduction 7.4.1

Here we work, usually, on a compact and convex subset Q of $\mathbb{R}^k$, $k\geq 1$ by considering a fixed point $\mathbf{x}_0\in Q$. Also, $f\in C^n(Q)$, $n\geq 1$ and μ for concreteness is taken to be a probability measure on Q. Using standard moment methods we obtain new strong upper bounds for the error $|\int_Q f\,d\mu - f(x_0)|$ under various typical circumstances. Thus, the presented results correspond to estimates of the pointwise convergence (on $\mathbb{R}^k$, $k \geq 1$) of a sequence of positive linear operators to the identity operator.

Earlier Censor (1971) produced some quantitative results similar to Shisha and Mond (1968), however in $\mathbb{R}^k$, $k \geq 1$.

We need the following:

Lemma 7.4.1. *Take Q a compact and convex subset of $\mathbb{R}^k$ and let $\mathbf{x}_0 := (x_{01}, \ldots, x_{0k}) \in Q$ be fixed and let μ be a probability measure on Q. Let $f \in C^n(Q)$ and suppose that each nth partial derivative $f_\alpha = \partial^\alpha f/\partial x^\alpha$, where $\alpha := (\alpha_1, \ldots, \alpha_k)$, $\alpha_i \in Z^+$, $i = 1, \ldots, k$ and $|\alpha| := \sum_{i=1}^k \alpha_i = n$, has, relative to Q and the ℓ_1-norm $\|\cdot\|$, a modulus of continuity $\omega_1(f_\alpha, h) \leq w$. Here h and w are fixed positive numbers. Then*

$$\left|\int_Q f\,d\mu - f(\mathbf{x}_0)\right| \leq \left|\sum_{j=0}^n \frac{1}{j!}\cdot\int_Q g_{\mathbf{x}}^{(j)}(0)\mu(d\mathbf{x})\right| + w\cdot\int_Q \phi_n(\|\mathbf{x}-\mathbf{x}_0\|)\mu(d\mathbf{x}), \qquad (7.4.1)$$

where $g_{\mathbf{x}}(t) := f(\mathbf{x}_0 + t(\mathbf{x}-\mathbf{x}_0))$, $t \geq 0$ and

$$\phi_n(x) := \int_0^{|x|}\left\lceil\frac{t}{h}\right\rceil\frac{(|x|-t)^{n-1}}{(n-1)!}dt, \quad (x \in \mathbb{R}).$$

PROOF: The jth derivative of $g_{\mathbf{z}}(t) = f(\mathbf{x}_0 + t\cdot(\mathbf{z}-\mathbf{x}_0))$ is given by

$$g_{\mathbf{z}}^{(j)}(t) = \left[\left(\sum_{i=1}^k (z_i - x_{0i})\cdot\frac{\partial}{\partial x_i}\right)^j f\right](x_{01} + t(z_1 - x_{01}), \ldots, x_{0k} + t(z_k - x_{0k})). \qquad (7.4.2)$$

Consequently

$$f(z_1, \ldots, z_k) = g_{\mathbf{z}}(1) = \sum_{j=0}^n \frac{g_{\mathbf{z}}^{(j)}(0)}{j!} + R_n(\mathbf{z}, 0) \qquad (7.4.3)$$

where

$$R_n(\mathbf{z}, 0) := \int_0^1\left(\int_0^{t_1}\cdots\left(\int_0^{t_{n-1}}(g_{\mathbf{z}}^{(n)}(t_n) - g_{\mathbf{z}}^{(n)}(0))dt_n\right)\cdots\right)dt_1.$$

We apply Lemma 7.1.1 to $(f_\alpha(\mathbf{x}_0 + t\cdot(\mathbf{z}-\mathbf{x}_0)) - f_\alpha(\mathbf{x}_0))$ as a function of $\mathbf{z}$, when $\omega_1(f_\alpha, h) \leq w :|f_\alpha(\mathbf{x}_0 + t\cdot(\mathbf{z}-\mathbf{x}_0)) - f_\alpha(\mathbf{x}_0)| \leq w\left\lceil\frac{t\|\mathbf{z}-\mathbf{x}_0\|}{h}\right\rceil$, all $t \geq 0. For \|\mathbf{z}-\mathbf{x}_0\| \neq 0$ it follows from

(7.4.2) using (7.1.14) that

$$|R_n(\mathbf{z},0)| \le \int_0^1 \int_0^{t_1} \cdots \int_0^{t_{n-1}} \Big(\sum_{|\alpha|=n} \frac{n!}{\alpha_1! \cdots \alpha_k!} \cdot |z_1 - x_{01}|^{\alpha_1} \cdots |z_k - x_{0k}|^{\alpha_k} \cdot w$$

$$\cdot \left\lceil \frac{t_n \|z - x_0\|}{h} \right\rceil dt_n) \cdots \Big) dt$$

$$= \sum_{|\alpha|=n} \frac{n!}{\alpha_1! \cdots \alpha_k!} \cdot \frac{\prod_{i=1}^k |z_i - x_{0i}|^{\alpha_i}}{\|\mathbf{z} - \mathbf{x}_0\|^n} \cdot w \cdot \phi_n(\|\mathbf{z} - \mathbf{x}_0\|)$$

$$= w\phi_n(\|\mathbf{z} - \mathbf{x}_0\|),$$

since $\|\mathbf{z} - \mathbf{x}_0\| = \sum_{i=1}^k |z_i - x_{0i}|$. Therefore

$$|R_n(\mathbf{z},0)| \le w\phi_n(\|\mathbf{z} - \mathbf{x}_0\|), \quad \text{for all } \mathbf{z} \in Q. \tag{7.4.4}$$

Also $g_{\mathbf{z}}(0) = f(\mathbf{x}_0)$. Integrating (7.4.3) relative to μ and using (7.4.4) we conclude our claim. □

The last is used in:

Theorem 7.4.1. *Take Q of the form $Q := \{\mathbf{x} \in \mathbb{R}^k : \|\mathbf{x}\| \le 1\}$, where $\|\cdot\|$ the ℓ_1-norm in $\mathbb{R}^k$, and let $\mathbf{x}_0 := (x_{01}, \dots, x_{0k}) \in Q$ be fixed. Let the measure μ satisfy $\mu(Q) = 1$ and $\int_Q \|\mathbf{x} - \mathbf{x}_0\| \mu(d\mathbf{x}) := d^*$. Also, let $f \in C^n(Q)$ and suppose that each of its nth partial derivatives has a modulus of continuity $\omega_1(f_\alpha, h) \le w$, where h and w are fixed positive numbers. Then we have the estimate*

$$\left| \int_Q f\, d\mu - f(\mathbf{x}_0) \right| \le \left| \sum_{j=1}^n \frac{1}{j!} \cdot \int_Q g_{\mathbf{x}}^{(j)}(0) \mu(d\mathbf{x}) \right| + w \cdot d^* \cdot \phi_n(1 + \|\mathbf{x}_0\|)/(1 + \|\mathbf{x}_0\|) \tag{7.4.5}$$

where $g_{\mathbf{x}}(t) := f(\mathbf{x}_0 + t \cdot (\mathbf{x} - \mathbf{x}_0))$, $t \ge 0$.

Remark 7.4.1. For $n = 1$, this yields the inequality ($x_0 \in Q$):

$$\left| \int_Q f\, d\mu - f(\mathbf{x}_0) \right| \le \left| \sum_{i=1}^k \frac{\partial f}{\partial x_i}(\mathbf{x}_0) \cdot \int_Q (x_i - x_{0i}) \mu(d\mathbf{x}) \right|$$
$$+ w \cdot d^* \cdot \phi_1(1 + \|\mathbf{x}_0\|)/(1 + \|\mathbf{x}_0\|).$$

Proof of Theorem 7.4.1: Let ν be the probability measure induced by μ and the mapping $x \to \|x - x_0\|$ into the interval $I := (0, 1 + \|x_0\|)$. Then $\int y\, d\nu = d^*$ and we need an upper bound of $\int \phi_n(y)\nu(dy)$. Here

$$\frac{\phi_n(y)}{y} \leq \frac{\phi_n(1+\|x_0\|)}{1+\|x_0\|},$$

thus

$$\int \phi_n(y)\nu(dy) \leq d^* \cdot \phi_n(1+\|x_0\|)/(1+\|x_0\|). \qquad \square$$

Next we give some explicit upper bounds.

Theorem 7.4.2. *Let μ be a probability measure on a convex subset Q of $\mathbb{R}^k$. Further, assume that*

$$\frac{1}{(n+1)} \cdot \left(\int \|\mathbf{x}-\mathbf{x}_0\|^{n+1}\mu(d\mathbf{x})\right)^{1/(n+1)} := h > 0,$$

where $\mathbf{x}_0$ is a fixed point of Q and $\|\cdot\|$ denotes the ℓ_1-norm in $\mathbb{R}^k$. Also, let $f \in C_B^n(Q)$ and suppose that each nth partial derivative has at h a modulus of continuity $\omega_1(f_\alpha, h) \leq w$, where w is a given positive number. Then

$$\left|\int_Q f\,d\mu - f(\mathbf{x}_0)\right| \leq \left|\sum_{j=1}^{n} \frac{1}{j!} \cdot \int_Q g_{\mathbf{x}}^{(j)}(0)\mu(d\mathbf{x})\right|$$

$$+\, wh^n \cdot \left[\frac{3}{2} \cdot \frac{(n+1)^n}{n!} + \frac{(n+1)^{n-1}}{8(n-1)!}\right], \tag{7.4.6}$$

where $g_{\mathbf{x}}(t) := f(\mathbf{x}_0 + t\cdot(\mathbf{x}-\mathbf{x}_0))$, $t \geq 0$.

PROOF: By making use of (7.4.1), (7.1.18), and Hölder's inequality. $\square$

As a related result we have:

Corollary 7.4.1. *Let the random vector $(X_1, \ldots, X_k)$ take values in a convex subset Q of $\mathbb{R}^k$, with distribution function μ and expectation $E(X_i) := x_{0i}$, $i = 1, \ldots, k$. Thus $\mathbf{x}_0 := (x_{01}, \ldots, x_{0k})$ is a fixed point of Q. Further, let $(\int \|\mathbf{x}-\mathbf{x}_0\|^2\mu(d\mathbf{x}))^{1/2} := \sigma$, where $\|\cdot\|$ denotes the ℓ_1-norm in $\mathbb{R}^k$. Be given $f \in C_B^1(Q)$ and suppose that each first partial derivative f_i has a modulus of continuity $\omega_1(f_i, \frac{1}{2}\sigma) \leq w$; $i = 1, \ldots, k$ where w is a given positive number. Then*

$$\left|\int_Q f\,d\mu - f(\mathbf{x}_0)\right| \leq (1.5625)\cdot w\sigma. \tag{7.4.7}$$

PROOF: Apply Theorem 7.4.2 with $n = 1$ and $h = \sigma/2$. □

As an application of the last corollary we give:

Example 7.4.1. Let $f \in C^1([0,1]^2)$, the two-dimensional Bernstein polynomials of f are defined by

$$B_{m,n}(f; x_1, x_2) := \sum_{k=0}^{m} \sum_{\ell=0}^{n} f\left(\frac{k}{m}, \frac{\ell}{n}\right) \cdot \binom{m}{k} \cdot \binom{n}{\ell} \cdot x_1^k \cdot (1-x_1)^{m-k} \cdot x_2^{\ell} \cdot (1-x_2)^{n-\ell},$$

all $(x_1, x_2) \in [0,1]^2$, it is known that $B_{m,n}(f) \to f$ uniformly on $[0,1]^2$. By giving the obvious probabilistic interpretation to the above positive linear operator (see Schurer and Steutel (1979)), as well as using Schwarz's inequality to estimate σ and applying (7.4.7) with $w := \max\{\omega_1(f_i, \frac{1}{4} \cdot (\frac{1}{\sqrt{m}} + \frac{1}{\sqrt{n}}))\colon i = 1,2\}$ one obtains:

$$|B_{m,n}(f; x_1, x_2) - f(x_1, x_2)| \le 1.5625 \cdot w \cdot \left(\sqrt{\frac{1}{m} \cdot x_1 \cdot (1-x_1)} + \sqrt{\frac{1}{n} \cdot x_2 \cdot (1-x_2)}\right)$$

$$\le 0.78125 \cdot w \cdot \left(\frac{1}{\sqrt{m}} + \frac{1}{\sqrt{n}}\right), \quad \text{all } (x_1, x_2) \in [0,1]^2. \tag{7.4.8}$$

The following Lemma 7.4.2 is needed in order to present a similar inequality over a rectangle $K \subset \mathbb{R}^2$ when we know only the expectations $\int (x_i - x_{0i})\mu(d\mathbf{x})$; $i = 1, 2$, where $\mathbf{x}_0 := (x_{01}, x_{02}) \in K$.

Preliminaries 7.4.1.

Let $K := [a,b] \times [c,d]$ and let $\phi\colon K \to \mathbb{R}$ be a continuous and convex function and be given $y := (y_1, y_2) \in K$. Consider probability measures μ on K satisfying $\int_K x_i \mu(dx) := y_i$, $i = 1, 2$. For later use, we would like to find

$$U(y) := U(\phi, y) := \sup_{\mu} \int_K \phi(x_1, x_2)\mu(dx). \tag{7.4.9}$$

Denote $\xi_1 := (b,d)$, $\xi_2 := (b,c)$, $\xi_3 := (a,c)$, $\xi_4 := (a,d)$,

$$D := [(\phi(\xi_1) + \phi(\xi_3)) - (\phi(\xi_2) + \phi(\xi_4))] \text{ and}$$

$$B := (b - y_1)/(b-a), \quad C := (d - y_2)/(d-c).$$

Further

$$T := [(1-B)\cdot\phi(\xi_2) + (B+C-1)\cdot\phi(\xi_3) + (1-C)\cdot\phi(\xi_4)],$$

$$\gamma := \min(1-B, 1-C) \quad \text{and} \quad \delta := \max(0, 1-B-C).$$

Lemma 7.4.2. *We have*

$$U(y) = \begin{cases} D\cdot\gamma + T, & \text{if } D > 0; \\ D\cdot\delta + T, & \text{if } D \le 0. \end{cases} \tag{7.4.10}$$

PROOF: From the well-known theory (Alfsen (1971); see Proposition 1.4.6 and Lemma 1.4.7, pp. 35–36), the optimal measure is carried by the four extreme points of K. Let z denote the mass at the vertex ξ_1, $0 \le z \le 1$. Then one can compute the masses p, q, r at ξ_2, ξ_3, ξ_4, respectively, from the given moment conditions:

$$(z+p+q+r=1;\quad zb+pb+qa+ra=y_1;\quad zd+pc+qc+rd=y_2). \tag{7.4.11}$$

Solving this system (7.4.11) we obtain

$$p = -z + (1-B);\quad q = z + ((B+C)-1);\quad r = -z + (1-C).$$

In order $p, q, r \ge 0$ we need $\delta \le z \le \gamma$. Moreover,

$$\theta := \int_{\{\xi_1,\xi_2,\xi_3,\xi_4\}} \phi(\xi)\mu(d\xi) = D\cdot z + T.$$

Hence,

$$U(y) = \sup_z \theta = \sup_z (D\cdot z + T) = \begin{cases} D\cdot\gamma + T, & \text{if } D > 0; \\ D\cdot\delta + T, & \text{if } D \le 0. \end{cases} \qquad \square$$

Now we are ready to state:

Proposition 7.4.1. *Let $K := [a,b]\times[c,d] \subset \mathbb{R}^2$ and a probability measure μ on K be such that $\int_K (x_i - x_{0i})\mu(d\mathbf{x}) := y_i$, $i = 1,2$ where the points $y := (y_1, y_2)$, $\mathbf{x}_0 := (x_{01}, x_{02}) \in K$ are given. Also, let $f \in C^n(K)$ and suppose that each of its nth partial derivatives has,* *relative to K and the ℓ_1-norm, a modulus of continuity $\omega_1(f_\alpha, h) \leq w$. Here h and w are fixed positive numbers. Then*

$$\left|\int_K f\,d\mu - f(\mathbf{x}_0)\right| \leq \left|\sum_{j=1}^{n} \frac{1}{j!} \cdot \int_K g_{\mathbf{x}}^{(j)}(0)\mu(d\mathbf{x})\right| + U(\phi, \mathbf{x}_0 + \mathbf{y}), \tag{7.4.12}$$

where $g_{\mathbf{x}}(t) := f(\mathbf{x}_0 + t \cdot (\mathbf{x} - \mathbf{x}_0))$, $t \geq 0$ and $\phi(\mathbf{x}) := w\phi_n(\|\mathbf{x} - \mathbf{x}_0\|)$.

PROOF: Immediate from Lemma 7.4.1 and the definition (7.4.9) of $U(\phi, y)$. □

As a complementary result we give:

Corollary 7.4.2. *Let $Q := \{\mathbf{x} \in \mathbb{R}^k\colon -\alpha \leq x_i \leq \alpha, i = 1, \ldots, k\}$, $\alpha > 0$ and μ a probability measure on Q. Also, let $f \in C^n(Q)$ and suppose that the nth order partial derivatives for $h > 0$ have modulus of continuity (relative to the ℓ_1-norm) at most $w > 0$. Denote $g_{\mathbf{x}}(t) := (\mathbf{x}_0 + t \cdot (\mathbf{x} - \mathbf{x}_0))$ for all $t \geq 0$.*

Then

$$\left|\int_Q f\,d\mu - f(\mathbf{x}_0)\right| \leq \left|\sum_{j=1}^{n} \frac{1}{j!} \cdot \int_Q g_{\mathbf{x}}^{(j)}(0)\mu(d\mathbf{x})\right| + w\phi_n(\|x_0\| + k\alpha). \tag{7.4.13}$$

PROOF: By Lemma 7.4.1 noting that $\phi_n(\|x - x_0\|) \leq \phi_n(\|x_0\| + k\alpha)$. □

Preliminaries 7.4.2

Let Q be a compact subset of $\mathbb{R}^k$ with the ℓ_1-norm. Consider the set

$$L_\alpha^{(n)} := \left\{f \in C^n(Q)\colon \omega_1\left(\frac{\partial^\beta f}{\partial x^\beta}, h\right) \leq Ah^\alpha; \text{ all } |\beta| = n\right\},$$

where h is a positive variable, $0 < \alpha \leq 1$ and A is a positive constant. As a related result, we have:

Theorem 7.4.3. *Let $f \in L_\alpha^{(n)}$ and n even. Consider a probability measure μ on Q such that*

$$\left(\int_Q \|z\|^{n+1}\mu(dz)\right)^{1/(n+1)} := D_{n+1} > 0.$$

Then

$$\left|\int_Q f\,d\mu - f(\mathbf{0})\right| \le \left|\sum_{j=1}^{n}\frac{1}{j!}\cdot\int_Q g_{\mathbf{z}}^{(j)}(0)\mu(d\mathbf{z})\right| + c\cdot(D_{n+1})^{n+\alpha}, \qquad (7.4.14)$$

where $c := A/[(1+\alpha)(2+\alpha)\cdots(n+\alpha)]$ *and* $g_{\mathbf{z}}(t) := f(t\cdot\mathbf{z})$, $t\ge 0$. *The above inequality is attained by the function* $r(z) := c\cdot\|z\|^{n+\alpha}$ *and the probability measure* μ *with mass* $1/2$ *at* z_0, $-z_0$ *and also by the measure with mass* 1 *at* z_0. *Here, we assume that* $\|z_0\| = D_{n+1}$, $\pm z_0 \in Q$ *and* $Q\subset\Delta := \{(z_1,\dots,z_1)\colon z_1\ge 0 \text{ or } z_1 < 0\}$.

PROOF: We have

$$f(z_1,\dots,z_k) = g(1) = \sum_{j=0}^{n}\frac{g^{(j)}(0)}{j!} + \int_0^1 (g^{(n)}(u) - g^{(n)}(0))\cdot\frac{(1-u)^{n-1}}{(n-1)!}\cdot du. \qquad (7.4.15)$$

Since $\omega_1(f_\beta,h)\le Ah^\alpha$; ($|\beta|) = n$), one has $|f_\beta(t\cdot\mathbf{z}) - f_\beta(\mathbf{0})| \le At^\alpha\cdot\|\mathbf{z}\|^\alpha$, for all $t\ge 0$.

Consequently, (compare the proof of Lemma 7.4.1),

$$|g^{(n)}(t) - g^{(n)}(0)| \le At^\alpha\|z\|^\alpha\cdot\left(\sum_{i=1}^{k}|z_i|\right)^n = At^\alpha\cdot\|z\|^{n+\alpha},$$

since $\|z\| = \sum_{i=1}^{k}|z_i|$. Denoting the remainder in (7.4.15) by $R_n(\mathbf{z},0)$, we get:

$$|R_n(z,0)| \le c\cdot\|z\|^{n+\alpha}. \qquad (7.4.16)$$

Integrating (7.4.15) relative to μ and using (7.4.16), one obtains (7.4.14). Here we used that

$$D_{n+\alpha} = \left(\int \|z\|^{n+\alpha}\mu(dz)\right)^{1/(n+\alpha)} \le D_{n+1}.$$

Further, note that the function $r(z) = c\cdot\|z\|^{n+\alpha} = c\cdot\left(\sum_{i=1}^{k}|z_i|\right)^{n+\alpha}$ is convex and since n is even, any nth order single partial derivative has the form $A\|z\|^\alpha$. Moreover,

$$|A\|x\|^\alpha - A\|y\|^\alpha| \le A(|\,\|x\| - \|y\|\,|)^\alpha \le A(\|x-y\|)^\alpha,$$

thus $r\in L_\alpha^{(n)}$ where $Q\subset\Delta$. Also, note that $g_{\mathbf{z}}(t) = r(t\cdot\mathbf{z}) = t^{n+\alpha}\cdot\|z\|^{n+\alpha}$ has $g_{\mathbf{z}}^{(j)}(0) = 0$ for all $0\le j\le n$. These remarks earlier imply the last assertion. □

CHAPTER EIGHT

OPTIMAL KOROVKIN TYPE INEQUALITIES UNDER CONVEXITY

8.1. On the Degree of Weak Convergence of a Sequence of Finite Measures to the Unit Measure Under Convexity

8.1.1. Introduction

The flavor of this section is conveyed by Proposition 8.1.1. It claims the equivalence of the weak convergence of a sequence of finite measures $\{\mu_j\}_{j\in\mathbb{N}}$ on $[a,b] \subset \mathbb{R}$ to the unit (Dirac) measure δ_{x_0}, where $x_0 \in (a,b)$, with the convergence of $\int f\, d\mu_j$ to $f(x_0)$, where $f \in C^m([a,b])$ for some $m \geq 0$ is such that $|f^{(m)}(t) - f^{(m)}(x_0)|$ is convex in t. For this restricted class of functions f we present quantitative estimates on the above weak convergence.

The inequalities shown here are usually the best possible and are stronger than the corresponding ones obtained from Shisha and Mond (1968), Mond and Vasudevan (1980), Gonska (1983), Anastassiou (1984) and others. Generalizations are given to $\mathbb{R}^k$, $k \geq 1$ and arbitrary real normed vector spaces where we are working on a compact and convex domain.

Here by using the method of optimal distance from Geometric Moment Theory, see Chapter 2, we obtain upper bounds and best upper bounds for $\left|\int_Q f\, d\mu - f(x_0)\right|$ under the convexity assumption, where Q is a compact and convex subset. These sometimes lead to sharp inequalities which are attained for particular μ and f. The presented results come from Anastassiou(3) (1987) and Anastassiou(1) (1986).

8.1.2. Preliminaries

We start with

Proposition 8.1.1. *Let m be an integer ≥ 1. Let $\{\mu_j\}_{j\in\mathbb{N}}$ be a sequence of measures on $[a,b] \subset \mathbb{R}$ with corresponding masses m_j: $0 < m_j \leq \tau$ and δ_{x_0} the unit (Dirac) measure at $x_0 \in (a,b)$. Then the following are equivalent:*

(i) $\mu_j \rightrightarrows \delta_{x_0}$ (*weakly*);

(ii) $\int f\,d\mu_j \to f(x_0)$ *for all* $f \in C^m([a,b])$

such that $|f^{(m)}(t) - f^{(m)}(x_0)|$ *is convex in* t.

Proof: (i) $\Rightarrow$ (ii) Obvious (Dunford and Schwartz (1957), p. 316). In fact (i) implies $\int f\,d\mu_j \to f(x_0)$ for all $f \in C([a,b])$.

(ii) $\Rightarrow$ (i) The set of functions $\{1, (t-x_0), (t-x_0)^2\}$ is a subset of $C^m([a,b])$ and for each of them $|f^{(m)}(t) - f^{(m)}(x_0)|$ is a convex function of t.

Therefore, by assumption, for the positive linear functionals $L_j(f) = \int f\,d\mu_j$ we have $L_j(f) \to f(x_0)$ for any $f \in \{1, (t-x_0), (t-x_0)^2\}$.

Since this triplet of functions is a Chebyshev system, by Korovkin's theorem for positive linear functionals Korovkin (1960), we get $\int f\,d\mu_j \to f(x_0)$ for all $f \in C([a,b])$. This implies $\mu_j \rightrightarrows \delta_{x_0}$ (weakly), see Dunford and Schwartz (1957). $\square$

The following result plays an important role in the proofs of this section.

Lemma 8.1.1. *Let* $(V, \|\cdot\|)$ *be a real normed vector space and* U *a star-shaped subset of* V *with respect to* $x_0 \in U$. *Let* w, h *be positive numbers such that* $h \le \|t - x_0\|$ *for each extreme point* $t \not\equiv x_0$ *of* U. *Consider a convex* $f\colon U \to \mathbb{R}$ *such that* $f(x_0) = 0$ *and*

$$|f(x) - f(y)| \le w \quad \text{if} \quad \|x - y\| \le h, \; x, y \in U. \tag{8.1.1}$$

Then the maximal function satisfying the above conditions is

$$\phi(t) = \frac{w}{h}\|t - x_0\|, \quad t \in U,$$

so that

$$f(t) \le \phi(t) \quad \text{for all} \quad t \in U.$$

Note. If U is convex, then in the lemma we require that the ball $B(x_0, h) \subset U$.

Proof: The function ϕ fulfills all the assumptions of the lemma. Namely, $\phi(x_0) = 0$ and for $x, y \in U$ with $\|x - y\| \le h$, we have $|\phi(x) - \phi(y)| \le w$. Also, one easily sees that ϕ is convex. Next, for $t \in U - \{x_0\}$ satisfying $\|t - x_0\| \le h$, consider $x \in U$ such that

$$t = \left(\frac{h - \|t - x_0\|}{h}\right) x_0 + \frac{\|t - x_0\|}{h} x.$$

Then $\|x - x_0\| = h$. Since f is convex and $f(x_0) = 0$, we have

$$f(t) = f\left(\left(\frac{h - \|t - x_0\|}{h}\right) x_0 + \frac{\|t - x_0\|}{h} x\right) \le \frac{\|t - x_0\|}{h} f(x)$$

and therefore $f(t) \le (\|t - x_0\|/h) f(x)$. Thus

$$f(t) \le \frac{\|t - x_0\|}{h} |f(x) - f(x_0)| \le \frac{\|t - x_0\|}{h} w$$

so

$$f(t) \le \frac{\|t - x_0\|}{h} w \quad \text{when } \|t - x_0\| \le h. \tag{8.1.2}$$

Now for $t \in U$ such that $\|t - x_0\| > h$, there is a finite sequence of points $x_1, \ldots, x_n$ on the line segment $\overline{tx_0}$ such that all of $\|x_0 - x_1\|$, $\|x_1 - x_2\|$, $\|x_2 - x_3\|, \ldots, \|x_n - t\|$ are $\le h$ and $\|x_0 - x_1\| + \|x_1 - x_2\| + \cdots + \|x_n - t\| = \|t - x_0\|$. Furthermore, the function $F(t) = f(t) - f(x_1)$ is convex, $F(x_1) = 0$ and fulfills (8.1.1). Since $\|t - x_1\| \le h$, by (8.1.2) we get $f(t) - f(x_1) \le (\|t - x_1\|/h) w$; similarly

$$f(x_1) - f(x_2) \le \frac{\|x_1 - x_2\|}{h} w, \ldots, f(x_n) - f(x_0) \le \frac{\|x_n - x_0\|}{h} w.$$

Adding up all these inequalities, we find $f(t) \le (\|t - x_0\|/h) w$ when $\|t - x_0\| > h$. The proof is now complete. □

8.1.3. One Dimensional Results

Theorem 8.1.1. *Let $r > 0$, μ a finite measure of mass m on an interval $[a, b]$, $x_0 \in (a, b)$. Set $c(x_0) = \max(x_0 - a, b - x_0)$ and*

$$\left(\int |t - x_0|^r \mu(dt)\right)^{1/r} = d_r(x_0), \tag{8.1.3}$$

and assume $d_r(x_0) > 0$. In order that μ exist, we also assume that $d_r^r(x_0) \le m \cdot (c(x_0))^r$. Next consider $f: [a, b] \to \mathbb{R}$ for which $|f(t) - f(x_0)|$ is convex in t and

$$|f(s) - f(t)| \le w \quad \text{when } s, t \in [a, b]; \quad |s - t| \le h. \tag{8.1.4}$$

Here $0 < h \leq \min(x_0 - a, b - x_0)$ *and* $w > 0$ *are fixed.*

A best upper bound is given by

$$\left|\int f\,d\mu - f(x_0)\right| - |m-1|\cdot|f(x_0)| \leq \begin{cases} wm^{1-(1/r)}\left(\dfrac{d_r(x_0)}{h}\right), & r \geq 1, \\[2ex] w(c(x_0))^{1-r}\dfrac{d_r^r(x_0)}{h}, & r \leq 1. \end{cases} \tag{8.1.5}$$

Remark 8.1.1. When $m = 1$, (8.1.5) implies

$$\left|\int f\,d\mu - f(x_0)\right| \leq \begin{cases} w\left(\dfrac{d_r(x_0)}{h}\right), & r \geq 1, \\[2ex] w(c(x_0))^{1-r}\dfrac{d_r^r(x_0)}{h}, & r \leq 1. \end{cases} \tag{8.1.6}$$

If $w = \omega_1(f,h)$ the modulus of continuity of f in $[a,b]$, and $r \geq 1$, (8.1.6) becomes

$$\left|\int f\,d\mu - f(x_0)\right| \leq \omega_1(f,h)\frac{d_r(x_0)}{h}, \tag{8.1.7}$$

which in case $d_r(x_0) = l \cdot h$, $l \geq 1$, turns out to be

$$\left|\int f\,d\mu - f(x_0)\right| \leq l \cdot \omega_1\left(f, \frac{1}{l}\cdot d_r(x_0)\right). \tag{8.1.8}$$

Note that inequality (8.1.7) is sharp when $r = 1$, namely, equality is attained by $f(t) = |t - x_0|$ where both of sides are $d_1(x_0)$.

Corollary 8.1.1. *For* $m = 1$ *and* $h = d_2(x_0) \leq \min(x_0 - a, b - x_0)$ *we have*

$$\left|\int f\,d\mu - f(x_0)\right| \leq \omega_1(f, d_2(x_0)). \tag{8.1.9}$$

This is also true for $f \in C_B(\mathbb{R})$ *(the space of real, bounded, continuous functions on* $(-\infty,\infty)$*) when* $h = d_2(x_0) < \infty$.

PROOF: Obvious from (8.1.7). □

PROOF OF THEOREM 8.1.1. Let $g(t) = f(t) - f(x_0)$. From Lemma 8.1.1 we have

$$|g(t)| \leq \frac{w}{h}|t - x_0|.$$

Thus

$$\left|\int f\,d\mu - f(x_0)\right| = \left|\int g\,d\mu + (m-1)f(x_0)\right| \le \int |g|d\mu + |m-1|\cdot|f(x_0)|,$$

i.e.,

$$\left|\int f\,d\mu - f(x_0)\right| \le |m-1|\cdot|f(x_0)| + \frac{w}{h}\int |t-x_0|\mu(dt). \tag{8.1.10}$$

Here, equality holds for $f(t) = (w/h)|t-x_0|$ which fulfills the assumptions of the theorem.

The best constant θ in (8.1.10) is given by

$$\theta = \sup_{\mu}\int |t-x_0|\mu(dt),$$

where μ ranges over all measures on $[a,b]$ of mass m satisfying (8.1.3).

Letting $\gamma = m^{-1}\mu$ we determine

$$U = \sup_{\gamma}\int |t-x_0|\gamma(dt),$$

where γ ranges over all probability measures on $[a,b]$ satisfying

$$\int |t-x_0|^r\cdot\gamma(dt) = d_r^r(x_0)/m.$$

Note that $0 \le |t-x_0| \le c(x_0) = \max(x_0-a, b-x_0)$. Taking the probability measure ρ induced by γ and the mapping $t \to |t-x_0|$ and denoting $u = |t-x_0|$, we seek

$$U = \sup_{\rho}\int u\,\rho(du) \quad (0 \le u \le c(x_0))$$

over all probability measures ρ such that

$$\int u^r\cdot\rho(du) = d_r^r(x_0)/m.$$

It follows (see Chapter 2, method of optimal distance) that

$$U = \phi(d_r^r(x_0)/m),$$

where

$$\Gamma_1 = \{(z,\phi(z)) : 0 \le z \le c^r(x_0)\}$$

is the upper boundary of the convex hull of the curve

$$\Gamma_0 = \{(u^r, u): 0 \le u \le c(x_0)\}.$$

When $r \ge 1$, Γ_0 is concave and

$$U = d_r(x_0)/m^{1/r},$$

while, when $r < 1$, Γ_0 is convex and

$$U = \frac{d_r^r(x_0)}{m}(c(x_0))^{1-r}.$$

As a result we get the best upper bound

$$\left|\int f\,d\mu - f(x_0)\right| \le |m-1| \cdot |f(x_0| + \frac{w}{h}\theta,$$

which completes the proof of the theorem. □

An application of Corollary 8.1.1 is

Corollary 8.1.2. *Let $f \in C_B[0,\infty)$ be such that $|f(t) - f(x_0)|$ is a convex function of t for a fixed $x_0 \ge 1$. Consider the Szász–Mirakjan operator applied to f at x_0:*

$$(U_n f)(x_0) = e^{-n \cdot x_0} \sum_{k=0}^{\infty} f\left(\frac{k}{n}\right) \frac{(n \cdot x_0)^k}{k!}.$$

Then

$$|(U_n f)(x_0) - f(x_0)| \le \omega_1\left(f, \left(\frac{x_0}{n}\right)^{1/2}\right).$$

PROOF: Consider $(X_j)_{j\in\mathbb{N}}$, Poisson (i.i.d.) random variables with parameter $x_0 \ge 1$, so that $E(X) = \mathrm{Var}(X) = x_0$. Put $S_n = \sum_{j=1}^{n} X_j$, $n \ge 1$; then $E(S_n/n) = x_0$ and $\mathrm{Var}(S_n/n) = x_0/n$. Note that $\sqrt{x_0/n} \le x_0$, so we can apply inequality (8.1.9) for $\mu = F_{S_n/n}$, the distribution function of S_n/n. □

For differentiable functions we have

Theorem 8.1.2. *Let $r > 0$, μ a finite measure on $[a,b] \subset \mathbb{R}$, $x_0 \in (a,b)$ and $c(x_0) = \max(x_0 - a, b - x_0)$. Put*

$$c_k(x_0) = \int (t - x_0)^k \mu(dt), \quad k = 0, 1, \dots, n;$$
$$d_r(x_0) = \left(\int |t - x_0|^r \cdot \mu(dt) \right)^{1/r}. \tag{8.1.11}$$

Let $f \in C^n[a,b]$, $n \geq 1$, and assume $|f^{(n)}(t) - f^{(n)}(x_0)|$ is convex in t and

$$|f^{(n)}(s) - f^{(n)}(t)| \leq w \ \text{ if } s, t \in [a,b] \text{ and } |s - t| \leq h. \tag{8.1.12}$$

Here $0 < h \leq \min(x_0 - a, b - x_0)$ and $w > 0$ are fixed.

Then

$$E(x_0) = \left| \int f\, d\mu - f(x_0) \right| - |f(x_0)| \cdot |c_0(x_0) - 1| - \sum_{k=1}^{n} \frac{|f^{(k)}(x_0)|}{k!} \cdot |c_k(x_0)|$$

$$\leq \begin{cases} \dfrac{w}{h(n+1)!} d_r^{n+1}(x_0) c_0(x_0)^{1-((n+1)/r)}, & r \geq n+1 \\[2ex] \dfrac{w}{h(n+1)!} d_r^r(x_0)(c(x_0))^{(n+1)-r}, & r \leq n+1. \end{cases} \tag{8.1.13}$$

Note. When $r = n + 1$ and $w = \omega_1(f^{(n)}, h)$,

$$E(x_0) \leq \frac{\omega_1(f^{(n)}, h)}{h(n+1)!} d_{n+1}^{n+1}(x_0), \tag{8.1.14}$$

which, for $h = d_{n+1}^{n+1}(x_0)/(n+1)!$, becomes

$$E(x_0) \leq \omega_1 \left(f^{(n)}, \frac{d_{n+1}^{n+1}(x_0)}{(n+1)!} \right). \tag{8.1.15}$$

Inequality (8.1.14) is sharp; equality is attained by the function

$$\tilde{f}(t) = \begin{cases} \dfrac{(t - x_0)^{n+1}}{(n+1)!}, & x_0 \leq t \leq b, \\[2ex] 0, & a \leq t \leq x_0, \end{cases}$$

when $b - x_0 \geq x_0 - a$, and by the function

$$\tilde{\tilde{f}}(t) = \begin{cases} \dfrac{(x_0 - t)^{n+1}}{(n+1)!}, & a \leq t \leq x_0, \\ \\ 0, & x_0 \leq t \leq b, \end{cases}$$

when $b - x_0 \leq x_0 - a$.

In the first case an optimal measure μ_{x_0} is of mass $c_0(x_0)$, supported by $\{x_0, b\}$ and in the second case it is of the same mass $c_0(x_0)$, supported by $\{x_0, a\}$. In both cases the corresponding masses are $[c_0(x_0) - (d_{n+1}(x_0)/c(x_0))^{n+1}]$ and $(d_{n+1}(x_0)/c(x_0))^{n+1}$.

Remark 8.1.2. When $r = n$ and $w = \omega_1(f^{(n)}, h)$, inequality (8.1.13) becomes

$$E(x_0) \leq \frac{\omega_1(f^{(n)}, h)}{h(n+1)!} d_n^n(x_0) c(x_0) = \frac{\omega_1(f^{(n)}, h)}{h} \left(\frac{d_n(x_0)}{c(x_0)}\right)^n \frac{(c(x_0))^{n+1}}{(n+1)!}. \tag{8.1.16}$$

This is also sharp and equality is attained as in (8.1.14).

Proof of Theorem 8.1.2.

$$f(t) = \sum_{k=0}^{n} \frac{f^{(k)}(x_0)}{k!}(t - x_0)^k + I_t, \tag{8.1.17}$$

where

$$I_t = \int_{x_0}^{t} \left(\int_{x_0}^{t_1} \cdots \left(\int_{x_0}^{t_{n-1}} (f^{(n)}(t_n) - f^{(n)}(x_0)) dt_n \right) \cdots \right) dt_1.$$

By Lemma 8.1.1

$$|f^{(n)}(t) - f^{(n)}(x_0)| \leq \frac{w}{h} \cdot |t - x_0|,$$

$$|I_t| \leq \frac{w}{h} \cdot \frac{|t - x_0|^{n+1}}{(n+1)!}.$$

From (8.1.17), integrating relative to μ, we get

$$\left| \int f \, d\mu - f(x_0) \right| \leq |f(x_0)| \cdot |c_0(x_0) - 1| + \sum_{k=1}^{n} \frac{|f^{(k)}(x_0)|}{k!} \cdot |c_k(x_0)| + \frac{w}{h(n+1)!} \int |t - x_0|^{n+1} \mu(dt).$$

We would like to find

$$\theta = \sup_{\mu} \int |t - x_0|^{n+1} \mu(dt)$$

over all measures μ on $[a,b]$ of mass $c_0(x_0)$ with $(\int |t-x_0|^r \mu(dt))^{1/r} = d_r(x_0)$, when $d_r(x_0) > 0$.

Equivalently, we want

$$U = \sup_{\gamma} \int |t-x_0|^{n+1} \gamma(dt) \qquad (\theta = c_0(x_0)U)$$

over all probability measures $\gamma = m^{-1}\mu$ such that

$$\int |t-x_0|^r \gamma(dt) = d_r^r(x_0)/c_0(x_0).$$

Note that $0 \leq |t-x_0| \leq c(x_0) = \max(x_0 - a, b - x_0)$. Let ρ be the probability measure induced by γ and the mapping $t \to |t-x_0|$ and let $u = |t-x_0|$; we want to find

$$U = \sup_{\rho} \int u^{n+1} \rho(du) \qquad (0 \leq u \leq c(x_0)),$$

where ρ runs over all probability measures on $[0, c(x_0)]$ such that

$$\int u^r \rho(du) = d_r^r(x_0)/c_0(x_0).$$

From the method of optimal distance, see Chapter 2, it follows that

$$U = \psi(d_r^r(x_0)/c_0(x_0)),$$

where $\{(z, \psi(z)) : 0 \leq z \leq c^r(x_0)\}$ is the upper boundary of the convex hull of the curve

$$G_0 = \{(u^r, u^{n+1}) : 0 \leq u \leq c(x_0)\}.$$

When $r \geq n+1$, G_0 is concave and

$$U = d_r^{n+1}(x_0)/(c_0(x_0))^{(n+1)/r},$$

while, when $r < n+1$, G_0 is convex and

$$U = \frac{d_r^r(x_0)}{c_0(x_0)}(c(x_0))^{(n+1-r)}.$$

Note that, for $r = n+1$, we find

$$U = d_{n+1}^{n+1}(x_0)/c_0(x_0).$$

Thus we get the upper bound

$$\left|\int f\,d\mu - f(x_0)\right| \le |f(x_0)|\cdot|c_0(x_0)-1| + \sum_{k=1}^{n}\frac{|f^{(k)}(x_0)|}{k!}\cdot|c_k(x_0)| + \frac{w}{h(n+1)!}\theta.$$

This completes the proof of the theorem. □

Corollary 8.1.3. *Let $x_0 \in (a,b)$ and $f \in C^1([a,b])$ be such that $|f'(t)-f'(x_0)|$ is a convex function of t. Let μ be a probability measure on $[a,b]$ for which $\int t\,\mu(dt) = x_0$ and*

$$\left(\int (t-x_0)^2\mu(dt)\right)^{1/2} = d_2(x_0) > 0.$$

If $d_2^2(x_0) \le 2\min(x_0-a, b-x_0)$, we get the sharp (attained) inequality

$$\left|\int f\,d\mu - f(x_0)\right| \le \omega_1(f', \frac{1}{2}d_2^2(x_0)). \tag{8.1.18}$$

And if $d_2(x_0) \le 2\min(x_0-a, b-x_0)$, we obtain the sharp inequality:

$$\left|\int f\,d\mu - f(x_0)\right| \le \omega_1(f', \frac{1}{2}d_2(x_0))d_2(x_0). \tag{8.1.19}$$

Corollary 8.1.4. *Let the random variable X have distribution μ, $E(X) = x_0$ and $\mathrm{Var}(X) = \sigma^2 < \infty$. Consider those $f \in C^1(\mathbb{R})$ for which $Ef(X) < \infty$ and $|f'(t)-f'(x_0)|$ is convex in t. Then we have the sharp inequality (attained by $f(t) = (t-x_0)^2$):*

$$|Ef(X) - f(x_0)| \le \min\left\{\omega_1\left(f', \frac{\sigma^2}{2}\right), \omega_1\left(f', \frac{\sigma}{2}\right)\sigma\right\}. \tag{8.1.20}$$

The next result will be used in Theorem 8.1.3.

Lemma 8.1.2. *Let $x_0 \in (a,b) \subset \mathbb{R}$, and let $c_1(x_0)$ and $c_0(x_0) > 0$, $d_1(x_0) > 0$ be given numbers. Consider all measures μ on $[a,b]$ of mass $c_0(x_0)$ such that $\int(t-x_0)\mu(dt) = c_1(x_0)$, $\int|t-x_0|\mu(dt) = d_1(x_0)$. Put*

$$U(x_0) = \sup_{\mu}\int\frac{(t-x_0)^2}{c_0(x_0)}\mu(dt).$$

Then

$$U(x_0) = (b - x_0)\left(\frac{d_1(x_0) + c_1(x_0)}{2c_0(x_0)}\right) + (x_0 - a)\left(\frac{d_1(x_0) - c_1(x_0)}{2c_0(x_0)}\right). \qquad (8.1.21)$$

An optimal measure is supported by $\{a, x_0, b\}$.

PROOF: Easy. □

Inequality (8.1.13) can be improved if we know more about μ. One result in this direction is

Theorem 8.1.3. *Under the hypothesis of Lemma* 8.1.2, *let* $f \in C^1([a,b])$ *with* $\omega_1(f', h) \leq w$, *where* w, h *are given positive numbers such that* $0 < h \leq \min(x_0 - a, b - x_0)$. *Suppose* $|f'(t) - f'(x_0)|$ *is a convex function of* t.

Then

$$\left|\int f\, d\mu - f(x_0)\right| \leq |f(x_0)| \cdot |c_0(x_0) - 1| + |f'(x_0)| \cdot |c_1(x_0)| + \frac{w}{2h} c_0(x_0) U(x_0). \qquad (8.1.22)$$

This inequality gives a best upper bound.

Inequality (8.1.22) is sharp, namely, it is attained when $c_0(x_0) = 1$, $c_1(x_0) = 0$, $\omega_1(f', h) = w$, $f(t) = (t - x_0)^2/2$ and μ is the probability measure supported by $\{a, x_0, b\}$ with masses

$$\frac{d_1(x_0)}{2(x_0 - a)}, \left[1 - \frac{d_1(x_0)}{2(x_0 - a)} - \frac{d_1(x_0)}{2(b - x_0)}\right], \frac{d_1(x_0)}{2(b - x_0)},$$

respectively.

PROOF: Easy. □

Note. When $n = r = 1$, inequality (8.1.22) is better than the corresponding inequality (8.1.13).

An application to Corollary 8.1.3 is

Corollary 8.1.5. *Let* $f \in C^1([0,1])$ *be such that* $|f'(t) - f'(x_0)|$ *is a convex function of* t, *let* $x_0 \in (0,1)$ *and consider the nth Bernstein operator applied to* f *at* x_0

$$(B_n f)(x_0) = \sum_{k=0}^{n} f(k/n) \binom{n}{k} x_0^k (1 - x_0)^{n-k}.$$

Then

(i)

$$|(B_nf)(x_0) - f(x_0)| \le \omega_1\left(f', \frac{x_0(1-x_0)}{2n}\right) \le \omega_1\left(f', \frac{1}{8n}\right), \tag{8.1.23}$$

(ii)

$$\begin{aligned}|(B_nf)(x_0) - f(x_0)| &\le \omega_1\left(f', \frac{1}{2}\sqrt{\frac{x_0(1-x_0)}{n}}\right) \cdot \sqrt{\frac{x_0(1-x_0)}{n}} \\ &\le \omega_1\left(f', \frac{1}{4\sqrt{n}}\right) \cdot \frac{1}{2\sqrt{n}}, \qquad x_0 \in \left[\frac{1}{5}, \frac{4}{5}\right].\end{aligned} \tag{8.1.24}$$

PROOF: (i) Let $(X_j)_{j\in\mathbb{N}}$ be Bernoulli (i.i.d.) random variables such that $P(X_j = 0) = 1 - x_0$, $P(X_j = 1) = x_0$; then $E(X) = x_0$, $\mathrm{Var}(X) = x_0(1-x_0)$. Put $S_n = \sum_{j=1}^n X_j$, $n \ge 1$; then $E(S_n/n) = x_0$ and $\mathrm{Var}(S_n/n) = x_0(1-x_0)/n < 2\min(x_0, 1-x_0)$. Now apply inequality (8.1.18) to $\mu = F_{S_n/n}$, the distribution function of S_n/n. Further, note that $\max_{0<x_0<1}(x_0(1-x_0)) = \frac{1}{4}$, being attained at $x_0 = \frac{1}{2}$.

(ii) Proved similarly. □

An application to Corollary 8.1.4 is

Corollary 8.1.6. *Let f be a real function, bounded and having a continuous bounded derivative on $(-\infty, \infty)$ and let $|f'(t) - f'(x_0)|$ be a convex function of t for some fixed $x_0 \in \mathbb{R}$. Consider the nth Weierstrass operator applied to f at x_0:*

$$(W_nf)(x_0) = \sqrt{n/\pi}\int_{-\infty}^{\infty} f(x)e^{-n(x-x_0)^2}\,dx.$$

Then

$$|(W_nf)(x_0) - f(x_0)| \le \min\left\{\omega_1\left(f', \frac{1}{4n}\right), \omega_1\left(f', \frac{1}{2\sqrt{2n}}\right) \cdot \frac{1}{\sqrt{2n}}\right\}. \tag{8.1.25}$$

PROOF: As that of Corollary 8.1.5. Here the random variable X has the normal distribution $(x_0, \frac{1}{2})$ with density $(1/\sqrt{\pi})e^{-(x-x_0)^2}$. Then apply inequality (8.1.20). □

Remark 8.1.3. In Theorems 8.1.1–2 and related results, when $x_0 = 0$ and $h \le \min(|a|, b)$, we can use instead of ω_1

$$\begin{aligned}\overline{\omega}_1(f^{(m)}, h) &= \sup\{|f^{(m)}(x) - f^{(m)}(y)| : x \cdot y \ge 0 \text{ and } |x-y| \le h\} \\ &< \omega_1(f^{(m)}, h), \quad \text{for any integer } m \ge 0.\end{aligned}$$

8.1.4. Multidimensional Results

Definition 8.1.1. For f a continuous real valued function on a compact subset Q of $\mathbb{R}^k$, $k \geq 1$, its modulus of continuity is

$$\omega_1(f,h) = \sup\{|f(\mathbf{x}) - f(\mathbf{y})|: \text{all } \mathbf{x},\mathbf{y} \in Q, \|\mathbf{x}-\mathbf{y}\| \leq h\},$$

where $\|\cdot\|$ is a norm in $\mathbb{R}^k$.

Theorem 8.1.4. *Let Q be a compact and convex subset of $\mathbb{R}^k$, $k \geq 1$, let $\mathbf{x}_0 = (x_{01}, \ldots, x_{0k}) \in Q$ be fixed and let μ be a probability measure on Q. Let $f \in C^n(Q)$, $n \geq 1$, and suppose that each nth partial derivative $f_\alpha = \partial^\alpha f/\partial x^\alpha$, where $\alpha = (\alpha_1, \ldots, \alpha_k)$, $\alpha_i \geq 0$, $i = 1, \ldots, k$, and $|\alpha| = \sum_{i=1}^k \alpha_i = n$ has, relative to Q and the l_1-norm, a modulus of continuity $\omega_1(f_\alpha, h) \leq w$, and each $|f_\alpha(\mathbf{x}) - f_\alpha(\mathbf{x}_0)|$ is a convex function of $\mathbf{x}$. Here h and w are given positive numbers, and h is chosen so that the ball in $\mathbb{R}^k$: $B(\mathbf{x}_0, h)$ is contained in Q. Then*

$$\left|\int_Q f\,d\mu - f(\mathbf{x}_0)\right| \leq \left|\sum_{j=1}^n \frac{1}{j!}\int_Q g_{\mathbf{x}}^{(j)}(0)\mu(d\mathbf{x})\right| + \frac{w}{h(n+1)!}\int_Q \|x - x_0\|^{n+1}\mu(d\mathbf{x}), \quad (8.1.26)$$

where $g_{\mathbf{x}}(t) = f(\mathbf{x}_0 + t(\mathbf{x}-\mathbf{x}_0))$, $t \geq 0$.

PROOF:

$$f(z_1, \ldots, z_k) = g_{\mathbf{z}}(1) = \sum_{j=0}^n \frac{g_{\mathbf{z}(0)}^{(j)}}{j!} + R_n(\mathbf{z}, 0), \quad (8.1.27)$$

where

$$g_{\mathbf{z}}^{(j)}(t) = \left[\left(\sum_{i=1}^k (z_i - x_{0i})\frac{\partial}{\partial x_i}\right)^j f\right](x_{01} + t(z_1 - x_{01}), \ldots, x_{0k} + t(z_k - x_{0k})) \quad (8.1.28)$$

is the jth derivative of $g_{\mathbf{z}}(t) = f(\mathbf{x}_0 + t(\mathbf{z}-\mathbf{x}_0))$ and

$$R_n(\mathbf{z},0) = \int_0^1 \left(\int_0^{t_1} \cdots \left(\int_0^{t_{n-1}} (g_{\mathbf{z}}^{(n)}(t_n) - g_{\mathbf{z}}^{(n)}(0))dt_n\right)\cdots\right)dt_1.$$

By Lemma 8.1.1 we get

$$|f_\alpha(\mathbf{x}_0 + t(\mathbf{z}-\mathbf{x}_0)) - f_\alpha(\mathbf{x}_0)| \leq \frac{w}{h}t\|\mathbf{z}-\mathbf{x}_0\|, \quad \text{all } t \geq 0.$$

It follows from (8.1.28) that

$$|R_n(\mathbf{x},0)| \leq \int_0^1 \left[\int_0^{t_1} \cdots \left[\int_0^{t_{n-1}} \left(\sum_{|\alpha|=n} \frac{n!\prod_{i=1}^k |z_i - x_{0i}|^{\alpha_i}}{\alpha_1!\cdots\alpha_k!}\frac{w}{h}\|\mathbf{z}-\mathbf{x}_0\| t_n\right)\cdot dt_n\right]\cdots\right]\cdot dt_1 = \frac{w}{h}\frac{\|\mathbf{z}-\mathbf{x}_0\|^{n+1}}{(n+1)!}.$$

Therefore

$$|R_n(\mathbf{z},0)| \leq \frac{w}{h}\frac{\|\mathbf{z}-\mathbf{x}_0\|^{n+1}}{(n+1)!}, \quad \text{for all } \mathbf{z} \in Q. \tag{8.1.29}$$

Note $g_{\mathbf{z}}(0) = f(\mathbf{x}_0)$. Integrating (8.1.27) relative to μ and using (8.1.29), (8.1.26) follows. □

Remark 8.1.4. Let n be even and let Q be the closed line segment in $\mathbb{R}^k$ $(k \geq 1)$ joining

$$\underbrace{(-1,-1,\ldots,-1)}_{k} \quad \text{to} \quad \underbrace{(1,1,\ldots 1)}_{k}.$$

Let $\mathbf{x}_0 = \mathbf{0}$, let $0 < h \leq 1$ and take $w = \max\{\omega_1(f_\alpha, h)\colon \text{all } \alpha \text{ such that } |\alpha| = n\}$. For $f(\mathbf{x}) = \|\mathbf{x}\|^{n+1}/(n+1)!$, $\|\cdot\|$ the l_1 norm in $\mathbb{R}^k$, equality is attained in (8.1.26).

Namely, all $f_\alpha(\mathbf{x}) = \|\mathbf{x}\|$ are convex functions so that $\omega_1(f_\alpha, h) = h$ and $g_{\mathbf{z}}^{(j)}(0) = 0$ for all $0 \leq j \leq n$.

Illustration 8.1.1. (i) For $Q = \{\mathbf{x} \in \mathbb{R}^k\colon \|\mathbf{x}\|_{l_1} \leq 1\}$ we have

$$\left|\int_Q f\,d\mu - f(\mathbf{x}_0)\right| \leq \left|\sum_{j=1}^n \frac{1}{j!}\int_Q g_{\mathbf{x}}^{(j)}(0)\mu(d\mathbf{x})\right| + \frac{w}{h}\frac{(1+\|\mathbf{x}_0\|)^{n+1}}{(n+1)!}, \tag{8.1.30}$$

where h is such that $B(\mathbf{x}_0, h) \subset Q$.

(ii) For $Q = \{\mathbf{x} \in \mathbb{R}^k\colon -\lambda \leq x_i \leq \lambda,\ i = 1,\ldots,k\}$, $\lambda > 0$, we get

$$\left|\int_Q f\,d\mu - f(\mathbf{x}_0)\right| \leq \left|\sum_{j=1}^n \frac{1}{j!}\int_Q g_{\mathbf{x}}^{(j)}(0)\mu(d\mathbf{x})\right| + \frac{w}{h}\frac{(\|\mathbf{x}_0\| + k\lambda)^{n+1}}{(n+1)!}, \tag{8.1.31}$$

where $\|\cdot\|$ is the l_1 norm in $\mathbb{R}^k$ and $B(\mathbf{x}_0, h) \subset Q$.

Corollary 8.1.7. *Let the random vector $(X_1,\ldots,X_k)$ take values in a convex subset Q of $\mathbb{R}^k$, with distribution function μ and expectations $E(X_i) = x_{0i}$, $i = 1,\ldots,k$. Thus*

$\mathbf{x}_0 = (x_{01}, \ldots, x_{0k})$ *is a fixed point of* Q. *Further, put* $\left(\int \|\mathbf{x} - \mathbf{x}_0\|^2 \mu(d\mathbf{x})\right)^{1/2} = \sigma$, *where* $\|\cdot\|$ *denotes the* l_1 *norm in* $\mathbb{R}^k$. *Let* f *have continuous first order partial derivatives on* Q, *let* f *and these derivatives be bounded on* Q *and let* $|(f_i(\mathbf{x}) - f_i(\mathbf{x}_0)|$ *be a convex function of* $\mathbf{x}$ *for* $i = 1, \ldots, k$, *where* f_i *is the ith first partial derivative of* f. *For* $h > 0$ *set*

$$\omega_1^*(f_i, h) = \max\{\omega_1(f_i, h) : i = 1, \ldots, k\}.$$

If $h = \sigma^2/2$ *and* $B(\mathbf{x}_0, \sigma^2/2) \subset Q$, *then*

$$\left|\int_Q f\, d\mu - f(\mathbf{x}_0)\right| \leq \omega_1^* \left(f_i, \frac{\sigma^2}{2}\right). \tag{8.1.32}$$

If $h = \sigma/2$ *and* $B(\mathbf{x}_0, \sigma/2) \subset Q$, then

$$\left|\int_Q f\, d\mu - f(\mathbf{x}_0)\right| \leq \omega_1^* \left(f_i, \frac{\sigma}{2}\right) \sigma. \tag{8.1.33}$$

PROOF: Apply Theorem 8.1.4 with $n = 1$ and $w = \omega_1^*(f_i, h)$. □

Remark 8.1.5. Let r, $C(\mathbf{x}_0)$, $D_r(\mathbf{x}_0)$ be given positive numbers. Assume the convex and compact set Q lies in the ball $0 \leq \|\mathbf{x} - \mathbf{x}_0\|_{l_1} \leq C(\mathbf{x}_0)$ and that the probability measure μ on Q satisfies

$$\left(\int \|\mathbf{x} - \mathbf{x}_0\|_{l_1}^r \cdot \mu(d\mathbf{x})\right)^{1/r} = D_r(\mathbf{x}_0).$$

Then, using standard moment methods, we find that the remainder term on the right-hand side of (8.1.26) is $\leq (w/h(n+1)!) \cdot D_r^{n+1}(\mathbf{x}_0)$ if $r \geq n+1$, and $\leq (w/h(n+1)!) \cdot D_r^r(\mathbf{x}_0) \cdot (C(\mathbf{x}_0))^{(n+1)-r}$ if $r \leq n+1$. Thus we have generalized (8.1.13) to higher dimension.

As a further result we have

Proposition 8.1.2. *Take* Q *a convex and compact subset of* $\mathbb{R}^k$, *let* $\mathbf{x}_0 = (x_{01}, \ldots, x_{0k}) \in Q$ *be fixed and let* μ *be a probability measure on* Q. *Let* $f \in C(Q)$, $|f(\mathbf{x}) - f(\mathbf{x}_0)|$ *being a convex function of* $\mathbf{x}$. *Assume* f *has, relative to* Q *and a norm* $\|\cdot\|$ *in* $\mathbb{R}^k$, *a modulus of continuity for which* $\omega_1(f, h) \leq w$. *Here* h, w *are given positive numbers such that the ball* $B(\mathbf{x}_0, h) \subset Q$. *Then*

$$\left|\int_Q f\, d\mu - f(\mathbf{x}_0)\right| \leq \frac{w}{h} \left(\int_Q \|\mathbf{x} - \mathbf{x}_0\| \mu(d\mathbf{x})\right). \tag{8.1.34}$$

This inequality is sharp; equality is attained by $f(\mathbf{x}) = \|\mathbf{x} - \mathbf{x}_0\|$ *when* $w = \omega_1(f, h)$.

PROOF: Obvious, using Lemma 8.1.1. □

Remark 8.1.6. Assume the first two sentences of Remark 8.1.5, except that instead of $\|\cdot\|_{l_1}$ take any norm in $\mathbb{R}^k$. Then, using standard moment methods, we obtain the best upper bound:

$$\left|\int_Q f\,d\mu - f(\mathbf{x}_0)\right| \leq \begin{cases} \dfrac{w}{h} D_r(\mathbf{x}_0), & \text{if } r \geq 1, \\ \\ \dfrac{w}{h} D_r^r(\mathbf{x}_0)(C(\mathbf{x}_0))^{1-r}, & \text{if } r \leq 1. \end{cases} \tag{8.1.35}$$

This is a generalization of inequality (8.1.5).

8.1.5. An Abstract Result

As a related result we give

Theorem 8.1.5. *Let* μ *be a finite measure of mass* m *on the nonempty convex and compact subset* Q *of the real normed vector space* $(V, \|\cdot\|)$. *Consider* $x_0 \in Q$ *and* $h > 0$ *so that the ball* $B(x_0, h) \subset Q$. *Also consider* $C(x_0) > 0$ *such that* $0 \leq \|x - x_0\| \leq C(x_0)$ *for all* $x \in Q$. *Assume that*

$$\left(\int_Q \|x - x_0\|^r \cdot \mu(dx)\right)^{1/r} = D_r(x_0), \tag{8.1.36}$$

where $r > 0$ *and* $D_r(x_0) > 0$ *are given. For the existence of* μ *we suppose that* $D_r^r(x_0) \leq m \cdot C^r(x_0)$.

Let $f: Q \to \mathbb{R}$ *be such that* $|f(x) - f(x_0)|$ *is convex in* x *and*

$$|f(x) - f(y)| \leq w \quad \text{when} \quad x, y \in Q: \|x - y\| \leq h, \tag{8.1.37}$$

where $w > 0$ *is given. Then we get the best upper bound*

$$\left|\int_Q f\,d\mu - f(x_0)\right| - |m - 1| \cdot |f(x_0)| \leq \begin{cases} w \cdot m^{1-(1/r)} \cdot \dfrac{D_r(x_0)}{h}, & r \geq 1; \\ \\ w \cdot (C(x_0))^{1-r} \cdot \dfrac{D_r^r(x_0)}{h}, & r \leq 1. \end{cases} \tag{8.1.38}$$

PROOF: Similar to Theorem 7.2.1, using Lemma 8.1.1. □

Example 8.1.1. Following the notation of Theorem 8.1.5: let Q be a non-empty convex and compact subset of the real normed vector space $(V, \|\cdot\|)$. Let $x_0 \in Q$ and $\{x_1, \ldots, x_n\}$ ($n \in \mathbb{N}$) be a finite set of distinct points in Q such that the ball

$$B\left(x_0; n^{1/r} \cdot \left(\sum_{i=1}^{n} \|x_0 - x_i\|^{-r}\right)^{-1/r}\right) \subset Q.$$

Let f be a real valued function on Q so that $|f(x) - f(x_0)|$ is convex in x. For $r \geq 1$, the rth Shepard metric interpolation operator (see Shepard (1968)) is given by

$$S_n^r(f, x_0) = S_{\{x_1,\ldots,x_n\}}^r(f, x_0)$$

$$:= \begin{cases} \displaystyle\sum_{i=1}^{n} f(x_i) \cdot \frac{\prod_{j=1, j\neq i}^{n} \|x_0 - x_j\|^r}{\sum_{\ell=1}^{n} \prod_{k=1,\, k\neq \ell}^{n} \|x_0 - x_k\|^r}, & \text{if } x_0 \notin \{x_1, \ldots, x_n\}; \\ \\ f(x_i), & \text{if otherwise.} \end{cases} \tag{8.1.39}$$

Obviously S_n^r is a positive linear operator on the space of the above functions f and $S_n^r(f, x_i) = f(x_i)$ for all $i = 1, \ldots, n$. In particular $S_n^r(1, x) = 1$ for all $x \in Q$. To $S_n^r(\cdot, x_0)$ corresponds a probability measure $\mu_n^{(x_0)}$ so that $S_n^r(f, x_0) = \int_Q f \, d\mu_n^{(x_0)}$. Thus, $S_n^r(\|x - x_0\|^r, x_0)$ is equal to

$$D_{r,n}^r(x_0) := \int_Q \|x - x_0\|^r \mu_n^{(x_0)}(dx) = n\left(\sum_{i=1}^{n} \|x_0 - x_i\|^{-r}\right)^{-1}, \tag{8.1.40}$$

which is the harmonic mean of $\|x_0 - x_i\|^r$, $i = 1, \ldots, n$.

Now by applying (8.1.38) of Theorem 8.1.5 we obtain:

$$|S_n^r(f, x_0) - f(x_0)| \leq \omega_1(f, D_{r,n}(x_0)) \tag{8.1.41}$$

which improves the corresponding result due to Gonska (Theorem 4.1(ii)) (1983).

8.2 On the Rate of Weak Convergence of Convex Type Finite Measures to the Unit Measure

8.2.1. Introduction

Let μ_n, $n \geq 1$, be a sequence of finite measures of mass m_n on the convex subset $M \neq \emptyset$ of the real normed vector space $(V, \|\cdot\|)$.

In Sections 7.2–7.3 we studied the rate of the weak convergence of μ_n to δ_{x_0}; the unit measure at $x_0 \in M$, where M is also compact.

Namely, among others, we presented the best possible constant $K_n(x_0)$ in the inequality

$$\left|\int_M f\,d\mu_n - f(x_0)\right| \leq |m_n - 1| \cdot |f(x_0)| + m_n \cdot K_n(x_0), \tag{8.2.1}$$

subject to the given moment condition

$$\left(\int_M \|x - x_0\|^r \mu_n(dx)\right)^{1/r} = D_{rn}(x_0), \qquad r > 0,$$

where f is a real valued measurable function on M such that

$$|f(x) - f(y)| \leq w \quad \text{when} \quad x, y \in M;\ \|x - y\| \leq h.$$

Here, $h > 0$ and $w > 0$ are fixed. The optimal constant $K_n(x_0) = K_n(m_n,\, r,\, D_{rn}(x_0),\, h,\, w,\, f(x_0))$ was determined independently of $f(x_0)$ by the use of standard moment methods; see in Chapter 2, method of optimal distance.

The most inequality (8.2.1) can be is almost attained for suitable μ_n and f. This interesting feature comes from the fact that in calculating $K_n(x_0)$ the involved part of the upper boundary of the convex hull of the related moment problem is part of its closure and not of the related moment graph.

In this section we present a sufficient condition so that inequality (8.2.1) is attained, that is, is sharp; see Theorems 8.2.1 and 8.2.2 and their corresponding Corollaries 8.2.1 and 8.2.2. Here M can be also unbounded and $D_{rn}(x_0)$ is assumed to be finite.

This condition supposes that $r \geq 1.7095114$, $D_{rn}(x_0) \geq r^{-1} \cdot (r+1)^{1-1/r} \cdot m_n^{1/r} \cdot h$, and that the measure μ_n is such that the cumulative distribution function corresponding to the probability measure $\rho_n = m_n^{-1}\mu_n \circ \tau^{-1}$, where $\tau(x) = \|x - x_0\|$, is concave. Here inequality (8.2.1) is attained by a step function and a measure μ_n so that the corresponding probability measure ρ_n is the weak*-homeomorphic image of an at most two-points-supported probability measure. Naturally, in the above concave case, inequality (8.2.1) is stronger than the earlier ones from Sections 7.2–7.3.

When f fulfills a Lipschitz type condition and the cumulative distribution function corresponding to ρ_n fulfills a higher order convexity condition, the best upper bounds to $|\int_M f\, d\mu_n - f(x_0)|$ are simplified considerably and the derived inequalities are again attained, i.e., sharp inequalities; see Theorems 8.2.3 and 8.2.4, improving a related result from Section 7.2. Our proofs are carried out by means coming from the convex moment theory; see Kemperman (1971) and Chapter 2. Also in this section we give applications to the concave positive linear operators arising from probabilistic distributions, e.g., the Weierstrass, logistic, and exponential operators; see subsection 8.2.3.

The presented estimates improve a lot the corresponding ones in the literature. The material of this section comes from Anastassiou (1988).

8.2.2. Main Results

Using convex moment theory methods we show the following results.

Theorem 8.2.1. *Let μ be a finite measure of mass m on the non-empty convex and compact subset M of the real normed vector space $(V, \|\cdot\|)$. Consider $x_0 \in M$ and $C(x_0) > 0$ such that $0 \le \|x - x_0\| \le C(x_0)$ for all $x \in M$. Let $\tau(x) = \|x - x_0\|$ and let the induced probability measure $\rho = m^{-1}\mu \circ \tau^{-1}$ on $[0, C(x_0)]$. Assume that the cumulative distribution function corresponding to ρ is concave in $(0, C(x_0)]$. Suppose further that*

$$\left(\int_M \|x - x_0\|^r \mu(dx)\right)^{1/r} = D_r(x_0), \tag{8.2.2}$$

where $0 < r \le 1$ or $r \ge (\ln 2/\ln 1.5) \approx 1.7095114$ and $D_r(x_0) > 0$ are given, with

$$D_r^r(x_0) \le m(r+1)^{-1}C^r(x_0). \tag{8.2.3}$$

Consider $h > 0$ as given. In the case of $r \geq 1.7095114$ we would assume that

$$\beta(\beta+1)\left(\frac{r+1}{2r}\right)h\left[1-\left(\frac{\beta-1}{\beta+1}\right)^{(r/(r+1))}\right] \leq C(x_0), \tag{8.2.4}$$

where $\beta = \lceil C(x_0)/h \rceil - 1$; $\lceil \cdot \rceil$ is the ceiling of the number.

Let $f: M \to \mathbb{R}$ be such that

$$|f(x) - f(y)| \leq w \text{ if } \|x - y\| \leq h, x, y \in M, \tag{8.2.5}$$

where $w > 0$ is given.

Then the best possible constant $K(x_0) = K(m, r, D_r(x_0), h, w, f(x_0))$ in the inequality

$$\left|\int_M f\,d\mu - f(x_0)\right| \leq |m-1|\,|f(x_0)| + mK(x_0) \tag{8.2.6}$$

is given as follows (independently of $f(x_0)$).

Let $\phi_1(u) = \int_0^u \lceil z/h \rceil dz$, $u > 0$,

(i) $K(x_0) = \left[1 + \left(\dfrac{\phi_1(C(x_0)) - C(x_0)}{(C(x_0))^{r+1}}\right) \cdot \left(\dfrac{r+1}{m}\right) \cdot D_r^r(x_0)\right] \cdot w,$

when $0 < r \leq 1$;

(ii) $K(x_0) = \psi((r+1) \cdot m^{-1} \cdot D_r^r(x_0))$,

when $r \geq 1.7095114$. Here $\psi(y)$ is defined in $[0, C^r(x_0)]$ as follows: For $k = 1, \ldots, \beta$ consider the numbers

$$y_{1k} = (k-1)^{(r/(r+1))} k^r (k+1)^{(r^2/(r+1))} \times h^r \left(\frac{r+1}{2r}\right)^r \left[1 - \left(\frac{k-1}{k+1}\right)^{(r/(r+1))}\right]^r$$

and

$$y_{2k} = k^r (k+1)^r h^r \left(\frac{r+1}{2r}\right)^r \left[1 - \left(\frac{k-1}{k+1}\right)^{(r/(r+1))}\right]^r. \tag{8.2.7}$$

Hence

$$\psi(y) = \begin{cases} w\left[\left(\dfrac{y_{21} - y}{y_{21}}\right) + y\dfrac{\phi_1(y_{21}^{1/r})}{y_{21}^{(1+1/r)}}\right]; & 0 \leq y \leq y_{21}, \\ w\left[\left(\dfrac{y_{2k} - y}{y_{2k} - y_{1k}}\right)\dfrac{\phi_1(y_{1k}^{1/r})}{y_{1k}^{1/r}} + \left(\dfrac{y - y_{1k}}{y_{2k} - y_{1k}}\right)\dfrac{\phi_1(y_{2k}^{1/r})}{y_{2k}^{1/r}}\right]; & \\ \quad y_{1k} \leq y \leq y_{2k},\ k = 2, \ldots, \beta, & \\ w\dfrac{\phi_1(y^{1/r})}{y^{1/r}}, & \text{elsewhere.} \end{cases} \tag{8.2.8}$$

Remark 8.2.1. (i) When $r \geq 1.7095114$ and $(r+1)m^{-1}D_r^r(x_0) \geq y_{21}$, inequality (8.2.6) is attained by $w \cdot \lceil \|x - x_0\|/h \rceil$ and a measure μ such that the cumulative distribution function corresponding to $\rho = m^{-1}\mu \circ \tau^{-1}$ is concave in $(0, C(x_0)]$.

The measure ρ is the weak*-homeomorphic image (see Kemperman (1971), pp. 133, 136 and Chapter 2) of an at most two-points-supported probability measure on $[y_{21}^{1/r}, C(x_0)]$.

(ii) When $0 < r \leq 1$ or $r \geq 1.7095114$ and $0 < (r+1)m^{-1}D_r^r(x_0) < y_{21}$, inequality (8.2.6) is almost attained in a similar way as in case (i). The last comes from the fact that in determining $K(x_0)$, the involved part of the upper boundary of the convex hull of the related moment problem (see proof of Theorem 8.2.1) is part of the closure and not of the related moment graph.

The next result is the result corresponding to Theorem 8.2.1 when the domain M is unbounded.

Theorem 8.2.2. *Let μ be a finite measure of mass m on the unbounded convex subset M of the real normed vector space $(V, \|\cdot\|)$. Let $\tau(x) = \|x - x_0\|$, $x_0 \in M$, and let the induced probability measure $\rho = m^{-1}\mu \circ \tau^{-1}$ on $[0, +\infty)$. Assume that the cumulative distribution function corresponding to ρ is concave in $(0, +\infty)$. Suppose further that*

$$\left(\int_M \|x - x_0\|^r \mu(dx) \right)^{1/r} = D_r(x_0) < \infty, \tag{8.2.9}$$

where $r \geq 1.7095114$ and $D_r(x_0) > 0$ are given.

Let $f: M \to \mathbb{R}$ be such that $|f(x) - f(y)| \leq w$ if $\|x - y\| \leq h$, $x, y \in M$, where $w, h > 0$ are given.

Then the best possible constant $K(x_0) = K(m, r, D_r(x_0), h, w, f(x_0))$ in the inequality

$$\left| \int_M f\, d\mu - f(x_0) \right| \leq |m-1|\,|f(x_0)| + mK(x_0) \tag{8.2.10}$$

is given as follows (independently of $f(x_0)$). We have

$$K(x_0) = \psi((r+1)m^{-1}D_r^r(x_0)), \tag{8.2.11}$$

where $\psi(y)$ is defined in $(0, +\infty)$ by (8.2.8), when we set $C(x_0) = +\infty$ and define y_{1k}, y_{2k} as in (8.2.7) for $k = 1, \ldots, +\infty$.

Inequality (8.2.10) is attained and almost attained as inequality (8.2.6) of Theorem 8.2.1 (in Remark 8.2.1 replace $C(x_0)$ by $+\infty$).

When f fulfills a Lipschitz type condition the best upper bounds we obtain look much simpler.

Theorem 8.2.3. *Let μ be a finite measure of mass m on the non-empty connected and compact subset M of the real normed vector space $(V, \|\cdot\|)$. Consider $x_0 \in M$ and $C(x_0) > 0$ such that $0 \le \|x - x_0\| \le C(x_0)$ for all $x \in M$. Let $\tau(x) = \|x - x_0\|$ and let the induced probability measure $\rho = m^{-1}\mu \circ \tau^{-1}$ on $[0, C(x_0)]$. Assume that the cumulative distribution function F corresponding to ρ possesses an $(s-1)$th derivative $F^{(s-1)}(x)$ $(s \ge 1)$ throughout the interval $(0, C(x_0)]$ and that further $(-1)^s F^{(s-1)}(x)$ is convex in $(0, C(x_0)]$. Suppose also that*

$$\left(\int_M \|x - x_0\|^r \mu(dx)\right)^{1/r} = D_r(x_0) > 0, \qquad r > 0 \tag{8.2.12}$$

with

$$D_r^r(x_0) \le \binom{r+s}{s}^{-1} mC^r(x_0). \tag{8.2.13}$$

Let $f\colon M \to \mathbb{R}$ be such that

$$|f(x) - f(y)| \le k\|x - y\|^\alpha, \qquad (k > 0, 0 < \alpha \le 1). \tag{8.2.14}$$

Then we find the following best upper bound,

$$\left|\int_M f\,d\mu - f(x_0)\right| - |m-1|\,|f(x_0)|$$
$$\le \begin{cases} k \cdot \binom{\alpha+s}{s}^{-1} \cdot \binom{r+s}{s} \cdot C(x_0)^{\alpha-r} \cdot D_r^r(x_0), & \alpha \ge r; \\ k \cdot \binom{\alpha+s}{s}^{-1} \cdot \binom{r+s}{s}^{\alpha/r} \cdot D_r^\alpha(x_0) \cdot m^{1-(\alpha/r)}, & \alpha \le r. \end{cases} \tag{8.2.15}$$

Remark 8.2.2. Inequality (8.2.15) is attained by $f(x) = k\|x - x_0\|^\alpha$ and a measure μ such that the probability measure $\rho = m^{-1}\mu \circ \tau^{-1}$ fulfills the convexity assumption of Theorem

8.2.3. This measure ρ is the weak*-homeomorphic image (see Kemperman (1971), pp. 133, 136 and Chapter 2) of an at most two-points-supported probability measure on $[0, C(x_0)]$.

Inequality (8.2.15) improves the corresponding inequality from Proposition 7.2.1.

The next result is the result corresponding to Theorem 8.2.3 when the domain M is unbounded.

Theorem 8.2.4. *Let μ be a finite measure of mass m on the unbounded connected subset M of the real normed vector space $(V, \|\cdot\|)$. Let $\tau(x) = \|x - x_0\|$, $x_0 \in M$, and let the induced probability measure $\rho = m^{-1}\mu \circ \tau^{-1}$ on $[0, +\infty)$. Assume that the cumulative distribution function F corresponding to ρ possesses an $(s-1)$th derivative $F^{(s-1)}(x)$ $(s \geq 1)$ throughout the interval $(0, +\infty)$ and that further $(-1)^s F^{(s-1)}(x)$ is convex in $(0, +\infty)$. Suppose also that*

$$\left(\int_M \|x - x_0\|^r \mu(dx)\right)^{1/r} = D_r(x_0) < \infty, \tag{8.2.16}$$

where $r > 0$ and $D_r(x_0) > 0$ are given. Let $f: M \to \mathbb{R}$ be such that

$$|f(x) - f(y)| \leq k\|x - y\|^\alpha \qquad (k > 0, 0 < \alpha \leq 1). \tag{8.2.17}$$

Here we would assume that $r \geq \alpha$. Then we get the best upper bound

$$\left|\int_M f\, d\mu - f(x_0)\right| - |m - 1| \cdot |f(x_0)|$$

$$\leq k \cdot \begin{pmatrix} \alpha + s \\ s \end{pmatrix}^{-1} \cdot \begin{pmatrix} r + s \\ s \end{pmatrix}^{\alpha/r} \cdot D_r^\alpha(x_0) \cdot m^{1-(\alpha/r)}. \tag{8.2.18}$$

Remark 8.2.3. Inequality (8.2.18) is attained by $f(x) = k\|x - x_0\|^\alpha$ and a measure μ such that the probability measure $\rho = m^{-1}\mu \circ \tau^{-1}$ fulfills the convexity assumption of Theorem 8.2.4. This measure ρ is the weak*-homeomorphic image (see Kemperman (1971), pp. 133, 136 and Chapter 2) of a one-point-supported probability measure on $[0, +\infty)$.

Inequality (8.2.6) becomes concrete in an important special case

Corollary 8.2.1. *Under the assumptions of Theorem* 8.2.1, *but without assumption* (8.2.4), *we consider the case of $m = 1$ and $r = 2$.*

(i) *If* $h = D_2(x_0)$ *and* $C(x_0) \geq (2.3366256) \cdot D_2(x_0)$, *then*

$$\left|\int_M f\,d\mu - f(x_0)\right| \leq (1.4268781) \cdot \omega_1(f, D_2(x_0)). \tag{8.2.19}$$

(ii) *If* $h = D_2(x_0)/2$ *and* $C(x_0) \geq \sqrt{3}\,D_2(x_0)$, *then*

$$\left|\int_M f\,d\mu - f(x_0)\right| \leq (2.2679492) \cdot \omega_1\left(f, \frac{D_2(x_0)}{2}\right). \tag{8.2.20}$$

Inequalities (8.2.19–20) *are attained and the numbers* 2.3366256, 1.4268781, 2.2679492 *are truncated irrationals.*

PROOF: It is similar to the proof of Theorem 8.2.1. The constraint on $C(x_0)$ in each case is instead of requirements (8.2.3–4) of Theorem 8.2.1. □

Also, inequality (8.2.10) can be simplified considerably in the same special case.

Corollary 8.2.2. *Under the assumptions of Theorem* 8.2.2 *we consider the case of* $m = 1$ *and* $r = 2$. *Then*

$$\left|\int_M f\,d\mu - f(x_0)\right| \leq \min\left\{(1.4268781) \cdot \omega_1(f, D_2(x_0)); (2.2679492) \cdot \omega_1\left(f, \frac{D_2(x_0)}{2}\right)\right\}, \tag{8.2.21}$$

which is an attained inequality.

PROOF: It is a straightforward application of Theorem 8.2.2. □

8.2.3. Applications

The results of this section are applications of Theorem 8.2.2 (see Corollary 8.2.2) to some important concave positive linear operators arising from Probability theory, i.e., Weierstrass, logistic, and exponential operators. Similar applications can be given for the rest of the theorems.

We need to introduce the notion of the concave operator.

Definition 8.2.1. Let μ_t be a finite measure of mass m_t on M ($t \in M$), which is a non-empty convex subset of the real normed vector space $(V, \|\cdot\|)$. Let $\tau(x) = \|x - t\|$ and let the induced probability measure $\rho_t = m_t^{-1}\mu_t \circ \tau^{-1}$ on $[0, c_t]$, where c_t is either finite or infinite for all $t \in M$. Assume that the cumulative distribution function corresponding to ρ_t is concave in $(0, c_t]$, for all $t \in M$. For $f \in C_B(M)$ we define

$$L(f,t) = \int_M f(x)\mu_t d(x), \quad \text{for all } \; t \in M.$$

Obviously L is a positive linear operator acting on $C_B(M)$. We shall call it *concave.*

Remark 8.2.4. Let the normal probability density function

$$f(x) = \frac{1}{\sigma\sqrt{2\pi}} e^{-(x-x_0)^2/2\sigma^2},$$

where $x \in \mathbb{R}$, x_0 is a fixed real number, and $\sigma > 0$, with a corresponding cumulative distribution function F and an associated probability measure μ. For $\tau(x) = |x - x_0|$ and $x > 0$ we have $\tau^{-1}([0,x]) = [x_0 - x, x_0 + x]$. Thus $\rho = \mu \circ \tau^{-1}$ is a probability measure on $[0, +\infty)$ and $\rho([0,x]) = \mu([x_0 - x, x_0 + x]) = F(x_0 + x) - F(x_0 - x)$.

Let G be the cumulative distribution function corresponding to ρ. Thus $G(x) = F(x_0 + x) - F(x_0 - x)$ and $G'(x) = F'(x_0 + x) + F'(x_0 - x) = f(x_0 + x) + f(x_0 - x)$. Hence $G'(x) = (2/\sigma\sqrt{2\pi})e^{-x^2/2\sigma^2}$ and $G''(x) = -G'(x)(x/\sigma^2)$. Therefore for $x \geq 0$ it holds that $G''(x) \leq 0$.

We have proved that G is a concave function in $[0, +\infty)$.

Based on Remark 8.2.4 we state

Theorem 8.2.5. *Let $f \in C_B(\mathbb{R})$ and let the Weierstrass operator W_n, $n \in \mathbb{N}$, acting on $C_B(\mathbb{R})$, be defined by*

$$(W_n f)(t) = \sqrt{\frac{n}{\pi}} \int_{-\infty}^{\infty} f(x) e^{-n(x-t)^2} dx.$$

Then

$$\|W_n f - f\|_\infty \leq \min\left\{(1.4268781) \cdot \omega_1\left(f, \frac{1}{\sqrt{2n}}\right); (2.2679492) \cdot \omega_1\left(f, \frac{1}{2\sqrt{2n}}\right)\right\}. \quad (8.2.22)$$

PROOF: Let X_j be real independent random variables having the normal distribution function $(t, 1/2)$, $t \in \mathbb{R}$. Thus $S_n = \sum_{j=1}^n X_j$, $n \geq 1$, has the normal distribution $(nt, n/2)$.

And S_n/n has the normal distribution function $(t, 1/2n)$, denoted by $F_{S_n/n}$. From Remark 8.2.4 we conclude that $F_{S_n/n}$ fulfills the assumption of Theorem 8.2.2. Obviously $(W_n f)(t) = \int_{-\infty}^{\infty} f(x) dF_{S_n/n}(x)$. Using Corollary 8.2.2 we establish the validity of inequality (8.2.22). □

From Remark 8.2.4 and the last proof we can conclude that the Weierstrass operator W_n is a concave positive linear operator.

Remark 8.2.5. Let X be a real random variable having the logistic distribution (see Feller (1966), pp. 52–53) with $E(X) = x_0 \in \mathbb{R}$ and $\mathrm{Var}(X) = \sigma^2$, $\sigma > 0$.

It has a cumulative distribution function

$$F(x) = (1 + e^{-\pi(x-x_0)/\sigma\sqrt{3}})^{-1}, \qquad x \in \mathbb{R} \tag{8.2.23}$$

with an associated probability measure μ. For $\tau(x) = |x - x_0|$ and $x > 0$ we have $\tau^{-1}([0, x]) = [x_0 - x, x_0 + x]$. Thus $\rho = \mu \circ \tau^{-1}$ is a probability measure on $[0, +\infty)$ and $\rho([0, x]) = \mu([x_0 - x, x_0 + x]) = F(x_0 + x) - F(x_0 - x)$.

Let G be the cumulative distribution function corresponding to ρ. Hence $G(x) = F(x_0 + x) - F(x_0 - x)$, giving us

$$G(x) = (1 + e^{-\pi x/\sigma\sqrt{3}})^{-1} - (1 + e^{\pi x/\sigma\sqrt{3}})^{-1}$$

which is a concave function in $[0, +\infty)$.

Based on Remark 8.2.5 we give

Lemma 8.2.1. *Let $\{X_n\}$, $n \geq 1$, be a sequence of real random variables having the logistic distribution with $E(X_n) = x_0 \in \mathbb{R}$ and $\mathrm{Var}(X_n) = \varepsilon_n^2$, $\varepsilon_n > 0$. It has a cumulative distribution function*

$$F_n(x) = (1 + e^{-\pi(x-x_0)/\varepsilon_n\sqrt{3}})^{-1}, \qquad x \in \mathbb{R}.$$

Let μ_n be the probability measure associated to F_n and $f \in C_B(\mathbb{R})$. Then

$$\left|\int_{-\infty}^{\infty} f\, d\mu_n - f(x_0)\right| \leq \min\left\{(1.4268781) \cdot \omega_1(f, \varepsilon_n); (2.2679492) \cdot \omega_1\left(f, \frac{\varepsilon_n}{2}\right)\right\}, \tag{8.2.24}$$

$$\mu_n \rightrightarrows \delta_{x_0} \text{ (weakly)}, \quad \text{as } n \to +\infty \text{ and } \varepsilon_n \to 0. \tag{8.2.25}$$

PROOF: It is a direct application of Remark 8.2.5 and Corollary 8.2.2. □

The logistic operator is introduced by

Definition 8.2.2. Consider the sequence of real random variables $\{X_{nt}\}$, $n \geq 1$, with a mean $t \in \mathbb{R}$ and a variance ε_n^2, $\varepsilon_n > 0$, having the logistic distribution function

$$F_{nt}(x) = (1 + e^{-\pi(x-t)/\varepsilon_n\sqrt{3}})^{-1}, \qquad x \in \mathbb{R}.$$

Let μ_{nt} be the probability measure associated to F_{nt}. For $f \in C_B(\mathbb{R})$ we define the operator

$$L_n(f,t) = \int_{-\infty}^{\infty} f(x)\mu_{nt}d(x), \quad \text{for all } t \in \mathbb{R}.$$

By Remark 8.2.5 $\{L_n\}$, $n \geq 1$, is a sequence of concave positive linear operators, called logistic.

The convergence of a sequence of logistic operators is described as follows:

Theorem 8.2.6. *Let $\{L_n\}$, $n \geq 1$, be a sequence of logistic operators and $f \in C_B(\mathbb{R})$. Then*

$$\|L_nf - f\|_\infty \leq \min\left\{(1.4268781)\cdot\omega_1(f,\varepsilon_n); (2.2679492)\cdot\omega_1\left(f,\frac{\varepsilon_n}{2}\right)\right\}, \tag{8.2.26}$$

$$\lim L_nf = f, \quad \textit{uniformly as } n \to \infty \textit{ and } \varepsilon_n \to 0. \tag{8.2.27}$$

PROOF: Use of Lemma 8.2.1. □

Remark 8.2.6. Let X_t, $t \in \mathbb{R}$, be the real random variable having an exponential probability density function

$$f_t(x) = \begin{cases} \alpha e^{-\alpha(x-t)}, & t \leq x \leq +\infty \\ 0, & -\infty \leq x < t, \end{cases} \tag{8.2.28}$$

where $\alpha > 0$. It has a cumulative distribution function

$$F_t(x) = \begin{cases} 1 - e^{-\alpha(x-t)}, & t \leq x \leq +\infty \\ 0, & -\infty \leq x < t \end{cases} \tag{8.2.29}$$

with an associated probability measure μ_t. One can find that $E(X_t) = t + \alpha^{-1}$, $E(X_t^2) = \alpha^{-2}[\alpha^2 t^2 + 2\alpha t + 2]$, and $\mathrm{Var}(X_t) = \alpha^{-2}$. Furthermore, it holds that

$$D_2(t) = \left(\int_{-\infty}^{\infty} (x-t)^2 \mu_t d(x)\right)^{1/2} = \sqrt{2}\,\alpha^{-1}. \tag{8.2.30}$$

For $\tau(x) = |x - t|$ and $x > 0$ we have $\tau^{-1}([0, x]) = [t - x, t + x]$. Thus $\rho_t = \mu_t \circ \tau^{-1}$ is a probability measure on $[0, +\infty)$ and $\rho_t([0, x]) = \mu_t([t - x, t + x]) = F_t(t + x) - F_t(t - x) = F_t(t + x)$. Let G_t be the cumulative distribution function corresponding to ρ_t. Then $G_t(x) = F_t(t + x)$, which implies that $G_t(x) = 1 - e^{-\alpha x}$ for all $x \geq 0$, all $t \in \mathbb{R}$. It is clear that G_t is a concave function in $[0, +\infty)$, for all $t \in \mathbb{R}$.

Based on Remark 8.2.6 we have

Lemma 8.2.2. *Let $\{X_{nt}\}$, $n \geq 1$, $t \in \mathbb{R}$, be a sequence of real random variables having an exponential distribution with probability density function*

$$f_{nt}(x) = \begin{cases} \varepsilon_n^{-1} e^{-\varepsilon_n^{-1}(x-t)}, & t \leq x \leq +\infty \\ 0, & -\infty \leq x < t, \end{cases}$$

where $\varepsilon_n > 0$.

Let F_{nt} be the cumulative distribution function of X_{nt}, where μ_{nt} is the probability measure associated to F_{nt} on $\mathbb{R}$ and $f \in C_B(\mathbb{R})$. Then

$$\left|\int_{-\infty}^{\infty} f\, d\mu_{nt} - f(t)\right| \leq \min\left\{(1.4268781)\cdot\omega_1(f, \sqrt{2}\varepsilon_n); (2.2679492)\cdot\omega_1\left(f, \frac{\sqrt{2}}{2}\varepsilon_n\right)\right\}, \tag{8.2.31}$$

$$\mu_{nt} \rightrightarrows \delta_t \ (\textit{weakly}), \quad \textit{as } n \to +\infty \textit{ and } \varepsilon_n \to 0. \tag{8.2.32}$$

Proof: It is a direct application of Remark 8.2.6 and Corollary 8.2.2. Here we should note that $D_2(t) = \sqrt{2}\varepsilon_n$. □

The exponential operator is introduced by

Definition 8.2.3. Let the sequence of real random variables $\{X_{nt}\}$, $n \geq 1$, $t \in \mathbb{R}$, have the exponential distribution with probability density function

$$f_{nt}(x) = \begin{cases} \varepsilon_n^{-1} e^{-\varepsilon_n^{-1}(x-t)}, & t \leq x \leq +\infty \\ 0, & -\infty \leq x < t, \end{cases}$$

where $\varepsilon_n > 0$. For $f \in C_B(\mathbb{R})$ we define the operator

$$L_n(f,t) = \varepsilon_n^{-1} e^{\varepsilon_n^{-1} t} \int_t^{+\infty} f(x) e^{-\varepsilon_n^{-1} x} dx, \quad \text{for all } t \in \mathbb{R}.$$

By Remark 8.2.6 $\{L_n\}$, $n \geq 1$, is a sequence of concave positive linear operators called exponential.

The convergence of a sequence of exponential operators is described as follows:

Theorem 8.2.7. *Let $\{L_n\}$, $n \geq 1$, be a sequence of exponential operators and $f \in C_B(\mathbb{R})$. Then*

$$\|L_n f - f\|_\infty \leq \min\left\{(1.4268781) \cdot \omega_1(f, \sqrt{2}\varepsilon_n); (2.2679492) \cdot \omega_1\left(f, \frac{\sqrt{2}}{2}\varepsilon_n\right)\right\}, \tag{8.2.33}$$

$$\lim L_n f = f, \quad \textit{uniformly as } n \to \infty \textit{ and } \varepsilon_n \to 0. \tag{8.2.34}$$

PROOF: Use of Lemma 8.2.2. □

8.2.4. Proofs of the Main Theorems

PROOF OF THEOREM 8.2.1: Let $g(x) = f(x) - f(x_0)$. From Lemma 7.1.1 we have

$$|g(x)| \leq \left\lceil \frac{\|x - x_0\|}{h} \right\rceil w, \quad \text{for all } x \in M.$$

Therefore

$$\begin{aligned} \left|\int_M f\, d\mu - f(x_0)\right| &= \left|\int_M g\, d\mu + (m-1) f(x_0)\right| \\ &\leq \int_M \left\lceil \frac{\|x - x_0\|}{h} \right\rceil \cdot w\mu(dx) + |m-1|\,|f(x_0)|. \end{aligned}$$

The last holds as an equality when

$$f(x) = \left\lceil \frac{\|x - x_0\|}{h} \right\rceil \cdot w, \quad \text{where } \left\lceil \frac{\|x - x_0\|}{h} \right\rceil \cdot w \text{ fulfills (8.2.5).}$$

Thus, the best constant $K(x_0)$ in (8.2.6) is given by

$$mK(x_0) = \sup_\mu \int_M \left\lceil \frac{\|x - x_0\|}{h} \right\rceil \cdot w\mu(dx),$$

where μ ranges over all measures on M of mass m which satisfy (8.2.2), so that the cumulative distribution function associated to $\rho = m^{-1}\mu \circ \tau^{-1}$ is concave in $(0, C(x_0)]$.

Calling $u = \|x - x_0\|$ and $\phi(u) = w\lceil u/h \rceil$, $u > 0$, we get

$$K(x_0) = \sup_{\rho} \int \phi(u)\rho(du) \qquad (0 \le u \le C(x_0)),$$

where the supremum is taken over all probability measures ρ, such that

$$\int u^r \rho(du) = m^{-1} \cdot D_r^r(x_0), \quad r > 0,$$

with the associated cumulative distribution function being concave in $(0, C(x_0)]$.

It follows from convex moment methods (see Kemperman (1971), pp. 128, 132, 133, 136, 137, 141, 143, 151, 158–160 and Chapter 2) that

$$K(x_0) = \psi((r+1) \cdot m^{-1} \cdot D_r^r(x_0)),$$

where $\Gamma_1 = \{(t, \psi(t)) : 0 \le t \le C^r(x_0)\}$ describes the upper boundary of the convex hull conv Γ_0 of the curve

$$\Gamma_0 = \{(u^r, \phi^*(u)) : 0 \le u \le C(x_0)\}.$$

Here

$$\phi^*(u) = \begin{cases} w\phi_1(u)/u, & 0 < u \le C(x_0) \\ 0, & u = 0, \end{cases}$$

where

$$\phi_1(u) = \int_0^u \left\lceil \frac{z}{h} \right\rceil dz, \qquad u > 0.$$

Condition (8.2.3) guarantees the existence of the optimal $K(x_0)$.

Setting $y = u^r$, that is $u = y^{1/r}$, we have

$$\Gamma_0 = \{(y, \phi^*(y^{1/r})) : 0 \le y \le C^r(x_0)\}.$$

Thus, in order to determine the best $K(x_0)$ it suffices to determine the upper boundary of the convex hull of the curve Γ_0 in the cases $0 < r \le 1$ and $r \ge 1.7095114$.

Case (I). $0 < r \le 1$.

Claim I. The upper boundary of conv Γ_0 is the line segment connecting the points $(0, w)$ and $(C^r(x_0),\ w\phi_1(C(x_0))/C(x_0))$. Thus we get

$$K(x_0) = \left[1 + \left(\frac{\phi_1(C(x_0)) - C(x_0)}{(C(x_0))^{r+1}}\right) \cdot \left(\frac{r+1}{m}\right) \cdot D_r^r(x_0)\right] \cdot w.$$

PROOF OF CLAIM I. It is enough to consider the function

$$\gamma(y) = \begin{cases} \phi_1(y^{1/r})/y^{1/r}, & 0 < y \leq \varepsilon^r \\ 0, & y = 0, \end{cases}$$

where $\varepsilon > 0$.

The upper boundary of the convex hull of the curve $\{(y, \gamma(y))\colon\ y \in [0, \varepsilon^r]\}$ is the line segment connecting $(0, 1)$ to $(\varepsilon^r, \phi_1(\varepsilon)/\varepsilon)$.

The above is established as follows: Note that $\phi_1(y^{1/r}) = ky^{1/r} - \frac{1}{2}k(k-1)h$, where $k = \lceil y^{1/r}/h \rceil$. Hence $\gamma(y) = 1$ for $0 < y \leq h^r$, so that $\lim_{y \to 0} \gamma(y) = 1$. One can easily prove that $\gamma(y)$ is a continuous and increasing function in $(0, \varepsilon^r]$.

Consequently, the slope $\alpha(y) = (\gamma(y) - 1)/y$ of the line segment $\overline{(0,1), (y, \gamma(y))}$ is also a continuous function in $(0, \varepsilon^r]$. Since $\alpha'(y) \geq 0$ throughout $((k-1)^r h^r,\ k^r h^r]$ for all $k = 1, \ldots, \lceil \varepsilon/h \rceil$, $\alpha(y)$ is an increasing function there.

Thus $\alpha(y)$ is an increasing function everywhere in $(0, \varepsilon^r]$ with $\alpha(y) = 0$ for $0 < y \leq h^r$. Therefore the curve $\{(y, \gamma(y))\colon\ y \in [0, \varepsilon^r]\}$ lies below the line segment connecting $(0, 1)$ to $(\varepsilon^r, \phi_1(\varepsilon)/\varepsilon)$.

Case (II). $r \geq 1.7095114$.

Claim II. The upper boundary of conv Γ_0, under assumption (8.2.4), is given by ψ which is defined by (8.2.8).

PROOF OF CLAIM II. It is enough to consider the function

$$\gamma(y) = \begin{cases} \phi_1(y^{1/r})/y^{1/r}, & 0 < y \leq \varepsilon^r \\ 0, & y = 0. \end{cases}$$

Here we would assume that

$$\zeta\cdot(\zeta+1)\cdot\left(\frac{r+1}{2r}\right)\cdot h\cdot\left[1-\left(\frac{\zeta-1}{\zeta+1}\right)^{r/(r+1)}\right]\le\varepsilon,\quad\text{where } \zeta=\left\lceil\frac{\varepsilon}{h}\right\rceil-1. \tag{8.2.35}$$

The upper boundary of the convex hull of the curve $\{(y,\gamma(y))\colon\ y\in[0,\varepsilon^r]\}$, under assumption (8.2.35), is given by $\{(y,\xi(y))\colon\ y\in[0,\varepsilon^r]\}$, where

$$\xi(y)=\begin{cases}\left(\dfrac{y_{21}-y}{y_{21}}\right)+\gamma(y_{21})\left(\dfrac{y}{y_{21}}\right), & 0\le y\le y_{21},\\[2ex] \left(\dfrac{y_{2k}-y}{y_{2k}-y_{1k}}\right)\gamma(y_{1k})+\left(\dfrac{y-y_{1k}}{y_{2k}-y_{1k}}\right)\gamma(y_{2k}), & y_{1k}\le y\le y_{2k};k=2,\ldots,\zeta,\\ \gamma(y), & \text{elsewhere.}\end{cases}$$

Here, for $k=1,\ldots,\zeta$ the numbers y_{1k}, y_{2k} are defined by (8.2.7) and the normalizing condition (8.2.35) says that

$$y_{2\zeta}\le\varepsilon^r.$$

This upper boundary is justified as follows:

Note that $\phi_1(y^{1/r})=ky^{1/r}-\frac{1}{2}k(k-1)h$, $k=\lceil y^{1/r}/h\rceil$. Hence $\gamma(y)=1$ for $0<y\le h^r$, so that $\lim_{y\to0}\gamma(y)=1$. One can easily prove that $\gamma(y)$ is a continuous and increasing function in $(0,\varepsilon^r]$. When $(k-1)^rh^r<y\le k^rh^r$, $\gamma(y)$ is given by

$$\gamma_k(y)=k-\frac{1}{2}k(k-1)hy^{-1/r}\quad\text{for all }\ k=1,\ldots,\zeta.$$

Therefore

$$\gamma_{k+1}(y)=(k+1)-\frac{1}{2}k(k+1)hy^{-1/r}.$$

Since

$$\gamma_k''(y)=-\left(\frac{r+1}{2r^2}\right)k(k-1)hy^{-(2+1/r)}<0$$

we have that $\gamma(y)$ is a concave function throughout $((k-1)^rh^r,\,k^rh^r]$, $k=1,\ldots,\zeta$. But

$$\gamma_k'(y)=\frac{k(k-1)}{2r}hy^{-(1+1/r)}$$

and

$$y_{k+1}'(y)=\frac{k(k+1)}{2r}hy^{-(1+1/r)}$$

give

$$\gamma_k'(kh) \neq \gamma_{k+1}'(kh) \quad \text{for all} \quad k = 1, \ldots, \zeta.$$

Consequently, $\gamma(y)$ is a piecewise concave function over $[0, \varepsilon^r]$.

For $k \geq 2$, the tangent line segment to $\gamma(y)$ over the two consecutive intervals $((k-1)^r h^r, k^r h^r]$, $(k^r h^r, (k+1)^r h^r]$ has the following points of tangency:

$$(y_{1k}, \gamma_k(y_{1k})), (y_{2k}, \gamma_{k+1}(y_{2k})), \quad \text{respectively.}$$

It has a slope

$$\tan \varphi_k = (\gamma_{k+1}(y_{2k}) - \gamma_k(y_{1k}))/(y_{2k} - y_{1k})$$

which is equal to $\gamma_k'(y_{1k}) = \gamma_{k+1}'(y_{2k})$. Here the points y_{1k}, y_{2k} are described by (8.2.7). And one can determine that

$$\tan \varphi_k = \frac{2^r r^r}{(r+1)^{(r+1)} h^r k^r [(k+1)^{r/(r+1)} - (k-1)^{r/(r+1)}]^{(r+1)}}$$

for all $k = 2, \ldots, \zeta$ and a fixed $r \geq 1.7095114$.

Next we would like to present the main steps in the proof that $\tan \varphi_k$ is decreasing in $k \geq 2$.

We have that

$$\left(\frac{k+2}{k-1}\right) \leq \left(\frac{k+1}{k}\right)^{(2r+1)}$$

is true for all integers $k \geq 2$ and a real $r \geq 1.7095114$. The last implies that

$$\chi(k) = \left[\left(\frac{k-1}{k+1}\right)^{r/(r+1)} + \left(\frac{k+2}{k}\right)^{r/(r+1)}\right]$$

is a decreasing function in integers $k \geq 2$ and by $\lim_{k \to +\infty} \chi(k) = 2$, we get that $\chi(k) \geq 2$ for all integers $k \geq 2$ and a real $r \geq 1.7095114$.

Consequently

$$k^{r/(r+1)}[(k+1)^{r/(r+1)} - (k-1)^{r/(r+1)}]$$

is increasing in $k \geq 2$, giving us that

$$k^r [(k+1)^{r/(r+1)} - (k-1)^{r/(r+1)}]^{(r+1)}$$

is increasing in $k \geq 2$. Therefore $\tan \varphi_k$ is decreasing in $k \geq 2$.

Consider now the tangent from the point $(0,1)$ to the graph of $(y, \gamma_2(y))$. It has a slope $(\gamma_2(y)-1)/y$ which should be equal to $\gamma_2'(y)$ at the point of tangency.

The corresponding y to the contact point will be $y_{21} = h^r(r+1)^r r^{-r}$, as in (8.2.7), with $y_2'(y_{21}) = r^r(r+1)^{-(r+1)}h^{-r}$.

Note that $h^r < y_{21} < 2^r h^r$ and that one can verify that $\tan \varphi_3 \leq \gamma_2'(y_{21})$ along with $y_{21} \leq y_{12}$, y_{12} as in (8.2.7), for all $r \geq 1.7095114$. We have that $y_{21} = y_{12}$ at $r = 1.7095114$. That is why we consider only $r \geq 1.7095114$.

Furthermore one can prove that

$$(k-1)^r h^r < y_{1k} < k^r h^r < y_{2k} < (k+1)^r h^r \quad \text{and} \quad y_{2k} < y_{1(k+1)}$$

for all $k = 2, \ldots, \zeta$ and a fixed $r \geq 1.7095114$.

By Figure 8.2.1 the proof of Claim II is completed and by this the proof of Theorem 8.2.1 is finished.

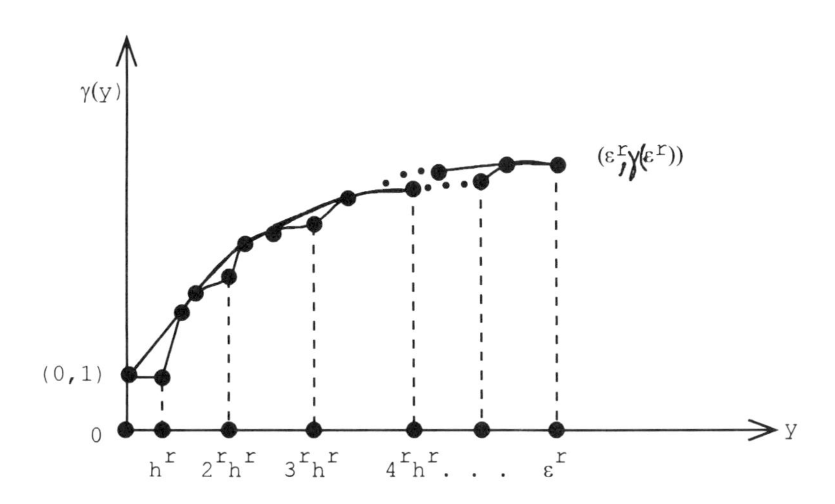

Figure 8.2.1. The convex hull of $(y, \gamma(y))$. □

PROOF OF THEOREM 8.2.2: Similar to Theorem 8.2.1; Case (ii). The convex hull of the related moment problem is the same; it is only extended over $\mathbb{R}_+$. Here we should note that

$$\begin{aligned}\int_M \left\lceil \frac{\|x-x_0\|}{h} \right\rceil \mu(dx) &< \int_M \left(\frac{\|x-x_0\|}{h} + 1 \right) \mu(dx) \\ &\leq 1 + h^{-1} \left(\int_M \|x-x_0\|^r \mu(dx) \right)^{1/r} \\ &= 1 + h^{-1} D_r(x_0) < \infty.\end{aligned}$$

Therefore $\int_M \lceil \|x-x_0\|/h \rceil \mu(dx) < \infty$, which implies $\int_M |f| d\mu < \infty$. □

PROOF OF THEOREM 8.2.3: Let $g(x) = f(x) - f(x_0)$. By assumption (8.2.14) we get

$$\begin{aligned}\left| \int_M f\,d\mu - f(x_0) \right| &= \left| \int_M g\,d\mu + (m-1) f(x_0) \right| \\ &\leq \int_M |g| d\mu + |m-1|\,|f(x_0)| \\ &\leq k \int_M \|x-x_0\|^\alpha \mu(dx) + |m-1|\,|f(x_0)|;\end{aligned}$$

i.e.,

$$\left| \int_M f\,d\mu - f(x_0) \right| \leq |m-1|\,|f(x_0)| + k \int_M \|x-x_0\|^\alpha \mu(dx).$$

The last inequality holds as an equality when $f(x) = k\|x-x_0\|^\alpha$ and here we would like to maximize its right-hand side over all μ fulfilling the assumption of the theorem. Thus, we wish to find

$$U(x_0) = \sup_\mu \int_M \|x-x_0\|^\alpha \mu(dx),$$

where μ ranges over all measures on M of mass m which satisfy (8.2.12), with the induced probability measure $\rho = m^{-1}\mu \circ \tau^{-1}$ on $[0, C(x_0)]$ as in the assumption of the theorem.

Calling $u = \|x-x_0\|$ we get

$$m^{-1} U(x_0) = \sup_\rho \int u^\alpha \rho(du) \qquad (0 \leq u \leq C(x_0)),$$

where ρ is such that

$$\int u^r \rho(du) = m^{-1} D_r^r(x_0), \qquad r > 0$$

and fulfills the convexity assumption of the theorem. From Kemperman (1971), pp. 137, 143, and Chapter 2, for $\xi(u) = u^\alpha$ we obtain $\xi^*(u) = \binom{\alpha+s}{s}^{-1} \cdot u^\alpha$, $0 \le u \le C(x_0)$, where ξ^* is according to the terminology of that paper.

Similarly, for $\phi(u) = u^r$ we obtain

$$\phi^*(u) = \binom{r+s}{s}^{-1} \cdot u^r, \qquad 0 \le u \le C(x_0).$$

Thus by applying convex moment methods (see Kemperman (1971), pp. 128, 132, 133, 136, 137, 141, 143, 144, 158, 159 and Chapter 2), the last moment problem is transferred to

$$m^{-1}U(x_0) = \binom{\alpha+s}{s}^{-1} \cdot \sup_\nu \int u^\alpha \nu(du)$$

over all probability measures ν on $[0, C(x_0)]$ such that

$$\int u^r \nu(du) = \binom{r+s}{s} \cdot m^{-1} \cdot D_r^r(x_0), \qquad r > 0. \tag{8.2.36}$$

Really, it is enough to find

$$\tilde{U}(x_0) = \sup_\nu \int u^\alpha \nu(du)$$

over all probability measures ν that fulfill (8.2.36).

Finally we need to consider the curve (see Kemperman (1971), pp. 151, 160 and Chapter 2)

$$\{(u^r, u^\alpha): 0 \le u \le C(x_0)\},$$

which is the same as the curve

$$\{(y, y^{\alpha/r}): 0 \le y \le C^r(x_0)\}.$$

Here when $\alpha \ge r$ the above curve is convex, while when $\alpha \le r$ the curve is concave.

Consequently, when $a \ge r$ we get

$$\tilde{U}(x_0) = (C(x_0))^{\alpha-r} \cdot \binom{r+s}{s} \cdot m^{-1} \cdot D_r^r(x_0)$$

and when $\alpha \leq r$ we get

$$\tilde{U}(x_0) = \binom{r+s}{s}^{\alpha/r} \cdot m^{-\alpha/r} \cdot D_r^\alpha(x_0).$$

Therefore

$$U(x_0) = \begin{cases} \binom{\alpha+s}{s}^{-1} \cdot C^{\alpha-r}(x_0) \cdot \binom{r+s}{s} \cdot D_r^r(x_0), & \alpha \geq r \\ \binom{\alpha+s}{s}^{-1} \cdot \binom{r+s}{s}^{(\alpha/r)} \cdot m^{1-(\alpha/r)} \cdot D_r^\alpha(x_0), & \alpha \leq r \end{cases}$$

which completes the proof of the theorem. □

PROOF OF THEOREM 8.2.4: Similar to Theorem 8.2.3, case of $r \geq \alpha$. The graph of the related moment problem is the same; it is only extended over $\mathbb{R}_+$. Here we should note that

$$\int_M \|x - x_0\|^\alpha \mu(dx) \leq \left(\int_M \|x - x_0\|^r \mu(dx)\right)^{\alpha/r} = D_r^\alpha(x_0) < \infty,$$

which implies $\int_M |f| d\mu < \infty$. □

8.3. On the Smooth Rate of Weak Convergence of Convex Type Finite Measures to the Unit Measure

8.3.1. Introduction

Let μ_j, $j \geq 1$, be a sequence of finite measures of mass c_{0j} on $[a,b] \subset \mathbb{R}$. Call

$$\int_{[a,b]} (x - x_0)^k \cdot \mu_j(dx) = c_{kj}(x_0), \qquad k = 1, \ldots, n,$$

where $x_0 \in [a,b]$ and $n \geq 1$ are fixed.

In Section 7.2 we studied the rate of the weak convergence of μ_j to δ_{x_0}, the unit measure at x_0. This study is continued here. Namely, among others there given that only the moment

$$\left(\int_{[a,b]} |x - x_n|^n \cdot \mu_j(dx)\right)^{1/n} = d_{nj}(x_0) \tag{8.3.1}$$

was prescribed, we showed the best possible $R_{nj}(x_0)$, so that the inequality

$$\left|\int_{[a,b]} f\,d\mu_j - f(x_0)\right| \le$$

$$|f(x_0)|\cdot|c_{0j}-1| + \sum_{k=1}^{n} \frac{|f^{(k)}(x_0)|}{k!}\cdot|c_{kj}(x_0)| + R_{nj}(x_0), \tag{8.3.2}$$

where $f \in C^n([a,b])$, can be attained by appropriate well-described f and μ_j of two points support. That is, the above inequality is sharp.

Note that the optimal constant is

$$R_{nj}(x_0) = w\cdot\phi_n(C(x_0))\cdot\left(\frac{d_{nj}(x_0)}{C(x_0)}\right)^n, \qquad n \ge 1,$$

where the first modulus of continuity

$$\omega_1(f^{(n)},h) \le w$$

with w, h given positive numbers, $0 < h \le b-a$, $C(x_0) = \max(x_0 - a, b - x_0)$, and

$$\phi_n(x) = \frac{1}{n!}\left(\sum_{i=0}^{\infty}(|x| - ih)_+^n\right), \qquad x \in \mathbb{R}$$

is a polynomial spline function.

The proving method there came from the standard theory of moments; see Kemperman (1968) and Chapter 2; method of optimal distance.

In this section, among others, we improve inequality (8.3.2) by presenting a smaller constant than $R_{nj}(x_0)$, for any $h \in (0, C(x_0))$ $(b-a/2 \le C(x_0) \le b-a)$; see Theorem 8.3.1.

The above is true under the assumption that μ_j fulfills a higher-order convexity condition, all $j \ge 1$. Namely, let $\tau(x) = |x - x_0|$ and the induced probability measure $\rho_j = c_{0j}^{-1}\mu_j \circ \tau^{-1}$ on $[0, C(x_0)]$.

We would assume that the corresponding to ρ_j cumulative distribution function F_j possesses an $(s-1)$st derivative $F_j^{(s-1)}(x)$ $(s \ge 1)$ throughout the interval $(0, C(x_0)]$ and that further $(-1)^s F_j^{(s-1)}(x)$ is convex in $(0, C(x_0)]$. When $\omega_1(f^{(n)},h) \le Ah^\alpha$, all $0 < h \le b-a$, $0 < \alpha \le 1$, $A > 0$: for the error

$$\left|\int_{[a,b]} f\,d\mu_j - f(x_0)\right|$$

we obtain a much more simplified upper bound; see Theorem 8.3.2.

The established inequalities are sharp, in fact they are attained by well-described $f \in C^n([a,b])$ and μ_j that fulfills the above convexity condition, so that the associated probability measure ρ_j is the weak*-homeomorphic image of a probability measure supported by $\{0, C(x_0)\}$.

The proving method here comes from the convex theory of moments; see Kemperman (1971) and Chapter 2.

At the end we present well-known probabilistic distributions that satisfy the assumed convexity condition. Note that

$$L_j(f, x_0) := \int_{[a,b]} f \, d\mu_{x0j},$$

all $f \in C([a,b])$, $j \geq 1$, is a positive linear functional. Thus our work was also motivated by the study of Korovkin (1960) for the convergence of a sequence of positive linear functionals to the unit functional at $x_0 \in [a,b]$.

The presented estimates improve the corresponding ones in the literature. The material of this section comes from Anastassiou(1) (1989).

8.3.2. Main Results

Using convex moment theory methods, we obtain the following results:

Theorem 8.3.1. *Let μ be a finite measure on $[a,b] \subset \mathbb{R}$ such that $\mu([a,b]) = c_0 > 0$. Consider $x_0 \in [a,b]$ and $0 \leq |x - x_0| \leq C(x_0) = \max(b - x_0,\ x_0 - a)$, all $x \in [a,b]$. Let $\tau(x) = |x - x_0|$ and the induced probability measure $\rho = c_0^{-1}\mu \circ \tau^{-1}$ on $[0, C(x_0)]$.*

Assume that the corresponding to ρ cumulative distribution function F possesses an $(s-1)$st derivative $F^{(s-1)}(x)$ $(s \geq 1)$ throughout the interval $(0, C(x_0)]$ and that further $(-1)^s F^{(s-1)}(x)$ is convex in $(0, C(x_0)]$.

Suppose also that

$$\left(\int_{[a,b]} |x - x_0|^n \mu(dx) \right)^{1/n} = d_n(x_0) > 0, \qquad n \geq 1$$

with

$$d_n^n(x_0) \leq \binom{n+s}{s}^{-1} \cdot c_0 \cdot C^n(x_0). \tag{8.3.3}$$

Call

$$\int_{[a,b]} (x - x_0)^k \mu(dx) = c_k(x_0), \qquad k = 1, \ldots, n.$$

Let $f \in C^n([a,b])$ such that the first modulus of continuity $\omega_1(f^{(n)}, h) \leq w$, where w, h are fixed positive numbers, $0 < h \leq b - a$.

Then we have the upper bounds

$$\left| \int_{[a,b]} \left(f(x) - \sum_{k=0}^{n} \frac{f^{(k)}(x_0)}{k!} \cdot (x - x_0)^k \right) \cdot \mu(dx) \right| \leq U(x_0) \tag{8.3.4}$$

and

$$\left| \int_{[a,b]} f \, d\mu - f(x_0) \right|$$
$$\leq |f(x_0)| \cdot |c_0 - 1| + \sum_{k=1}^{n} \frac{|f^{(k)}(x_0)|}{k!} \cdot |c_k(x_0)| + U(x_0), \tag{8.3.5}$$

where

$$U(x_0) = w \cdot \left(\prod_{j=1}^{s} (n+j) \right) \cdot \left(\frac{d_n(x_0)}{C(x_0)} \right)^n \cdot \frac{\phi_{n+s}(C(x_0))}{C^s(x_0)}. \tag{8.3.6}$$

Here

$$\phi_n(x) = \int_0^{|x|} \left(\int_0^{x_1} \cdots \left(\int_0^{x_{n-1}} \left\lceil \frac{x_n}{h} \right\rceil dx_n \right) \cdots \right) dx_1$$
$$= \frac{1}{n!} \cdot \left(\sum_{i=0}^{\infty} (|x| - ih)_+^n \right), \qquad x \in \mathbb{R}, \ n \geq 1 \tag{8.3.7}$$

is a polynomial spline function, where $\lceil \cdot \rceil$ is the ceiling of the number.

Inequalities (8.3.4) and (8.3.5), when $0 < h < C(x_0)$ $(b - a/2 \leq C(x_0) \leq b - a)$, improve earlier corresponding results; see Theorem 7.2.3 and Corollary 7.2.2.

Remark 8.3.1. Inequalities (8.3.4) and (8.3.5) are sharp, namely, are attained. The optimal measure μ_{x_0} is supported by $[x_0, b]$ or $[a, x_0]$ and is such that the corresponding to $\rho = c_0^{-1} \mu_{x_0} \circ \tau^{-1}$ cumulative distribution function fulfills the assumption of the theorem.

The probability measure ρ is the weak*-homeomorphic image (see Kemperman (1971), pp. 133, 136 and Chapter 2) of a probability measure on $[0, C(x_0)]$, supported by $\{0, C(x_0)\}$ with masses

$$\left[1 - \binom{n+s}{s} \cdot \left(\frac{d_n(x_0)}{C(x_0)}\right)^n \cdot c_0^{-1}\right]$$

and

$$\binom{n+s}{s} \cdot \left(\frac{d_n(x_0)}{C(x_0)}\right)^n \cdot c_0^{-1},$$

respectively.

When $b - x_0 \geq x_0 - a$, the optimal function is $w \cdot \phi_n((x - x_0)_+)$ and when $b - x_0 \leq x_0 - a$, the optimal function is $w \cdot \phi_n((x_0 - x)_+)$.

Each of these optimal functions is the limit of a sequence of functions f having continuous nth derivatives satisfying $\omega_1(f^{(n)}, h) \leq w$ and $f^{(k)}(x_0) = 0$, $k = 0, 1, \ldots, n$, and such that the difference of the two sides of (8.3.4), (8.3.5) tends to 0.

In fact both of the above optimal functions are of the form

$$\tilde{f}(t) = \begin{cases} w \cdot \phi_n(t), & t \in [0, C(x_0)], \\ 0, & t \in [K(x_0), 0], \end{cases}$$

where $K(x_0) = \max(x_0 - b, a - x_0)$ and either $t = x - x_0$ or $t = x_0 - x$.

And actually $\lim_{N \to +\infty} f_{nN}(t) = \tilde{f}(t)$, all $t \in [K(x_0), C(x_0)]$, where

$$f_{nN}(t) = w \cdot \int_0^t \left(\int_0^{t_1} \cdots \left(\int_0^{t_{n-1}} f_{0N}(t_n) dt_n\right) \cdots\right) dt_1.$$

Here, for $k = 0, 1, \ldots, \lceil C(x_0)/h \rceil - 1$ and $N \geq 1$: f_{0N} is the continuous function defined as

$$f_{0N}(t) = \begin{cases} 0, & \text{if } K(x_0) \leq t < 0; \\ \dfrac{Nwt}{2h} + kw\left(1 - \dfrac{N}{2}\right), & \text{if } kh \leq t \leq \left(k + \dfrac{2}{N}\right) h; \\ (k+1)w, & \text{if } \left(k + \dfrac{2}{N}\right) h < t \leq (k+1)h; \\ \left\lceil \dfrac{C(x_0)}{h} \right\rceil \cdot w, & \text{if } \left(\left\lceil \dfrac{C(x_0)}{h} \right\rceil - 1 + \dfrac{2}{N}\right) h < t \leq C(x_0). \end{cases}$$

Observe that $f_{0N}(t)$ fulfills $\omega_1(f_{0N}, h) \leq w$ and further

$$\lim_{N \to +\infty} f_{0N}(t) = \begin{cases} \lceil t/h \rceil \cdot w, & t \in [0, C(x_0)]; \\ 0, & t \in [K(x_0), 0]. \end{cases}$$

PROOF OF THEOREM 8.3.1: We have ($a \leq x \leq b$)

$$f(x) = \sum_{k=0}^{n} \frac{f^{(k)}(x_0)}{k!} \cdot (x - x_0)^k + R_n(x, x_0), \tag{8.3.8}$$

where

$$R_n(x, x_0) = \int_{x_0}^{x} \left(\int_{x_0}^{x_1} \left(\cdots \left(\int_{x_0}^{x_{n-1}} \left(f^{(n)}(x_n) - f^{(n)}(x_0) \right) \right) dx_n \right) dx_{n-1} \cdots dx_1 .$$

By $\omega_1(f^{(n)}, h) \leq w$ and from Corollary 7.1.1 we have

$$|f^{(n)}(x_n) - f^{(n)}(x_0)| \leq w \cdot \left\lceil \frac{|x_n - x_0|}{h} \right\rceil .$$

Call $\phi_0(t) = \lceil t/h \rceil$. Then note that

$$\phi_k(x) = \int_0^x \phi_{k-1}(t) dt, \quad \text{all } x \geq 0, \ k \geq 1.$$

And by the change of variable technique, one can prove that

$$|R_n(x, x_0)| \leq w \cdot \phi_n(|x - x_0|), \quad \text{all } x \in [a, b], \ n \geq 1. \tag{8.3.9}$$

We would like to find

$$U(x_0) = \sup_{\mu} w \cdot \int_{[a,b]} \phi_n(|x - x_0|) \mu(dx)$$

provided that μ fulfills

$$\int_{[a,b]} |x - x_0|^n \cdot \mu(dx) = d_n^n(x_0), \qquad n \geq 1.$$

Here μ is a finite measure on $[a, b]$ such that $\mu([a, b]) = c_0 > 0$ and the induced probability measure $\rho = c_0^{-1} \mu \circ \tau^{-1}$ on $[0, C(x_0)]$ fulfills the convexity assumption of the theorem. Letting $u = |x - x_0|$ we can reformulate our moment problem into

$$c_0^{-1} U(x_0) = \sup_{\rho} w \cdot \int_{[0, C(x_0)]} \phi_n(u) \rho(du)$$

such that

$$\int_{[0,C(x_0)]} u^n \rho(du) = d_n^n(x_0) c_0^{-1},$$

where ρ is a convex (as above) probability measure on $[0, C(x_0)]$.

Now we apply convex moment methods from Kemperman (1971) and Chapter 2. In the terminology of that article (see p. 137, the case of $x_0 = 0$), for $u > 0$ we get

$$\frac{1}{s!} u^s \phi^*(u) = \frac{1}{(s-1)!} \cdot \int_0^u (u-x)^{s-1} \phi_n(x) dx = \phi_{n+s}(u).$$

That is

$$\phi^*(u) = \frac{s! \phi_{n+s}(u)}{u^s}, \quad \text{all } u > 0; \ s, n \geq 1.$$

And $\phi^*(0) = \phi_n(0) = 0$. Since $\phi_{n+s}^{(k)} = \phi_{n+s-k}$, $k = 1, \ldots, s$, by L'Hospital's rule we get $\lim_{u \to 0_+} \phi^*(u) = 0$. Hence,

$$\phi^*(u) = \begin{cases} \dfrac{s! \phi_{n+s}(u)}{u^s}, & 0 < u \leq C(x_0), \\ 0, & u = 0, \end{cases}$$

is continuous at zero, $n \geq 1$.

In the same terminology for $\psi(u) = u^n$, we get that

$$\psi^*(u) = \binom{n+s}{s}^{-1} \cdot u^n, \quad \text{all } u \in [0, C(x_0)], \ n \geq 1.$$

By the same convex methods (see Kemperman (1971), pp. 136, 137, 151, 158, 159 and Chapter 2) we want

$$w^{-1} c_0^{-1} U(x_0) = \sup_v \int_{[0,C(x_0)]} \phi^*(u) v(du),$$

the supremum taken over all probability measures v on $[0, C(x_0)]$ such that

$$\int_{[0,C(x_0)]} u^n v(du) = \binom{n+s}{s} d_n^n(x_0) c_0^{-1}.$$

We know that $\phi_n(x)$ is a convex function on $\mathbb{R}$ and strictly increasing on $\mathbb{R}_+$, $n \geq 1$ (see Remark 7.1.3). Furthermore $\phi_n(u^{1/n})$ is convex in $u \geq 0$, $n \geq 1$ (see proof of Theorem 7.2.3).

Again from Kemperman (1971), p. 137 and Chapter 2, the case of $x_0 = 0$ there, we have that

$$\phi^*(u) = \int_0^1 \phi_n((1-t)u)st^{s-1}dt, \quad \text{all } u > 0,\ n \geq 1.$$

Hence $\phi^*(u)$ is a strictly increasing function in $u > 0$.

One can easily prove now that also $\phi^*(u^{1/n})$ is a strictly increasing and convex function in $u > 0$, $n \geq 1$.

For our moment problem in its final form we need to consider the graph of

$$\{(u^n, \phi^*(u)) : 0 \leq u \leq C(x_0)\},$$

which is the same as the graph of

$$\Gamma_0 = \{(y, \phi^*(y^{1/n})) : 0 \leq y \leq C^n(x_0)\}.$$

The upper boundary of the convex hull of Γ_0 is the line segment

$$\overline{(0,0), \left(C^n(x_0), \frac{s!\phi_{n+s}(C(x_0))}{C^s(x_0)}\right)}.$$

By assumption (8.3.3) of the theorem we have

$$\binom{n+s}{s} \cdot d_n^n(x_0)c_0^{-1} \leq C^n(x_0).$$

Standard moment methods now imply that (see Kemperman (1968) and (1971), pp. 151, 160 and Chapter 2; method of optimal distance)

$$w^{-1}c_0^{-1}U(x_0) = \binom{n+s}{s} \cdot \left(\frac{d_n(x_0)}{C(x_0)}\right)^n \cdot \frac{c_0^{-1}s!\phi_{n+s}(C(x_0))}{C^s(x_0)}$$

giving us

$$U(x_0) = w \cdot \binom{n+s}{s} \cdot \left(\frac{d_n(x_0)}{C(x_0)}\right)^n \cdot \frac{s!\phi_{n+s}(C(x_0))}{C^s(x_0)}. \tag{8.3.10}$$

Integrating (8.3.8) against μ we obtain

$$\int_{[a,b]} \left(f(x) - \sum_{k=0}^{n} \frac{f^{(k)}(x_0)}{k!}(x-x_0)^k\right) \cdot \mu(dx) = \int_{[a,b]} R_n(x, x_0)\mu(dx)$$

and

$$\int_{[a,b]} f\,d\mu - f(x_0) = f(x_0)(c_0 - 1) + \sum_{k=1}^{n} \frac{f^{(k)}(x_0)}{k!} c_k(x_0) + \int_{[a,b]} R_n(x, x_0)\mu(dx).$$

Hence by (8.3.9) we get

$$\left| \int_{[a,b]} \left(f(x) - \sum_{k=0}^{n} \frac{f^{(k)}(x_0)}{k!}(x - x_0)^k \right) \cdot \mu(dx) \right| \le w \cdot \int_{[a,b]} \phi_n(|x - x_0|)\mu(dx)$$

and

$$\left| \int_{[a,b]} f\,d\mu - f(x_0) \right| \le |f(x_0)|\,|c_0 - 1| + \sum_{k=1}^{n} \frac{|f^{(k)}(x_0)|}{k!} |c_k(x_0)| + w \cdot \int_{[a,b]} \phi_n(|x - x_0|)\mu(dx).$$

But it holds that

$$w \cdot \int_{[a,b]} \phi_n(|x - x_0|)\mu(dx) \le U(x_0),$$

where $U(x_0)$ is given by (8.3.10). The proof of the theorem is now complete. □

Comment 8.3.1. Inequality (7.1.18) says that

$$\phi_n(x) \le \left[\frac{|x|^{n+1}}{(n+1)!h} + \frac{|x|^n}{2n!} + \frac{h|x|^{n-1}}{8(n-1)!} \right), \quad \text{all } x \in \mathbb{R},\ n \ge 1.$$

Thus

$$\phi_{n+s}(C(x_0)) \le \left(\frac{(C(x_0))^{n+s+1}}{(n+s+1)!h} + \frac{(C(x_0))^{n+s}}{2(n+s)!} + \frac{h(C(x_0))^{n+s-1}}{8(n+s-1)!} \right),$$

which implies that

$$\frac{(n+s)!}{n!} \cdot \frac{\phi_{n+s}(C(x_0))}{(C(x_0))^{n+s}} \le \left[\frac{C(x_0)}{n!(n+s+1)h} + \frac{1}{2n!} + \frac{(n+s)h}{8n!C(x_0)} \right].$$

Hence for the remainder of Theorem 8.3.1 (see (8.3.6)) we have

$$U(x_0) \le \left[\frac{C(x_0)}{n!(n+s+1)h} + \frac{1}{2n!} + \frac{(n+s)h}{8n!C(x_0)} \right] w\, d_n^n(x_0). \tag{8.3.11}$$

Letting $h = r\, d_n(x_0)$, $r > 0$, we obtain

$$U(x_0) \le \left[\frac{C(x_0)}{(n+s+1)r} + \frac{d_n(x_0)}{2} + \frac{(n+s)r\, d_n^2(x_0)}{8C(x_0)}\right] \cdot \frac{d_n^{n-1}(x_0)}{n!} \cdot \omega_1(f^{(n)}, r\, d_n(x_0)). \quad (8.3.12)$$

When $f^{(n)}$ fulfills a Lipschitz type condition, the upper bounds we obtain look much simpler.

Theorem 8.3.2. *Let μ be a finite measure on $[a,b] \subset \mathbb{R}$ such that $\mu([a,b]) = c_0 > 0$. Consider $x_0 \in [a,b]$ and $0 \le |x - x_0| \le C(x_0) = \max(b - x_0,\ x_0 - a)$, all $x \in [a,b]$. Let $\tau(x) = |x - x_0|$ and the induced probability measure $\rho = c_0^{-1}\mu \circ \tau^{-1}$ on $[0, C(x_0)]$. Assume that the corresponding to ρ cumulative distribution function F possesses an $(s-1)$st derivative $F_{(x)}^{(s-1)}$ $(s \ge 1)$ throughout the interval $(0, C(x_0)]$ and that further $(-1)^s F_{(x)}^{(s-1)}$ is convex in $(0, C(x_0)]$.*

Suppose also that

$$\left(\int_{[a,b]} |x - x_0|^n \mu(dx)\right)^{1/n} = d_n(x_0) > 0, \qquad n \ge 1$$

with

$$d_n^n(x_0) \le \binom{n+s}{s}^{-1} \cdot c_0 \cdot C^n(x_0). \quad (8.3.13)$$

Call

$$\int_{[a,b]} (x - x_0)^k \mu(dx) = c_k(x_0), \qquad k = 1, \ldots, n.$$

Let $f \in C^n([a,b])$ such that $\omega_1(f^{(n)}, h) \le Ah^\alpha$, all $0 < h \le b - a$; $0 < \alpha \le 1$, $A > 0$. Then we find the upper bounds

$$\left|\int_{[a,b]} \left(f(x) - \sum_{k=0}^{n} \frac{f^{(k)}(x_0)}{k!}(x - x_0)^k\right) \cdot \mu(dx)\right| \le B\tilde{U}(x_0) \quad (8.3.14)$$

and

$$\left|\int_{[a,b]} f\, d\mu - f(x_0)\right| \le |f(x_0)|\, |c_0 - 1| + \sum_{k=1}^{n} \frac{|f^{(k)}(x_0)|}{k!} |c_k(x_0)| + B\tilde{U}(x_0), \quad (8.3.15)$$

where

$$B = A / \prod_{i=1}^{n} (\alpha + i)$$

and

$$\tilde{U}(x_0) = \binom{n+s}{s} \cdot \binom{n+\alpha+s}{s}^{-1} \cdot d_n^n(x_0) \cdot C^\alpha(x_0), \qquad n \geq 1. \tag{8.3.16}$$

Remark 8.3.2. Inequalities (8.3.14) and (8.3.15) are sharp, namely, are attained. The optimal measure μ_{x_0} is supported by $[x_0, b]$ or $[a, x_0]$ and is such that the corresponding to $\rho = c_0^{-1}\mu_{x_0} \circ \tau^{-1}$ cumulative distribution function fulfills the assumption of the theorem.

That probability measure ρ is the weak*-homeomorphic image (see Kemperman (1971), pp. 133, 136 and Chapter 2) of a probability measure on $[0, C(x_0)]$, supported by $\{0, C(x_0)\}$ with masses

$$\left[1 - \binom{n+s}{s} \cdot \left(\frac{d_n(x_0)}{C(x_0)}\right)^n \cdot c_0^{-1}\right]$$

and

$$\binom{n+s}{s} \cdot \left(\frac{d_n(x_0)}{C(x_0)}\right)^n \cdot c_0^{-1},$$

respectively. When $b - x_0 \geq x_0 - a$, the optimal function is

$$\tilde{f}(x) = \begin{cases} B(x-x_0)^{n+\alpha}, & x_0 \leq x \leq b, \\ 0, & a \leq x \leq x_0. \end{cases}$$

When $b - x_0 \leq x_0 - a$, the optimal function is

$$\tilde{\tilde{f}}(x) = \begin{cases} 0, & x_0 \leq x \leq b, \\ B(x_0 - x)^{n+\alpha}, & a \leq x \leq x_0. \end{cases}$$

Here note that $\tilde{f}^{(k)}(x_0) = \tilde{\tilde{f}}^{(k)}(x_0) = 0$, all $k = 0, 1, \ldots, n$. Furthermore

$$\tilde{f}^{(n)}(x) = \begin{cases} A(x-x_0)^\alpha, & x_0 \leq x \leq b, \\ 0, & a \leq x \leq x_0, \end{cases}$$

$$\tilde{\tilde{f}}^{(n)}(x) = \begin{cases} 0, & x_0 \leq x \leq b, \\ A(-1)^n(x_0 - x)^\alpha, & a \leq x \leq x_0. \end{cases}$$

Using $|\,|x|^\alpha - |y|^\alpha| \leq |\,|x| - |y|\,|^\alpha \leq |x-y|^\alpha$, all $x, y \in \mathbb{R}$, $0 < \alpha \leq 1$, we get

$$\omega_1(\tilde{f}^{(n)}, h) \leq Ah^\alpha \quad \text{and} \quad \omega_1(\tilde{\tilde{f}}^{(n)}, h) \leq Ah^\alpha,$$

fulfilling the assumption of the theorem.

PROOF OF THEOREM 8.3.2: We have ($a \leq x \leq b$)

$$f(x) = \sum_{k=0}^{n} \frac{f^{(k)}(x_0)}{k!}(x - x_0)^k + R_n(x, x_0), \tag{8.3.17}$$

where

$$R_n(x, x_0) = \int_{x_0}^{x} \left(\int_{x_0}^{x_1} \left(\cdots \left(\int_{x_0}^{x_{n-1}} (f^{(n)}(x_n) - f^{(n)}(x_0)) \right) dx_n \right) \cdots \right) dx_1.$$

By $\omega_1(f^{(n)}, h) \leq Ah^\alpha$ we have

$$|f^{(n)}(x_n) - f^{(n)}(x_0)| \leq A|x_n - x_0|^\alpha.$$

And one can prove that

$$|R_n(x, x_0)| \leq A|x - x_0|^{n+\alpha} / \prod_{i=1}^{n}(\alpha + i).$$

Setting $B = A / \prod_{i=1}^{n}(\alpha + i)$ we have

$$|R_n(x, x_0)| \leq B|x - x_0|^{n+\alpha}, \quad \text{all } x \in [a, b],\ n \geq 1. \tag{8.3.18}$$

We would like to find

$$\tilde{U}(x_0) = \sup_{\mu} \int_{[a,b]} |x - x_0|^{n+\alpha} \mu(dx)$$

provided that μ fulfills

$$\int_{[a,b]} |x - x_0|^n \mu(dx) = d_n^n(x_0), \qquad n \geq 1.$$

Here μ is a finite measure on $[a, b]$ such that $\mu([a, b]) = c_0 > 0$ and the induced probability measure $\rho = c_0^{-1} \mu \circ \tau^{-1}$ on $[0, C(x_0)]$ fulfills the convexity assumption of the theorem.

Letting $u = |x - x_0|$ we can reformulate our moment problem into

$$c_0^{-1} \tilde{U}(x_0) = \sup_{\rho} \int_{[0,C(x_0)]} u^{n+\alpha} \rho(du)$$

such that

$$\int_{[0,C(x_0)]} u^n \rho(du) = d_n^n(x_0) c_0^{-1},$$

where ρ is a convex (as above) probability measure on $[0, C(x_0)]$.

Now we apply convex moment methods from Chapter 2. In the terminology of Chapter 2 and Kemperman (1971), p. 137; the case of $x_0 = 0$, we get that $\phi(u) = u^{n+\alpha}$ is transformed into

$$\phi^*(u) = \binom{n+\alpha+s}{s}^{-1} \cdot u^{n+\alpha}$$

and $\psi(u) = u^n$ is transformed into $\psi^*(u) = \binom{n+s}{s}^{-1} \cdot u^n$, for all $u \in [0, C(x_0)]$, $n \geq 1$.

By the same convex methods (see Kemperman (1971), pp. 136, 137, 151, 158, 159 and Chapter 2) we want

$$c_0^{-1} \cdot \tilde{U}(x_0) \cdot \binom{n+\alpha+s}{s} = \sup_v \int_{[0,C(x_0)]} u^{n+\alpha} v(du),$$

the supremum taken over all probability measures v on $[0, C(x_0)]$ such that

$$\int_{[0,C(x_0)]} u^n v(du) = \binom{n+s}{s} \cdot d_n^n(x_0) \cdot c_0^{-1}. \tag{8.3.19}$$

Essentially we need to find

$$\tilde{\tilde{U}}(x_0) = \sup_v \int_{[0,C(x_0)]} u^{n+\alpha} v(du)$$

subject to (8.3.19).

For our moment problem in its final form we need to consider the graph of

$$\{(u^n, u^{n+\alpha}) : 0 \leq u \leq C(x_0)\},$$

which is the same as the graph of

$$\Gamma_0 = \{(y, y^{(n+\alpha/n)}) : 0 \leq y \leq C^n(x_0)\}.$$

The last curve is obviously convex. And the upper boundary of the convex hull of Γ_0 is the line segment

$$\overline{(0,0), (C^n(x_0), C^{n+\alpha}(x_0))}.$$

By assumption (8.3.13) of the theorem we have

$$\binom{n+s}{s} \cdot d_n^n(x_0) \cdot c_0^{-1} \leq C^n(x_0).$$

Standard moment methods now imply that (see Kemperman (1968), (1971), pp. 151, 160 and Chapter 2, method of optimal distance)

$$\tilde{\tilde{U}}(x_0) = \binom{n+s}{s} \cdot d_n^n(x_0) \cdot c_0^{-1} \cdot C^\alpha(x_0),$$

giving us

$$\tilde{U}(x_0) = \binom{n+s}{s} \cdot \binom{n+\alpha+s}{s}^{-1} \cdot d_n^n(x_0) \cdot C^\alpha(x_0). \tag{8.3.20}$$

Integrating (8.3.17) against μ we obtain

$$\int_{[a,b]} \left(f(x) - \sum_{k=0}^{n} \frac{f^{(k)}(x_0)}{k!}(x-x_0)^k \right) \cdot \mu(dx) = \int_{[a,b]} R_n(x,x_0)\mu(dx)$$

and

$$\int_{[a,b]} f\,d\mu - f(x_0) = f(x_0)(c_0 - 1) + \sum_{k=1}^{n} \frac{f^{(k)}(x_0)}{k!} c_k(x_0) + \int_{[a,b]} R_n(x,x_0)\mu(dx).$$

Hence by (8.3.18) we get

$$\left| \int_{[a,b]} \left(f(x) - \sum_{k=0}^{n} \frac{f^{(k)}(x_0)}{k!}(x-x_0)^k \right) \cdot \mu(dx) \right| \leq B \cdot \int_{[a,b]} |x-x_0|^{n+\alpha}\mu(dx)$$

and

$$\left| \int_{[a,b]} f\,d\mu - f(x_0) \right| \leq |f(x_0)|\,|c_0 - 1| + \sum_{k=1}^{n} \frac{|f^{(k)}(x_0)|}{k!} |c_k(x_0)| + B \cdot \int_{[a,b]} |x-x_0|^{n+\alpha}\mu(dx).$$

But it holds that

$$\int_{[a,b]} |x-x_0|^{n+\alpha}\mu(dx) \leq \tilde{U}(x_0),$$

where $\tilde{U}(x_0)$ is given by (8.3.20). The proof of the theorem is now complete. □

8.3.3. Applications

The assumed convexity condition in our results is fulfilled by many commonly used probability distributions.

Example 8.3.1. Let $[a,b] \subset \mathbb{R}$, $x_0 \in [a,b)$ such that $b - x_0 \geq x_0 - a$. Let $\alpha > 0$ and set $k(x_0) = (1 - e^{-\alpha(b-x_0)})^{-1}$. Consider the *truncated exponential probability density function*

$$f_{x_0}(x) = \begin{cases} k(x_0)\alpha e^{-\alpha(x-x_0)}, & x_0 \leq x \leq b, \\ 0, & a \leq x < x_0. \end{cases}$$

Then the corresponding cumulative distribution function $F_{x_0}(x)$ (μ_{x_0} is the associated probability measure) is given by

$$F_{x_0}(x) = \begin{cases} k(x_0)(1 - e^{-\alpha(x-x_0)}), & x_0 \leq x \leq b, \\ 0, & a \leq x < x_0, \end{cases}$$

and it has the property that $(-1)^s F_{x_0}^{(s-1)}(x)$ is convex in $[x_0, b]$ for any integer $s \geq 1$.

Let $G(x)$ be the cumulative distribution function of $\rho = \mu_{x_0} \circ \tau^{-1}$, where $\tau(x) = |x - x_0|$. For $0 \leq x \leq b - x_0$ we have $\tau^{-1}([0,x]) = [x_0 - x, x_0 + x] \cap [a,b]$ and

$$\begin{aligned} \rho([0,x]) &= \mu_{x_0}([x_0 - x, x_0 + x] \cap [a,b]) \\ &= F_{x_0}(x_0 + x) - F_{x_0}(\max(a, x_0 - x)) = F_{x_0}(x_0 + x), \end{aligned}$$

by $F_{x_0}(x_0) = 0$. Therefore $G(x) = \rho([0,x]) = k(x_0)(1 - e^{-\alpha x})$. (Obviously ρ is a probability measure on $[0, b - x_0]$.) We observe that $(-1)^s G^{(s-1)}(x)$ is convex in $[0, b - x_0]$ for any integer $s \geq 1$.

Now let $f \in C([a,b])$, then

$$\int_{[a,b]} f\, d\mu_{x_0} = \frac{\alpha}{e^{-\alpha x_0} - e^{-\alpha b}} \cdot \int_{x_0}^{b} f(x) e^{-\alpha x} dx,$$

where $x_0 \in [a,b)$ such that $b - x_0 \geq x_0 - a$.

Example 8.3.2. (Case of $s = 1$). Let $[a,b] \subset \mathbb{R}$, $x_0 \in [a,b]$, and $C(x_0) = \max(b - x_0,$

$x_0 - a)$. Consider the *two sides-truncated normal probability density function on* $\mathbb{R}$, $\sigma > 0$,

$$f_{x_0}(x) = \begin{cases} 0, & -\infty < x < a, \\ \dfrac{K(x_0)}{\sigma\sqrt{2\pi}} \cdot e^{-(x-x_0)^2/2\sigma^2}, & a \le x \le b, \\ 0, & b < x < +\infty, \end{cases}$$

where

$$K(x_0) = \left(\int_a^b \frac{e^{-(x-x_0)^2/2\sigma^2}}{\sigma\sqrt{2\pi}} \cdot dx \right)^{-1} > 0.$$

Here F_{x_0} is the corresponding cumulative distribution function and μ_{x_0} the associated probability measure which is essentially supported by $[a, b]$.

For $\tau(x) = |x - x_0|$, all $x \in [a, b]$, we have that $\tau^{-1}([0, x]) = [x_0 - x,\ x_0 + x] \cap [a, b]$, for all $x \in [0, C(x_0)]$. Thus $\rho = \mu_{x_0} \circ \tau^{-1}$ is a probability measure on $[0, C(x_0)]$ and $\rho([0, x]) = \mu_{x_0}([x_0 - x,\ x_0 + x] \cap [a, b])$.

Let G be the corresponding to ρ cumulative distribution function, that is, $G(x) = \rho([0, x])$, all $x \in [0, C(x_0)]$.

(i) Assume $b - x_0 \ge x_0 - a$. Let $0 \le x < x_0 - a$, then

$$\rho([0, x]) = \mu_{x_0}([x_0 - x, x_0 + x]) = F_{x_0}(x_0 + x) - F_{x_0}(x_0 - x),$$

that is,

$$G(x) = F_{x_0}(x_0 + x) - F_{x_0}(x_0 - x)$$

and

$$G'(x) = F'_{x_0}(x_0 + x) + F'_{x_0}(x_0 - x) = f_{x_0}(x_0 + x) + f_{x_0}(x_0 - x).$$

Hence

$$G'(x) = \frac{2K(x_0)}{\sigma\sqrt{2\pi}} e^{-x^2/2\sigma^2} > 0$$

and $G''(x) = -G'(x)(x/\sigma^2)$. Thus for $x \in [0, x_0 - a)$ we get $G''(x) \le 0$. We have proved that G is concave in $[0, x_0 - a)$.

Let now $x_0 - a \le x \le b - x_0$, then

$$\rho([0, x]) = \mu_{x_0}([x_0 - x, x_0 + x] \cap [a, b]) = \mu_{x_0}([a, x_0 + x]) = F_{x_0}(x_0 + x).$$

That is, $G(x) = F_{x_0}(x_0 + x)$ and $G'(x) = F'_{x_0}(x_0 + x) = f_{x_0}(x_0 + x)$. Hence

$$G'(x) = \frac{K(x_0)}{\sigma\sqrt{2\pi}} e^{-x^2/2\sigma^2} > 0 \quad \text{and} \quad G''(x) = -G'(x)\frac{x}{\sigma^2} \leq 0.$$

We have proved that G is concave in $[x_0 - a, b - x_0]$.

Since

$$\begin{aligned} \lim_{x \to (x_0 - a)} -G'(x) &= \frac{2K(x_0)}{\sigma\sqrt{2\pi}} e^{-(x_0-a)^2/2\sigma^2} \\ &> \frac{K(x_0)}{\sigma\sqrt{2\pi}} e^{-(x_0-a)^2/2\sigma^2} = G'(x_0 - a). \end{aligned}$$

We have established that G is *concave* in $[0, b - x_0]$.

(ii) Assume $b - x_0 < x_0 - a$. Let $0 \leq x < b - x_0$, then

$$\rho([0, x]) = \mu_{x_0}([x_0 - x, x_0 + x]) = F_{x_0}(x_0 + x) - F_{x_0}(x_0 - x),$$

that is,

$$G(x) = F_{x_0}(x_0 + x) - F_{x_0}(x_0 - x)$$

and

$$G'(x) = F'_{x_0}(x_0 + x) + F'_{x_0}(x_0 - x) = f_{x_0}(x_0 + x) + f_{x_0}(x_0 - x).$$

Hence

$$G'(x) = \frac{2K(x_0)}{\sigma\sqrt{2\pi}} e^{-x^2/2\sigma^2} > 0$$

and $G''(x) = -G'(x)(x/\sigma^2)$. Thus for $x \in [0, b - x_0)$ we get $G''(x) \leq 0$. We have proved that G is concave in $[0, b - x_0)$.

Let $b - x_0 \leq x \leq x_0 - a$, then

$$\begin{aligned} \rho([0, x]) &= \mu_{x_0}([x_0 - x, x_0 + x] \cap [a, b]) = \mu_{x_0}([x_0 - x, b]) \\ &= F_{x_0}(b) - F_{x_0}(x_0 - x) = 1 - F_{x_0}(x_0 - x). \end{aligned}$$

Thus $G(x) = 1 - F_{x_0}(x_0 - x)$ and

$$G'(x) = F'_{x_0}(x_0 - x) = f_{x_0}(x_0 - x) = \frac{K(x_0)}{\sigma\sqrt{2\pi}} e^{-x^2/2\sigma^2},$$

that is,

$$G'(x) = \frac{K(x_0)}{\sigma\sqrt{2\pi}} e^{-x^2/2\sigma^2} > 0$$

and

$$G''(x) = -G'(x)\frac{x}{\sigma^2} \leq 0.$$

We have proved that G is concave in $[b - x_0, x_0 - a]$.

Since

$$\begin{aligned} \lim_{x \to (b-x_0)_-} G'(x) &= \frac{2K(x_0)}{\sigma\sqrt{2\pi}} e^{-(b-x_0)^2/2\sigma^2} \\ &> \frac{K(x_0)}{\sigma\sqrt{2\pi}} e^{-(b-x_0)^2/2\sigma^2} = G'(b - x_0). \end{aligned}$$

we have established that G is *concave* in $[0, x_0 - a]$.

Now let $f \in C([a,b])$, then

$$\int_{[a,b]} f \, d\mu_{x_0} = \frac{K(x_0)}{\sigma\sqrt{2\pi}} \cdot \int_a^b f(x) \cdot e^{-(x-x_0)^2/2\sigma^2} \cdot dx,$$

where $x_0 \in [a, b]$.

CHAPTER NINE

OPTIMAL KOROVKIN TYPE INEQUALITIES FOR CONVOLUTION TYPE OPERATORS

9.1. Sharp Inequalities for Convolution Operators

9.1.1. Introduction

The positive linear convolution-type operators we consider here are given by:

Definition 9.1.1. Let $f \in C([-2a, 2a])$, $a > 0$, and let μ be a probability measure on $[-a, a]$. For every $x \in [-a, a]$, we set

$$L(f, x) = \int_{-a}^{a} ([f(x + y) + f(x - y)]/2)\mu(dy). \tag{9.1.1}$$

We call L a *positive linear convolution operator* from $C([-2a, 2a])$ into $C([-a, a])$. Let $f \in C^n([-2a, 2a])$, $a > 0$, $n \geq 0$ even, $x_0 \in [-a, a]$. Also let L be a positive linear convolution operator, defined above from $C^n([-2a, 2a])$ into $C([-a, a])$. In this section, using moment methods (see Chapter 2), we find best upper bounds for $|L(f, x_0) - f(x_0)|$, leading to attainable sharp inequalities involving the second modulus of continuity of $f^{(n)}$ or an upper bound of it. Our inequalities are attained by conveniently chosen n-times continuously differentiable functions and finitely supported probability measures. These inequalities estimate the degree of convergence of positive linear convolution operators, such as (9.1.1), to the unit operator. The presented results here come from Anastassiou (2) (1989).

For what follows we need

Lemma 9.1.1. *For all $x, t \in \mathbb{R}$, we have*

$$\big|\, |x + t|^\alpha + |x - t|^\alpha - 2|x|^\alpha \big| \leq 2|t|^\alpha, \qquad 0 \leq \alpha \leq 2. \tag{9.1.2}$$

Proof: (9.1.2) is trivial for $\alpha = 0$ and for $t = 0$. For $t \neq 0$, (9.1.2) becomes

$$\left| \left|\frac{x}{t} + 1\right|^\alpha + \left|\frac{x}{t} - 1\right|^\alpha - 2\left|\frac{x}{t}\right|^\alpha \right| \leq 2, \qquad 0 < \alpha \leq 2.$$

Thus, it is enough to prove that $\left| |y+1|^\alpha + |y-1|^\alpha - 2|y|^\alpha \right| \le 2$ for all $y \in \mathbb{R}$, $0 < \alpha \le 2$.

Case I. $0 < \alpha \le 1$. We observe that

$$\begin{aligned} \left| |y+1|^\alpha + |y-1|^\alpha - 2|y|^\alpha \right| &\le \left| |y+1|^\alpha - |y|^\alpha \right| + \left| |y-1|^\alpha - |y|^\alpha \right| \\ &\le \left| |y+1| - |y| \right|^\alpha + \left| |y-1| - |y| \right|^\alpha \le 2. \end{aligned}$$

The middle inequality follows from the subadditivity of $|x|^\alpha$.

Case II. $1 < \alpha \le 2$. From the convexity everywhere of $|y|^\alpha$, we obtain

$$|y+1|^\alpha + |y-1|^\alpha - 2|y|^\alpha \ge 0.$$

Hence, it is enough to prove

$$|y+1|^\alpha + |y-1|^\alpha - 2|y|^\alpha \le 2 \quad \text{for all } y \in \mathbb{R}.$$

Without loss of generality we may assume $0 \le y \le 1$.

Consequently, what remains to be proved is that

$$(1+y)^\alpha + (1-y)^\alpha - 2y^\alpha \le 2 \quad \text{for all } 0 \le y \le 1.$$

For $0 \le y \le 1$, let $g(y) = (1+y)^\alpha + (1-y)^\alpha - 2y^\alpha - 2$. It is enough to prove, for $0 < y < 1$, that $g'(y) \le 0$. But by the subadditivity of $|x|^{\alpha-1}$, if $0 < y < 1$, then $(1+y)^{\alpha-1} \le (1-y)^{\alpha-1} + (2y)^{\alpha-1} \le (1-y)^{\alpha-1} + 2y^{\alpha-1}$ and hence $g'(y) \le 0$. □

Corollary 9.1.1. *For all $x, t \in \mathbb{R}$ we have*

$$\left| |x+2t|^\alpha - 2|x+t|^\alpha + |x|^\alpha \right| \le 2|t|^\alpha, \qquad 0 \le \alpha \le 2. \tag{9.1.3}$$

PROOF: Apply (9.1.2) with x replaced by $x+t$. □

We recall (see Lorentz (1966), p. 47).

Definition 9.1.2. For $f \in C([\alpha, \beta])$, $-\infty < \alpha < \beta < \infty$, the *second modulus of continuity* of f in $[\alpha, \beta]$ is given by

$$\omega_2(f,h) := \sup_{\substack{\alpha \le x \le x+2t \le \beta \\ t \le h}} |f(x) - 2f(x+t) + f(x+2t)|, \tag{9.1.4}$$

$$0 \leq h \leq \frac{\beta - \alpha}{2}.$$

Proposition 9.1.1. *The function* $|t|^\alpha$ $(0 \leq \alpha \leq 2)$ *has in* $[-\gamma, \gamma]$ $(\gamma > 0)$ *the second modulus of continuity*

$$\omega_2(|t|^\alpha, h) = 2h^\alpha, \qquad 0 < h \leq \gamma. \tag{9.1.5}$$

PROOF: Apply (9.1.3). Equality holds for $x = -h$, $t = h$. □

9.1.2. Preliminaries

In this section we consider $f \in C^n([-2a, 2a])$, $a > 0$, n is even and ≥ 0, and we study inequalities involving $\omega_2(f^{(n)}, h)$, $0 \leq h \leq a$, or an upper bound on $\omega_2(f^{(n)}, h)$.

For fixed $x_0 \in [-a, a]$, by using Taylor's formula with Cauchy remainder for $f(x_0 + y)$, $f(x_0 - y)$ $(n \geq 2$ even), we obtain

$$(\Delta_y^2 f)(x_0) = 2 \cdot \sum_{\rho=1}^{n/2} \frac{f^{(2\rho)}(x_0)}{(2\rho)!} y^{2\rho} + \int_0^y (\Delta_t^2 f^{(n)})(x_0) \cdot \frac{(y-t)^{n-1}}{(n-1)!} dt, \tag{9.1.6}$$

where

$$(\Delta_y^2 f)(x_0) := f(x_0 + y) + f(x_0 - y) - 2f(x_0) \tag{9.1.7}$$

$$(\text{thus } (\Delta_{-y}^2 f)(x_0) = (\Delta_y^2 f)(x_0)) \tag{9.1.8}$$

is the so-called central second order difference. Now we integrate (9.1.6) relative to the variable y with respect to a probability measure μ over $[-a, a]$. We make a change of variable in the remainder of (9.1.6) and we use the fact that n is even. Also, we denote

$$c_{2\rho} := \int_{-a}^{a} y^{2\rho} \mu(dy). \tag{9.1.9}$$

Thus

$$L(f, x_0) - f(x_0) - \sum_{\rho=1}^{n/2} \frac{f^{(2\rho)}(x_0)}{(2\rho)!} c_{2\rho} = U_n, \tag{9.1.10}$$

where

$$U_n := \frac{1}{2} \cdot \int_{-a}^{a} \left(y^n \cdot \int_0^1 (\Delta_{\theta y}^2 f^{(n)})(x_0) \cdot \frac{(1-\theta)^{n-1}}{(n-1)!} d\theta \right) \mu(dy). \tag{9.1.11}$$

Using (9.1.4) it follows that

$$|U_n| \leq V_n, \tag{9.1.12}$$

where

$$V_n := \frac{1}{2} \cdot \int_{-a}^{a} \left(y^n \cdot \int_0^1 \omega_2(f^{(n)}, \theta|y|) \cdot \frac{(1-\theta)^{n-1}}{(n-1)!} d\theta \right) \mu(dy). \tag{9.1.13}$$

Note that $U_n = V_n$ when $x_0 = 0$ and $f^{(n)}(y) = |y|^\alpha$, with $0 \leq \alpha \leq 2$ (by Proposition 9.1.1).

It is helpful to introduce

$$\omega_2(t) := \omega_2(f^{(n)}, |t|), \qquad 0 \leq |t| \leq a, \tag{9.1.14}$$

and, if $n \geq 1$, the even functions

$$G_j(y) := \int_0^y \omega_2(t) \cdot \frac{(y-t)^{j-1}}{(j-1)!} dt, \qquad j = 1, 2, \ldots, n. \tag{9.1.15}$$

Further, let $G_0(y) := \omega_2(y)$. Observe that

$$V_n = \frac{1}{2} \cdot \int_{-a}^{a} G_n(y) \mu(dy) \tag{9.1.16}$$

and $G_j'(y) = G_{j-1}(y)$, $j = 1, 2, \ldots, n$, since $\omega_2(f^{(n)}, h)$ is continuous.

9.1.3. Main Results

Using the previous notations and assumptions, we have

Proposition 9.1.2. *Let μ be a probability measure on $[-a, a]$, $a > 0$, and define*

$$d_r := \left[\int_{-a}^{a} |y|^r \mu(dy) \right]^{1/r}, \qquad r > 0.$$

Let $n \geq 2$ be even, and consider all $f \in C^n([-2a, 2a])$ such that

$$\omega_2(f^{(n)}, |y|) \leq 2A|y|^\alpha, \qquad 0 \leq |y| \leq a, \ 0 < \alpha \leq 2, \ A > 0.$$

Then, for $x_0 \in [-a, a]$,

$$\left| L(f, x_0) - f(x_0) - \sum_{\rho=1}^{n/2} \frac{f^{(2\rho)}(x_0)}{(2\rho)!} c_{2\rho} \right| \leq A d_{n+2}^{n+\alpha} / [(\alpha+1)(\alpha+2) \cdots (\alpha+n)]. \tag{9.1.17}$$

The last inequality is attained (therefore is sharp), when $x_0 = 0$, by the function

$$f_*(y) := A|y|^{\alpha+n}/[(\alpha+1)(\alpha+2)\cdots(\alpha+n)] \tag{9.1.18}$$

and the associated probability measure μ having mass $1/2$ at the two points $\pm d_{n+2}$.

PROOF: We easily find that

$$G_n(y) \le 2A|y|^{\alpha+n}/[(\alpha+1)\cdots(\alpha+n)] \qquad (0 \le |y| \le a) \tag{9.1.19}$$

and thus

$$V_n \le A d_{n+\alpha}^{n+\alpha}/[(\alpha+1)\cdots(\alpha+n)]. \tag{9.1.20}$$

As $d_{n+\alpha} \le d_{n+2}$ and $|U_n| \le V_n$, we get

$$|U_n| \le A d_{n+2}^{n+\alpha}/[(\alpha+1)\cdots(\alpha+n)]. \tag{9.1.21}$$

Using (9.1.10), we finally obtain (9.1.17). Also, note that $f_*^{(n)}(y) = A|y|^\alpha$; hence, by Proposition 9.1.1, $\omega_2(f_*^{(n)}, |y|) = 2A|y|^\alpha$. Also, $f_*^{(k)}(0) = 0$, for $k = 0, \ldots, n$. This proves the desired equality. □

The next results assume that $f \in C^n([-2a, 2a])$ $(a > 0)$, and that $\omega_2(f^{(n)}, |t|) \le g(t)$ $(0 \le |t| \le a)$, where g is a given, arbitrary, bounded, even positive function which is Borel measurable. We set

$$\hat{G}_n(y) := \int_0^y g(t)(y-t)^{n-1}/(n-1)!dt.$$

Theorem 9.1.1. *Let ψ be a function on $[0, a]$ such that $\psi(0) = 0$, which is continuous and strictly increasing. Let a probability measure μ exist on $[-a, a]$ with*

$$\psi^{-1}\left(\int_{-a}^{a} \psi(|y|)\mu(dy)\right) = d. \tag{9.1.22}$$

Suppose ($n \ge 2$ even) that

$$\mathcal{H}_n(u) := \hat{G}_n(\psi^{-1}(u)) \tag{9.1.23}$$

is concave on $[0, \psi(a)]$. Then, for every $x_0 \in [-a, a]$,

$$E(x_0) := \left| L(f, x_0) - f(x_0) - \sum_{\rho=1}^{n/2} \frac{f^{(2\rho)}(x_0)}{(2\rho)!} \cdot c_{2\rho} \right| \le \frac{1}{2} \cdot \hat{G}_n(d). \tag{9.1.24}$$

The above inequality is attained (and hence is sharp) when $x_0 = 0$, $f(y) = \frac{1}{2} \cdot \hat{G}_n(y)$ *(implying* $\omega_2(f^{(n)}, |t|) = g(t)$*), and* $\mu = \delta_d$. *If* $g(t) = 2 \cdot A \cdot |t|^\alpha$, $0 < \alpha \leq 2$, $A > 0$, *one can choose* f *as* f_* *of* (9.1.18).

PROOF: From (9.1.10), (9.1.12) and (9.1.16), we need to prove only that

$$\int_{-a}^{a} \hat{G}_n(y)\mu(dy) \leq \hat{G}_n(d).$$

Note that both $\hat{G}_n(y)$ and $\psi(|y|)$ are even functions on $[-a, a]$. It follows from the moment method of optimal distance that (see Chapter 2)

$$\sup_{\mu} \int_{-a}^{a} \hat{G}_n(y)\mu(dy) = \hat{G}_n(d),$$

since, by the concavity of $\mathcal{H}_n$, the set

$$\Gamma_1 := \{(u, \mathcal{H}_n(u)) : 0 \leq u \leq \psi(a)\}$$

is the upper boundary of the convex hull of the curve

$$\Gamma_0 := \{(\psi(y), \hat{G}_n(y)) : 0 \leq y \leq a\}. \qquad \square$$

The next theorem generalizes Theorem 9.1.1.

Theorem 9.1.2. *Let* μ *be a probability measure on* $[-a, a]$ *and consider the upper concave envelope* $\mathcal{H}_n^*(u)$ *of* $\mathcal{H}_n(u)$ *of* (9.1.23) ($n \geq 0$ *even*). *Also, consider* $E(x_0)$ *of* (9.1.24).

Then

$$E(x_0) \leq \frac{1}{2} \cdot \mathcal{H}_n^*(\psi(d)), \tag{9.1.25}$$

where $x_0 \in [-a, a]$.

When $\mathcal{H}_n$ *is concave, the right-hand side of* (9.1.25) *equals* $\frac{1}{2} \cdot \hat{G}_n(d)$. *If, moreover,* $\omega_2(\frac{1}{2} \cdot g, |t|) = g(t)$, *the inequality is attained as in Theorem* 9.1.1. *Otherwise* (9.1.25) *is still (non-trivially, that is with* $\mu \neq \delta_{x_0}$*) attained, when* $x_0 = 0$ *and* $\omega_2(\frac{1}{2} \cdot g, |t|) = g(t)$, *by the same function* $f(y) = \frac{1}{2} \cdot \hat{G}_n(y)$ *and a two points supported probability measure* μ. *In particular, when* $\mathcal{H}_n$ *is convex,*

$$\mathcal{H}_n^*(\psi(d)) = (\psi(d)/\psi(a)) \cdot \hat{G}_n(a), \qquad n > 0.$$

PROOF: As in Theorem 9.1.1. □

Let g be an arbitrary continuous even positive function on $[-a,a]$ (allowing $g(0)=0$). Let ψ be a continuous, strictly increasing function on $[0,a]$ with $\psi(0)=0$. Let $\hat{G}_n$ be as above.

Next we give, without proof, sufficient conditions for $\mathcal{H}_n(u)=\hat{G}_n(\psi^{-1}(u))$ to be convex (concave) on $[0,\psi(a)]$, $n>0$ being even.

Theorem 9.1.3. (i) *Assume that $\psi \in C^n((0,a))$, $n \geq 0$ even, satisfies*

$$\psi^{(k)}(0) \geq 0 \quad \textit{for } k=0,\dots,n-1. \tag{9.1.26}$$

Suppose further, that

$$g(y)/\psi^{(n)}(y) \textit{ is non-decreasing on each interval where } \psi^{(n)} \textit{ is positive.} \tag{9.1.27}$$

Then the function $\mathcal{H}_n=\hat{G}_n\circ\psi^{-1}$ is convex. In particular, $\hat{G}_n(y)/\psi(y)$ is non-decreasing.

(ii) *Assume that $\psi \in C^n((0,a))$, $n \geq 0$ even, satisfies*

$$\psi^{(k)}(0) \leq 0 \quad \textit{for } k=0,\dots,n-1. \tag{9.1.28}$$

Suppose, further that

$$g(y)/\psi^{(n)}(y) \textit{ is non-increasing on each interval where } \psi^{(n)} \textit{ is positive.} \tag{9.1.29}$$

Then the function $\mathcal{H}_n=\hat{G}_n\circ\psi^{-1}$ is concave. In particular, $\hat{G}_n(y)/\psi(y)$ is non-increasing.

Corollary 9.1.2. *Let $1<m\leq n-1$ be such that $\psi^{(m)}(y)$ is non-increasing as long as it is positive. Suppose further that $\psi^{(k)}(0)\geq 0$ for $1\leq k\leq m-1$. Then $\mathcal{H}_n$ is convex for $n\geq 4$ even.*

Proposition 9.1.3. *If $n \geq 2$ is even and $\psi''(y)\leq 0$ $(0<y\leq a)$, then $\mathcal{H}_n$ is convex on $[0,\psi(a)]$.*

9.2. Sharp Inequalities for Non-Positive Generalized Convolution Operators

9.2.1. Introduction

Jakimowski and Ramanujan (1964) gave a Korovkin type theorem for a specific sequence of non-positive linear operators. Similar results for non-positive linear operators have been extremely rare.

In this section a sequence of non-positive generalized convolution type linear operators is introduced. The rate of convergence of this sequence to the unit operator is given through sharply attained inequalities in several cases. Using the moment method of optimal distance (see Chapter 2) we are able to simplify a lot the remainders of the appearing inequalities, especially under convexity/concavity. A related Korovkin type theorem is presented at the end. The appearing results here come from Anastassiou(3) (1989).

9.2.2. Background

Let $(\mu_N)_{N\in\mathbb{N}}$ be a sequence of probability measures on $[a,b]\subset\mathbb{R}$. Here $0\in(a,b)$ and $|a|\le b$.

For $r\in\mathbb{N}$ and $n\in\mathbb{Z}_+$ we set

$$\alpha_j=\begin{cases}(-1)^{r-j}\cdot\binom{r}{j}\cdot j^{-n}, & j=1,\dots,r\\ 1-\sum_{j=1}^{r}(-1)^{r-j}\cdot\binom{r}{j}\cdot j^{-n}, & j=0.\end{cases}$$

That is, $\sum_{j=0}^{r}\alpha_j=1$. Let $[\alpha,\beta]\subset\mathbb{R}$ such that

$$[a,b]\subset\left[\frac{\alpha}{r},\frac{\beta}{r}\right].$$

Obviously $0\in(\alpha,\beta)$.

For $x_0\in[\alpha-ra,\beta-rb]$ we have that $x_0+j\cdot[a,b]\subset[\alpha,\beta]$, all $j=0,1,\dots,r$. Note that $b-a\le(\beta-\alpha)/r$. Let $(L_N)_{N\in\mathbb{N}}$ be the sequence of linear operators from $C^n([\alpha,\beta])$ into $C([\alpha-ra,\beta-rb])$ defined by

$$(L_Nf)(x_0):=\int_{[a,b]}\left(\sum_{j=0}^{r}\alpha_j\cdot f(x_0+j\cdot y)\right)\cdot\mu_N(dy),\tag{9.2.1}$$

for all $f \in C^n([\alpha,\beta])$, all $x_0 \in [\alpha - ra, \beta - rb]$. These operators are of *convolution type* and in general *are not* positive, as we see from the following examples.

Example 9.2.1. Let $r = 2$, $n = 3$. Then $\alpha_0 = \frac{23}{8}$, $\alpha_1 = -2$, $\alpha_2 = \frac{1}{8}$. Consider $f(x) = x^2 \geq 0$ and $x_0 = 0$. We get

$$(L_N x^2)(0) = -\frac{3}{2} \cdot \int_{[a,b]} y^2 \mu_N(dy) \leq 0.$$

Example 9.2.2. Let r, n be even numbers. Consider $f(x) = x^n \geq 0$ and again $x_0 = 0$. Let

$$c_{nN} = \int_{[a,b]} y^n \mu_N(dy) \geq 0$$

and note that

$$\sum_{j=1}^{r} (-1)^{r-j} \cdot \binom{r}{j} = -1.$$

Then

$$\begin{aligned}(L_N x^n)(0) &= \int_{[a,b]} \left(\sum_{j=0}^{r} \alpha_j \cdot (0 + j \cdot y)^n \right) \cdot \mu_N(dy) \\ &= c_{nN} \cdot \left(\sum_{j=0}^{r} \alpha_j \cdot j^n \right) = c_{nN} \cdot \left(\sum_{j=1}^{r} (-1)^{r-j} \binom{r}{j} \right) \\ &= -c_{nN} \leq 0,\end{aligned}$$

i.e., $(L_N x^n)(0) \leq 0$.

Definition 9.2.1. (See Lorentz (1966), p. 47.) In here the rth modulus of smoothness of $f^{(n)}$ is given by

$$\omega_r(f^{(n)}, h) := \sup_{|t| \leq h} \|\Delta_t^r f^{(n)}(x)\|_{\infty,[\alpha,\beta]}.$$

Here $0 \leq h \leq (\beta - \alpha)/r$ and

$$\Delta_t^r f^{(n)}(x) := \sum_{j=0}^{r} (-1)^{r-j} \cdot \binom{r}{j} \cdot f^{(n)}(x + j \cdot t).$$

9.2.3. Results

The main result of this section is

Theorem 9.2.1. *Here we consider $n \in \mathbb{N}$. Let*

$$\delta_k = \sum_{j=0}^{r} \alpha_j \cdot j^k, \qquad k = 1, \ldots, n$$

and

$$c_{kN} = \int_{[a,b]} y^k \mu_N(dy), \quad \text{all } N \in \mathbb{N}.$$

Also, consider the even function

$$G_n(y) = \int_0^{|y|} \frac{(|y| - t)^{n-1}}{(n-1)!} \cdot \omega_r(f^{(n)}, t) \cdot dt,$$

all $y \in [a,b]$; $f \in C^n([\alpha, \beta])$. Then

$$|(L_N f)(x_0) - f(x_0)| \le \sum_{k=1}^{n} \frac{|f^{(k)}(x_0)|}{k!} \cdot |\delta_k| \cdot |c_{kN}|$$

$$+ \int_{[a,b]} G_n(y) \mu_N(dy), \quad \text{all } x_0 \in [\alpha - ra, \beta - rb], \tag{9.2.2}$$

and

$$\|L_N f - f\|_{\infty, [\alpha - ra, \beta - rb]} \le \sum_{k=1}^{n} \frac{\|f^{(k)}\|_{\infty, [\alpha - ra, \beta - rb]}}{k!} \cdot |\delta_k| \cdot |c_{kN}|$$

$$+ \int_{[a,b]} G_n(y) \mu_N(dy). \tag{9.2.3}$$

Corollary 9.2.1. *Let $G_0(y) = \omega_r(f, |y|)$, all $y \in [a,b]$; $f \in C([\alpha, \beta])$. Then*

$$|(L_N f)(x_0) - f(x_0)| \le \int_{[a,b]} G_0(y) \mu_N(dy), \tag{9.2.4}$$

all $x_0 \in [\alpha - ra, \beta - rb]$, and

$$\|L_N f - f\|_{\infty, [\alpha - ra, \beta - rb]} \le \int_{[a,b]} G_0(y) \mu_N(dy). \tag{9.2.5}$$

Proof of Theorem 9.2.1: Since μ_N is a probability measure on $[a,b]$ and $\sum_{j=0}^{r} \alpha_j = 1$ we have $L_N c = c$, all $c \in \mathbb{R}$, so that

$$(L_N f)(x_0) - f(x_0) = \sum_{j=0}^{r} \alpha_j \cdot \int_{[a,b]} [f(x_0 + j \cdot y) - f(x_0)] \cdot \mu_N(dy).$$

For $f \in C^n([\alpha, \beta])$ and $y \in [a, b]$ we obtain

$$f(x_0 + j \cdot y) = \sum_{k=0}^{n-1} \frac{f^{(k)}(x_0)}{k!} \cdot (j \cdot y)^k + \int_0^{j \cdot y} \frac{(j \cdot y - x)^{n-1}}{(n-1)!} f^{(n)}(x_0 + x) \cdot dx$$

$$= \sum_{k=0}^{n-1} \frac{f^{(k)}(x_0)}{k!} \cdot (j \cdot y)^k + j^n \cdot \int_0^y \frac{(y-t)^{n-1}}{(n-1)!} \cdot f^{(n)}(x_0 + j \cdot t) \cdot dt.$$

Multiplying by α_j and summing up, we get

$$\sum_{j=0}^{r} \alpha_j \cdot [f(x_0 + j \cdot y) - f(x_0)] = \sum_{k=1}^{n} \frac{f^{(k)}(x_0)}{k!} \cdot \delta_k \cdot y^k + \int_0^y \frac{(y-t)^{n-1}}{(n-1)!} \cdot \tau(t) \cdot dt. \tag{9.2.6}$$

Here

$$\tau(t) := \sum_{j=0}^{r} \alpha_j \cdot j^n \cdot f^{(n)}(x_0 + j \cdot t) - \delta_n \cdot f^{(n)}(x_0),$$

and note that

$$-\sum_{j=1}^{r} (-1)^{r-j} \cdot \binom{r}{j} = (-1)^r \cdot \binom{r}{0}.$$

Thus, we see that

$$\tau(t) = \sum_{j=0}^{r} \alpha_j \cdot j^n \cdot [f^{(n)}(x_0 + j \cdot t) - f^{(n)}(x_0)]$$

$$= \sum_{j=1}^{r} (-1)^{r-j} \cdot \binom{r}{j} \cdot [f^{(n)}(x_0 + j \cdot t) - f^{(n)}(x_0)]$$

$$= \sum_{j=1}^{r} (-1)^{r-j} \cdot \binom{r}{j} \cdot f^{(n)}(x_0 + j \cdot t) - f^{(n)}(x_0) \cdot \left(\sum_{j=1}^{r} (-1)^{r-j} \cdot \binom{r}{j} \right)$$

$$= \sum_{j=1}^{r} (-1)^{r-j} \cdot \binom{r}{j} \cdot f^{(n)}(x_0 + j \cdot t) + (-1)^r \cdot \binom{r}{0} \cdot f^{(n)}(x_0 + o \cdot t)$$

$$= \sum_{j=0}^{r} (-1)^{r-j} \cdot \binom{r}{j} \cdot f^{(n)}(x_0 + j \cdot t) = \Delta_t^r f^{(n)}(x_0).$$

That is, $\tau(t) = \Delta_t^r f^{(n)}(x_0)$.

Hence,

$$|\tau(t)| \leq \omega_r(f^{(n)}, |t|), \quad \text{all } t \in [a, b],$$

independently of x_0.

Integrating (9.2.6) relative to μ_N we find that

$$(L_N f)(x_0) - f(x_0) = \sum_{k=1}^{n} \frac{f^{(k)}(x_0)}{k!} \cdot \delta_k \cdot c_{kN} + R_n,$$

where

$$|R_n| \leq \int_{[a,b]} G_n(y) \mu_N(dy).$$

Inequality (9.2.2) is sharp. □

Theorem 9.2.2. *Inequality* (9.2.2) *at* $x_0 = 0$ *is attained by* $f(x) = x^{r+n}$, $r, n \in \mathbb{N}$, *and a probability measure* μ_N *supported on* $[0, b]$.

Inequality (9.2.4) is also sharp.

Corollary 9.2.2. *Inequality* (9.2.4) *at* $x_0 = 0$ *is attained by* $f(x) = x^r$, $r \in \mathbb{N}$, *and a probability measure* μ_N *supported on* $[0, b]$.

Proof of Theorem 9.2.2: Since

$$f^{(n)}(x) = (r+n)(r+n-1)\cdots(r+1)x^r$$

we get

$$\begin{aligned} \Delta_t^r f^{(n)}(x) &= (r+n)(r+n-1)\cdots(r+1) \cdot \Delta_t^r x^r \\ &= (r+n)(r+n-1)\cdots(r+1) r! t^r, \qquad t > 0 \end{aligned}$$

(see Schumaker (1981), p. 54). That is,

$$\omega_r(f^{(n)}, t) = (r+n)(r+n-1)\cdots(r+1) r! t^r, \quad \text{all } 0 \leq t \leq (\beta - \alpha)/r.$$

Therefore

$$G_n(y) = \int_0^{|y|} \frac{(|y| - t)^{n-1}}{(n-1)!} (r+n)(r+n-1)\cdots(r+1) r! t^r dt = r! |y|^{r+n},$$

i.e.,

$$G_n(y) = r! |y|^{r+n}.$$

And we have $f^{(k)}(0) = 0$, $k = 0, 1, \ldots, n$.

Thus, the right-hand side of inequality (9.2.2) in this case equals

$$r! \cdot \int_{[a,b]} |y|^{r+n} \mu_N(dy).$$

Also,

$$\begin{aligned}
(L_N x^{r+n})(0) &= \int_{[a,b]} \left(\sum_{j=0}^{r} \alpha_j \cdot (j \cdot y)^{r+n} \right) \cdot \mu_N(dy) \\
&= \int_{[a,b]} \left(\sum_{j=1}^{r} (-1)^{r-j} \cdot \binom{r}{j} \cdot j^r \cdot y^{r+n} \right) \cdot \mu_N(dy) \\
&= \left(\sum_{j=0}^{r} (-1)^{r-j} \cdot \binom{r}{j} \cdot j^r \right) \cdot \int_{[a,b]} y^{r+n} \mu_N(dy) \\
&= (\Delta_1^r x^r)(0) \cdot \int_{[a,b]} y^{r+n} \mu_N(dy) = r! \cdot \int_{[a,b]} y^{r+n} \mu_N(dy)
\end{aligned}$$

(see Schumaker (1981), p. 54). So we find that

$$|(L_N x^{r+n})(0)| = r! \cdot \left| \int_{[a,b]} y^{r+n} \mu_N(dy) \right|,$$

which is the left-hand side of inequality (9.2.2) for this case.

Since $\mu_N([a,0]) = 0$, we have that

$$\int_{[a,b]} |y|^{r+n} \mu_N(dy) = \int_{[0,b]} y^{r+n} \mu_N(dy) = \left| \int_{[a,b]} y^{r+n} \mu_N(dy) \right|,$$

proving the claim of the theorem. □

Next for

$$\int_{[a,b]} G_n(y) \mu_N(dy), \qquad n \in \mathbb{Z}_+,$$

the remainders of inequalities (9.2.2) and (9.2.4), we find best upper bounds in two different important cases.

Here the function G_n can be defined for an arbitrary modulus of smoothness.

Theorem 9.2.3. *Let ψ be a continuous and strictly increasing function on $[0,b]$ such that $\psi(0) = 0$, and let*

$$\psi^{-1} \left(\int_{[a,b]} \psi(|y|) \mu_N(dy) \right) = d_N, \qquad N \geq 1.$$

Consider the function $H_n = G_n \circ \psi^{-1}$ *on* $[0, \psi(b)]$, $n \in \mathbb{Z}^+$. *Then we find that the best upper bounds are*

$$\int_{[a,b]} G_n(y)\mu_N(dy) \le \begin{cases} G_n(d_N), & \text{if } H_n \text{ is concave.} \\ (\psi(d_N)/\psi(b)) \cdot H_n(\psi(b)), & \text{if } H_n \text{ is convex.} \end{cases} \tag{9.2.7}$$

In general, we obtain

Corollary 9.2.3. *Consider the upper concave envelope* $H_n^*(u)$ *of* $H_n(u)$. *We find the best upper bound, which is (also attained)*

$$\int_{[a,b]} G_n(y)\mu_N(dy) \le H_n^*(\psi(d_N)), \qquad n \in \mathbb{Z}_+. \tag{9.2.8}$$

Remark 9.2.1. When H_n, $n \in \mathbb{Z}_+$ is concave or convex, we have that $H_n^*(\psi(d_N))$ equals $G_n(d_N)$ or $(\psi(d_N)/\psi(b)) \cdot H_n(\psi(b))$, respectively.

PROOF OF THEOREM 9.2.3: (i) When H_n is concave, note that both $G_n(y)$ and $\psi(|y|)$ are even functions on $[a, b]$, where $|a| \le b$. It follows from the moment method of optimal distance (see Chapter 2) that

$$\sup_{\mu_N} \int_{[a,b]} G_n(y)\mu_N(dy) = \sup_{\mu_N} \int_{[0,b]} G_n(y)\mu_N(dy) = G_n(d_N),$$

since by concavity of H_n the set $\Gamma_1 := \{(u, H_n(u))\colon 0 \le u \le \psi(b)\}$ describes the upper boundary of the convex hull conv Γ_0 of the curve

$$\Gamma_0 := \{(\psi(y), G_n(y)) \colon 0 \le y \le b\}.$$

(ii) When H_n is convex the proof is similar. □

That in Theorem 9.2.3 it is not strange to assume that H_n is concave or convex is justified by the following: Let g be a modulus of smoothness and consider

$$G_n(y) := \int_0^{|y|} \frac{(|y| - t)^{n-1}}{(n-1)!} \cdot g(t) \cdot dt, \quad \text{all } y \in [a, b],\ n \in \mathbb{N}.$$

Then we have

Lemma 9.2.1. (i) *Let* $\psi \in C^n((0,b))$ *such that* $\psi^{(k)}(0) \leq 0$ *for* $k = 1,\ldots,n-1$ *and* $g(y)/\psi^{(n)}(y)$ *is non-increasing, whenever* $\psi^{(n)}(y) > 0$. *Then* $H_n := G_n \circ \psi^{-1}$ *is a concave function.*

(ii) *Let* $\psi \in C^n((0,b))$ *such that* $\psi^{(k)}(0) \geq 0$, $k = 1,\ldots,n-1$ *and* $g(y)/\psi^{(n)}(y)$ *is non-decreasing, whenever* $\psi^{(n)}(y) > 0$. *Then* $H_n := G_n \circ \psi^{-1}$ *is a convex function.*

For the remainder of inequality (9.2.2) we find the following simple upper bound without any special assumptions.

Theorem 9.2.4. *Let*

$$\left(\int_{[a,b]} |y|^{n+1}\mu_N(dy)\right)^{1/n+1} = \tau_N > 0, \qquad n \in \mathbb{N}$$

and

$$G_n^*(y) := \int_0^{|y|} \frac{(|y|-t)^{n-1}}{(n-1)!} \cdot \omega_1(f^{(n)}, t)dt,$$

all $y \in [a,b]$, *where* $\omega_1(f^{(n)}, t)$ *is the first modulus of continuity of* $f^{(n)}$, $f \in C^n([\alpha,\beta])$. *Then*

$$\int_{[a,b]} G_n(y)\mu_N(dy) \leq 2^r \cdot G_n^*(\tau_N), \qquad r \in \mathbb{N}. \tag{9.2.9}$$

PROOF: We have $\omega_r(f^{(n)}, |y|) \leq 2^{r-1} \cdot \omega_1(f^{(n)}, |y|)$, all $y \in [a,b]$ (see Schumaker (1981), p. 55). Furthermore,

$$\omega_1(f^{(n)}, |y|) \leq \overline{\omega}_1(|y|) \leq 2 \cdot \omega_1(f^{(n)}, |y|),$$

all $y \in [a,b]$, where $\overline{\omega}_1$ is a concave first modulus of continuity (see Lorentz (1966), p. 45).

Thus,

$$\omega_r(f^{(n)}, |y|) \leq 2^{r-1} \cdot \overline{\omega}_1(|y|) \leq 2^r \cdot \omega_1(f^{(n)}, |y|), \quad \text{all } y \in [a,b].$$

Set

$$\overline{G}_n(y) := \int_0^{|y|} \frac{(|y|-t)^{n-1}}{(n-1)!} \cdot \overline{\omega}_1(t) \cdot dt, \quad \text{all } y \in [a,b].$$

Hence,

$$\begin{aligned} G_n(y) &= \int_0^{|y|} \frac{(|y|-t)^{n-1}}{(n-1)!} \cdot \omega_r(f^{(n)}, t) \cdot dt \\ &\leq 2^{r-1} \cdot G_n^*(y) \leq 2^{r-1} \cdot \overline{G}_n(y) \leq 2^r \cdot G_n^*(y), \quad \text{all } y \in [a,b]. \end{aligned}$$

The function $\psi(y) = y^{n+1}$ on $[0, b]$ is continuous, strictly increasing, and $\psi(0) = 0$. And $\psi^{(n)}(y) = (n+1)!y > 0$, all $y \in (0, b]$, along with $\psi^{(k)}(0) = 0$, $k = 1, \ldots, n-1$. Since $\overline{\omega}_1(y)$ is concave on $[0, b]$, this implies that $\overline{\omega}_1(y)/y$ is decreasing in $y > 0$, so that $\overline{\omega}_1(y)/\psi^{(n)}(y)$ is decreasing on $(0, b]$.

Thus by Lemma 9.2.1 (i) we get that $\overline{H}_n = \overline{G}_n \circ \psi^{-1}$ is a concave function on $[0, b^{n+1}]$; and by Theorem 9.2.3 we obtain

$$\int_{[a,b]} \overline{G}_n(y)\mu_N(dy) \leq \overline{G}_n(\tau_N),$$

giving us

$$\begin{aligned}\int_{[a,b]} G_n(y)\mu_N(dy) &\leq 2^{r-1} \cdot \int_{[a,b]} \overline{G}_n(y)\mu_N(dy) \\ &\leq 2^{r-1} \cdot \overline{G}_n(\tau_N) \leq 2^r \cdot G_n^*(\tau_N).\end{aligned}$$

The proof of the theorem is now complete. □

A related Korovkin type theorem has as follows.

Theorem 9.2.5. *Let*

$$\int_{[a,b]} |y|\mu_N(dy) = l_N > 0. \tag{9.2.10}$$

Then

$$\|L_N f - f\|_{\infty,[\alpha-ra,\beta-rb]} \leq 2^r \cdot \omega_1(f, l_N), \quad \text{all } f \in C([\alpha, \beta]). \tag{9.2.11}$$

That is, when $l_N \to 0$ as $N \to \infty$, we get

$$L_N f \xrightarrow{u} f \text{ on } [\alpha - ra, \beta - rb], \quad \text{all } f \in C([\alpha, \beta]). \tag{9.2.12}$$

Remark 9.2.2. Since $[\alpha - ra, \beta - rb] \subset [\alpha - a, \beta - b]$, the optimal result (9.2.12) is for $r = 1$. When $r = 1$ we have

$$(L_N f)(x_0) = \int_{[a,b]} f(x_0 + y)\mu_N(dy).$$

Proof of Theorem 9.2.5: We have $\omega_r(f, |y|) \leq 2^{r-1} \cdot \omega_1(f, |y|)$, all $y \in [a, b]$ (see Schumaker (1981), p. 55). Furthermore

$$\omega_1(f, |y|) \leq \overline{\omega}_1(|y|) \leq 2 \cdot \omega_1(f, |y|),$$

all $y \in [a,b]$, where $\overline{\omega}_1$ is a concave first modulus of continuity (see Lorentz (1966), p. 45). Thus

$$\omega_r(f,|y|) \le 2^{r-1} \cdot \overline{\omega}_1(|y|) \le 2^r \cdot \omega_1(f,|y|), \quad \text{all } y \in [a,b].$$

From inequality (9.2.5) we get

$$\|L_N f - f\|_{\infty,[\alpha-ra,\beta-rb]} \le \int_{[a,b]} \omega_r(f,|y|)\mu_N(dy) \le 2^{r-1} \cdot \int_{[a,b]} \omega_1(f,|y|)\mu_N(dy)$$

$$\le 2^{r-1} \cdot \int_{[a,b]} \overline{\omega}_1(|y|)\mu_N(dy). \tag{9.2.13}$$

Note that $|y|$, $\overline{\omega}_1(|y|)$ are even functions on $[a,b]$, $|a| \le b$, and μ_N fulfills (9.2.10). In particular $\overline{\omega}_1(|y|)$ is a concave function on $[0,b]$. Therefore by applying the moment method of optimal distance (see Chapter 2) we find that

$$\sup_{\mu_N} \int_{[a,b]} \overline{\omega}_1(|y|)\mu_N(dy) = \sup_{\mu_N} \int_{[0,b]} \overline{\omega}_1(|y|)\mu_N(dy)$$

$$= \overline{\omega}_1(l_N) \le 2 \cdot \omega_1(f,l_N) \qquad (l_N \le b). \tag{9.2.14}$$

Clearly from (9.2.13) and (9.2.14) we obtain (9.2.11). □

CHAPTER TEN

10.1. OPTIMAL KOROVKIN TYPE INEQUALITIES FOR POSITIVE LINEAR STOCHASTIC OPERATORS

10.1.1. Introduction

The rate of convergence of a sequence of positive linear stochastic operators $\{T_j\}_{j\in\mathbb{N}}$ to the unit operator I on spaces of stochastic processes X is studied. This is mainly done for stochastic processes that are smooth over a compact and convex index set $Q \subset \mathbb{R}^k$, $k \geq 1$. Nearly best upper bounds are given for $|E(T_jX)(x_0) - (EX)(x_0)|$, $x_0 \in Q$, where E is the expectation operator and T_j are E-commutative. These lead to strong and elegant inequalities many times sharp, which involve the first modulus of continuity of EX_α, where X_α is a (partial) derivative of X. The case of Q being a compact convex subset of a real normed vector space is also met and there the upper bound is the best possible.

This work was motivated by Weba (1990) where it is assumed that T_j are E-commutative and stochastically simple. He was the first to produce Korovkin type inequalities involving stochastic processes. According to his work, if a stochastic process $X(t,\omega)$, $t \in Q$—a compact convex subset of a real normed vector space, $\omega \in \underline{0}$—a probability space, is to be approximated by positive linear operators T_j, then the maximal error in the qth mean is $(q \geq 1)$

$$\|T_jX - X\| := \sup_{t\in Q}(E|(T_jX)(t,\omega) - X(t,\omega)|^q)^{1/q}.$$

So Weba establishes upper bounds for $\|T_jX - X\|$ involving his own natural general first modulus of continuity of X with several interesting applications.

We meet the pointwise case of $q = 1$. Without stochastic simplicity of T_j we find nearly best and best upper bounds for $|E(T_jX)(x_0) - (EX)(x_0)|$, $x_0 \in Q$. The proving method here is again the moment method of optimal distance (see Chapter 2). These upper bounds usually lead to sharp inequalities that are attained by concrete simple T_j and X.

An interesting application to multidimensional stochastic Bernstein operators is presented at the end. The material of this chapter comes from Anastassiou(1) (1991).

PART A: ONE-DIMENSIONAL RESULTS

10.1.2. Preliminaries

Let $(\underline{0}, A, P)$ denote a probability space and $L^1(\underline{0}, A, P)$ be the space of all real-valued random variables $Y = Y(\omega)$ with

$$\int_{\underline{0}} |Y(\omega)| P(d\omega) < \infty.$$

Let $X = X(t, \omega)$ denote a stochastic process with index set $[a,b] \subset \mathbb{R}$ and real state space $(\mathbb{R}, \mathcal{B})$, where $\mathcal{B}$ is the σ-field of Borel subsets of $\mathbb{R}$. Here $C([a,b])$ is the space of continuous real-valued functions on $[a,b]$ and $B([a,b])$ is the space of bounded real-valued functions on $[a,b]$. Also $C_{\underline{0}}([a,b]) := C([a,b],\ L^1(\underline{0}, A, P))$ is the space of L^1-continuous stochastic processes in t and

$$B_{\underline{0}}([a,b]) := \left\{ X\colon \sup_{t\in[a,b]} \int_{\underline{0}} |X(t,\omega)| P(d\omega) < \infty \right\};$$

obviously $C_{\underline{0}}([a,b]) \subset B_{\underline{0}}([a,b])$. Consider the subspace $C^n_{\underline{0}}([a,b]) := \{X\colon$ there exists $X^{(k)}(t,\omega) \in C_{\underline{0}}([a,b])$ and it is continuous in t for each $\omega \in \underline{0}$, $k = 0, 1, \ldots, n\}$, i.e., for every $\omega \in \underline{0}$ we have $X(t,\omega) \in C^n([a,b])$. $(EX)(t) = \int_{\underline{0}} X(t,\omega) P(d\omega)$ is the expectation operator, defined for all $t \in [a,b]$.

Consider the linear operator

$$T\colon C_{\underline{0}}([a,b]) \hookrightarrow B_{\underline{0}}([a,b]).$$

If $X \in C_{\underline{0}}([a,b])$ is nonnegative and TX, too, then T is called positive. If $ET = TE$, then T is called E-commutative. Here

$$\phi_n(x) := \int_0^{|x|} \left\lceil \frac{t}{h} \right\rceil \frac{(|x| - t)^{n-1}}{(n-1)!} dt \qquad (x \in \mathbb{R}),$$

where $\lceil\cdot\rceil$ is the ceiling of the number. Note that $\phi_n(x)$ is continuous, convex on $\mathbb{R}$ and strictly increasing on $\mathbb{R}^+$ $(n \geq 1)$ (see Remark 7.1.3).

10.1.3. Results

Following Preliminaries 10.1.2.

Theorem 10.1.1. *Consider the positive E-commutative linear operator*

$$T: C_{\underline{0}}([a,b]) \hookrightarrow B_{\underline{0}}([a,b]).$$

Let

$$c(t_0) = \max(t_0 - a, b - t_0) \quad \textit{and}$$

$$d_n(t_0) = ((T(|t - t_0|^n))(t_0))^{1/n}, \qquad n \in \mathbb{N}. \tag{10.1.1}$$

Let $X \in C^n_{\underline{0}}([a,b])$ such that $\omega_1(EX^{(n)}, h) \leq w$, where ω_1 is the first modulus of continuity and w, h are fixed positive numbers, $0 < h \leq b - a$. Then we get the upper bound

$$|(E(TX))(t_0) - (EX)(t_0)| \leq |(EX)(t_0)| \cdot |(T(1))(t_0) - 1|$$
$$+ \sum_{k=1}^{n} \frac{|(EX^{(k)})(t_0)|}{k!} \cdot |(T(t - t_0)^k)(t_0)| + w\phi_n(c(t_0)) \cdot \left(\frac{d_n(t_0)}{c(t_0)}\right)^n \tag{10.1.2}$$

for each $t_0 \in [a,b]$.

Theorem 10.1.2. *Inequality* (10.1.2) *is sharp.*

PROOF OF THEOREM 10.1.1: We have by Taylor's formula

$$X(t,\omega) = \sum_{k=0}^{n} \frac{X^{(k)}(t_0,\omega)}{k!}(t - t_0)^k$$
$$+ \int_{t_0}^{t} (X^{(n)}(s,\omega) - X^{(n)}(t_0,\omega)) \frac{(t-s)^{n-1}}{(n-1)!} ds,$$

for all $\omega \in \underline{0}$; $t, t_0 \in [a,b]$. Hence

$$(EX)(t) = \sum_{k=0}^{n} \frac{(EX^{(k)})(t_0)}{k!}(t - t_0)^k$$
$$+ E\left(\int_{t_0}^{t} (X^{(n)}(s,\omega) - X^{(n)}(t_0,\omega)) \frac{(t-s)^{n-1}}{(n-1)!} ds\right)$$

and by Fubini's Theorem we obtain

$$(EX)(t) = \sum_{k=0}^{n} \frac{(EX^{(k)})(t_0)}{k!}(t - t_0)^k + \mathcal{R}_n(t, t_0),$$

where the remainder

$$\mathcal{R}_n(t, t_0) := \int_{t_0}^{t} ((EX^{(n)})(s) - (EX^{(n)})(t_0)) \cdot \frac{(t-s)^{n-1}}{(n-1)!} ds.$$

From Corollary 7.1.1 we have

$$|(EX^{(n)})(s) - (EX^{(n)})(t_0)| \leq w \left\lceil \frac{|s - t_0|}{h} \right\rceil.$$

Now one can easily prove that

$$|\mathcal{R}_n(t, t_0)| \leq w\phi_n(|t - t_0|), \quad \text{all } t \in [a, b]. \tag{10.1.3}$$

Furthermore,

$$(T(EX))(t_0) = \sum_{k=0}^{n} \frac{(EX^{(k)})(t_0)}{k!} \cdot (T(t - t_0)^k)(t_0) + (T(\mathcal{R}_n(t, t_0)))(t_0).$$

From [Weber (1990), pp. 3–5] we have the following results

(i) $C([a,b]) \subset C_{\underline{0}}([a,b])$,

(ii) if $X \in C_{\underline{0}}([a,b])$, then $EX \in C([a,b])$, and

(iii) if T is E-commutative then T maps the subspace $C([a,b])$ into $B([a,b])$.

One can rewrite

$$\begin{aligned}(T(EX))(t_0) - (EX)(t_0) &= (EX)(t_0) \cdot [(T(1))(t_0) - 1] \\ &+ \sum_{k=1}^{n} \frac{(EX^{(k)})(t_0)}{k!} \cdot (T(t - t_0)^k)(t_0) + (T(\mathcal{R}_n(t, t_0)))(t_0).\end{aligned}$$

Consequently by E-commutativity of T

$$\begin{aligned}&|E[(TX)(t_0, \omega) - X(t_0, \omega)]| \\ &\leq |(EX)(t_0)| \cdot |(T(1))(t_0) - 1| + \sum_{k=1}^{n} \frac{|(EX^{(k)})(t_0)|}{k!} \\ &\cdot |(T(t - t_0)^k)(t_0)| + |(T(\mathcal{R}_n(t, t_0)))(t_0)|.\end{aligned}$$

But by the positivity of T and (10.1.3) we obtain

$$|(T(\mathcal{R}_n(t,t_0)))(t_0)| \le (T(|\mathcal{R}_n(t,t_0)|))(t_0) \le w(T(\phi_n(|t-t_0|)))(t_0).$$

That is,

$$\begin{aligned}&|(E(TX))(t_0)-(EX)(t_0)|\\ &\le |(EX)(t_0)|\cdot|(T(1))(t_0)-1| + \sum_{k=1}^{n}\frac{|(EX^{(k)})(t_0)|}{k!}\\ &\cdot|(T(t-t_0)^k)(t_0)| + w(T(\phi_n(|t-t_0|)))(t_0).\end{aligned}$$

Since $(T(\cdot))(t_0)$ is a positive linear functional on $C([a,b])$, it follows from the Riesz representation theorem that

$$(T(f))(t_0) = \int_{[a,b]} f(t)\mu_{t_0}(dt), \quad \text{all } f \in C([a,b]),$$

where μ_{t_0} is a finite nonnegative measure on $[a,b]$. Thus, as in the proof of Theorem 7.2.3 we get (by applying the moment method of optimal distance, see Chapter 2)

$$(T(\phi_n(|t-t_0|)))(t_0) \le \phi_n(c(t_0))\cdot\left(\frac{d_n(t_0)}{c(t_0)}\right)^n.$$

The last establishes inequality (10.1.2). □

PROOF OF THEOREM 10.1.2: Let $(\underline{\Omega}, A, P)$ be an arbitrary probability space and $X(t,\omega) = f(t)\cdot\varphi(\omega)$; $f \in C^n([a,b])$, φ is a real valued random variable such that $E\varphi = \rho \in \mathbb{R}-\{0\}$. Obviously $X(t,\omega) \in C^n_{\underline{\Omega}}([a,b])$. We have $X^{(k)}(t,\omega) = f^{(k)}(t)\cdot\varphi(\omega)$ and $(EX^{(k)})(t) = f^{(k)}(t)\cdot\rho$, $k=0,1,\ldots,n$, all $t \in [a,b]$. Thus inequality (10.1.2) reduces to

$$\begin{aligned}&|(Tf)(t_0)-f(t_0)| \le |f(t_0)|\cdot|(T(1))(t_0)-1|\\ &+\sum_{k=1}^{n}\frac{|f^{(k)}(t_0)|}{k!}\cdot|(T(t-t_0)^k)(t_0)| + w\cdot|\rho|^{-1}\cdot\phi_n(c(t_0))\cdot\left(\frac{d_n(t_0)}{c(t_0)}\right)^n\end{aligned}$$

all $t_0 \in [a,b]$.

Hence inequality (10.1.2) is sharp. This is attained in the sense of Theorem 7.2.3 and Corollary 7.2.2. That is achieved by $w\cdot|\rho|^{-1}\cdot\phi_n((t-t_0)_+)\cdot\varphi(\omega)$ and a nonnegative measure μ_{t_0} supported by $\{t_0, b\}$, when $t_0 - a \le b - t_0$, also attained by $w\cdot|\rho|^{-1}\cdot\phi_n((t_0-t)_+)\cdot\varphi(\omega)$

and a nonnegative measure μ_{t_0} supported by $\{t_0, a\}$ when $t_0 - a \geq b - t_0$: in each case with masses

$$\left((T(1))(t_0) - \left(\frac{d_n(t_0)}{c(t_0)}\right)^n\right) \quad \text{and} \quad \left(\frac{d_n(t_0)}{c(t_0)}\right)^n,$$

respectively. Here μ_{t_0} fulfills

$$(T(f))(t_0) = \int_{[a,b]} f(t)\mu_{t_0}(dt), \quad \text{all } f \in C([a,b]).$$

The optimal operator T is a positive E-commutative linear operator described as follows:

Define

$$(T(1))(t_0) := c_0(t_0) \geq 0$$

and

$$(T(|t - t_0|^n))(t_0) := d_n^n(t_0) \geq 0$$

such that $d_n^n(t_0) \leq c_0(t_0) \cdot c^n(t_0)$, all $t_0 \in [a,b]$. Furthermore, define

$$\begin{aligned}(TX)(t_0,\omega) := &\left(c_0(t_0) - \left(\frac{d_n(t_0)}{c(t_0)}\right)^n\right) \cdot X(t_0,\omega) \\ &+ \left(\frac{d_n(t_0)}{c(t_0)}\right)^n \cdot X(\gamma(t_0),\omega), \quad \text{all } (t_0,\omega) \in [a,b] \times \underline{0},\end{aligned}$$

where

$$\gamma(t_0) := \begin{cases} a, & \text{if } t_0 - a \geq b - t_0 \\ b, & \text{if } t_0 - a \leq b - t_0. \end{cases}$$

Obviously T is a positive linear operator.

Additionally

$$\begin{aligned}(E(TX))(t_0,\omega) = (E(TX))(t_0) = &\left(c_0(t_0) - \left(\frac{d_n(t_0)}{c(t_0)}\right)^n\right) \cdot (EX)(t_0) \\ &+ \left(\frac{d_n(t_0)}{c(t_0)}\right)^n \cdot (EX)(\gamma(t_0)) = (T(EX))(t_0,\omega),\end{aligned}$$

i.e., $ET = TE$. □

A refinement of Theorem 10.1.1 has as follows:

Theorem 10.1.3. *Consider the positive E-commutative linear operator*

$$T: C_{\underline{0}}([a,b]) \hookrightarrow B_{\underline{0}}([a,b]).$$

Let $T(1)(t_0) = m_{t_0}$ *and*

$$[m_{t_0}^{-1}(T(|t-t_0|^{n+1}))(t_0)]^{1/n+1} = d_{n+1}(t_0) > 0, \qquad n \in \mathbb{N}. \tag{10.1.4}$$

Let $X \in C_{\underline{0}}^n([a,b])$ *such that* $\omega_1(EX^{(n)}, rd_{n+1}(t_0)) \le w$, *where* r, w *are given positive numbers. Then*

$$|(E(TX))(t_0) - (EX)(t_0)| \le |(EX)(t_0)| \cdot |m_{t_0} - 1|$$

$$+ \sum_{k=1}^{n} \frac{|(EX^{(k)})(t_0)|}{k!} \cdot |(T(t-t_0)^k)(t_0)| + \frac{wm_{t_0}}{rn!} \cdot \left[\frac{nr^2}{8} + \frac{r}{2} + \frac{1}{n+1}\right] \cdot d_{n+1}^n(t_0), \tag{10.1.5}$$

for each $t_0 \in [a,b]$.

Corollary 10.1.1. *Inequality* (10.1.5) *is sharp.*

PROOF OF THEOREM 10.1.3: From inequality (7.1.8) we obtain

$$\phi_n(|t-t_0|) \le \left(\frac{|t-t_0|^{n+1}}{(n+1)!h} + \frac{|t-t_0|^n}{2n!} + \frac{h|t-t_0|^{n-1}}{8(n-1)!}\right).$$

Hence

$$T(\phi_n(|t-t_0|))(t_0) \le \Big(\frac{T(|t-t_0|^{n+1})(t_0)}{(n+1)!h} + \frac{T(|t-t_0|^n)(t_0)}{2n!} + \frac{h}{8(n-1)!} \cdot T(|t-t_0|^{n-1})(t_0)\Big).$$

Thus so far (following the proof of Theorem 10.1.1)

$$|(E(TX))(t_0) - (EX)(t_0)| \le |(EX)(t_0)| \cdot |m_{t_0} - 1| + \sum_{k=1}^{n} \frac{|EX^{(k)}(t_0)|}{k!} \cdot |(T(t-t_0)^k)(t_0)| + \mathcal{R}^*,$$

where

$$\mathcal{R}^* := w \cdot \Big[\frac{T(|t-t_0|^{n+1})(t_0)}{(n+1)!h} + \frac{T(|t-t_0|^n)(t_0)}{2n!} + \frac{h}{8(n-1)!} \cdot T(|t-t_0|^{n-1})(t_0)\Big].$$

From Riesz representation theorem and Hölder's inequality we have

$$T(|t-t_0|^n)(t_0) \le (T(|t-t_0|^{n+1})(t_0))^{n/n+1} \cdot (m_{t_0})^{1/n+1}$$

and

$$T(|t-t_0|^{n-1})(t_0) \le (T(|t-t_0|^{n+1})(t_0))^{(n-1)/(n+1)} \cdot (m_{t_0})^{2/n+1}.$$

Hence,

$$\begin{aligned}
\mathcal{R}^* &\le \Bigg[\frac{T(|t-t_0|^{n+1})(t_0)}{(n+1)!h} + \frac{(T(|t-t_0|^{n+1})(t_0))^{n/n+1} \cdot (m_{t_0})^{1/n+1}}{2n!} \\
&\quad + \frac{h}{8(n-1)!} \cdot (T(|t-t_0|^{n+1})(t_0))^{(n-1)/(n+1)} \cdot (m_{t_0})^{2/n+1}\Bigg] \\
&= w \cdot \left[\frac{d_{n+1}^{n+1}(t_0) m_{t_0}}{(n+1)!h} + \frac{d_{n+1}^{n}(t_0) m_{t_0}}{2n!} + \frac{h d_{n+1}^{n-1}(t_0) m_{t_0}}{8(n-1)!}\right] \\
&= w m_{t_0} \cdot \left[\frac{d_{n+1}^{n}(t_0)}{(n+1)!r} + \frac{d_{n+1}^{n}(t_0)}{2n!} + \frac{r d_{n+1}^{n}(t_0)}{8(n-1)!}\right] \\
&= \frac{w m_{t_0} d_{n+1}^{n}(t_0)}{rn!} \cdot \left[\frac{nr^2}{8} + \frac{r}{2} + \frac{1}{n+1}\right],
\end{aligned}$$

where $h = rd_{n+1}(t_0)$, $r > 0$. Inequality (10.1.5) has been established. □

PROOF OF COROLLARY 10.1.1: Let $(\underline{0}, A, P) = ([-1,1], \mathcal{B}, \lambda/2)$, where λ is the Lebesgue measure. Let $X(t,\omega) = 2|t|^{n+1} \cdot |\omega|$, $(t,\omega) \in [a,b] \times [-1,1]$, $0 \in [a,b]$. Obviously $X(t,\omega) \in C_{\underline{0}}^n([a,b])$. We have

$$X^{(k)}(t,\omega) = 2(n+1)n(n-1)\cdots(n+2-k) \cdot |t|^{n+1-k} \cdot |\omega|$$

and

$$(EX^{(k)})(t) = (n+1)n(n-1)\cdots(n+2-k) \cdot |t|^{n+1-k},$$

all $k = 0,1,\ldots,n$. In particular, $(EX)(t) = |t|^{n+1}$ and $(EX^{(n)})(t) = (n+1)!|t|$. Take $t_0 = 0$, assume $m_0 = 1$ and let $r \downarrow 0$.

Assume also that $\omega_1((EX^{(n)})(t), rd_{n+1}(0)) = w$. Obviously the left-hand side of inequality (10.1.5) reduces to $(T(|t|^{n+1}))(0)$. And the right-hand side of (10.1.5) becomes

$$\begin{aligned}
&\frac{\omega_1((n+1)!|t|, rd_{n+1}(0))}{rn!} \cdot \left[\frac{nr^2}{8} + \frac{r}{2} + \frac{1}{n+1}\right] \cdot d_{n+1}^{n}(0) \\
&= (n+1) d_{n+1}^{n+1}(0) \cdot \left[\frac{nr^2}{8} + \frac{r}{2} + \frac{1}{n+1}\right] \to d_{n+1}^{n+1}(0) = (T(|t|^{n+1}))(0).
\end{aligned}$$

The last establishes our claim. □

Next we meet an interesting special case

Theorem 10.1.4. *Consider the positive E-commutative linear operator*

$$T\colon C_{\underline{0}}([a,b]) \hookrightarrow B_{\underline{0}}([a,b]).$$

Let

$$T(1)(t_0) = m_{t_0}, \quad r > 0, \quad [(T((t-t_0)^2))(t_0)]^{1/2} = d_2(t_0) > 0 \tag{10.1.6}$$

and $(T(t-t_0))(t_0) = 0$, *for some* $t_0 \in [a,b]$. *Let* $X \in C^1_{\underline{0}}([a,b])$. *Then*

$$|(E(TX))(t_0) - (EX)(t_0)| - |(EX)(t_0)| \cdot |m_{t_0} - 1|$$
$$\leq \begin{cases} \omega_1(EX', rd_2(t_0)) \cdot \dfrac{d_2(t_0)}{8r} \cdot (2 + r\sqrt{m_{t_0}})^2, & \text{when } 0 < r \leq 2/\sqrt{m_{t_0}}, \\ \\ \sqrt{m_{t_0}} \cdot \omega_1(EX', rd_2(t_0)) \cdot d_2(t_0), & \text{when } r \geq 2/\sqrt{m_{t_0}}. \end{cases} \tag{10.1.7}$$

Corollary 10.1.2. *When* $m_{t_0} = 1$ *and* $r = 1/2$ *we obtain*

$$|E(TX)(t_0) - (EX)(t_0)| \leq 1.5625\omega_1\left(EX', \frac{1}{2}d_2(t_0)\right) \cdot d_2(t_0). \tag{10.1.8}$$

Corollary 10.1.3. *Inequality* (10.1.7) *is sharp.*

PROOF: Similar to Corollary (10.1.1). □

PROOF OF THEOREM 10.1.4: As in the proof of Theorem 10.1.1 for $n = 1$ we obtain

$$|(E(TX))(t_0) - (EX)(t_0)| \leq |(EX)(t_0)| \cdot |m_{t_0} - 1|$$
$$+ \omega_1(EX', h) \cdot (T(\phi_1(|t-t_0|)))(t_0).$$

From inequality (7.1.17) we have

$$\phi_1(|t-t_0|) \leq \left(\frac{(t-t_0)^2}{2h} + \frac{|t-t_0|}{2} + \frac{h}{8}\right).$$

Thus

$$(T(\phi_1(|t-t_0|)))(t_0) \leq \left[\frac{(T((t-t_0)^2))(t_0)}{2h} + \frac{(T(|t-t_0|))(t_0)}{2} + \frac{hm_{t_0}}{8}\right]. \tag{10.1.9}$$

From Riesz representation theorem and Hölder's inequality we obtain

$$(T(|t - t_0|))(t_0) \leq \sqrt{m_{t_0}} \cdot d_2(t_0).$$

Here we choose $h = rd_2(t_0)$. Applying these on the right-hand side of (10.1.9) we find

$$\begin{aligned}(T(\phi_1(|t - t_0|)))(t_0) &\leq d_2(t_0) \cdot \left(\frac{1}{2r} + \frac{\sqrt{m_{t_0}}}{2} + \frac{rm_{t_0}}{8}\right) \\ &= \frac{d_2(t_0)}{8r} \cdot (2 + r\sqrt{m_{t_0}})^2.\end{aligned}$$

Therefore,

$$\begin{aligned}|(E(TX))(t_0) - (EX)(t_0)| &\leq |(EX)(t_0)| \cdot |m_{t_0} - 1| \\ &+ \omega_1(EX', rd_2(t_0)) \cdot \frac{d_2(t_0)}{8r} \cdot (2 + r\sqrt{m_{t_0}})^2. \end{aligned} \tag{10.1.10}$$

Observe that $g(r) := (2 + r\sqrt{m_{t_0}})^2/8r$ is increasing in r iff $r \geq 2/\sqrt{m_{t_0}}$ and note that $g(2/\sqrt{m_{t_0}}) = \sqrt{m_{t_0}}$. Consequently

$$\begin{aligned}&|(E(TX))(t_0) - (EX)(t_0)| - |(EX)(t_0)| \cdot |m_{t_0} - 1| \\ &\leq \omega_1\left(EX', \frac{2}{\sqrt{m_{t_0}}} d_2(t_0)\right) \cdot d_2(t_0) \cdot g\left(\frac{2}{\sqrt{m_{t_0}}}\right) \\ &\leq \omega_1(EX', rd_2(t_0)) \cdot d_2(t_0) \cdot \sqrt{m_{t_0}}, \quad \text{for any } r \geq 2/\sqrt{m_{t_0}}. \end{aligned} \tag{10.1.11}$$

The validity of inequality (10.1.7) now is clear from inequalities (10.1.10) and (10.1.11). □

PART B: MULTIDIMENSIONAL RESULTS

10.1.4. Preliminaries

Let $(\underline{\Omega}, A, P)$ denote a probability space. Let $X = X(z, \omega)$ denote a stochastic process with index set Q, a compact convex subset of $\mathbb{R}^k$, $k \geq 1$ with l_1-norm, and real state space $(\mathbb{R}, \mathcal{B})$. Here $C(Q)$ is the space of continuous real-valued functions on Q and $B(Q)$ is the space of bounded real-valued functions on Q. Also $C_{\underline{\Omega}}(Q)$ is the space of L^1-continuous stochastic processes in z and

$$B_{\underline{\Omega}}(Q) = \left\{X : \sup_{z \in Q} \int_{\underline{\Omega}} |X(z, \omega)| P(d\omega) < \infty\right\};$$

obviously $C_{\underline{0}}(Q) \subset B_{\underline{0}}(Q)$.

Consider the subspace $C_{\underline{0}}^{n}(Q) = \{X$: there exists $X_\alpha = \vartheta^\alpha X/\vartheta x^\alpha \in C_{\underline{0}}(Q)$, $[\alpha = (\alpha_1, \ldots, \alpha_k)$, $\alpha_i \in \mathbb{Z}^+$, $i = 1, \ldots, k$, $|\alpha| = \sum_{i=1}^{k} \alpha_i = \rho$, $0 \le \rho \le n]$, and it is continuous in z for each $\omega \in \underline{0}$, all $\alpha\}$, i.e., for every $\omega \in \underline{0}$ we have $X(z, \omega) \in C^n(Q)$, $n \in \mathbb{N}$

$$(EX)(z) = \int_{\underline{0}} X(z, \omega) P(d\omega)$$

is the expectation operator, defined for all $z \in Q$.

Consider the linear operator

$$T: C_{\underline{0}}(Q) \hookrightarrow B_{\underline{0}}(Q)$$

which is positive and E-commutative as defined in Preliminaries 10.1.2.

Definition 10.1.1. Let $(V, \|\cdot\|)$ be a real normed vector space and Q a compact convex subset of V. For a continuous function $f: Q \to \mathbb{R}$ we define its first modulus of continuity

$$\omega_1(f, h) := \sup\{|f(x) - f(y)|\text{: all } x, y \in Q, \|x - y\| \le h\}.$$

This has all the basic properties of the standard first modulus of continuity defined on $(\mathbb{R}, |\cdot|)$, where $Q = [a, b] \subset \mathbb{R}$.

In the following we need

Lemma 10.1.1. *Let T be such that $(T(1))(x_0) = 1$, where $x_0 = (x_{01}, \ldots, x_{0k}) \in Q$. Let $X \in C_{\underline{0}}^{n}(Q)$ and suppose that relative to the l_1-norm $\omega_1(EX_\alpha, h) \le w$ for all α such that $|\alpha| = n$, where h, w are fixed positive numbers.*

Then

$$\begin{aligned} |(E(TX))(x_0) - (EX)(x_0)| &\le \sum_{j=1}^{n} \frac{|E(TG^{(j)}(0, z))(x_0)|}{j!} \\ &+ w \cdot (T(\phi_n(\|z - x_0\|)))(x_0), \end{aligned} \tag{10.1.12}$$

where $G(t, z, \omega) = X(x_0 + t(z - x_0), \omega)$, all $t \ge 0$, $z \in Q$, $\omega \in \underline{0}$ and $\|\cdot\| = \|\cdot\|_{l_1}$.

Proof: The jth derivative with respect to t of

$$G(t, z, \omega) = X(x_0 + t(z - x_0), \omega)$$

is given by

$$G^{(j)}(t,z,\omega) = \left[\left(\sum_{i=1}^{k}(z_i - x_{0i})\frac{\vartheta}{\vartheta x_i}\right)^j X\right]$$
$$(x_{01} + t(z_1 - x_{01}), \ldots, x_{0k} + t(z_k - x_{0k}), \omega).$$

Hence

$$X(z_1,\ldots,z_k,\omega) = G(1,z,\omega) = \sum_{j=0}^{n}\frac{G^{(j)}(0,z,\omega)}{j!} + \mathcal{R}_n(0,z,\omega),$$

where

$$\mathcal{R}_n(0,z,\omega) := \int_0^1\left(\int_0^{t_1}\cdots\left(\int_0^{t_{n-1}}(G^{(n)}(t_n,z,\omega) - G^{(n)}(0,z,\omega))dt_n\right)\cdots\right)dt_1.$$

From Lemma 7.1.1 we have

$$|(EX_\alpha)(x_0 + t(z - z_0)) - (EX_\alpha)(x_0)| \le w\left\lceil\frac{t\|z - x_0\|}{h}\right\rceil, \quad \text{all } t \ge 0,$$

all α such that $|\alpha| = n$. Furthermore,

$$(EX)(z_1,\ldots,z_k) = (EG)(1,z) = \sum_{j=0}^{n}\frac{(EG^{(j)})(0,z)}{j!} + E\mathcal{R}_n(0,z).$$

Note that

$$E\mathcal{R}_n(0,z) = \int_0^1\left(\int_0^{t_1}\cdots\left(\int_0^{t_{n-1}}((EG^{(n)})(t_n,z) - (EG^{(n)})(0,z))dt_n\right)\cdots\right)dt_1$$

by Fubini's Theorem and $X \in C^n_{\underline{0}}(Q)$. When $z \ne x_0$ we get

$$\begin{aligned}
|E\mathcal{R}_n(0,z)| &\le \int_0^1\int_0^{t_1}\cdots\int_0^{t_{n-1}}\left(\sum_{|\alpha|=n}\frac{n!}{\alpha_1!\cdots\alpha_k!}\right.\\
&\qquad \left.\cdot|z_1 - x_{01}|^{\alpha_1}\cdots|z_k - x_{0k}|^{\alpha_k}\cdot w\cdot\left(\left\lceil\frac{t_n\|z - x_0\|}{h}\right\rceil dt_n\right)\cdots\right)dt_1\\
&= \sum_{|\alpha|=n}\frac{n!}{\alpha_1!\cdots\alpha_k!}\cdot\frac{\prod_{i=1}^k|z_i - x_{0i}|^{\alpha_i}}{\|z - x_0\|^n}\cdot w\cdot\phi_n(\|z - x_0\|)\\
&= w\phi_n(\|z - x_0\|),
\end{aligned}$$

where

$$\|z - x_0\| = \sum_{i=1}^{k}|z_i - x_{0i}|.$$

Thus

$$|E\mathcal{R}_n(0,z)| \le w\phi_n(\|z-x_0\|), \quad \text{all } z \in Q.$$

Also $EG(0,z) = EX(x_0)$. Hence

$$(EX)(z) - (EX)(x_0) = \sum_{j=1}^{n} \frac{(EG^{(j)})(0,z)}{j!} + E\mathcal{R}_n(0,z).$$

Therefore

$$\begin{aligned}
&|(T(EX))(x_0) - (EX)(x_0)| \\
&\quad \le \sum_{j=1}^{n} \frac{|(T(EG^{(j)}(0,z)))(x_0)|}{j!} + |T(E\mathcal{R}_n(0,z))(x_0)| \\
&\quad \le \sum_{j=1}^{n} \frac{|(T(EG^{(j)}(0,z)))(x_0)|}{j!} + (T|E\mathcal{R}_n(0,z)|)(x_0) \\
&\quad \le \sum_{j=1}^{n} \frac{|(T(EG^{(j)}(0,z)))(x_0)|}{j!} + w\cdot(T(\phi_n(\|z-x_0\|)))(x_0).
\end{aligned}$$

Inequality (10.1.12) is established. □

When Q is a ball the remainder term in inequality (10.1.12) becomes quite elegant.

Theorem 10.1.5. *Let $Q = \{x \in \mathbb{R}: \|x\| \le 1\}$, where $\|\cdot\|$ is the l_1-norm in $\mathbb{R}^k$. Let T be such that $(T(1))(x_0) = 1$ and $(T(\|x-x_0\|))(x_0) = d^*_{x_0}$, where $x_0 = (x_{01},\dots,x_{0k}) \in Q$. Consider $X \in C^n_{\underline{0}}(Q)$, $n \in \mathbb{N}$, so that $\omega_1(EX_\alpha,h) \le w$ for all α such that $|\alpha| = n$, where h,w are fixed positive numbers. Then*

$$|(E(TX))(x_0) - (EX)(x_0)| \le \sum_{j=1}^{n} \frac{|(E(TG^{(j)}(0,z)))(x_0)|}{j!} + wd^*_{x_0}\cdot\frac{\phi_n(1+\|x_0\|)}{(1+\|x_0\|)}, \tag{10.1.13}$$

where $G(t,z,\omega) = X(x_0 + t(z-x_0),\omega)$, all $t \ge 0$, $z \in Q$, $\omega \in \underline{0}$.

Proof: We use inequality (10.1.12) taking into account the following: Note $\|x-x_0\| \le 1+\|x_0\|$. Let $x \ne x_0$, from the convexity of ϕ_n, $n \in \mathbb{N}$ we obtain

$$\frac{\phi_n(\|x-x_0\|)}{\|x-x_0\|} \le \frac{\phi_n(1+\|x_0\|)}{(1+\|x_0\|)}.$$

Hence

$$\phi_n(\|x - x_0\|) \leq \frac{\phi_n(1 + \|x_0\|)}{(1 + \|x_0\|)} \cdot \|x - x_0\|, \quad \text{all } x \in Q.$$

By positivity of T we get

$$(T(\phi_n(\|x - x_0\|)))(x_0) \leq \frac{\phi_n(1 + \|x_0\|)}{(1 + \|x_0\|)} \cdot d^*_{x_0}.$$

The last establishes inequality (10.1.13). □

A refinement of inequality (10.1.12) has as follows:

Theorem 10.1.6. *Let Q be a compact convex subset of $(\mathbb{R}^k, \|\cdot\|_{l_1})$. Let T be such that $(T(1))(x_0) = 1$ and*

$$\frac{1}{(n+1)} \cdot ((T(\|z - x_0\|^{n+1}))(x_0))^{1/(n+1)} = h > 0,$$

where $x_0 = (x_{01}, \ldots, x_{0k}) \in Q$, $n \in \mathbb{N}$. Consider $X \in C^n_{\underline{0}}(Q)$, so that $\omega_1(EX_\alpha, h) \leq w$ for all α such that $|\alpha| = n$, where w is a fixed positive number. Then

$$|(E(TX))(x_0) - (EX)(x_0)| \leq \sum_{j=1}^{n} \frac{|(E(TG^{(j)}(0, z)))(x_0)|}{j!}$$

$$+ wh^n \cdot \left(\frac{3}{2} \cdot \frac{(n+1)^n}{n!} + \frac{(n+1)^{n-1}}{8(n-1)!}\right), \tag{10.1.14}$$

where $G(t, z, \omega) = X(x_0 + t(z - x_0), \omega)$, all $t \geq 0$, $z \in Q$, $\omega \in \underline{0}$.

Proof: We use inequality (10.1.12) taking into account inequality (7.1.18), positivity of T, and Hölder's inequality. □

Corollary 10.1.4. *Case of $n = 1$. Additionally, we assume that $(T(z_i - x_{0i}))(x_0) = 0$, all $i = 1, \ldots, k$, where $z = (z_1, \ldots, z_k) \in Q$. Then*

$$|(E(TX))(x_0) - (EX)(x_0)| \leq 1.5625 \cdot w \cdot ((T(\|z - x_0\|^2))(x_0))^{1/2}. \tag{10.1.15}$$

Proof: Easy. □

When Q is a cube inequality (10.1.12) leads to an inequality with a specific remainder.

Proposition 10.1.1. *Let $Q = \{x \in \mathbb{R}^k \colon -a \le x_i \le a,\ i = 1, \ldots, k\}$, $a > 0$. Let T be such that $(T(1))(x_0) = 1$, where $x_0 = (x_{01}, \ldots, x_{0k}) \in Q$. Consider $X \in C_{\underline{0}}^n(Q)$ and assume that relative to the l_1-norm $\omega_1(EX_\alpha, h) \le w$ for all α such that $|\alpha| = n$, where h, w are given positive numbers. Then*

$$|(E(TX))(x_0) - (EX)(x_0)| \le \sum_{j=1}^{n} \frac{|E(TG^{(j)}(0,z))(x_0)|}{j!} + w\phi_n(\|x_0\| + ka), \qquad (10.1.16)$$

where $G(t, z, \omega) = X(x_0 + t(z - x_0), \omega)$, all $t \ge 0$, $z \in Q$, $\omega \in \underline{0}$.

Proof: By Lemma 10.1.1, the positivity of T and

$$\phi_n(\|x - x_0\|) \le \phi_n(\|x_0\| + ka). \qquad \square$$

PART C: ABSTRACT RESULTS

10.1.5. Preliminaries

Let $(\underline{0}, A, P)$ denote a probability space. Let $X = X(t, \omega)$ denote a stochastic process with index set K, a compact convex subset of $(V, \|\cdot\|)$ a real normed vector space, and real state space $(\mathbb{R}, \mathcal{B})$.

Here $C(K)$ is the space of continuous real-valued functions on K and $B(K)$ is the space of bounded real valued functions on K. Also $C_{\underline{0}}(K)$ is the space of L^1-continuous stochastic processes in t and

$$B_{\underline{0}}(K) = \left\{ X \colon \sup_{t \in K} \int_{\underline{0}} |X(t, \omega)| P(d\omega) < \infty \right\};$$

obviously $C_{\underline{0}}(K) \subset B_{\underline{0}}(K)$.

From [Weba (1990), p. 4, Lemma 2.1] if $X(t, \omega) \in C_{\underline{0}}(K)$ then $(EX)(t) \in C(K)$, where E is the expectation operator as defined in Preliminaries 10.1.2.

Consider the linear operator

$$T \colon C_{\underline{0}}(K) \hookrightarrow B_{\underline{0}}(K)$$

which is positive and E-commutative as defined in Preliminaries 10.1.2. Again from [Weba (1990), p. 5, Lemma 3.4] since T is a E-commutative operator it maps $C(K)$ into $B(K)$.

The following is a very general result.

Theorem 10.1.7. *Consider $t_0 \in K$ and $C(t_0) > 0$ such that $0 \le \|t - t_0\| \le C(t_0)$ for all $t \in K$. Let T be as in Preliminaries 10.1.5 so that $(T(1))(t_0) = m_{t_0} > 0$ and*

$$(T(\|t - t_0\|^r)(t_0))^{1/r} = D_r(t_0) > 0, \qquad r > 0$$

with $D_r^r(t_0) \le m_{t_0} \cdot C^r(t_0)$. Let $X \in C_{\underline{0}}(K)$ such that $\omega_1(EX, h) \le w$, where h, w are given positive numbers. Then the best possible constant $K(t_0)$ in the inequality

$$|(E(TX))(t_0) - (EX)(t_0)| \le |(EX)(t_0)| \cdot |m_{t_0} - 1| + m_{t_0} K(t_0) \tag{10.1.17}$$

is given as follows. Set

$$n(t_0) = \left\lceil \frac{C(t_0)}{h} \right\rceil, \qquad k(t_0) = \left\lceil \frac{D_r(t_0)}{h \cdot m_{t_0}^{1/r}} \right\rceil$$

(it is $1 \le k(t_0) \le n(t_0)$).

(i) *$K(t_0) = n(t_0) \cdot w$, when $k(t_0) = n(t_0)$, i.e., when $D_r(t_0)/m_{t_0}^{1/r} > C(t_0) - h$.*

(ii) *$K(t_0) = [1 + (1/m_{t_0}) \cdot (D_r(t_0)/h)^r \cdot (n(t_0) - 1)^{1-r}] \cdot w$, when $r \le 1$ and $k(t_0) < n(t_0)$.*

(iii) *$K(t_0) = (k(t_0) + \theta_{k(t_0)}) \cdot w \le [1 + D_r(t_0)/(h \cdot m_{t_0}^{1/r})] \cdot w$, when $r \ge 1$ and $k(t_0) < n(t_0)$.*

Here

$$\theta_{k(t_0)} = \frac{[(D_r^r(t_0)/m_{t_0}) - (k(t_0) - 1)^r \cdot h^r]}{[k^r(t_0) \cdot h^r - (k(t_0) - 1)^r \cdot h^r]}.$$

Corollary 10.1.5. *When $r \ge 1$ we get the upper bound*

$$\begin{aligned} |(E(TX))(t_0) - (EX)(t_0)| &\le |(EX)(t_0)| \cdot |m_{t_0} - 1| \\ &+ \omega_1(EX, h) \cdot \left(m_{t_0} + \frac{D_r(t_0)}{h} \cdot m_{t_0}^{1-(1/r)} \right). \end{aligned} \tag{10.1.18}$$

Corollary 10.1.6. *Let $r \ge 1$, $(T(1))(t_0) = 1$ and $h = D_r(t_0)$. Then*

$$|(E(TX))(t_0) - (EX)(t_0)| \le 2\omega_1(EX, D_r(t_0)). \tag{10.1.19}$$

PROOF OF THEOREM 10.1.7: From Lemma 7.1.1 we have

$$|(EX)(t)-(EX)(t_0)| \le w\left\lceil\frac{\|t-t_0\|}{h}\right\rceil,$$

all $t, t_0 \in K$. By Riesz representation theorem,

$$(Tf)(t_0)=\int_K f(t)\mu_{t_0}(dt), \quad \text{all } f\in C(K),$$

where μ_{t_0} is a nonnegative measure on K. Note that $\mu_{t_0}(K)=(T(1))(t_0)=m_{t_0}$. Since (EX) is a continuous function on K

$$(T(EX))(t_0)=\int_K (EX)(t)\mu_{t_0}(dt).$$

Therefore

$$\begin{aligned}|(T(EX))(t_0)-(EX)(t_0)\cdot T(1)(t_0)| &= \left|\int_K ((EX)(t)-(EX)(t_0))\cdot\mu_{t_0}(dt)\right|\\ &\le \int_K |(EX)(t)-(EX)(t_0)|\cdot\mu_{t_0}(dt)\\ &\le w\cdot\int_K\left\lceil\frac{\|t-t_0\|}{h}\right\rceil\cdot\mu_{t_0}(dt).\end{aligned}$$

Furthermore, by E-commutativity of T we get that

$$\begin{aligned}|(E(TX))(t_0)-(EX)(t_0)| \le |(EX)(t_0)|\cdot|m_{t_0}-1|\\ +w\cdot\int_K\left\lceil\frac{\|t-t_0\|}{h}\right\rceil\cdot\mu_{t_0}(dt).\end{aligned}$$

As in Theorem 7.2.1 we determine the smallest possible constant $K(t_0)$ on

$$w\cdot\int_K\left\lceil\frac{\|t-t_0\|}{h}\right\rceil\cdot\mu_{t_0}(dt)\le m_{t_0}\cdot K(t_0).$$ □

10.1.6. Application: Convergence of Multidimensional Stochastic Bernstein Operators

The next topic has to do with an application of Corollary 10.1.4. Let Q be the following compact convex subset of $(\mathbb{R}^k, \|\cdot\|_{l_1})$, $k\ge 1$,

$$Q=\left\{(z_1,\ldots,z_k)\in\mathbb{R}^k\colon 0\le z_i\le 1 \text{ for each } i \text{ and } \sum_{i=1}^k z_i\le 1\right\}.$$

Given $N \in \mathbb{N}$ the Nth multidimensional stochastic Bernstein operator B_N maps $X \in C_{\underline{0}}(Q)$ onto the stochastic process

$$(B_N X)(z_1, \dots, z_k, \omega) = \sum_{(m_1,\dots,m_k)\in M} X\left(\frac{m_1}{N}, \dots, \frac{m_k}{N}, \omega\right) \cdot F_{(m_1,\dots,m_k)}(z_1, \dots, z_k),$$

where summation is taken over the set $M = \{(m_1, \dots, m_k)\colon\ m_i \in \mathbb{Z}_+$ for each i and $\sum_{i=1}^k m_i \leq N\}$ and

$$F_{(m_1,\dots,m_k)}(z_1, \dots, z_k) = \frac{N!}{(\prod_{i=1}^k m_i!) \cdot (N - \sum_{i=1}^k m_i)!} \times \left(\prod_{i=1}^k z_i^{m_i}\right) \cdot \left(1 - \sum_{i=1}^k z_i\right)^{N - \sum_{i=1}^k m_i}$$

is defined for each $(z_1, \dots, z_k) \in Q$.

Observe that $B_N X \in C_{\underline{0}}(Q)$. Furthermore B_N is a positive E-commutative linear operator with the properties

$$(B_n 1)(x_0) = 1, \qquad (B_N z_i)(x_0) = x_{0i}$$

(i.e., $(B_N(z_i - x_{0i}))(x_0) = 0$), also

$$(B_N(z_i^2))(x_0) = x_{0i}^2 + \frac{1}{N} x_{0i}(1 - x_{0i}),$$

all $i = 1, \dots, k$, for each $x_0 = (x_{01}, \dots, x_{0k}) \in Q$.

Theorem 10.1.8. *Let $X \in C_{\underline{0}}^1(Q)$. Call*

$$w := \max\left\{\omega_1\left(EX_\alpha, \frac{k}{4\sqrt{N}}\right) :\ \textit{all } \alpha \textit{ such that } |\alpha| = 1\right\}.$$

Then

$$\sup_{x_0 \in Q} |(E(B_N X))(x_0) - (EX)(x_0)| \leq 0.78125 \cdot w \cdot \frac{k}{\sqrt{N}}, \quad \textit{all } N \in \mathbb{N}. \tag{10.1.20}$$

Proof: Let $x_0 \in Q$ and note that

$$\|z - x_0\|^2 = \left(\sum_{i=1}^k |z_i - x_{0i}|\right)^2 = \sum_{i=1}^k (z_i - x_{0i})^2 + 2 \cdot \sum_{1=i\neq j=1}^k |z_i - x_{0i}| \cdot |z_j - x_{0j}|.$$

Thus

$$
\begin{aligned}
&(B_N(\|z - x_0\|^2))(x_0)\\
&= \sum_{i=1}^{k}(B_N(z_i - x_{0i})^2)(x_0) + 2\cdot \sum_{1=i\neq j=1}^{k} (B_N(|z_i - x_{0i}|\cdot|z_j - x_{0j}|))(x_0)\\
&\le \sum_{i=1}^{k}(B_N(z_i - x_{0i})^2)(x_0)\\
&\quad + 2\cdot \sum_{1=i\neq j=1}^{k} [((B_N(z_i - x_{0i})^2)(x_0))^{1/2}\cdot((B_N(z_j - x_{0j})^2)(x_0))^{1/2}]\\
&= \sum_{i=1}^{k}\frac{1}{N}x_{0i}(1 - x_{0i})\\
&\quad + 2\cdot \sum_{1=i\neq j=1}^{k}\left[\left(\frac{1}{N}x_{0i}(1 - x_{0i})\right)^{1/2}\cdot\left(\frac{1}{N}x_{0j}(1 - x_{0j})\right)^{1/2}\right]\\
&= \frac{1}{N}\cdot\sum_{i=1}^{k}x_{0i}(1 - x_{0i}) + \frac{2}{N}\cdot\sum_{1=i\neq j=1}^{k}[x_{0i}(1 - x_{0i})x_{0j}(1 - x_{0j})]^{1/2}\\
&\le \frac{1}{N}\cdot\sum_{i=1}^{k}\frac{1}{4} + \frac{2}{N}\cdot\sum_{1=i\neq j=1}^{k}\left(\frac{1}{16}\right)^{1/2}\\
&= \frac{k}{4N} + \frac{k(k-1)}{4N} = \frac{k^2}{4N}.
\end{aligned}
$$

We have applied above the special properties of B_N and Schwartz's inequality for positive linear functionals. We have found that

$$\theta := (B_N(\|z - x_0\|^2))(x_0) \le \frac{k^2}{4N}, \quad \text{for each } x_0 \in Q.$$

That is, $\theta^{1/2} \le k/2\sqrt{N}$ and $h := \frac{1}{2}\theta^{1/2} \le k/4\sqrt{N}$. Obviously

$$\omega_1(EX_\alpha, h) \le \omega_1\left(EX_\alpha, \frac{k}{4\sqrt{N}}\right) \le w$$

all α such that $|\alpha| = 1$. Clearly inequality (10.1.15) implies inequality (10.1.20). $\square$

CHAPTER ELEVEN

11.1. OPTIMAL KOROVKIN TYPE INEQUALITIES FOR POSITIVE LINEAR OPERATORS USING AN EXTENDED COMPLETE TSCHEBYCHEV SYSTEM

11.1.1. Introduction

Let $[a,b] \subset \mathbb{R}$ and let $\{L_j\}_{j\in\mathbb{N}}$ be a sequence of positive linear operators from $C^{n+1}([a,b])$ to $C([a,b])$, $n \geq 0$. As mentioned earlier in Chapter 7, the convergence of L_j to the unit operator I is closely related to the weak convergence of a sequence of positive finite measures μ_j to the unit measure δ_t, $t \in [a,b]$. Very general estimates with rates are given for the error $|\int_{[a,b]} f\, d\mu_j - f(t)|$, where $f \in C^{n+1}([a,b])$, in the presence of an extended complete Tschebychev system. These lead to sharp or nearly sharp inequalities of Korovkin type and are connected to the theory of best L_1 approximation by generalized polynomials.

The material of this chapter comes from Anastassiou(4) (1989).

The following introductory notions come from Widder (1928), which reference will be of constant help throughout this chapter.

Let the functions $f, u_0, u_1, \ldots, u_n \in C^{n+1}([a,b])$, $n \geq 0$, and let the Wronskians

$$W_i(x) := W[u_0(x), u_1(x), \ldots, u_i(x)] := \begin{vmatrix} u_0(x) & u_1(x) & \cdots & u_i(x) \\ u_0'(x) & u_1'(x) & \cdots & u_i'(x) \\ \vdots & & & \\ u_0^{(i)}(x) & u_1^{(i)}(x) & \cdots & u_i^{(i)}(x) \end{vmatrix}, \quad i = 0, 1, \ldots, n$$

and assume that all $W_i(x)$ are positive throughout $[a,b]$.

Obviously the functions

$$\phi_0(x) := W_0(x) := u_0(x), \quad \phi_1(x) := \frac{W_1(x)}{(W_0(x))^2},$$
$$\phi_i(x) := \frac{W_i(x)W_{i-2}(x)}{(W_{i-1}(x))^2}, \qquad i = 2, 3, \ldots, n$$

are positive everywhere on $[a,b]$.

Consider the linear differential operator of order $i \geq 0$

$$L_i f(x) := \frac{W[u_0(x), u_1(x), \ldots, u_{i-1}(x), f(x)]}{W_{i-1}(x)} \tag{11.1.1}$$

$i = 1, \ldots, n+1$; $L_0 f(x) := f(x)$, all $x \in [a,b]$. Here $W[u_0(x), u_1(x), \ldots, u_{i-1}(x), f(x)]$ denotes the Wronskian of $u_0, u_1, \ldots, u_{i-1}, f$. Note that for $i = 1, \ldots, n+1$ we have

$$L_i f(x) = \phi_0(x)\phi_1(x)\cdots\phi_{i-1}(x)\frac{d}{dx}\frac{1}{\phi_{i-1}(x)}\frac{d}{dx}\frac{1}{\phi_{i-2}(x)}\frac{d}{dx} \cdots \frac{d}{dx}\frac{1}{\phi_1(x)}\frac{d}{dx}\frac{f(x)}{\phi_0(x)}.$$

Consider also the functions

$$g_i(x,t) := \frac{1}{W_i(x)} \cdot \begin{vmatrix} u_0(t) & u_1(t) & \cdots & u_i(t) \\ u_0'(t) & u_1'(t) & \cdots & u_i'(t) \\ \vdots & & & \\ u_0^{(i-1)}(t) & u_1^{(i-1)}(t) & \cdots & u_i^{(i-1)}(t) \\ u_0(x) & u_1(x) & \cdots & u_i(x) \end{vmatrix} \tag{11.1.2}$$

$$i = 1, 2, \ldots, n; \quad g_0(x,t) := \frac{u_0(x)}{u_0(t)}, \quad \text{all } x, t \in [a,b].$$

Note that $g_i(x,t)$ as a function of x, is a linear combination of $u_0(x)$, $u_1(x), \ldots, u_i(x)$ and furthermore it holds

$$\begin{aligned} g_i(x,t) &= \frac{\phi_0(x)}{\phi_0(t)\cdots\phi_i(t)} \int_t^x \phi_1(x_1) \int_t^{x_1} \cdots \int_t^{x_{i-2}} \phi_{i-1}(x_{i-1}) \int_t^{x_{i-1}} \phi_i(x_i)\,dx_i\,dx_{i-1}\cdots dx_1 \\ &= \frac{1}{\phi_0(t)\cdots\phi_i(t)} \int_t^x \phi_0(s)\cdots\phi_i(s)\,g_{i-1}(x,s)\,ds, \quad \text{all } i = 1, 2, \ldots, n. \end{aligned}$$

This chapter's results are mainly motivated by the following result (see Karlin and Studden (1966), p. 376)

Theorem. *Let $u_0, u_1, \ldots, u_n \in C^n([a,b])$, $n \geq 0$. Then $\{u_i\}_{i=0}^n$ is an extended complete Tschebyshev* (E.C.T.) *system on $[a,b]$ iff $W_i(x)$ are positive everywhere on $[a,b]$, all $i = 0, 1, \ldots, n$.*

Remark. The definition of an E.C.T. system is quite complicated and lengthy, see Karlin and Studden (1966), pp. 1, 4, 5, 6, 375. For the convenience of the reader we use the statement of the last theorem as its definition.

Here, we would like to set

$$N_n(x,t) := \int_t^x g_n(x,s)ds$$

and

$$E_n(x,t) := f(x) - \sum_{i=0}^{n} L_i f(t) \cdot g_i(x,t) - L_{n+1} f(t) \cdot N_n(x,t) \tag{11.1.3}$$

all $x, t \in [a,b]$, $n \geq 0$.

Let L be a positive linear operator from $C^{n+1}([a,b])$ into $C([a,b])$, $n \geq 0$. It follows from the Riesz representation theorem, for every $t \in [a,b]$ there is a positive finite measure μ_t such that

$$L(f,t) = \int_{[a,b]} f(x)\mu_t(dx), \quad \text{all } f \in C^{n+1}([a,b]).$$

In Section 7.2, we met the special case of $u_i(x) := x^i$, $i = 0, 1, \ldots, n$. Therefore, it is still of interest to investigate for strong upper bounds to

$$|L(f,t) - f(t)| = \left| \int_{[a,b]} f(x)\mu_t(dx) - f(t) \right|$$

using general (E.C.T.)'s. In this chapter we present upper bounds to

$$\int_{[a,b]} |E_n(x,t)|\mu(dx); \qquad \left| \int_{[a,b]} f(x)\mu(dx) - f(t) \right|,$$

where μ is a positive finite measure on $[a,b]$ and t is a fixed point in $[a,b]$.

These bounds lead to sharp or nearly sharp inequalities in the natural very general "environment" of an (E.C.T.) extended complete Tschebychev system for different standard cases.

Here the convergence rates are given by the first modulus of continuity $\omega_1(L_{n+1}f, h)$, $0 < h \leq b - a$. Equivalently, the presented results estimate in the very general (E.C.T.) setting the rate of weak convergence of a sequence of positive finite measures to the unit measure at a fixed point.

At the end of the chapter we give concrete examples of systems of functions $\{u_i\}_{i=0}^n$ fulfilling the assumptions of the theorems.

11.1.2. Main Results

In the following for special functions u_0, u_1 we obtain sharp inequalities.

Theorem 11.1.1. *Let μ be a positive finite measure of mass m on $[a,b] \subset \mathbb{R}$ and t a fixed point in (a,b), such that*

$$\int_{[a,b]} |x-t|\mu(dx) = d > 0. \tag{11.1.4}$$

Let the functions $f(x)$, $u_0(x)$, $u_1(x), \ldots, u_n(x)$ belong to $C^{n+1}([a,b])$, $n \geq 0$, and let the Wronskians $W_0(x)$, $W_1(x), \ldots, W_n(x)$ be positive throughout $[a,b]$.

Assume that $u_0(x) = c > 0$ and $u_1(x)$ is a concave function for $x \leq t$ and a convex function for $x \geq t$. Define

$$\tilde{G}_n(x,t) := \left| \int_t^x g_n(x,s) \left\lceil \frac{|s-t|}{h} \right\rceil ds \right|, \qquad x,t \in [a,b], \tag{11.1.5}$$

where $0 < h \leq b-a$ is given and $\lceil \cdot \rceil$ is the ceiling of the number; $n \geq 0$. Assume that the first modulus of continuity $\omega_1(L_{n+1}f, h) \leq w$, where $w > 0$ is given.

Consider the error function

$$E_n(x,t) := f(x) - f(t) - \sum_{i=1}^n L_i f(t) \cdot g_i(x,t) - L_{n+1} f(t) \cdot N_n(x,t).$$

Then we have the upper bounds

$$\int_{[a,b]} |E_n(x,t)|\mu(dx) \leq w \cdot \max\left\{ \frac{\tilde{G}_n(b,t)}{b-t}, \frac{\tilde{G}_n(a,t)}{t-a} \right\} \cdot d, \tag{11.1.6}$$

and

$$\left| \int_{[a,b]} f\, d\mu - f(t) \right| \leq |m-1| \cdot |f(t)| + \sum_{i=1}^n |L_i f(t)| \cdot \left| \int_{[a,b]} g_i(x,t)\mu(dx) \right|$$

$$+ |L_{n+1} f(t)| \cdot \left| \int_{[a,b]} N_n(x,t)\mu(dx) \right| + w \cdot \max\left\{ \frac{\tilde{G}_n(b,t)}{b-t}, \frac{\tilde{G}_n(a,t)}{t-a} \right\} \cdot d; \quad n \geq 0. \tag{11.1.7}$$

Sharpness of inequalities (11.1.6) and (11.1.7) is proved next.

Theorem 11.1.2. *Let $c(t) := \max(t-a, b-t)$, where $t \in (a,b)$ is fixed and let $0 < h \leq b-a$. For*

$$k = 0, 1, \ldots, \left\lceil \frac{c(t)}{h} \right\rceil - 1$$

and $N \geq 1$ *define the continuous function* f_N *as follows:*

$$f_N(y) := \begin{cases} \frac{Nwy}{2h} + kw\left(1 - \frac{N}{2}\right), & \text{if } kh \leq y \leq \left(k + \frac{2}{N}\right)h; \\ (k+1)w, & \text{if } \left(k + \frac{2}{N}\right)h < y \leq (k+1)h; \\ \left\lceil \frac{c(t)}{h} \right\rceil w, & \text{if } \left(\left\lceil \frac{c(t)}{h} \right\rceil - 1 + \frac{2}{N}\right)h < y \leq c(t). \end{cases} \tag{11.1.8}$$

Observe that

$$\lim_{N \to +\infty} f_N(y) = \left\lceil \frac{y}{h} \right\rceil w, \qquad 0 \leq y \leq c(t).$$

Define

$$\tilde{G}_{nN}(x,t) := \left| \int_t^x g_n(x,s) f_N(|s-t|) ds \right|, \quad \text{all } x, t \in [a,b],\ n \geq 0,\ N \geq 1. \tag{11.1.9}$$

Then (as $N \to +\infty$*) inequalities (11.1.6) and (11.1.7) of Theorem 11.1.1 are attained, i.e., they are sharp.*

Namely:

(i) *Assume that*

$$\frac{\tilde{G}_n(b,t)}{b-t} \geq \frac{\tilde{G}_n(a,t)}{t-a} \quad \text{and} \quad d \leq m(b-t).$$

The optimal elements are the function

$$f(x) = \begin{cases} \tilde{G}_{nN}(x,t), & t \leq x \leq b; \\ 0, & a \leq x \leq t, \end{cases}$$

with $\omega_1(L_{n+1}f, h) \leq w$*, and* μ *which is the positive measure of mass* m *with masses* $[m - (d/b-t)]$ *and* $(d/b-t)$ *at* t *and* b*, respectively.*

(ii) *Assume that*

$$\frac{\tilde{G}_n(b,t)}{b-t} \leq \frac{\tilde{G}_n(a,t)}{t-a} \quad \text{and} \quad d \leq m(t-a).$$

The optimal elements are the function

$$f(x) = \begin{cases} 0, & t \leq x \leq b; \\ \tilde{G}_{nN}(x,t), & a \leq x \leq t, \end{cases}$$

with $\omega_1(L_{n+1}f, h) \leq w$*, and* μ *which is the positive measure of mass* m *with masses* $[m - (d/t-a)]$ *and* $(d/t-a)$ *at* t *and* a*, respectively.*

The following relates to best L_1-approximation by generalized polynomials.

Corollary 11.1.1. *Inequality* (11.1.6) *of Theorem* 11.1.1 *implies*

$$\min_{\substack{(c_0,c_1,\ldots,c_n,c_{n+1}) \\ c_i\in\mathbb{R},i=0,1,\ldots,n+1}} \int_{[a,b]} \left| f(x) - \sum_{i=0}^{n} c_i g_i(x,t) - c_{n+1} N_n(x,t) \right| \cdot \mu(dx)$$

$$\leq w \cdot \max\left\{ \frac{\tilde{G}_n(b,t)}{b-t}, \frac{\tilde{G}_n(a,t)}{t-a} \right\} \cdot d, \quad n \geq 0. \tag{11.1.10}$$

Remark 11.1.1. Given that $d = \int_{[a,b]} |x-t|\mu(dx) < \infty$, where μ is a positive nonfinite measure on $[a,b]$, inequality (11.1.6) of Theorem 11.1.1 and inequality (11.1.10) are still valid.

In general we obtain

Theorem 11.1.3. *Let μ be a positive finite measure of mass m on $[a,b] \subset \mathbb{R}$ and t a fixed point in $[a,b]$, such that*

$$\left(\int_{[a,b]} |x-t|^{n+2} \mu(dx) \right)^{1/(n+2)} = h, \tag{11.1.11}$$

where $0 < h \leq b-a$ is given; $n \geq 0$.

Let the functions $f(x)$, $u_0(x)$, $u_1(x), \ldots, u_n(x)$ belong to $C^{n+1}([a,b])$ and let the Wronskians $W_0(x)$, $W_1(x), \ldots, W_n(x)$ be positive throughout $[a,b]$. Assume that the first modulus of continuity $\omega_1(L_{n+1}f, h) \leq w$, where $w > 0$ is given.

Consider the error function

$$E_n(x,t) := f(x) - \sum_{i=0}^{n} L_i f(t) \cdot g_i(x,t) - L_{n+1} f(t) \cdot N_n(x,t).$$

Then we have the upper bounds

$$\int_{[a,b]} |E_n(x,t)|\mu(dx) \leq w \cdot (m^{1/(n+2)} + 1) \cdot \left(\int_{[a,b]} |N_n(x,t)|^{(n+2/n+1)} \mu(dx) \right)^{(n+1/n+2)} \tag{11.1.12}$$

and

$$\left| \int_{[a,b]} f\, d\mu - f(t) \right| \leq |f(t)| \cdot \left| \int_{[a,b]} g_0(x,t)\mu(dx) - 1 \right| +$$

$$+\sum_{i=1}^{n}|L_if(t)|\cdot\left|\int_{[a,b]}g_i(x,t)\mu(dx)\right|+|L_{n+1}f(t)|\cdot\left|\int_{[a,b]}N_n(x,t)\mu(dx)\right| \tag{11.1.13}$$

$$+w\cdot(m^{1/(n+2)}+1)\cdot\left(\int_{[a,b]}|N_n(x,t)|^{(n+2/n+1)}\cdot\mu(dx)\right)^{(n+1/n+2)} \quad ; \quad n\geq 0.$$

A more general connection to best L_1 approximation by generalized polynomials has as follows:

Corollary 11.1.2. *Inequality* (11.1.12) *of Theorem* 11.1.3 *implies*

$$\min_{\substack{(c_0,c_1,\ldots,c_n,c_{n+1})\\ c_i\in\mathbb{R},i=0,1,\ldots,n+1}}\int_{[a,b]}\left|f(x)-\sum_{i=0}^{n}c_ig_i(x,t)-c_{n+1}N_n(x,t)\right|\cdot\mu(dx)$$

$$\leq w\cdot(m^{1/(n+2)}+1)\cdot\left(\int_{[a,b]}|N_n(x,t)|^{(n+2/n+1)}\cdot\mu(dx)\right)^{(n+1/n+2)}, \quad n\geq 0. \tag{11.1.14}$$

The following theorem improves Theorem 11.1.3 under a Lipschitz condition.

Theorem 11.1.4. *Let μ be a positive finite measure of mass m on $[a,b]\subset\mathbb{R}$ and t a fixed point in $[a,b]$, such that*

$$\left(\int_{[a,b]}|x-t|^{n+2}\cdot\mu(dx)\right)^{1/(n+2)}=h, \tag{11.1.15}$$

where $0<h\leq b-a$ is given; $n\geq 0$.

Let the functions $f(x)$, $u_0(x)$, $u_1(x),\ldots,u_n(x)$ belong to $C^{n+1}([a,b])$ and let the Wronskians $W_0(x)$, $W_1(x),\ldots,W_n(x)$ be positive throughout $[a,b]$. Assume that the first modulus of continuity $\omega_1(L_{n+1}f,\delta)\leq A\delta^\alpha$, all $0<\delta\leq b-a$, $A>0$, $0<\alpha\leq 1$.

Consider the error function

$$E_n(x,t):=f(x)-\sum_{i=0}^{n}L_if(t)\cdot g_i(x,t)-L_{n+1}f(t)\cdot N_n(x,t).$$

Then we find the upper bounds ($n\geq 0$)

$$\int_{[a,b]}|E_n(x,t)|\mu(dx)$$

$$
\leq \begin{cases} A \cdot h^{\alpha} \cdot \left(\int_{[a,b]} |N_n(x,t)|^{(n+2/n+1)} \cdot \mu(dx) \right)^{(n+1/n+2)}, & m \leq 1; \\ \\ A \cdot h^{\alpha} \cdot m^{(1-\alpha/n+2)} \cdot \left(\int_{[a,b]} |N_n(x,t)|^{(n+2/n+1)} \cdot \mu(dx) \right)^{(n+1/n+2)}, & m \geq 1. \end{cases} \tag{11.1.16}
$$

Remark 11.1.2. We see that when $\omega_1(L_{n+1}f, \delta) \leq A\delta^{\alpha}$, inequality (11.1.16) improves the corresponding results from inequalities (11.1.12) and (11.1.13) of Theorem 11.1.3.

11.1.3. Examples

1) The system of functions $u_i(x) = x^i$, $i = 0, 1, \ldots, n$, defined on $[a, b]$, fulfills the assumptions of Theorems 11.1.1 and 11.1.3.

In particular $L_i f(t) = f^{(i)}(t)$, $g_i(x,t) = (x-t)^i / i!$, $t \in [a,b]$, see Widder (1928), p. 133.

2) According to Widder (1928), p. 135 consider $\phi_0(x) = 1$, $\phi_i(x) = \cosh ix$, $i = 1, \ldots, n$ defined on $[a, b]$, $t = 0 \in (a, b)$.

Note that $\phi_i(0) = 1$, $i = 0, 1, \ldots, n$, $g_0(x, s) = 1$ and

$$
g_i(x, 0) = \int_0^x \phi_1(s) \cdots \phi_i(s) g_{i-1}(x, s) ds, \qquad i = 1, \ldots, n.
$$

In particular, $g_1(x, 0) = \sinh x$. Then the system of functions $u_i(x) = g_i(x, 0)$, $i = 0, 1, \ldots, n$ fulfills the assumptions of Theorems 11.1.1 and 11.1.3.

Namely $u_0(x) = 1$, $u_1(x) = \sinh x$ and clearly $u_1(x)$ is a concave function for $x \leq 0$, and a convex function for $x \geq 0$.

3) The system of functions

$$
\{u_i(x)\}_{i=0}^{n} = \{1, (-1)^{i-1} \sin ix, (-1)^i \cos ix\}_{i=1}^{(n/2)}
$$

defined on $[a, b]$, $t = 0 \in (a, b)$, n even, fulfills the assumptions of Theorem 11.1.3.

In particular (see Widder (1928), p. 151(11))

$$
g_{2i}(x, 0) = \frac{2^i}{(2i)!}[1 - \cos x]^i,
$$

$$
g_{2i+1}(x, 0) = \frac{2^i}{(2i+1)!}[1 - \cos x]^i \sin x
$$

and

$$L_{2i+1} = D(D^2+1^2)(D^2+2^2)\cdots(D^2+i^2),$$
$$L_{2i+2}f(0) = D^2(D^2+1^2)(D^2+2^2)\cdots(D^2+i^2)f(0),$$

where D indicates the operation of differentiation.

4) Let $\phi_0(x) = 1$, $\phi_i(x) = e^{\varphi(i)\cdot x}$, $i = 1, \ldots, n$ be defined on $[a,b]$, with $\varphi(i) \neq 0$, e.g., $\varphi(i) = i$, $\varphi(i) = -i^{-1}$.

Then (see Widder (1928), p. 135) we have

$$g_i(x,t) = \frac{1}{\phi_0(t)\cdots\phi_i(t)} \int_t^x \phi_0(s)\cdots\phi_i(s) g_{i-1}(x,s)ds, \quad i = 1, \ldots, n, \quad g_0(x,t) = 1, \quad t \in [a,b].$$

From the same reference we get that, the system of functions $u_i(x) = g_i(x,t)$, $i = 0, 1, \ldots, n$ fulfills the assumption of Theorem 11.1.3.

11.1.4. Auxiliary Results

The next results are by themselves independently interesting.

Lemma 11.1.1. *Let g be a differentiable real valued function on $[a,b]^2 \subset \mathbb{R}^2$ with $g(x,x) = 0$ all $x \in [a,b]$, and φ be a bounded real valued function on $[a,b]$.*

Define

$$G(x,t) := \int_t^x g(x,s)\varphi(s)ds, \quad \textit{all } x,t \in [a,b].$$

Then

$$\frac{\partial G(x,t)}{\partial x} = \int_t^x \frac{\partial g(x,s)}{\partial x}\varphi(s)ds.$$

PROOF: Easy. □

As a consequence we obtain

Lemma 11.1.2. *Let*

$$G_n(x,t) := \int_t^x g_n(x,s) \cdot \left\lceil \frac{|s-t|}{h} \right\rceil ds, \quad \textit{all } x,t \in [a,b],$$

where $0 < h \leq b-a$ and $\lceil \cdot \rceil$ is the ceiling of the number.

Then

$$\frac{\partial G_n(x,t)}{\partial x} = \int_t^x \frac{\partial g_n(x,s)}{\partial x} \cdot \left\lceil \frac{|s-t|}{h} \right\rceil ds, \quad n \geq 1. \tag{11.1.17}$$

and

$$\frac{\partial^2 G_n(x,t)}{\partial x^2} = \int_t^x \frac{\partial^2 g_n(x,s)}{\partial x^2} \cdot \left\lceil \frac{|s-t|}{h} \right\rceil ds, \quad n \geq 2. \tag{11.1.18}$$

PROOF: See Widder (1928), p. 132(6) and apply Lemma 11.1.1 once/twice. □

The last result is used next

Lemma 11.1.3. *Assume that $u_0(x) = c > 0$ and $u_1(x)$ is a convex function for $x \geq t$. Let*

$$G_n(x,t) := \int_t^x g_n(x,s) \left\lceil \frac{|s-t|}{h} \right\rceil ds, \quad \textit{all } x,t \in [a,b], \textit{ where } 0 < h \leq b-a, n \geq 0.$$

Then $G_n(x,t) > 0$ for $x > t$, $G_n(t,t) = 0$, and it is a strictly increasing function in $x \geq t$, also $G_n(x,t)$ is a continuous function in $x \in [a,b]$.

Furthermore, $G_n(x,t)$ is a strictly convex function in $x \geq t$, $n \geq 1$, and $G_0(x,t)$ is a convex function in $x \geq t$.

PROOF: From $W_0(x) = \phi_0(x) = u_0(x) = c > 0$ and $W_1(x) = W[u_0(x), u_1(x)] = cu_1'(x) > 0$, $u_1(x)$ is a strictly increasing function everywhere on $[a,b]$. Hence $\phi_1(x) = W_1(x)/(W_0(x))^2 = u_1'(x)/c > 0$.

By assumption $u_1(x)$ is a convex function in $x \geq t$ implying that $u_1'(x)$ is an increasing function there, that is, $\phi_1(x)$ is increasing in $x \geq t$.

Recall that

$$g_n(x,t) = \frac{1}{\phi_1(t)\cdots\phi_n(t)} \int_t^x \phi_1(x_1) \int_t^{x_1} \cdots \int_t^{x_{n-2}} \phi_{n-1}(x_{n-1}) \int_t^{x_{n-1}} \phi_n(x_n) dx_n dx_{n-1} \cdots dx_1$$

and $g_n(x,t) > 0$ $(x > t)$, $g_n(t,t) = 0$; $n \geq 1$, with $g_0(x,t) = 1$.

Consequently

$$\frac{\partial g_n(x,t)}{\partial x} = \frac{\phi_1(x)}{\phi_1(t)\cdots\phi_n(t)} \int_t^x \phi_2(x_1) \cdots \int_t^{x_{n-2}} \phi_n(x_{n-1}) dx_{n-1} \cdots dx_1.$$

From $\phi_i(x) > 0$, $i = 1, \ldots, n$, $n \geq 2$ and $\phi_1(x)$ being an increasing function we have that $\partial g_n(x,t)/\partial x$ is a strictly increasing function in $x \geq t$, note that

$$\frac{\partial g_n(x,t)}{\partial x} > 0 \ (x > t), \quad \frac{\partial g_n(t,t)}{\partial x} = 0.$$

Hence $g_n(x,t)$ is a strictly convex function in $x \geq t$, $n \geq 2$ and clearly $g_1(x,t)$ is convex in $x \geq t$. One can easily prove that $G_n(x,t)$ is a continuous function in $x \in [a,b]$, $n \geq 0$.

From Lemma 11.1.2

$$\frac{\partial^i G_n(x,t)}{\partial x^i} = \int_t^x \frac{\partial^i g_n(x,s)}{\partial x^i} \cdot \left\lceil \frac{s-t}{h} \right\rceil ds, \quad (x \geq t, n \geq 2), \; i = 1,2.$$

It is clear that $G_n(x,t)$ is a strictly increasing function in $x \geq t$, $n \geq 2$.

By strict convexity of $g_n(x,s)$ in $x \geq s$ we get $\partial^2 g_n(x,s)/\partial x^2 > 0$ $(x > s)$, which leads to

$$\frac{\partial^2 G_n(x,t)}{\partial x^2} > 0 \; (x > t), \quad \frac{\partial^2 G_n(t,t)}{\partial x^2} = 0.$$

Thus $G_n(x,t)$ is a strictly convex function in $x \geq t$, $n \geq 2$.

Since $g_0(x,t) = 1$, all $x,t \in [a,b]$, one has

$$G_0(x,t) = \int_t^x \left\lceil \frac{s-t}{h} \right\rceil ds, \quad (x \geq t).$$

As an integral of an increasing function $G_0(x,t)$ is a convex function in $x \geq t$, also is strictly increasing in $x \geq t$. Note that

$$g_1(x,t) = \phi_1^{-1}(t) \cdot \int_t^x \phi_1(s) ds.$$

From $\partial g_1(x,t)/\partial x = \phi_1(x)/\phi_1(t)$ and ϕ_1 an increasing function, we have that $\partial g_1(x,t)/\partial x$ is increasing in $x \geq t$. Obviously, $\partial g_1(x,t)/\partial x > 0$ for all $x \in [a,b]$.

Let s be such that $t \leq s \leq x_1 < x_2$, then

$$\frac{\partial g_1(x_2,s)}{\partial x} \cdot \left\lceil \frac{s-t}{h} \right\rceil \geq \frac{\partial g_1(x_1,s)}{\partial x} \cdot \left\lceil \frac{s-t}{h} \right\rceil.$$

Adding

$$\int_t^{x_1} \frac{\partial g_1(x_2,s)}{\partial x} \cdot \left\lceil \frac{s-t}{h} \right\rceil ds \geq \int_t^{x_1} \frac{\partial g_1(x_1,s)}{\partial x} \cdot \left\lceil \frac{s-t}{h} \right\rceil ds$$

and

$$\int_{x_1}^{x_2} \frac{\partial g_1(x_2,s)}{\partial x} \cdot \left\lceil \frac{s-t}{h} \right\rceil ds > 0,$$

one gets

$$\int_t^{x_2} \frac{\partial g_1(x_2,s)}{\partial x} \cdot \left\lceil \frac{s-t}{h} \right\rceil ds > \int_t^{x_1} \frac{\partial g_1(x_1,s)}{\partial x} \cdot \left\lceil \frac{s-t}{h} \right\rceil ds.$$

The last inequality and Lemma 11.1.2 (11.1.17) imply that $\partial G_1(x,t)/\partial x$ is strictly increasing in $x \geq t$, which in turn implies that $G_1(x,t)$ is a strictly convex function in $x \geq t$.

Since

$$\frac{\partial G_1(x,t)}{\partial x} > 0 \ (x > t), \ \frac{\partial G_1(t,t)}{\partial x} = 0$$

we conclude that $G_1(x,t)$ is a strictly increasing function in $x \geq t$. □

The counterpart of Lemma 11.1.3 has as follows:

Lemma 11.1.4. *Assume that $u_0(x) = c > 0$ and $u_1(x)$ is a concave function for $x \leq t$. When $x \leq t$, $x, t \in [a,b]$, we have*

$$G_n(x,t) = \int_t^x g_n(x,s) \cdot \left\lceil \frac{t-s}{h} \right\rceil ds, \quad \textit{where } 0 < h \leq b-a, n \geq 0.$$

If n is odd, then $G_n(x,t)$ is a strictly decreasing and a strictly convex function in $x \leq t$, $G_n(x,t) > 0$ for $x < t$. If n is even, then $G_n(x,t)$ is a strictly increasing and a strictly concave function in $x \leq t$. Furthermore, $G_0(x,t)$ is a strictly increasing and a concave function in $x \leq t$. Also $G_n(x,t) < 0$ $(x < t)$ for n zero or even; with $G_n(t,t) = 0$ all $n \geq 0$.

PROOF: By assumption $u_1(x)$ is a concave function in $x \leq t$ implying that $u_1'(x)$ is a decreasing function there, that is, $\phi_1(x)$ is decreasing in $x \leq t$. See that for $n \geq 1$

$$\begin{aligned}
\frac{\partial g_n(x,t)}{\partial x} &= \frac{\phi_1(x)}{\phi_1(t)\cdots\phi_n(t)} \cdot \int_t^x \phi_2(x_1) \int_t^{x_1} \cdots \int_t^{x_{n-2}} \phi_n(x_{n-1}) dx_{n-1} \cdots dx_1 \\
&= (-1)^{n-1} \frac{\phi_1(x) \cdot B(x,t)}{\phi_1(t)\cdots\phi_n(t)},
\end{aligned}$$

where

$$\begin{aligned}
B(x,t) &:= \int_x^t \phi_2(x_1) \int_{x_1}^t \cdots \int_{x_{n-2}}^t \phi_n(x_{n-1}) dx_{n-1} \cdots dx_1 > 0 \ (x < t), \\
B(t,t) &:= 0.
\end{aligned}$$

Since $B(x,t)$ is a strictly decreasing function in $x \leq t$, we get that $\phi_1(x) \cdot B(x,t)$ is also strictly decreasing in $x \leq t$.

When $n > 1$ is odd then

$$\frac{\partial g_n(x,t)}{\partial x} > 0 \ (x < t), \ \frac{\partial g_n(t,t)}{\partial x} = 0$$

and it is a strictly decreasing function in $x \leq t$. When n is even then

$$\frac{\partial g_n(x,t)}{\partial x} < 0 \ (x < t), \ \frac{\partial g_n(t,t)}{\partial x} = 0$$

and it is a strictly increasing function in $x \leq t$.

We have proved that for n odd, $g_n(x,t) < 0$ $(x < t)$, $g_n(t,t) = 0$ and it is strictly concave in $x \leq t$ for $n > 1$; clearly $g_1(x,t)$ is concave in $x \leq t$. Also for n even $g_n(x,t) > 0$ $(x < t)$, $g_n(t,t) = 0$ and it is strictly convex in $x \leq t$.

From Lemma 11.1.2

$$\frac{\partial^i G_n(x,t)}{\partial x^i} = \int_t^x \frac{\partial^i g_n(x,s)}{\partial x^i} \cdot \left\lceil \frac{t-s}{h} \right\rceil ds, \ (x \leq t, n \geq 2), \ i = 1,2.$$

It is clear that when $n > 2$ is odd, then $G_n(x,t)$ is a strictly decreasing and a strictly convex function in $x \leq t$. And when n is even, then $G_n(x,t)$ is a strictly increasing and a strictly concave function in $x \leq t$. Note that for $n \geq 1$ odd, $G_n(x,t) > 0$ and for n zero or even, $G_n(x,t) < 0$, where $x < t$, with $G_n(t,t) = 0$ all $n \geq 0$.

One can easily see that

$$G_0(x,t) = \int_t^x \left\lceil \frac{t-s}{h} \right\rceil ds$$

is a concave and a strictly increasing function in $x \leq t$.

From $\partial g_1(x,t)/\partial x = \phi_1(x)/\phi_1(t)$ and ϕ_1 a decreasing function, we have that $\partial g_1(x,t)/\partial x$ is decreasing in $x \leq t$. Obviously $\partial g_1(x,t)/\partial x > 0$ for all $x \in [a,b]$.

Let s be such that $x_1 < x_2 \leq s \leq t$, then

$$\frac{\partial g_1(x_1,s)}{\partial x} \cdot \left\lceil \frac{t-s}{h} \right\rceil \geq \frac{\partial g_1(x_2,s)}{\partial x} \cdot \left\lceil \frac{t-s}{h} \right\rceil.$$

Adding

$$\int_{x_2}^t \frac{\partial g_1(x_1,s)}{\partial x} \cdot \left\lceil \frac{t-s}{h} \right\rceil ds \geq \int_{x_2}^t \frac{\partial g_1(x_2,s)}{\partial x} \cdot \left\lceil \frac{t-s}{h} \right\rceil ds$$

and

$$\int_{x_1}^{x_2} \frac{\partial g_1(x_1,s)}{\partial x} \cdot \left\lceil \frac{t-s}{h} \right\rceil ds > 0$$

one has

$$\int_{x_1}^t \frac{\partial g_1(x_1,s)}{\partial x} \cdot \left\lceil \frac{t-s}{h} \right\rceil ds > \int_{x_2}^t \frac{\partial g_1(x_2,s)}{\partial x} \cdot \left\lceil \frac{t-s}{h} \right\rceil ds$$

or

$$\int_t^{x_1} \frac{\partial g_1(x_1,s)}{\partial x} \cdot \left\lceil \frac{t-s}{h} \right\rceil ds < \int_t^{x_2} \frac{\partial g_1(x_2,s)}{\partial x} \cdot \left\lceil \frac{t-s}{h} \right\rceil ds.$$

The last inequality and Lemma 11.1.2 (11.1.17) imply that $\partial G_1(x,t)/\partial x$ is strictly increasing in $x \le t$, which says that $G_1(x,t)$ is a strictly convex function in $x \le t$. Since

$$\frac{\partial G_1(x,t)}{\partial x} < 0 \ (x < t), \ \frac{\partial G_1(t,t)}{\partial x} = 0$$

we conclude that $G_1(x,t)$ is a strictly decreasing function in $x \le t$. □

Lemmas 11.1.3 and 11.1.4 enable us to conclude:

Lemma 11.1.5. *Assume that $u_0(x) = c > 0$ and $u_1(x)$ is a concave function for $x \le t$ and a convex function for $x \ge t$.*

Let $\tilde{G}_n(x,t) := |G_n(x,t)|$, where

$$G_n(x,t) := \int_t^x g_n(x,s) \cdot \left\lceil \frac{|s-t|}{h} \right\rceil ds, \ \text{all } x,t \in [a,b], \ 0 < h \le b-a, \ n \ge 0.$$

Then for $n \ge 1$, $\tilde{G}_n(x,t)$ is a strictly decreasing function in $x \le t$ and a strictly increasing function in $x \ge t$, furthermore it is a continuous and a strictly convex function in $x \in [a,b]$.

Especially $\tilde{G}_0(x,t)$ possesses the above properties, with the exception that it is a convex function in $x \in [a,b]$. In particular, $\tilde{G}_n(x,t) > 0$ for $x \ne t$, with $\tilde{G}_n(t,t) = 0$, all $n \ge 0$.

Lemma 11.1.5 implies the following lemma, which is used in the proof of Theorem 11.1.1.

Lemma 11.1.6. *Under the assumptions of Lemma 11.1.5, for fixed $t \in (a,b)$, we have that*

$$\tilde{G}_n(x,t) \le \max\left\{ \frac{\tilde{G}_n(b,t)}{b-t}, \frac{\tilde{G}_n(a,t)}{t-a} \right\} \cdot |x-t|, \tag{11.1.19}$$

all $x \in [a,b]$; for all $n \ge 1$.

Equality can be true only at $x = t$ and at $x = a$ or b.

The above inequality is also true for $n = 0$, but equality can hold elsewhere not only at the points t, a or b.

PROOF: When $t < x < b$ by strict convexity of $\tilde{G}_n(x,t)$, $n \geq 1$, we get

$$\frac{\tilde{G}_n(x,t)}{x-t} < \frac{\tilde{G}_n(b,t)}{b-t}, \qquad (\tilde{G}_n(t,t) = 0).$$

Thus

$$\tilde{G}_n(x,t) < \left(\frac{\tilde{G}_n(b,t)}{b-t}\right) \cdot (x-t) \leq \max\left\{\frac{\tilde{G}_n(b,t)}{b-t}, \frac{\tilde{G}_n(a,t)}{t-a}\right\} \cdot (x-t).$$

And when $a < x < t$, again by strict convexity of $\tilde{G}_n(x,t)$ we get

$$\frac{\tilde{G}_n(a,t)}{a-t} < \frac{\tilde{G}_n(x,t)}{x-t}.$$

Hence

$$\tilde{G}_n(x,t) < \left(\frac{\tilde{G}_n(a,t)}{t-a}\right) \cdot (t-x) \leq \max\left\{\frac{\tilde{G}_n(b,t)}{b-t}, \frac{\tilde{G}_n(a,t)}{t-a}\right\} \cdot (t-x). \qquad \square$$

The next result is used in the proof of Theorem 11.1.4.

Lemma 11.1.7. *Let μ be a positive finite measure of mass $m \leq 1$ on $[a,b] \subset \mathbb{R}$ and t a fixed point in $[a,b]$. Then*

$$\left(\int_{[a,b]} |x-t|^r \mu(dx)\right)^{1/r}$$

is an increasing function in $r > 0$.

Proof: Similar to the proof of the related result in Feller (1971), p. 155(c). □

11.1.5. Proofs of Main Results

PROOF OF THEOREM 11.1.1: From Widder (1928), p. 138, Theorem II we have

$$f(x) = f(t) + \sum_{i=1}^{n} L_i f(t) \cdot g_i(x,t) + \int_t^x g_n(x,s) \cdot L_{n+1} f(s) ds,$$

all $x \in [a,b]$, fixed $t \in (a,b)$; $n \geq 0$. And from (11.1.3) we see that

$$f(x) = f(t) + \sum_{i=1}^{n} L_i f(t) \cdot g_i(x,t) + L_{n+1} f(t) \cdot N_n(x,t)$$
$$+ \int_t^x g_n(x,s) \cdot (L_{n+1} f(s) - L_{n+1} f(t)) ds.$$

Thus

$$E_n(x,t) = \int_t^x g_n(x,s) \cdot (L_{n+1}f(s) - L_{n+1}f(t))ds, \qquad n \geq 0.$$

By $\omega_1(L_{n+1}f, h) \leq w$ and from Corollary 7.1.1 we get

$$|L_{n+1}f(s) - L_{n+1}f(t)| \leq w \cdot \left\lceil \frac{|s-t|}{h} \right\rceil.$$

In the proofs of Lemmas 11.1.(3–4) we find

$$\begin{aligned}
&g_n(x,t) > 0, \quad x > t; \quad n \geq 1,\\
&g_n(x,t) < 0, \quad x < t; \quad n \text{ odd},\\
&g_n(x,t) > 0, \quad x < t; \quad n \text{ even},\\
&g_n(t,t) = 0, \quad n \geq 1 \quad g_0(x,t) = 1.
\end{aligned}$$

Let $x \leq t$ and n even, then

$$\begin{aligned}
|E_n(x,t)| &= \left| \int_x^t g_n(x,s) \cdot (L_{n+1}f(s) - L_{n+1}f(t)) \cdot ds \right|\\
&\leq \int_x^t g_n(x,s) \cdot |L_{n+1}f(s) - L_{n+1}f(t)| \cdot ds\\
&\leq w \cdot \int_x^t g_n(x,s) \cdot \left\lceil \frac{|s-t|}{h} \right\rceil \cdot ds\\
&= w \cdot \left| \int_t^x g_n(x,s) \cdot \left\lceil \frac{|s-t|}{h} \right\rceil \cdot ds \right|.
\end{aligned}$$

I.e.,

$$|E_n(x,t)| \leq w \cdot \tilde{G}_n(x,t)$$

for $x \leq t$ and n even.

Let $x \leq t$ and n odd, then

$$\begin{aligned}
|E_n(x,t)| &= \left| \int_x^t (-g_n(x,s)) \cdot (L_{n+1}f(s) - L_{n+1}f(t)) \cdot ds \right|\\
&\leq \int_x^t (-g_n(x,s)) \cdot |L_{n+1}f(s) - L_{n+1}f(t)| \cdot ds\\
&\leq w \cdot \int_x^t (-g_n(x,s)) \cdot \left\lceil \frac{|s-t|}{h} \right\rceil \cdot ds\\
&= w \cdot \left| \int_t^x g_n(x,s) \cdot \left\lceil \frac{|s-t|}{h} \right\rceil \cdot ds \right|\\
&= w \cdot \tilde{G}_n(x,t).
\end{aligned}$$

I.e.,

$$|E_n(x,t)| \le w \cdot \tilde{G}_n(x,t)$$

for $x \le t$ and n odd.

The last inequality is also true for $x \ge t$, all $n \ge 1$ and true for $n = 0$. Thus we have established that

$$|E_n(x,t)| \le w \cdot \tilde{G}_n(x,t),$$

all $x \in [a,b]$; $n \ge 0$.

Using inequality (11.1.19) from Lemma 11.1.6 we obtain

$$|E_n(x,t)| \le w \cdot \max\left\{\frac{\tilde{G}_n(b,t)}{b-t}, \frac{\tilde{G}_n(a,t)}{t-a}\right\} \cdot |x-t|, \tag{11.1.20}$$

all $x \in [a,b]$, fixed $t \in (a,b)$; $n \ge 0$.

An integration of inequality (11.1.20) against μ produces inequality (11.1.6).

Inequality (11.1.7) is established through

$$\left|\int_{[a,b]} f\,d\mu - f(t)\right| \le \left|\int_{[a,b]} (f(x)-f(t)) \cdot \mu(dx)\right| + |m-1| \cdot |f(t)|$$

and

$$(f(x)-f(t)) = \sum_{i=1}^{n} L_i f(t) \cdot g_i(x,t) + L_{n+1} f(t) \cdot N_n(x,t) + E_n(x,t). \qquad \square$$

PROOF OF THEOREM 11.1.2: Since

$$\lim_{N\to+\infty} (g_n(x,s) \cdot f_N(|s-t|)) = g_n(x,s) \cdot \left\lceil \frac{|s-t|}{h} \right\rceil w,$$

by bounded convergence theorem we have

$$\lim_{N\to+\infty} \int_t^x g_n(x,s) \cdot f_N(|s-t|)ds = \int_t^x g_n(x,s) \cdot \left\lceil \frac{|s-t|}{h} \right\rceil w\,ds.$$

Thus

$$\lim_{N\to+\infty} \left|\int_t^x g_n(x,s) \cdot f_N(|s-t|)ds\right| = w \cdot \left|\int_t^x g_n(x,s) \cdot \left\lceil \frac{|s-t|}{h} \right\rceil ds\right|$$

i.e.,

$$\lim_{N\to+\infty} \tilde{G}_{nN}(x,t) = w \cdot \tilde{G}_n(x,t).$$

By calling

$$G_{nN}(x,t) := \int_t^x g_n(x,s) \cdot f_N(|s-t|)ds,$$

we have for n odd that

$$\tilde{G}_{nN}(x,t) = G_{nN}(x,t) > 0, \qquad x \neq t$$

and for n zero or even that

$$\tilde{G}_{nN}(x,t) = \begin{cases} -G_{nN}(x,t) > 0, & a \le x < t, \\ G_{nN}(x,t) > 0, & t < x \le b. \end{cases}$$

In particular $\tilde{G}_{nN}(t,t) = G_{nN}(t,t) = 0$, all $n \geq 0$.

Let $n \geq 1$, from Widder (1928), p. 132(6) we have

$$\frac{\partial^i g_n(t,t)}{\partial x^i} = \begin{cases} 0, & i = 0, 1, \ldots, n-1 \\ 1, & i = n. \end{cases}$$

Applying Leibnitz's formula repeatedly, we find

$$\frac{\partial^i}{\partial x^i} G_{nN}(x,t) = \int_t^x \frac{\partial^i g_n(x,s)}{\partial x^i} f_N(|s-t|)ds, \qquad i = 0, 1, \ldots, n$$

and

$$\frac{\partial^{n+1}}{\partial x^{n+1}} G_{nN}(x,t) = \int_t^x \frac{\partial^{n+1} g_n(x,s)}{\partial x^{n+1}} f_N(|s-t|)ds + f_N(|x-t|), \quad \text{all } x \in [a,b].$$

Hence

$$\frac{\partial^i}{\partial x^i} G_{nN}(t,t) = 0, \qquad i = 0, 1, \ldots, n+1.$$

And one can easily see that

$$\frac{\partial^i}{\partial x^i} \tilde{G}_{nN}(t,t) = 0, \qquad i = 0, 1, \ldots, n+1.$$

Since L_i is a linear differential operator of order i, $i = 1, \ldots, n+1$; $L_0 f(t) = f(t)$ (see (11.1.1)), we get $L_i \tilde{G}_{nN}(t,t) = 0$, $i = 0, 1, \ldots, n+1$; $n \geq 0$.

From Widder (1928), p. 132 we have

$$L_{n+1} G_{nN}(x,t) = f_N(|x-t|), \quad \text{all } x \in [a,b]; n \geq 0.$$

Hence for n odd we find that

$$L_{n+1}\tilde{G}_{nN}(x,t) = f_N(|x-t|), \quad \text{all } x \in [a,b].$$

And for n zero or even we find that

$$L_{n+1}\tilde{G}_{nN}(x,t) = \begin{cases} -f_N(t-x), & a \le x \le t; \\ f_N(x-t), & t \le x \le b. \end{cases}$$

Now consider the case (i) of this theorem with f and μ described as there. Then note that

$$L_{n+1}f(x) = \begin{cases} f_N(x-t), & t \le x \le b; \\ 0, & a \le x \le t. \end{cases}$$

Hence one can easily see that

$$\omega_1(L_{n+1}f, h) \le w$$

and $L_i f(t) = 0$, $i = 0, 1, \ldots, n+1$.

Consequently the left-hand sides of inequalities (11.1.6) and (11.1.7) equal to $(\tilde{G}_{nN}(b,t)/b-t)d$ which, as $N \to +\infty$, converges to $w(\tilde{G}_n(b,t)/b-t)d$, i.e., to the right-hand side of these inequalities, proving them as attained.

Finally consider the case (ii) of the theorem with f and μ described as there. Then note that

$$L_{n+1}f(x) = \begin{cases} 0, & t \le x \le b; \\ (-1)^{n+1} f_N(t-x), & a \le x \le t. \end{cases}$$

Again one can easily see that

$$\omega_1(L_{n+1}f, h) \le w$$

and

$$L_i f(t) = 0, \qquad i = 0, 1, \ldots, n+1.$$

Consequently the left-hand sides of inequalities (11.1.6) and (11.1.7) equal to $(\tilde{G}_{nN}(a,t)/t-a)d$ which, as $N \to +\infty$, converges to $w(\tilde{G}_n(a,t)/t-a)d$, i.e., to the right-hand side of these inequalities proving them again attained. □

PROOF OF THEOREM 11.1.3: From Widder (1928), p. 138, Theorem II we have

$$f(x) = \sum_{i=0}^{n} L_i f(t) \cdot g_i(x,t) + \int_t^x g_n(x,s) \cdot L_{n+1} f(s) \cdot ds,$$

all $x, t \in [a,b]$; $n \geq 0$.

And from (11.1.3) we see that

$$f(x) = \sum_{i=0}^{n} L_i f(t) \cdot g_i(x,t) + L_{n+1} f(t) \cdot N_n(x,t) + \int_t^x g_n(x,s) \cdot (L_{n+1} f(s) - L_{n+1} f(t)) \cdot ds.$$

Thus

$$E_n(x,t) = \int_t^x g_n(x,s) \cdot (L_{n+1} f(s) - L_{n+1} f(t)) \cdot ds, \quad n \geq 0.$$

By $\omega_1(L_{n+1} f, h) \leq w$ and from Corollary 7.1.1 we get

$$|L_{n+1} f(s) - L_{n+1} f(t)| \leq w \cdot \left\lceil \frac{|s-t|}{h} \right\rceil,$$

where $\lceil \cdot \rceil$ is the ceiling of the number.

Let $x \leq t$ and n even, then

$$\begin{aligned} |E_n(x,t)| &= \left| \int_x^t g_n(x,s) \cdot (L_{n+1} f(s) - L_{n+1} f(t)) \cdot ds \right| \\ &\leq \int_x^t g_n(x,s) \cdot |L_{n+1} f(s) - L_{n+1} f(t)| \cdot ds \\ &\leq w \cdot \int_x^t g_n(x,s) \cdot \left\lceil \frac{|s-t|}{h} \right\rceil \cdot ds \\ &\leq w \cdot \left\lceil \frac{|x-t|}{h} \right\rceil \cdot \left| \int_t^x g_n(x,s) ds \right|. \end{aligned}$$

I.e.,

$$|E_n(x,t)| \leq w \cdot \left\lceil \frac{|x-t|}{h} \right\rceil \cdot |N_n(x,t)|$$

for $x \leq t$ and n even.

Let $x \leq t$ and n odd, then

$$\begin{aligned} |E_n(x,t)| &= \left| \int_x^t (-g_n(x,s)) \cdot (L_{n+1} f(s) - L_{n+1} f(t)) \cdot ds \right| \\ &\leq \int_x^t (-g_n(x,s)) \cdot |L_{n+1} f(s) - L_{n+1} f(t)| \cdot ds \\ &\leq w \cdot \int_x^t (-g_n(x,s)) \cdot \left\lceil \frac{|s-t|}{h} \right\rceil \cdot ds \\ &\leq w \cdot \left\lceil \frac{|x-t|}{h} \right\rceil \cdot \left| \int_t^x g_n(x,s) ds \right|. \end{aligned}$$

I.e.,

$$|E_n(x,t)| \le w \cdot \left\lceil \frac{|x-t|}{h} \right\rceil \cdot |N_n(x,t)|$$

for $x \le t$ and n odd.

The last inequality is also true for $x \ge t$, all $n \ge 1$ and true for $n = 0$. Thus we have established that

$$|E_n(x,t)| \le w \cdot \left\lceil \frac{|x-t|}{h} \right\rceil \cdot |N_n(x,t)|, \tag{11.1.21}$$

all $x, t \in [a,b]$; $n \ge 0$.

Integrating inequality (11.1.21) against μ ($\mu([a,b]) = m$) we get

$$\begin{aligned}
\int_{[a,b]} |E_n(x,t)|\mu(dx) &\le w \cdot \int_{[a,b]} \left\lceil \frac{|x-t|}{h} \right\rceil \cdot |N_n(x,t)| \cdot \mu(dx) \\
&\le w \cdot \int_{[a,b]} \left(1 + \frac{|x-t|}{h}\right) \cdot |N_n(x,t)| \cdot \mu(dx) \\
&= w \cdot \left[\int_{[a,b]} |N_n(x,t)| \cdot \mu(dx) + \frac{1}{h} \cdot \int_{[a,b]} |x-t| \cdot |N_n(x,t)| \cdot \mu(dx)\right] \\
&\le w \cdot \left[m^{1/(n+2)} + \frac{1}{h} \cdot \left(\int_{[a,b]} |x-t|^{n+2} \cdot \mu(dx)\right)^{1/(n+2)}\right] \\
&\quad \cdot \left(\int_{[a,b]} |N_n(x,t)|^{(n+2/n+1)} \cdot \mu(dx)\right)^{(n+1/n+2)} \\
&= w \cdot (m^{1/(n+2)} + 1) \cdot \left(\int_{[a,b]} |N_n(x,t)|^{(n+2/n+1)} \cdot \mu(dx)\right)^{(n+1/n+2)}.
\end{aligned}$$

The last inequality and equality come by applying Hölder's inequality twice and by the choice of h (see (11.1.11), respectively. Therefore we have proved inequality (11.1.12).

Inequality (11.1.13) is established from

$$\begin{aligned}
\left|\int_{[a,b]} f(x)\mu(dx) - f(t)\right| \le |f(t)| \cdot \left|\int_{[a,b]} g_0(x,t)\mu(dx) - 1\right| \\
+ \sum_{i=1}^{n} |L_i f(t)| \cdot \left|\int_{[a,b]} g_i(x,t)\mu(dx)\right| \\
+ |L_{n+1} f(t)| \cdot \left|\int_{[a,b]} N_n(x,t)\mu(dx)\right| \\
+ \int_{[a,b]} |E_n(x,t)|\mu(dx); \quad L_0 f(t) = f(t). \qquad \square
\end{aligned}$$

Proof of Theorem 11.1.4: As in the proof of Theorem 11.1.3 we have

$$E_n(x,t) = \int_t^x g_n(x,s) \cdot (L_{n+1}f(s) - L_{n+1}f(t)) \cdot ds,$$

all $x, t \in [a,b]$; $n \geq 0$.

The Lipschitz condition $\omega_1(L_{n+1}f, \delta) \leq A\delta^\alpha$ implies that

$$|L_{n+1}f(s) - L_{n+1}f(t)| \leq A|s-t|^\alpha, \quad \text{all } s,t \in [a,b].$$

Let $x \leq t$ and n zero or even, then

$$\begin{aligned} |E_n(x,t)| &= \left| \int_x^t g_n(x,s) \cdot (L_{n+1}f(s) - L_{n+1}f(t)) \cdot ds \right| \\ &\leq \int_x^t g_n(x,s) \cdot |L_{n+1}f(s) - L_{n+1}f(t)| \cdot ds \\ &\leq A \int_x^t g_n(x,s)|s-t|^\alpha ds \leq A|x-t|^\alpha \left| \int_t^x g_n(x,s)ds \right|. \end{aligned}$$

I.e.,

$$|E_n(x,t)| \leq A|x-t|^\alpha |N_n(x,t)|$$

for $x \leq t$ and n zero or even.

Let $x \leq t$ and n odd, then

$$\begin{aligned} |E_n(x,t)| &= \left| \int_x^t (-g_n(x,s)) \cdot (L_{n+1}f(s) - L_{n+1}f(t)) \cdot ds \right| \\ &\leq \int_x^t (-g_n(x,s)) \cdot |L_{n+1}f(s) - L_{n+1}f(t)| \cdot ds \\ &\leq A \int_x^t (-g_n(x,s))|s-t|^\alpha ds \leq A|x-t|^\alpha \left| \int_t^x g_n(x,s)ds \right|. \end{aligned}$$

I.e.,

$$|E_n(x,t)| \leq A|x-t|^\alpha |N_n(x,t)|$$

for $x \leq t$ and n odd.

The last inequality is also true for $x \geq t$, all $n \geq 0$. Thus we have established that

$$|E_n(x,t)| \leq A|x-t|^\alpha |N_n(x,t)|, \tag{11.1.22}$$

all $x, t \in [a,b]$; $n \geq 0$.

Integrating inequality (11.1.22) against μ ($\mu([a,b]) = m$) we get

$$\int_{[a,b]} |E_n(x,t)|\mu(dx) \le A \int_{[a,b]} |x-t|^\alpha \cdot |N_n(x,t)| \cdot \mu(dx)$$
$$\le A \cdot D_\alpha(t) \cdot \left(\int_{[a,b]} |N_n(x,t)|^{(n+2/n+1)} \cdot \mu(dx) \right)^{(n+1/n+2)},$$

where

$$D_\alpha(t) := \left(\int_{[a,b]} |x-t|^{\alpha(n+2)} \cdot \mu(dx) \right)^{1/(n+2)}.$$

The last inequality comes by the use of Hölder's inequality. I.e., we have obtained that

$$\int_{[a,b]} |E_n(x,t)|\mu(dx) \le A \cdot D_\alpha(t) \cdot \left(\int_{[a,b]} |N_n(x,t)|^{(n+2/n+1)} \cdot \mu(dx) \right)^{(n+1/n+2)}. \quad (11.1.23)$$

Case of $m \le 1$: By Lemma 11.1.7, since $\alpha(n+2) \le n+2$, we have

$$\left(\int_{[a,b]} |x-t|^{\alpha(n+2)} \cdot \mu(dx) \right)^{1/\alpha(n+2)} \le \left(\int_{[a,b]} |x-t|^{n+2} \cdot \mu(dx) \right)^{1/(n+2)},$$

that is,

$$D_\alpha(t) \le h^\alpha. \quad (11.1.24)$$

Now the first part of inequality (11.1.16) is established by (11.1.23) and (11.1.24).

Case of $m \ge 1$: We observe that

$$\left(\int_{[a,b]} |x-t|^{\alpha(n+2)} \cdot \mu(dx) \right)^{1/\alpha(n+2)} = m^{1/\alpha(n+2)} \cdot \left(\int_{[a,b]} |x-t|^{\alpha(n+2)} \cdot \frac{\mu}{m}(dx) \right)^{1/\alpha(n+2)}$$
$$\le m^{1/\alpha(n+2)} \cdot \left(\int_{[a,b]} |x-t|^{n+2} \cdot \frac{\mu}{m}(dx) \right)^{1/(n+2)}$$
$$= m^{((1-\alpha)/\alpha(n+2))} \cdot h.$$

Here we used again Lemma 11.1.7. That is we obtain

$$D_\alpha(t) \le m^{(1-\alpha/n+2)} \cdot h^\alpha. \quad (11.1.25)$$

Finally, inequalities (11.1.23) and (11.1.25) imply the second part of inequality (11.1.16). □

CHAPTER TWELVE

12.1. A GENERAL "K-ATTAINED" INEQUALITY RELATED TO THE WEAK CONVERGENCE OF PROBABILITY MEASURES TO THE UNIT MEASURE

12.1.1. Introduction

Here the rate of weak convergence of a sequence of probability measures to the unit measure at a fixed point t is given by a sharp attained general inequality using an extended complete Tschebychev system (E.C.T.). This involves a very flexible measure of smoothness, the generalized reduced K-functional, which has all the basic properties of the usual K-functional but depends on t.

The material of this chapter comes from Anastassiou: Memphis State University, Memphis, TN, Report Series 88-1.

The following introductory notions come again from Widder (1928), which is an essential reference for this chapter.

Let the functions g, $u_0, u_1, \ldots, u_n \in C^{n+1}([a,b])$, $n \geq 0$, $[a,b] \subset \mathbb{R}$, such that $u_0(x) = c > 0$ and $u_1(x)$ is a concave function for $x \leq t$ and a convex function for $x \geq t$, where t is a fixed point in (a,b).

Let the Wronskians

$$W_i(x) := W(u_0(x), u_1(x), \ldots, u_i(x)) := \begin{vmatrix} u_0(x) & u_1(x) & \cdots & u_i(x) \\ u_0'(x) & u_1'(x) & \cdots & u_i'(x) \\ \vdots & & & \\ u_0^{(i)}(x) & u_1^{(i)}(x) & \cdots & u_i^{(i)}(x) \end{vmatrix},$$

$$i = 0, 1, \ldots, n$$

and assume that all $W_i(x)$ are positive everywhere on $[a,b]$. Clearly the functions

$$\phi_0(x) := W_0(x) := u_0(x), \qquad \phi_1(x) := \frac{W_1(x)}{(W_0(x))^2},$$

$$\phi_i(x) := \frac{W_i(x) W_{i-2}(x)}{(W_{i-1}(x))^2}, \qquad i = 2, 3, \ldots, n$$

are positive everywhere on $[a,b]$.

Consider the linear differential operator of order $i \geq 0$

$$L_i g(x) := \frac{W[u_0(x), u_1(x), \ldots, u_{i-1}(x), g(x)]}{W_{i-1}(x)} \tag{12.1.1}$$

$i = 1, \ldots, n+1$; $L_0 g(x) := g(x)$ all $x \in [a,b]$. Here $W[u_0(x), u_1(x), \ldots, u_{i-1}(x), g(x)]$ denotes the Wronskian of $u_0, u_1, \ldots, u_{i-1}, g$.

Consider also the functions

$$g_i(x,t) := \frac{1}{W_i(x)} \cdot \begin{vmatrix} u_0(t) & u_1(t) & \cdots & u_i(t) \\ u_0'(t) & u_1'(t) & \cdots & u_i'(t) \\ \vdots & & & \\ u_0^{(i-1)}(t) & u_1^{(i-1)}(t) & \cdots & u_i^{(i-1)}(t) \\ u_0(x) & u_1(x) & \cdots & u_i(x) \end{vmatrix}, \tag{12.1.2}$$

$i = 1, 2, \ldots, n$; $g_0(x,t) := 1$, all $x, t \in [a,b]$. Note that $g_i(x,t)$ as a function of x, it is a linear combination of $u_0(x)$, $u_1(x), \ldots, u_i(x)$ and furthermore it holds

$$g_i(x,t) = \frac{1}{\phi_1(t)\cdots\phi_i(t)} \cdot \int_t^x \phi_1(x_1) \int_t^{x_1} \cdots \int_t^{x_{i-2}} \phi_{i-1}(x_{i-1}) \int_t^{x_{i-1}} \phi_i(x_i) dx_i dx_{i-1} \cdots dx_1,$$

all $i = 1, 2, \ldots, n$.

From Widder (1928), p. 138, Theorem II we have that

$$g(x) = g(t) + \sum_{i=1}^{n} L_i g(t) \cdot g_i(x,t) + R_n(x), \tag{12.1.3}$$

where

$$R_n(x) := L_{n+1} g(\xi) \cdot \int_t^x g_n(x,s) ds,$$

ξ is between x, t; all $x \in [a,b]$, t is a fixed point in (a,b), $n \geq 0$.

Here we denote by

$$N_n(x,t) := \int_t^x g_n(x,s) ds, \qquad \tilde{N}_n(x,t) := |N_n(x,t)|; \quad n \geq 0. \tag{12.1.4}$$

For a definition of an (E.C.T.) system, see Chapter 11.

Next we would like to introduce the following measure of smoothness, which plays a central role in this chapter.

Definition 12.1.1. Let $f \in C([a,b])$. Let $n \geq 0$ and $t \in (a,b)$ be fixed. We define the *generalized reduced K-functional* as:

$$\tilde{K}^{(t)}_{n+1}(f,h) := K(f,h) := K(h) := \inf_g (\|f-g\|_\infty + h \cdot \|L_{n+1}g\|_\infty), \quad h \geq 0. \qquad (12.1.5)$$

Here g ranges through the functions $g \in C^{n+1}([a,b])$ satisfying

$$L_i g(t) = 0, \qquad i = 1, \ldots, n,$$

where $\|\cdot\|_\infty$ denotes the sup norm.

The notion of $K(f,h)$ generalizes to a wide class of E.C.T. systems the notion of reduced K-functional (see Anastassiou(2) (1985)), where there we have the special case of $u_i(x) = x^i$, $i = 0, 1, \ldots, n$. E.g., consider any E.C.T. system in $C^{n+1}([a,b])$ containing $u_0(x) = c > 0$, $u_1(x) = \sinh x$; where here $t = 0 \in (a,b)$. And $K(f,h)$ is similar to the K-functional of Peetre (see Peetre (1963); Schumaker (1981), p. 60) where we do not impose the restrictions $L_i g(t) = 0$, $i = 1, \ldots, n$; there is

$$L_{n+1} = \frac{d^{n+1}}{dx^{n+1}}.$$

The quantity $K(f,h)$ measures how well f can be approximated by a smooth function g. It is the suitable, very flexible measure for the approximation of small functions f with large derivatives in the sup norm.

In Theorem 12.1.1 we present the basic properties of the introduced generalized reduced K-functional, proving that it behaves nicely like the other K-functionals.

In Theorem 12.1.2 we present a general sharp attained inequality involving $\tilde{K}^{(t)}_{n+1}(f,h)$, which is the main goal of this chapter. This inequality gives an estimate in the very general E.C.T. setting for the rate of weak convergence of a sequence of probability measures to the unit measure at a fixed point.

Finally, Theorem 12.1.3 contains an independent result in the uniform approximation of continuous functions by generalized polynomials.

12.1.2. Main Results

Theorem 12.1.1. *The generalized reduced K-functional has the following properties:* (i)

Subadditive in terms of f. And in terms of h is (ii) *continuous,* (iii) *nonnegative,* (iv) *monotonely increasing,* (v) *concave, hence* (vi) *subadditive.* (vii) $\tilde{K}_{n+1}^{(t)}(f,0) = 0$.

Remark 12.1.1. Thus $\tilde{K}_{n+1}^{(t)}(f,h)$ has all the basic properties of the usual Peetre functional $K_{n+1}(f,h)$. In particular $\tilde{K}_1^{(t)} = K_1^{(t)} = K_1$, by $u_0(x)$ constant and $L_1 = d/dx$.

Also, since $L_i c = 0$, $c \in \mathbb{R}$, all $i \geq 1$ we have that $\tilde{K}_{n+1}^{(t)}(c,h) = 0$, all $h \geq 0$.

PROOF OF THEOREM 12.1.1: (i) Let $g_1, g_2 \in C^{n+1}([a,b])$ such that $L_i g_1(t) = L_i g_2(t) = 0$, $i = 1,\ldots,n$, i.e., $L_i(g_1 + g_2)(t) = 0$, $i = 1,\ldots,n$. Therefore

$$\begin{aligned} K(f_1+f_2,h) &\leq \|(f_1+f_2)-(g_1+g_2)\|_\infty + h\cdot\|L_{n+1}(g_1+g_2)\|_\infty \\ &\leq (\|f_1-g_1\|_\infty + h\cdot\|L_{n+1}g_1\|_\infty) + (\|f_2-g_2\|_\infty + h\cdot\|L_{n+1}g_2\|_\infty), \end{aligned}$$

for all g_1, g_2 of the above class. Obviously one has

$$K(f_1+f_2,h) \leq K(f_1,h) + K(f_2,h).$$

(ii)(a) K is an upper-semicontinuous function in h, i.e.,

$$\limsup_{h\to c} K(h) \leq K(c).$$

PROOF: Observe that

$$\begin{aligned} \limsup_{h\to c} K(h) &\leq \limsup_{h\to c}(\|f-g\|_\infty + h\cdot\|L_{n+1}g\|_\infty) \\ &= \lim_{h\to c}(\|f-g\|_\infty + h\cdot\|L_{n+1}g\|_\infty) = (\|f-g\|_\infty + c\cdot\|L_{n+1}g\|_\infty), \end{aligned}$$

for all $g \in C^{n+1}([a,b])$ satisfying $L_i g(t) = 0$, $i = 1,\ldots,n$. The result is now obvious.

(b) K is right continuous in h.

PROOF: Since K is increasing in h we get $K(h) \leq K(h+\varepsilon)$, $\varepsilon > 0$. From (a) for $\varepsilon \to 0$ we have

$$\limsup_{h+\varepsilon\to h} K(h+\varepsilon) \leq K(h).$$

Furthermore it holds

$$K(h) \leq \liminf_{h+\varepsilon\to h} K(h+\varepsilon)$$

along with

$$\liminf_{h+\varepsilon\to h} K(h+\varepsilon) \le \limsup_{h+\varepsilon\to h} K(h+\varepsilon).$$

Therefore

$$\begin{aligned}\liminf_{h+\varepsilon\to h} K(h+\varepsilon) &= \limsup_{h+\varepsilon\to h} K(h+\varepsilon)\\ &= \lim_{h+\varepsilon\to h} K(h+\varepsilon) = K(h),\end{aligned}$$

which proves the claim.

(c) It holds $K(h-) \le K(h) \le K(h-)$, i.e., K is left continuous in h.

Proof: Clearly $K(h-) \le K(h)$, by K being increasing in h. If $0 < \varepsilon < h$ then $h/h-\varepsilon > 1$. Now see that

$$K(h) \le (\|f-g\|_\infty + h\cdot\|L_{n+1}g\|_\infty) \le \left(\frac{h}{h-\varepsilon}\right)\cdot(\|f-g\|_\infty + (h-\varepsilon)\cdot\|L_{n+1}g\|_\infty),$$

for all $g \in C^{n+1}([a,b])$ such that $L_i g(t) = 0$, $i = 1,\ldots,n$. Therefore

$$K(h) \le \left(\frac{h}{h-\varepsilon}\right)\cdot K(h-\varepsilon),$$

i.e.,

$$K(h) \le K(h-),$$

which proves the claim.

We have established that K is continuous in h.

(iii) obvious.

(iv) For $h_1 \le h_2$ we get

$$K(h_1) \le (\|f-g\|_\infty + h_1\cdot\|L_{n+1}g\|_\infty) \le (\|f-g\|_\infty + h_2\cdot\|L_{n+1}g\|_\infty),$$

for all $g \in C^{n+1}([a,b])$ such that $L_i g(t) = 0$, $i = 1,\ldots,n$. Thus $K(h_1) \le K(h_2)$.

(v) Let $0 \le \lambda \le 1$. We observe that

$$\begin{aligned}\lambda K(h_1) + (1-\lambda)K(h_2) &\le \lambda(\|f-g\|_\infty + h_1\cdot\|L_{n+1}g\|_\infty)\\ &\quad + (1-\lambda)(\|f-g\|_\infty + h_2\cdot\|L_{n+1}g\|_\infty)\\ &= \|f-g\|_\infty + (\lambda h_1 + (1-\lambda)h_2)\cdot\|L_{n+1}g\|_\infty,\end{aligned}$$

for all $g \in C^{n+1}([a,b])$ such that $L_i g(t) = 0$, $i = 1, \ldots, n$. Thus

$$\lambda K(h_1) + (1-\lambda)K(h_2) \leq K(\lambda h_1 + (1-\lambda)h_2),$$

proving the claim.

(vii) From Lemmas 12.1.(2–3) following later, we get that

$$\tilde{g}(x) := N_n(x,t) = \int_t^x g_n(x,s)ds \in C^{n+1}([a,b]), \qquad n \geq 0,$$

is a strictly monotone function.

Thus for $x \neq y$ we get that $\tilde{g}(x) \neq \tilde{g}(y)$, that is, $\tilde{g}(x)$ separates points. Furthermore from Widder (1928), p. 138 we have that

$$L_{n+1}\tilde{g}(x) = 1,$$
$$\tilde{g}^{(i)}(t) = 0, \qquad i = 0,1,2,\ldots,n.$$

So that $L_i\tilde{g}(t) = 0$, $i = 1, \ldots, n$.

Consider now the algebra A generated by $\{1, \tilde{g}(x)\}$ i.e., $A = \langle\{1, \tilde{g}(x)\}\rangle$. A typical element of A is a finite linear combination of non-negative powers of $\tilde{g}$. Note that A is a linear subspace of $C^{n+1}([a,b]) \subset C([a,b])$, which contains the constant functions and separates points. Therefore by the Stone–Weierstrass Theorem we get that

$$\overline{A} = C([a,b])$$

i.e., for $f \in C([a,b])$; $\forall \varepsilon > 0$ there exists $e \in A$ such that $\|f - e\|_\infty < \varepsilon$. Since $u_0(x)$ is constant we get that $L_i(1) = 0$, all $i = 1, \ldots, n$.

And since $\tilde{g}^{(i)}(t) = 0$, all $i = 0,1,2,\ldots,n$ we have that $(\tilde{g}^\ell)^{(i)}(t) = 0$, $i = 0,1,2,\ldots,n$ $(\ell \geq 1)$. That is, $L_i(\tilde{g}^\ell)(t) = 0$, $i = 1, \ldots, n$. These imply $L_i e(t) = 0$, $i = 1, \ldots, n$ for all $e \in A$. Hence A is contained in the class of functions $g \in C^{n+1}([a,b])$ such that

$$L_i g(t) = 0, \qquad i = 1, \ldots, n.$$

Consequently,

$$0 \leq \tilde{K}_{n+1}^{(t)}(f,0) = \inf_g(\|f-g\|_\infty) \leq \inf_{e\in A}(\|f-e\|_\infty) = 0.$$

We have established that

$$\tilde{K}_{n+1}^{(t)}(f,0) = 0. \qquad \square$$

Theorem 12.1.2. *Let μ be a probability measure on $[a,b] \subset \mathbb{R}$ such that*

$$\int_{[a,b]} |x-t|\mu(dx) = d_1(t) > 0, \tag{12.1.6}$$

where t is a fixed point of (a,b).

Let the functions $u_0(x), u_1(x), \ldots, u_n(x)$ belong to $C^{n+1}([a,b])$, $n \geq 0$, and let the Wronskians $W_0(x), W_1(x), \ldots, W_n(x)$ be positive throughout $[a,b]$ (i.e., $\{u_i\}_{i=0}^n$ is an E.C.T. system).

Assume that $u_0(x) = c > 0$ and $u_1(x)$ is a concave function for $x \leq t$ and a convex function for $x \geq t$.

Define

$$\Lambda_n(t) := \max\left\{\frac{\tilde{N}_n(b,t)}{b-t}, \frac{\tilde{N}_n(a,t)}{t-a}\right\}, \tag{12.1.7}$$

where

$$\tilde{N}_n(x,t) := \left|\int_t^x g_n(x,s)ds\right|, \quad \text{all } x \in [a,b];\ n \geq 0.$$

Consider $f \in C([a,b])$. Then

$$\left|\int_{[a,b]} f\,d\mu - f(t)\right| \leq 2 \cdot \tilde{K}_{n+1}^{(t)}\left(f; \frac{\Lambda_n(t)\cdot d_1(t)}{2}\right), \quad n \geq 0. \tag{12.1.8}$$

For special choices of $d_1(t)$ the above inequality is *attained* (i.e., *it is sharp*) by the function $N_n(x,t)$ and the probability measure μ_0 described as follows:

(i) If $\frac{\tilde{N}_n(b,t)}{b-t} \geq \frac{\tilde{N}_n(a,t)}{t-a}$ and $d_1(t) \leq b-t$, then the optimal probability measure μ_0 is supported at $\{t,b\}$ with masses $(1-\frac{d_1(t)}{b-t})$, $(\frac{d_1(t)}{b-t})$, respectively.

(ii) If $\frac{\tilde{N}_n(b,t)}{b-t} \leq \frac{\tilde{N}_n(a,t)}{t-a}$ and $d_1(t) \leq t-a$, then the optimal probability measure μ_0 is supported at $\{t,a\}$ with masses $(1-\frac{d_1(t)}{t-a})$, $(\frac{d_1(t)}{t-a})$, respectively.

Remark 12.1.2. According to Widder (1928), p. 139 when $u_i(x) = x^i$, $i = 0,1,2,\ldots,n$ we get

$$N_n(x,t) = \frac{(x-t)^{n+1}}{(n+1)!} \quad \text{and} \quad \tilde{N}_n(x,t) = \frac{|x-t|^{n+1}}{(n+1)!}.$$

Hence

$$\tilde{N}_n(b,t) = \frac{(b-t)^{n+1}}{(n+1)!}, \qquad \tilde{N}_n(a,t) = \frac{(t-a)^{n+1}}{(n+1)!}$$

and

$$\Lambda_n(t) = \max\left\{\frac{(b-t)^n}{(n+1)!}, \frac{(t-a)^n}{(n+1)!}\right\}.$$

That is,

$$\Lambda_n(t) = \frac{(c(t))^n}{(n+1)!},$$

where $c(t) := \max(b-t, t-a)$; $n \geq 0$. Here note that $\tilde{K}_{n+1}^{(t)} = K_{n+1}^{(t)}$: the reduced K-functional introduced in Anastassiou(2) (1985). In this special case inequality (12.1.8) reduces to the known attained inequality (see Anastassiou(2) (1985)),

$$\left|\int_{[a,b]} f\,d\mu - f(t)\right| \leq 2 \cdot K_{n+1}^{(t)}\left(f; \frac{d_1(t)\cdot(c(t))^n}{2(n+1)!}\right),$$

$t \in (a,b)$; $n \geq 0$.

Proof of Theorem 12.1.2: For $f \in C([a,b])$ and $g \in C^{n+1}([a,b])$ we have $f(x) - f(t) = (f(x) - g(x)) + (g(x) - g(t)) + (g(t) - f(t))$. Integrating relative to μ we obtain

$$\begin{aligned}\left|\int_{[a,b]} f\,d\mu - f(t)\right| &\leq \int_{[a,b]} |f(x) - g(x)|\mu(dx) \\ &\quad + |g(t) - f(t)| + \int_{[a,b]} |g(x) - g(t)|\mu(dx) \\ &\leq 2 \cdot \|f - g\|_\infty + \int_{[a,b]} |g(x) - g(t)|\mu(dx).\end{aligned}$$

Now assume $L_i g(t) = 0$ for $i = 1, \ldots, n$, then from (12.1.3) we have

$$g(x) - g(t) = L_{n+1}g(\xi) \cdot \int_t^x g_n(x,s)ds,$$

where ξ is between x, t. Hence $|g(x) - g(t)| \leq \|L_{n+1}g\|_\infty \cdot \tilde{N}_n(x,t)$. And from Lemma 12.1.5 following later, we obtain

$$|g(x) - g(t)| \leq \|L_{n+1}g\|_\infty \cdot \max\left\{\frac{\tilde{N}_n(b,t)}{b-t}, \frac{\tilde{N}_n(a,t)}{t-a}\right\} \cdot |x - t| \quad \text{for all } x \in [a,b].$$

Consequently,

$$\left|\int_{[a,b]} f\,d\mu - f(t)\right| \leq 2 \cdot \|f - g\|_\infty + \|L_{n+1}g\|_\infty \cdot \Lambda_n(t) \cdot d_1(t),$$

where $\Lambda_n(t)$, $d_1(t)$ as in (12.1.7), (12.1.6), respectively. Therefore we have proved (12.1.8)

$$\left|\int_{[a,b]} f\,d\mu - f(t)\right| \le 2\cdot \tilde{K}_{n+1}^{(t)}\left(f; \frac{\Lambda_n(t)\cdot d_1(t)}{2}\right).$$

Furthermore, consider $\tilde{f}(x) := N_n(x,t) \in C^{n+1}([a,b])$. This satisfies $L_{n+1}\tilde{f}(x) = 1$ with $\tilde{f}^{(i)}(t) = 0$, $i = 0,1,2,\dots,n$, so that $L_i\tilde{f}(t) = 0$, $i = 1,\dots,n$. Hence

$$\tilde{K}_{n+1}^{(t)}(\tilde{f},h) \le h\cdot \|L_{n+1}\tilde{f}\|_\infty = h.$$

That is,

$$2\cdot \tilde{K}_{n+1}^{(t)}\left(\tilde{f}; \frac{\Lambda_n(t)\cdot d_1(t)}{2}\right) \le \Lambda_n(t)\cdot d_1(t).$$

I.e., inequality (12.1.8) for $f = \tilde{f}$ leads to

$$\left|\int_{[a,b]} \tilde{f}\,d\mu\right| \le \Lambda_n(t)\cdot d_1(t). \tag{12.1.9}$$

Here we observe that

(i)

$$\begin{aligned}\left|\int_{[a,b]} \tilde{f}\,d\mu_0\right| &= \left|\int_{[a,b]} N_n(x,t)\mu_0(dx)\right| \\ &= \left(\frac{N_n(b,t)}{b-t}\right)\cdot d_1(t) = \left(\frac{\tilde{N}_n(b,t)}{b-t}\right)\cdot d_1(t) \\ &= \Lambda_n(t)\cdot d_1(t),\end{aligned}$$

i.e., inequality (12.1.9) is attained.

(ii)

$$\begin{aligned}\left|\int_{[a,b]} \tilde{f}\,d\mu_0\right| &= \left|\left(\frac{N_n(a,t)}{t-a}\right)\cdot d_1(t)\right| \\ &= \left(\frac{\tilde{N}_n(a,t)}{t-a}\right)\cdot d_1(t) = \Lambda_n(t)\cdot d_1(t),\end{aligned}$$

i.e., inequality (12.1.9) is again attained.

We have proved in both cases that inequality (12.1.8) is attained, i.e., it is sharp. □

From Lemmas 12.1.(2–3) following later and the proof of Theorem 12.1.1 (vii) we obtain the following independent result.

Theorem 12.1.3. *Assume that $u_0(x) = c > 0$ and $u_1(x)$ is a concave function for $x \leq t$ and a convex function for $x \geq t$; $x, t \in [a, b]$, t is a fixed point. Let*

$$\varphi_n(x) := N_n(x,t) := \int_t^x g_n(x,s)ds, \quad n \geq 0.$$

Consider the algebra $A_n = \langle\{1, \varphi_n(x)\}\rangle$, which is a linear subspace of $C^{n+1}([a,b])$. Then $\overline{A}_n = C([a,b])$, $n \geq 0$, in the uniform norm. Furthermore

$$L_i e(t) = 0, \quad i = 1, \ldots, n \ \text{ for all } e \in A_n; \ n \geq 1.$$

12.1.3. Auxiliary Results

The next results are by themselves independently interesting.

Lemma 12.1.1. *Let*

$$N_n(x,t) := \int_t^x g_n(x,s)ds, \quad \text{all } x, t \in [a,b].$$

Then

$$\frac{\partial N_n(x,t)}{\partial x} = \int_t^x \frac{\partial g_n(x,s)}{\partial x} ds, \quad n \geq 1 \tag{12.1.10}$$

and

$$\frac{\partial^2 N_n(x,t)}{\partial x^2} = \int_t^x \frac{\partial^2 g_n(x,s)}{\partial x^2} ds, \quad n \geq 2. \tag{12.1.11}$$

PROOF: See Widder (1928), p. 132(6) and apply Leibnitz's formula once/twice. □

The previous result is used in the following.

Lemma 12.1.2. *Assume that $u_0(x) = c > 0$ and $u_1(x)$ is a convex function for $x \geq t$. Let*

$$N_n(x,t) := \int_t^x g_n(x,s)ds, \quad \text{all } x, t \in [a,b], \ n \geq 0.$$

Then $N_n(x,t) > 0$ for $x > t$, $N_n(t,t) = 0$, and it is a strictly increasing function in $x \geq t$, also $N_n(x,t)$ is a continuous function in $x \in [a,b]$.

Furthermore, $N_n(x,t)$ is a strictly convex function in $x \geq t$, $n \geq 1$, and $N_0(x,t) = x - t$ is a trivially convex function in $x \geq t$.

PROOF: From $W_0(x) = \phi_0(x) = u_0(x) = c > 0$ and $W_1(x) = cu_1'(x) > 0$, $u_1(x)$ is a strictly increasing function everywhere on $[a,b]$. Hence $\phi_1(x) = W_1(x)/(W_0(x))^2 = u_1'(x)/c > 0$. By assumption $u_1(x)$ is a convex function in $x \geq t$ implying that $u_1'(x)$ is an increasing function there, that is, $\phi_1(x)$ is increasing in $x \geq t$.

Note that $g_n(x,t) > 0$ $(x > t)$, $g_n(t,t) = 0$; $n \geq 1$ with $g_0(x,t) = 1$. We have

$$\frac{\partial g_n(x,t)}{\partial x} = \frac{\phi_1(x)}{\phi_1(t)\cdots\phi_n(t)} \cdot \int_t^x \phi_2(x_1)\cdots\int_t^{x_{n-2}} \phi_n(x_{n-1})dx_{n-1}\cdots dx_1.$$

From $\phi_i(x) > 0$, $i = 1, \ldots, n$, $n \geq 2$ and $\phi_1(x)$ being an increasing function we have that $\partial g_n(x,t)/\partial x$ is a strictly increasing function in $x \geq t$, note that $\partial g_n(x,t)/\partial x > 0$ $(x > t)$, $\partial g_n(t,t)/\partial x = 0$.

Therefore $g_n(x,t)$ is a strictly convex function in $x \geq t$, $n \geq 2$ and clearly $g_1(x,t)$ is convex in $x \geq t$.

Obviously $N_n(x,t)$ is a continuous function in $x \in [a,b]$, $n \geq 0$.

From Lemma 12.1.1

$$\frac{\partial^i N_n(x,t)}{\partial x^i} = \int_t^x \frac{\partial^i g_n(x,s)}{\partial x^i} ds, \quad (x \geq t, n \geq 2),\ i = 1,2.$$

It is clear that $N_n(x,t)$ is a strictly increasing function in $x \geq t$, $n \geq 2$.

By strict convexity of $g_n(x,s)$ in $x \geq s$ we get $\partial^2 g_n(x,s)/\partial x^2 > 0$ $(x > s)$, which leads to

$$\frac{\partial^2 N_n(x,t)}{\partial x^2} > 0 \ (x > t), \quad \frac{\partial^2 N_n(t,t)}{\partial x^2} = 0.$$

Thus $N_n(x,t)$ is a strictly convex function in $x \geq t$, $n \geq 2$.

Since $g_0(x,t) = 1$, all $x,t \in [a,b]$, $N_0(x,t) = x - t$, $(x \geq t)$. Obviously $N_0(x,t)$ is a trivially convex and strictly increasing function in $x \geq t$. Note that

$$g_1(x,t) = \phi_1^{-1}(t) \cdot \int_t^x \phi_1(s)ds.$$

From $\partial g_1(x,t)/\partial x = \phi_1(x)/\phi_1(t)$ and ϕ_1 an increasing function, we have that $\partial g_1(x,t)/\partial x$ is increasing in $x \geq t$. Obviously $\partial g_1(x,t)/\partial x > 0$ for all $x \in [a,b]$.

Let s be such that $t \leq s \leq x_1 < x_2$, then

$$\frac{\partial g_1(x_2,s)}{\partial x} \geq \frac{\partial g_1(x_1,s)}{\partial x}.$$

Adding

$$\int_t^{x_1} \frac{\partial g_1(x_2,s)}{\partial x} ds \geq \int_t^{x_1} \frac{\partial g_1(x_1,s)}{\partial x} ds$$

and

$$\int_{x_1}^{x_2} \frac{\partial g_1(x_2,s)}{\partial x} ds > 0,$$

one has

$$\int_t^{x_2} \frac{\partial g_1(x_2,s)}{\partial x} ds > \int_t^{x_1} \frac{\partial g_1(x_1,s)}{\partial x} ds.$$

The last inequality and Lemma 12.1.1 (12.1.10) imply that $\partial N_1(x,t)/\partial x$ is strictly increasing in $x \geq t$, which in turn implies that $N_1(x,t)$ is a strictly convex function in $x \geq t$. Since

$$\frac{\partial N_1(x,t)}{\partial x} > 0 \ (x > t), \quad \frac{\partial N_1(t,t)}{\partial x} = 0$$

we conclude that $N_1(x,t)$ is a strictly increasing function in $x \geq t$. □

The counterpart of Lemma 12.1.2 follows:

Lemma 12.1.3. *Assume that $u_0(x) = c > 0$ and $u_1(x)$ is a concave function for $x \leq t$. Let*

$$N_n(x,t) := \int_t^x g_n(x,s)ds, \quad \text{all } x,t \in [a,b]\text{: } x \leq t, \ n \geq 0.$$

If n is odd, then $N_n(x,t)$ is a strictly decreasing and a strictly convex function in $x \leq t$, $N_n(x,t) > 0$ for $x < t$.

If n is even, then $N_n(x,t)$ is a strictly increasing and a strictly concave function in $x \leq t$. Furthermore $N_0(x,t) = x - t$ is a strictly increasing and a trivially concave function in $x \leq t$.

Also $N_n(x,t) < 0$ $(x < t)$ for n zero or even; with $N_n(t,t) = 0$ all $n \geq 0$.

PROOF: By assumption $u_1(x)$ is a concave function in $x \leq t$ implying that $u_1'(x)$ is a decreasing function there, that is, $\phi_1(x)$ is decreasing in $x \leq t$.

Observe that for $n \geq 1$

$$\begin{aligned}\frac{\partial g_n(x,t)}{\partial x} &= \frac{\phi_1(x)}{\phi_1(t)\cdots\phi_n(t)} \cdot \int_t^x \phi_2(x_1) \int_t^{x_1} \cdots \int_t^{x_{n-2}} \phi_n(x_{n-1})dx_{n-1}\cdots dx_1 \\ &= (-1)^{n-1} \cdot \frac{\phi_1(x)\cdot B(x,t)}{\phi_1(t)\cdots\phi_n(t)},\end{aligned}$$

where

$$B(x,t) := \int_x^t \phi_2(x_1) \int_{x_1}^t \cdots \int_{x_{n-2}}^t \phi_n(x_{n-1}) dx_{n-1} \cdots dx_1 > 0 \ (x < t),$$
$$B(t,t) := 0.$$

Since $B(x,t)$ is a strictly decreasing function in $x \leq t$, we have that $\phi_1(x) \cdot B(x,t)$ is also strictly decreasing in $x \leq t$. When $n > 1$ is odd then $\partial g_n(x,t)/\partial x > 0$ $(x < t)$, $\partial g_n(t,t)/\partial x = 0$ and it is a strictly decreasing function in $x \leq t$.

When n is even then $\partial g_n(x,t)/\partial x < 0$ $(x < t)$, $\partial g_n(t,t)/\partial x = 0$ and it is a strictly increasing function in $x \leq t$.

We have proved that for n odd, $g_n(x,t) < 0$ $(x < t)$, $g_n(t,t) = 0$ and it is strictly concave in $x \leq t$ for $n > 1$; clearly $g_1(x,t)$ is concave in $x \leq t$. Also for n even, $g_n(x,t) > 0$ $(x < t)$, $g_n(t,t) = 0$ and it is strictly convex in $x \leq t$.

From Lemma 12.1.1

$$\frac{\partial^i N_n(x,t)}{\partial x^i} = \int_t^x \frac{\partial^i g_n(x,s)}{\partial x^i} ds, \ (x \leq t, n \geq 2), \ i = 1,2.$$

It is clear that when $n > 2$ is odd, then $N_n(x,t)$ is a strictly decreasing and a strictly convex function in $x \leq t$.

And when n is even, then $N_n(x,t)$ is a strictly increasing and a strictly concave function in $x \leq t$. Note that for $n \geq 1$ odd, $N_n(x,t) > 0$ and for n zero or even, $N_n(x,t) < 0$, where $x < t$, with $N_n(t,t) = 0$ all $n \geq 0$. Obviously $N_0(x,t) = x - t$ is a trivially concave and a strictly increasing function in $x \leq t$.

From $\partial g_1(x,t)/\partial x = \phi_1(x)/\phi_1(t)$ and ϕ_1 a decreasing function, we get that $\partial g_1(x,t)/\partial x$ is decreasing in $x \leq t$. Obviously, $\partial g_1(x,t)/\partial x > 0$ for all $x \in [a,b]$. Let s be such that $x_1 < x_2 \leq s \leq t$, then

$$\frac{\partial g_1(x_1,s)}{\partial x} \geq \frac{\partial g_1(x_2,s)}{\partial x}.$$

Adding

$$\int_{x_2}^t \frac{\partial g_1(x_1,s)}{\partial x} ds \geq \int_{x_2}^t \frac{\partial g_1(x_2,s)}{\partial x} ds$$

and

$$\int_{x_1}^{x_2} \frac{\partial g_1(x_1,s)}{\partial x} ds > 0$$

one has

$$\int_{x_1}^{t} \frac{\partial g_1(x_1,s)}{\partial x} ds > \int_{x_2}^{t} \frac{\partial g_1(x_2,s)}{\partial x} ds$$

or

$$\int_{t}^{x_1} \frac{\partial g_1(x_1,s)}{\partial x} ds < \int_{t}^{x_2} \frac{\partial g_1(x_2,s)}{\partial x} ds.$$

The last inequality and Lemma 12.1.1 (12.1.10) imply that $\partial N_1(x,t)/\partial x$ is strictly increasing in $x \leq t$, which says that $N_1(x,t)$ is a strictly convex function in $x \leq t$. Since

$$\frac{\partial N_1(x,t)}{\partial x} < 0 \ (x < t), \ \frac{\partial N_1(t,t)}{\partial x} = 0$$

we conclude that $N_1(x,t)$ is a strictly decreasing function in $x \leq t$. □

Lemmas 12.1.2 and 12.1.3 enable us to conclude

Lemma 12.1.4. *Assume that* $u_0(x) = c > 0$ *and* $u_1(x)$ *is a concave function for* $x \leq t$ *and a convex function for* $x \geq t$.

Let $\tilde{N}_n(x,t) := |N_n(x,t)|$, *where*

$$N_n(x,t) := \int_t^x g_n(x,s)ds, \quad \textit{all } x,t \in [a,b], \ n \geq 0.$$

Then for $n \geq 1$, $\tilde{N}_n(x,t)$ *is a strictly decreasing function in* $x \leq t$ *and a strictly increasing function in* $x \geq t$, *furthermore it is a continuous and a strictly convex function in* $x \in [a,b]$.

Obviously $\tilde{N}_0(x,t) = |x-t|$ *possesses the above properties, but is a convex function in* $x \in [a,b]$. *In particular,* $\tilde{N}_n(x,t) > 0$ *for* $x \neq t$, *with* $\tilde{N}_n(t,t) = 0$, *all* $n \geq 0$.

Lemma 12.1.4 implies the next result which is used in the proof of Theorem 12.1.2.

Lemma 12.1.5. *Under the assumptions of Lemma* 12.1.4 *for fixed* $t \in (a,b)$, *we obtain that*

$$\tilde{N}_n(x,t) \leq \max\left\{\frac{\tilde{N}_n(b,t)}{b-t}, \frac{\tilde{N}_n(a,t)}{t-a}\right\} \cdot |x-t|,$$

all $x \in [a,b]$; *for all* $n \geq 1$. *Equality can be true only at* $x = t$ *and at* $x = a$ *or* b.

The above inequality, for $n = 0$, becomes identity.

PROOF: When $t < x < b$ by strict convexity of $\tilde{N}_n(x,t)$, $n \geq 1$; $\tilde{N}_n(t,t) = 0$ we have

$$\frac{\tilde{N}_n(x,t)}{x-t} < \frac{\tilde{N}_n(b,t)}{b-t}.$$

Hence

$$\begin{aligned}\tilde{N}_n(x,t) &< \left(\frac{\tilde{N}_n(b,t)}{b-t}\right) \cdot (x-t) \\ &\leq \max\left\{\frac{\tilde{N}_n(b,t)}{b-t}, \frac{\tilde{N}_n(a,t)}{t-a}\right\} \cdot (x-t).\end{aligned}$$

And when $a < x < t$, again by strict convexity of $\tilde{N}_n(x,t)$ we get

$$\frac{\tilde{N}_n(a,t)}{a-t} < \frac{\tilde{N}_n(x,t)}{x-t}.$$

Thus

$$\begin{aligned}\tilde{N}_n(x,t) &< \left(\frac{\tilde{N}_n(a,t)}{t-a}\right) \cdot (t-x) \\ &\leq \max\left\{\frac{\tilde{N}_n(b,t)}{b-t}, \frac{\tilde{N}_n(a,t)}{t-a}\right\} \cdot (t-x). \qquad \square\end{aligned}$$

CHAPTER THIRTEEN

13.1A GENERAL STOCHASTIC INEQUALITY INVOLVING BASIC MOMENTS

13.1.1. Introduction

Let $(\underline{\Omega}, F, P)$ be an arbitrary probability space and $(X_n)_{n\in\mathbb{N}}$: $\underline{\Omega} \to \mathbb{R}$ be a sequence of real, independent, not necessarily identically distributed random variables with distribution functions F_{X_n}, and $S_n = \sum_{i=1}^{n} X_i$.

Here is presented a general limit theorem with applications, for the weak convergence of the sums $\varphi(n) \cdot S_n$ (where $\varphi: \mathbb{N} \to \mathbb{R}^+$, $\varphi(n) \to 0$, $n \to \infty$) to a suitable limiting random variable X. These convergence results are given quantitatively: by employing a higher order measure of smoothness for the involved continuous real valued function f defined on the real axis $\mathbb{R}$, which measure of smoothness is evaluated at some mixture of associated moments and the function φ.

The presented inequality is sharper than the main inequality due to Butzer and Hahn(1) (1978), see relation (3.3), for the very typical case of $\beta = 1$.

The corresponding improvement is by a factor $2 \cdot r^{-1}$, $r \in \mathbb{N} - \{1, 2\}$ and our proof is carried out by quite different means.

This chapter's material comes from Anastassiou (1983).

13.1.2. Background

Let $C = C(\mathbb{R})$ be the space of real valued continuous functions defined on $\mathbb{R}$ and $C_B = C_B(\mathbb{R})$ the subspace of those bounded and uniformly continuous, endowed with norm

$$\|f\| = \sup_{x\in\mathbb{R}} |f(x)|.$$

For $r \in Z^+$ define

$$C_B^r = \{f \in C_B; f', f'', \ldots, f^{(r)} \in C_B\}.$$

To measure the smoothness of functions we use: the K-functional, which for $t \geq 0$ is defined by

$$K_r(t, f, C_B) = \inf_{g \in C_B^r} \{\|f - g\| + t\|g^{(r)}\|\}, \qquad r \in \mathbb{N} \tag{13.1.1}$$

(for the properties of K-functional see Butzer and Berens (1967).

Further, the rth modulus of continuity which is defined as, see also Butzer and Berens (1967)).

$$\omega_r(t, f; C_B) := \sup_{|h| \leq t} \left\| \sum_{k=0}^{r} (-1)^{r-k} \binom{r}{k} \cdot f(u + k \cdot h) \right\|, \qquad r \in \mathbb{N}. \tag{13.1.2}$$

The following fact is true, see Butzer and Berens (1967), pp. 192 and 258: for $f \in C_B$, $t \geq 0$ there are universal constants $C_{1,r}, C_{2,r} > 0$ such that

$$C_{1,r}\omega_r(t^{(1/r)}, f; C_B) \leq K_r(t, f; C_B) \leq C_{2,r} \cdot \omega_r(t^{(1/r)}, f; C_B), \tag{13.1.3}$$

establishing the equivalence of two measures of smoothness.

For $r \in \mathbb{N}$, $0 < \alpha \leq r$ another important class of functions is

$$\text{Lip}(\alpha, r, C_B) = \{f \in C_B; \omega_r(t, f, C_B) \leq L_f t^{\alpha}\}. \tag{13.1.4}$$

Note that for $\alpha = r' + \beta$, $r' \leq r - 1$, $0 < \beta \leq 1$ one has the fact, see Junggeburth et al (1973):

$$f \in \text{Lip}(r' + \beta, r, C_B) \Longleftrightarrow f^{(r')} \in \text{Lip}(\beta, r - r', C_B). \tag{13.1.5}$$

If z is any real random variable the operator $V_z : C_B \to C_B$, used by Trotter (1959), is defined by

$$(V_z f)(y) := \int f(x + y) dF_z(x), \qquad y \in \mathbb{R}. \tag{13.1.6}$$

For the properties of V_z see Feller (1966), p. 248ff. Namely, V_z has the contraction property

$$\|V_z f\| \leq \|f\|, \tag{13.1.7}$$

for any f which is F_z-integrable.

Also, the operators V_{z_1}, V_{z_2} commute and $V_{z_1} = V_{z_2}$ when z_1, z_2 are identically distributed (i.d.). For the independent real random variables $z_1, z_2, \ldots, z_n$ we get

$$V_{\sum_{i=1}^n z_i} f = V_{z_1} \circ \cdots \circ V_{z_n}(f) \tag{13.1.8}$$

for all f integrable with respect to $F_{\sum_{i=1}^{n} z_i}$ (by Fubini's theorem).

If $z'_1, \ldots, z'_n$ are independent, then for all f integrable with respect to $F_{\sum_{i=1}^{n}(z_i+z'_i)}$ we have

$$\left\| V_{\sum_{i=1}^{n} z_i} f - V_{\sum_{i=1}^{n} z'_i} f \right\| \leq \sum_{i=1}^{n} \|V_{z_i} f - V_{z'_i} f\| \tag{13.1.9}$$

(by the use of (13.1.7) and (13.1.8)).

Let $(X_i)_{i \in \mathbb{N}}$ be a sequence of real independent r.v., the corresponding limiting r.v. X is assumed to be φ-decomposable into the independent components $z_{i,n}$, $(1 \leq i \leq n)$, $n \in \mathbb{N}$ (define $\varphi: \mathbb{N} \to \mathbb{R}^+$ such that $\varphi(n) = 0(1)$ as $n \to +\infty$, then X is called *φ-decomposable* if $F_X = F_{\varphi(n) \cdot \sum_{i=1}^{n} z_{i,n}}$, for all $n \in \mathbb{N}$).

We set

$$G_{i,n}(x) := (F_{x_i} - F_{z_{i,n}})(x), \tag{13.1.10}$$

$$\mu_{j,i,n} := \int_{\mathbb{R}} x^j \cdot dG_{i,n}(x), \quad V_{j,i,n} := \int_{\mathbb{R}} |x|^j \cdot d|G_{i,n}(x)|,$$

where $|G_{i,n}(x)|$ denotes the total variation of $G_{i,n}(x)$ over $(-\infty, x]$. Here we need to assume that the $s \leq r$, $r \in \mathbb{N}$ moments of $G_{i,n}(x)$ are finite.

Also, we consider the r.v. $T_n := \varphi(n) \cdot (\sum_{i=1}^{n} x_i)$.

We will use the following, see Lemma 7.1.1.

Lemma 13.1.1. *Let D be a subset of the real normed vector space $V = (V, \|\cdot\|)$ which is star-shaped relative to the fixed point x_0 and consider $f: D \to \mathbb{R}$ with the properties:*

$$f(x_0) = 0 \tag{13.1.11}$$

and

$$\|s - t\| \leq h \Rightarrow |f(s) - f(t)| \leq w; \qquad w, h > 0.$$

Then there is a maximal such function $\emptyset$, namely

$$\emptyset(t) = \lceil \|t - x_0\|/h \rceil \cdot w, \tag{13.1.12}$$

where $\lceil \cdot \rceil$ denotes the ceiling of the number.

Note 13.1.1. Let $h > 0$ be fixed. We often use the even function defined by

$$\emptyset_n(x) := \int_0^{|x|} \left\lceil \frac{t}{h} \right\rceil \cdot \frac{(x-t)^{n-1}}{(n-1)!} \cdot dt, \qquad (x \in \mathbb{R}; n \in \mathbb{N}).$$

We find, see relation (7.1.18), that:

$$\emptyset_n(x) \leq \left(\frac{|x|^{n+1}}{(n+1)!h} + \frac{|x|^n}{2n!} + \frac{h|x|^{n-1}}{8(n-1)!} \right) \tag{13.1.13}$$

with equality true only at $x = 0$.

13.1.3. Results

The following theorem is our main result.

Theorem 13.1.1. *Let $r \in \mathbb{N}$ and*

$$V_{r,i,n} := \int_{\mathbb{R}} |x|^r d|G_{i,n}(x)| < \infty; \qquad i, n \in \mathbb{N}. \tag{13.1.14}$$

Further, assume that there are constants G_j such that

$$\varphi(n)^{j-r} \cdot \left(\sum_{i=1}^{n} |\mu_{j,i,n}| \right) \leq G_j \cdot \left(\sum_{i=1}^{n} V_{r,i,n} \right), \quad (0 \leq j \leq r-1). \tag{13.1.15}$$

Then for any $f \in \mathrm{Lip}(r, r, C_B)$, we obtain

$$\|V_{T_n} f - V_X f\| \leq \left(C_f + \frac{2 \cdot L_f}{r!} \right) \cdot (\varphi(n))^r \cdot \left(\sum_{i=1}^{n} V_{r,i,n} \right), \tag{13.1.16}$$

where

$$C_f := \sum_{j=0}^{r-1} C_j \cdot \frac{\|f^{(j)}\|}{j!}$$

and L_f is the Lipschitz constant of $f^{(r-1)}$.

PROOF: Since $f \in C_B^{(r-1)}$, for a fixed y we have

$$f(x+y) = \sum_{j=0}^{r-1} \frac{f^{(j)}(y)}{j!} \cdot x^j$$

$$+ \int_y^{x+y} \left(f_{(t)}^{(r-1)} - f_{(y)}^{(r-1)} \right) \cdot \frac{(x+y-t)^{r-2}}{(r-2)!} \cdot dt. \tag{13.1.17}$$

Because of $|\mu_{j,i,n}| < \infty$ for $0 \le j \le r-1$, $i, n \in \mathbb{N}$ by the definition of Trotter operator (13.1.6) and (13.1.17) we obtain

$$(V_{\varphi(n)X_i}f)(y) - (V_{\varphi(n)\cdot z_{i,n}}f)(y) = \int_{\mathbb{R}} f(\varphi(n)\cdot x + y)dG_{i,n}(x)$$

$$= \sum_{j=0}^{r-1} \frac{f^{(j)}(y)}{j!} \cdot (\varphi(n))^j \cdot \mu_{j,i,n} + \int K_{r-1,i,n}(x,y)dG_{i,n}(x) \tag{13.1.18}$$

where

$$K_{r-1,i,n}(x,y) := \int_y^{(\varphi(n)\cdot x+y)} (f^{(r-1)}_{(t)} - f^{(r-1)}_{(y)}) \cdot \frac{(\varphi(n)\cdot x + y - t)^{r-2}}{(r-2)!} \cdot dt. \tag{13.1.19}$$

For given $h_{i,n} > 0$ consider the first modulus of continuity $w_{i,n} := \omega_1(h_{i,n}, f^{(r-1)}; C_B)$. Then for the function

$$\Phi_{(r-1)}(t) := f^{(r-1)}_{(t)} - f^{(r-1)}_{(y)}$$

we see

$$\Phi_{(r-1)}(y) = 0$$

and by Lemma 13.1.1 we get

$$|\Phi_{(r-1)}(t)| \le w_{i,n} \cdot \left\lceil \frac{|t-y|}{h_{i,n}} \right\rceil. \tag{13.1.20}$$

Using (13.1.20) to estimate (13.1.19) we obtain

$$|K_{r-1,i,n}(x,y)| \le \frac{w_{i,n}}{(r-1)!} \cdot (\varphi(n)\cdot|x|)^{r-1} \cdot \left(1 + \frac{\varphi(n)\cdot|x|}{r\cdot h_{i,n}}\right). \tag{13.1.21}$$

However

$$\|V_{\varphi(n)X_i}f - V_{\varphi(n)z_{i,n}}f\| \le \sum_{j=0}^{r-1} \frac{\|f^{(j)}\|}{j!} \cdot (\varphi(n))^j \cdot |\mu_{j,i,n}| + L_{r-1,i,n}, \tag{13.1.22}$$

where

$$L_{r-1,i,n} := \int \|K_{r-1,i,n}(x,y)\|_{(y)} \cdot d|G_{i,n}(x)|. \tag{13.1.23}$$

Furthermore,

$$L_{r-1,i,n} \le \int \frac{w_{i,n}}{(r-1)!}(\varphi(n))^{r-1} \cdot |x|^{r-1} \cdot \left(1 + \frac{\varphi(n)\cdot|x|}{r\cdot h_{i,n}}\right) \cdot d|G_{i,n}(x)|$$

$$= w_{i,n} \cdot \frac{\varphi(n)^{r-1}}{(r-1)!} \cdot \left[V_{r-1,i,n} + \frac{\varphi(n)}{r \cdot h_{i,n}} \cdot V_{r,i,n} \right] =: (*). \tag{13.1.24}$$

Let $m_{i,n}$ be the mass of the total variation $|G_{i,n}|$.

To prove the theorem it is enough to consider those $m_{i,n} \neq 0$. By Hölder's inequality we get

$$V_{r-1,i,n} \leq V_{r,i,n}^{(r-1)/r} \cdot m_{i,n}^{(1/r)}$$

and we are using it in (13.1.24).

Therefore

$$(*) \leq w_{i,n} \cdot \frac{\varphi(n)^{r-1}}{(r-1)!} \cdot V_{r,i,n}^{(r-1)/r} \left[m_{i,n}^{(1/r)} + \frac{\varphi(n)}{r \cdot h_{i,n}} \cdot V_{r,i,n}^{(1/r)} \right].$$

Choosing

$$h_{i,n} := \frac{\varphi(n)}{r \cdot m_{i,n}^{(1/r)}} \cdot (V_{r,i,n}^{(1/r)})$$

we obtain

$$L_{r-1,i,n} \leq 2 \cdot w_{i,n} \cdot m_{i,n}^{(1/r)} \cdot \frac{\varphi(n)^{r-1}}{(r-1)!} \cdot V_{r,i,n}^{(r-1)/r}, \tag{13.1.25}$$

where $m_{i,n} := |G_{i,n}|(\mathbb{R})$ is finite.

Eventually,

$$\|V_{\varphi(n)X_i} f - V_{\varphi(n)\cdot z_{i,n}} f\| \leq \sum_{j=0}^{r-1} \frac{\|f^{(j)}\|}{j!} \cdot (\varphi(n))^j \cdot |\mu_{j,i,n}|$$

$$+ 2 \cdot w_{i,n} \cdot m_{i,n}^{(1/r)} \cdot \frac{\varphi(n)^{r-1}}{(r-1)!} \cdot V_{r,i,n}^{(r-1)/r}. \tag{13.1.26}$$

Using (13.1.9) and

$$w_{i,n} = \omega_1(h_{i,n}, f^{(r-1)}; C_B) \leq L_f \cdot h_{i,n} = L_f \cdot \frac{\varphi(n)}{r \cdot m_{i,n}^{(1/r)}} \cdot (V_{r,i,n}^{(1/r)})$$

and setting

$$\lambda_{j,i,n} := \frac{\|f^{(j)}\|}{j!} \cdot (\varphi(n))^j \cdot |\mu_{j,i,n}|$$

we find the following

$$\|V_{T_n} f - V_X f\| \leq \sum_{i=1}^{n} \|V_{\varphi(n)\cdot X_i} f - V_{\varphi(n)\cdot z_{i,n}} f\|$$

$$
\begin{aligned}
&\leq \sum_{j=0}^{r-1}\left(\sum_{i=1}^{n}\lambda_{j,i,n}\right) \\
&\qquad + 2\cdot\frac{\varphi(n)^{r-1}}{(r-1)!}\left(\sum_{i=1}^{n} w_{i,n}\cdot m_{i,n}^{(1/r)}\cdot V_{r,i,n}^{(r-1)/r}\right) \\
&= \sum_{j=0}^{r-1}\left(\sum_{i=1}^{n}\lambda_{j,i,n}\right) \\
&\qquad + 2\cdot\frac{\varphi(n)^{r}}{r!}\cdot L_f\cdot\left(\sum_{i=1}^{n} V_{r,i,n}\right) \\
&\leq \left(C_f+\frac{2\cdot L_f}{r!}\right)\cdot\varphi(n)^r\cdot\left(\sum_{i=1}^{n} V_{r,i,n}\right)
\end{aligned}
$$

from (13.1.15). □

Remark 13.1.1. The coefficient 2 in the right-hand side of (13.1.16) could be replaced by the finer one

$$\left[1+\frac{1}{2}+\frac{(r-1)}{8\cdot r}\right], \qquad r\in\mathbb{N}$$

which is always less than 2 and is obtained in a similar proof by using relation (13.1.13) (instead of (13.1.21) and Hölder's inequality).

However in applications we keep the factor 2 as more comfortable.

As an application of Theorem 13.1.1 we have

Theorem 13.1.2. *Let $r \in \mathbb{N}$, $0 < \alpha \leq r$ with $V_{r,i,n} < \infty$ $(i, n \in \mathbb{N})$, as well as $\mu_{j,i,n} = 0$, $(0 \leq j \leq r-1;\ i, n \in \mathbb{N})$.*

Then for $f \in C_B$ we get

$$\|V_{T_n}f - V_X f\| \leq 2\cdot C_{2,r}\cdot\omega_r\left(\left[\frac{\varphi(n)^r}{r!}\cdot\left(\sum_{i=1}^{n}V_{r,i,n}\right)\right]^{(1/r)}, f; C_B\right). \tag{13.1.27}$$

Especially for $f \in \mathrm{Lip}(\alpha, r, C_B)$ we obtain

$$\|V_{T_n}f - V_X f\| \leq 2\cdot C_{2,r}\cdot L_f\cdot(\varphi(n))^{\alpha}\cdot\left(\frac{\sum_{i=1}^{n}V_{r,i,n}}{r!}\right)^{(\alpha/r)}. \tag{13.1.28}$$

Proof: Consider $f \in C_B$, $g \in C_B^r$ as arbitrary.

Then

$$\|V_{T_n}f - V_X f\| \le \|V_{T_n}f - V_{T_n}g\| + \|V_{T_n}g - V_X g\| + \|V_X g - V_X f\| =: A_1 + A_2 + A_3,$$

by Trotter's operator contraction property (13.1.7)

$$A_1 + A_3 \le 2\|f - g\|.$$

Since

$$g^{(r-1)} \in C_B^1 \Rightarrow g \in \mathrm{Lip}(r, r, C_B) \ \text{ with } L_g = \|g^{(r)}\|.$$

By Theorem 13.1.1 we obtain

$$\begin{aligned}
A_2 = \|V_{T_n}g - V_X g\| &\le \left(\frac{2}{r!}\right) \cdot \|g^{(r)}\| \cdot \varphi(n)^r \cdot \left(\sum_{i=1}^{n} V_{r,i,n}\right) \\
\Rightarrow \|V_{T_n}f - V_X f\| &\le 2 \cdot \left(\|f - g\| + \frac{\varphi(n)^r}{r!} \cdot \left(\sum_{i=1}^{n} V_{r,i,n}\right) \cdot \|g^{(r)}\|\right) \\
&\le 2 \cdot K_r \left(\left[\frac{\varphi(n)^r}{r!} \cdot \sum_{i=1}^{n} V_{r,i,n}\right], f; C_B\right) \\
&\le 2 \cdot C_{2,r} \cdot \omega_r \left(\left[\frac{\varphi(n)^r}{r!} \cdot \sum_{i=1}^{n} V_{r,i,n}\right]^{(1/r)}, f; C_B\right). \quad \square
\end{aligned}$$

Next consider X_i as i.d. and also $z_{i,n}$ as i.d. with respect to i and n.

Corollary 13.1.1. *Let $r \in \mathbb{N}$, $V_r := \int |x|^r d|G(x)| < \infty$ and $\mu_j := \int x^j dG(x) = 0$, $0 \le j \le r-1$. Then for any $f \in \mathrm{Lip}(r, r, C_B)$ we get*

$$\|V_{T_n}f - V_X f\| \le \left(\frac{2 \cdot L_f}{r!}\right) \cdot (n \cdot \varphi(n)^r \cdot V_r). \tag{13.1.29}$$

Corollary 13.1.2. *Let $r \in \mathbb{N}$, $0 < \alpha \le r$ and suppose that $V_r < \infty$ and $\mu_j = 0$.*

(i) *Then for any $f \in C_B$ we get*

$$\|V_{T_n}f - V_X f\| \le 2 \cdot C_{2,r} \cdot \omega_r \left(\left[\frac{n \cdot \varphi(n)^r}{r!} \cdot V_r\right]^{(1/r)}, f; C_B\right). \tag{13.1.30}$$

(ii) *Also, for any $f \in \mathrm{Lip}(a, r, C_B)$ we obtain*

$$\|V_{T_n}f - V_X f\| \le 2 \cdot C_{2,r} \cdot L_f \cdot \left[\varphi(n) \cdot \left(\frac{n \cdot V_r}{r!}\right)^{(1/r)}\right]^{\alpha}. \tag{13.1.31}$$

Conclusion 13.1.1. Many other applications of Theorem 13.1.1 could follow, in fact all the related results in Butzer and Hahn(1), (2), (1978) and Hahn (1982) are improved by the favorable factor $2r^{-1}$, $r \geq 2$.

These are general probabilistic limit theorems with rates having applications: to stable limit laws, such as the central limit theorem, also to the weak law of large numbers. Other applications of them are given to the estimation of the speed of convergence of sequences of well-known positive linear operators (probabilistically arising) to the identity operator: such as the Bernstein polynomials, the Baskakov operators, the Weierstrass operators, the Szász–Mirakjan and Hahn operators.

CHAPTER FOURTEEN

14.1. MISCELLANEOUS SHARP INEQUALITIES AND KOROVKIN-TYPE CONVERGENCE THEOREMS INVOLVING SEQUENCES OF BASIC MOMENTS

14.1.1. Introduction

Here we present a generalization of a theorem due to Korovkin (1958) to sequences of arbitrary probability measures on $[0, \pi]$. Korovkin's result is concerned with the convergence of certain ratios of the Fourier coefficients of a sequence of density functions. Earlier, Stark (1972) gave a different generalization of this Korovkin theorem.

Analogous characterizations are given for the same type of ratios of the hyperbolic coefficients (respectively, the Laplace transforms) of a sequence of probability measures on $\mathbb{R}$ (respectively, on $\mathbb{R}^+$). In the course of the proofs we present various inequalities, on subsets of $\mathbb{R}$, leading to several sharp estimates. A number of related applications are given.

The material of this chapter comes from Anastassiou(3) (1985).

14.1.2. Fourier–Stieltjes Coefficients

Lemma 14.1.1. *For $k, l \geq 2$, $k, l \in \mathbb{N}$ there exists a positive constant $C(k,l) \geq [k^2(k^2 - 1)]/[l^2(l^2-1)]$ such that*

$$[k^2(1-\cos t) - (1-\cos kt)]$$

$$\leq C(k,l)[l^2(1-\cos t) - (1-\cos lt)], \quad \textit{all } t \in [0,\pi]. \tag{14.1.1}$$

PROOF: Since $|\sin nt| \leq n|\sin t|$, $n \in \mathbb{N}$, we have that for $t \in (0, \pi]$

$$n^2(1-\cos t) - (1-\cos nt) = \left(n^2 - \frac{\sin^2(nt/2)}{\sin^2(t/2)}\right) \cdot (1-\cos t) \geq 0.$$

The function

$$\varphi(t) := \frac{k^2(1-\cos t) - (1-\cos kt)}{l^2(1-\cos t) - (1-\cos lt)}$$

on $[0,\pi]$ satisfies

$$\lim_{t\to 0}\varphi(t) = \frac{k^2(k^2-1)}{l^2(l^2-1)} > 0,$$

so that $\varphi(t)$ is strictly positive and continuous, hence bounded. □

Remark 14.1.1. We conjecture that

$$C(k,l) = k^2(k^2-1)l^{-2}(l^2-1)^{-1} \quad \text{if } 2 \le l \le k.$$

It is correct for $k \le 5$ and the corresponding inequality is sharp.

Definition 14.1.1. Let μ be a probability measure on $[0,\pi]$. Its Fourier–Stieltjes coefficients are defined by

$$p_k := \int_0^\pi \cos kt\mu(dt) \qquad (k = 0,1,2,\ldots). \tag{14.1.2}$$

If $p_1 = 1$, then $\mu = \delta_0$.

Lemma 14.1.2. *Let μ be a probability measure on $[0,\pi]$ with Fourier–Stieltjes coefficients p_k, $k \in \mathbb{Z}^+$ and $p_1 \neq 1$.*

Then

$$\left[k^2 - \left(\frac{1-p_k}{1-p_1}\right)\right] \le C(k,l)\cdot\left[l^2 - \left(\frac{1-p_l}{1-p_1}\right)\right], \tag{14.1.3}$$

where $k,l \ge 2$, $k,l \in \mathbb{N}$.

PROOF: Integrate (14.1.1) relative to μ and divide both sides by $(1-p_1)$. □

Theorem 14.1.1. *Let $l \in \mathbb{N}$, $l \ge 2$. If $\{\mu_n\}_{n\in\mathbb{N}}$ is a sequence of probability measures on $[0,\pi]$ with Fourier–Stieltjes coefficients p_{kn} such that $p_{1n} \neq 1$ and $\lim_{n\to\infty}((1-p_{ln})/(1-p_{1n})) = l^2$, then $\lim_{n\to\infty}((1-p_{kn})/(1-p_{1n})) = k^2$ for all $k \ge 2$, $k \in \mathbb{N}$.*

PROOF: Use (14.1.3). □

Remark 14.1.2. In the sequel $k,l \in \mathbb{N}$ and $k,l \ge 2$. Let $g(s)$ be the characteristic function (Fourier transform) of a probability measure μ on $[0,\pi]$. We have that

$$\mathrm{Re}(1-g(s)) = \int_0^\pi (1-\cos st)\mu(dt), \qquad s \in \mathbb{R}.$$

Then by applying (14.1.1) we get

$$[k^2(1-\operatorname{Re} g(1))-(1-\operatorname{Re} g(k))] \leq C(k,l)[l^2(1-\operatorname{Re} g(1))-(1-\operatorname{Re} g(l))].$$

As an illustration, let $g(s)=|f(s)|^2$, where f is a characteristic function. Then also g is a characteristic function. Therefore

$$[k^2(1-|f(1)|^2)-(1-|f(k)|^2)] \leq C(k,l)[l^2(1-|f(1)|^2)-(1-|f(l)|^2)].$$

Consequently, if a sequence $\{f_n\}_{n\in\mathbb{N}}$ of characteristic functions with $|f_n(1)|<1$ satisfies

$$\lim_{n\to\infty}\left(\frac{1-|f_n(k)|^2}{1-|f_n(1)|^2}\right)=k^2 \quad \text{for one } k\in\mathbb{N}, k\geq 2$$

then for all such k.

14.1.3. Hyperbolic Coefficients

Now we proceed to a similar type of result.

Lemma 14.1.3. *Let $k\geq l>1$, $B(k,l):=l^2(l^2-1)k^{-2}(k^2-1)^{-1}$. Then*

$$[(\operatorname{Cos} hlt-1)-l^2(\operatorname{Cos} ht-1)] \leq B(k,l)[(\operatorname{Cos} hkt-1)-k^2(\operatorname{Cos} ht-1)] \tag{14.1.4}$$

for all $t\in\mathbb{R}$.

Here the constant $B(k,l)$ cannot be improved, i.e., (14.1.4) *is sharp.*

Proof: Easy, namely, by writing (14.1.4) as

$$\sum_{j=2}^{\infty}\frac{t^{2j}}{(2j)!}(l^{2j}-l^2) \leq B(k,l)\cdot\sum_{j=2}^{\infty}\frac{t^{2j}}{(2j)!}(k^{2j}-k^2).$$

The last assertion comes from

$$A_\gamma(t):=[(\operatorname{Cos} h\gamma t-1)-\gamma^2(\operatorname{Cos} ht-1)]\geq 0, \qquad \gamma>1$$

and

$$\lim_{t\to 0}(A_l/A_k)=B(k,l). \qquad \square$$

Definition 14.1.2. Let $\{\mu_n\}_{n\in\mathbb{N}}$ be a sequence of probability measures on $\mathbb{R}$ such that the following integrals exist:

$$\tilde{p}_{k,n} := \int_{\mathbb{R}} \operatorname{Cos} hkt\mu_n(dt) \qquad (k \in \mathbb{R}). \tag{14.1.5}$$

We shall call the numbers $\tilde{p}_{k,n}$ hyperbolic coefficients of the measure μ_n.

Next we have:

Theorem 14.1.2. *Let $\{\mu_n\}_{n\in\mathbb{N}}$ be a sequence of probability measures on $\mathbb{R}$ with $\tilde{p}_{k,n} < \infty$ and $\tilde{p}_{1,n} \neq 1$. Let $k > 1$ and suppose that*

$$\lim_{n\to\infty} \frac{\tilde{p}_{k,n} - 1}{\tilde{p}_{1,n} - 1} = k^2,$$

then

$$\lim_{n\to\infty} \frac{\tilde{p}_{l,n} - 1}{\tilde{p}_{1,n} - 1} = l^2 \ \text{ for all } 1 < l \leq k.$$

PROOF: Immediate from an integration of (14.1.4) with respect to μ_n. □

14.1.4. Laplace Transforms

The following result is a characterization of the convexity of functions and leads to some applications as it is Proposition 14.1.1.

Lemma 14.1.4. *If $f: (0,\infty) \to \mathbb{R}$ then $h(x) = f(x)/x$ is convex iff $\varphi(k) = (k^2 f(x) - f(kx))/(k^2 - k)$ is non-increasing in $k > 1$, for each $x > 0$. Equivalently if $\psi(k) = (kf(x) - f(kx))/(k^2 - k)$ is non-increasing in $k > 1$, for each $x > 0$.*

PROOF: Let us first assume that h is convex. Consider $1 < \lambda < k$. We have

$$\lambda x = \left(\frac{k-\lambda}{k-1}\right) x + \left(\frac{\lambda - 1}{k-1}\right) kx, \quad \text{with} \quad \frac{k-\lambda}{k-1} + \frac{\lambda-1}{k-1} = 1,$$

where both $(k-\lambda)/(k-1)$, $(\lambda-1)/(k-1) > 0$. Since h is convex, one has

$$h(\lambda x) \leq \left(\frac{k-\lambda}{k-1}\right) h(x) + \left(\frac{\lambda-1}{k-1}\right) h(kx). \tag{14.1.6}$$

Substituting $h(x) = f(x)/x$ one obtains precisely $\varphi(\lambda) \geq \varphi(k)$. Next, assume that φ is non-increasing for each $x > 0$. This is equivalent to (14.1.6). Now suppose $0 < x_1 < x_2$ and let $\alpha = x_2/x_1 > 1$. Applying (14.1.6) with $\lambda := (1+\alpha)/2$, $k = \alpha$, and $x = x_1$ one has

$$\begin{aligned} h\left(\frac{x_1+x_2}{2}\right) &= h\left(\left(\frac{1+\alpha}{2}\right)\cdot x_1\right) \\ &\leq \left(\frac{\alpha-((1+\alpha)/2)}{\alpha-1}\right)\cdot h(x_1) + \left(\frac{((1+\alpha)/2)-1}{\alpha-1}\right)\cdot h(\alpha x_1) \\ &= \frac{1}{2}h(x_1)+\frac{1}{2}h(x_2). \end{aligned}$$

Therefore h is convex.

Next we prove that $\varphi(k)$ being non-increasing in $k > 1$ for each $x > 0$ is equivalent to $\psi(k)$ being non-increasing. That is, we want to show that for $1 < \lambda < k$

$$\frac{[\lambda^2 f(x) - f(\lambda x)]}{(\lambda^2-\lambda)} - \frac{[k^2 f(x) - f(kx)]}{(k^2-k)} \geq 0$$

is equivalent to

$$\frac{[\lambda f(x) - f(\lambda x)]}{(\lambda^2-\lambda)} - \frac{[k f(x) - f(kx)]}{(k^2-k)} \geq 0.$$

In fact, the difference between the two left-hand sides equals

$$\frac{(\lambda^2-\lambda)f(x)}{(\lambda^2-\lambda)} - \frac{(k^2-k)f(x)}{(k^2-k)} = 0.$$ □

Remark 14.1.3. Let $f: (0,\infty) \to \mathbb{R}$ such that $f(x)/x$ is convex. Then by Lemma 14.1.4 one has for $1 < \lambda < k$, all $x > 0$, the two equivalent inequalities

$$(k^2 f(x) - f(kx)) \leq \left(\frac{k^2-k}{\lambda^2-\lambda}\right)\cdot(\lambda^2 f(x) - f(\lambda x)) \tag{14.1.7}$$

and

$$(kf(x) - f(kx)) \leq \left(\frac{k^2-k}{\lambda^2-\lambda}\right)\cdot(\lambda f(x) - f(\lambda x)). \tag{14.1.8}$$

Quantities such as $[\lambda f(x) - f(\lambda x)]$ can be regarded as a measure of linearity for f.

Proposition 14.1.1. *For $k > \lambda > 1$ and $x > 0$ we get the following two equivalent sharp inequalities:*

$$k^2(1-e^{-x}) - (1-e^{-kx}) < \left(\frac{k^2-k}{\lambda^2-\lambda}\right)\cdot(\lambda^2(1-e^{-x}) - (1-e^{-\lambda x})) \tag{14.1.9}$$

and

$$k(1-e^{-x})-(1-e^{-kx})<\left(\frac{k^2-k}{\lambda^2-\lambda}\right)\cdot(\lambda(1-e^{-x})-(1-e^{-\lambda x})). \tag{14.1.10}$$

PROOF: Note that the function $f(x)=1-e^{-x}$, $x\in(0,+\infty)$, fulfills

$$\frac{f(x)}{x}=\int_0^1 e^{-\theta x}d\theta$$

showing that $f(x)/x$ is convex. By Lemma 14.1.4 we have (14.1.7) and (14.1.8). These are precisely (14.1.9) and (14.1.10). The sharpness follows from

$$\begin{aligned}\lim_{x\to 0}\frac{k^2(1-e^{-x})-(1-e^{-kx})}{\lambda^2(1-e^{-x})-(1-e^{-\lambda x})}&=\lim_{x\to 0}\frac{k(1-e^{-x})-(1-e^{-kx})}{\lambda(1-e^{-x})-(1-e^{-\lambda x})}\\&=\left(\frac{k^2-k}{\lambda^2-\lambda}\right). \qquad \square\end{aligned}$$

Some applications of the last proposition are Theorem 14.1.3 and Proposition 14.1.2 following.

Definition 14.1.3. Let μ be a probability measure on $\mathbb{R}^+$. For $\lambda\geq 0$ its Laplace transform is defined by

$$\varphi(\lambda):=\int_0^\infty e^{-\lambda t}\mu(dt). \tag{14.1.11}$$

Lemma 14.1.5. *Let $k\geq\lambda>1$ and let $\{\mu_n\}_{n\in\mathbb{N}}$ be a sequence of probability measures on $\mathbb{R}^+$ with existing Laplace transforms φ_n such that $\varphi_n(1)\neq 1$ all $n\in\mathbb{N}$. Then one has the equivalent inequalities*

$$\left[k^2-\left(\frac{1-\varphi_n(k)}{1-\varphi_n(1)}\right)\right]\leq\left(\frac{k^2-k}{\lambda^2-\lambda}\right)\cdot\left[\lambda^2-\left(\frac{1-\varphi_n(\lambda)}{1-\varphi_n(1)}\right)\right] \tag{14.1.12}$$

and

$$\left[k-\left(\frac{1-\varphi_n(k)}{1-\varphi_n(1)}\right)\right]\leq\left(\frac{k^2-k}{\lambda^2-\lambda}\right)\cdot\left[\lambda-\left(\frac{1-\varphi_n(\lambda)}{1-\varphi_n(1)}\right)\right]. \tag{14.1.13}$$

PROOF: Integrate (14.1.9) and (14.1.10) relative to μ_n and divide by $(1-\varphi_n(1))>0$. $\square$

Theorem 14.1.3. *Let $\lambda>1$. Then*

$$\lim_{n\to\infty}\left(\frac{1-\varphi_n(\lambda)}{1-\varphi_n(1)}\right)=\lambda^2 \qquad (\textit{respectively } \lambda)$$

implies that

$$\lim_{n\to\infty}\left(\frac{1-\varphi_n(k)}{1-\varphi_n(1)}\right)=k^2 \qquad (\textit{respectively } k) \textit{ for all } k\ge\lambda.$$

PROOF: Apply (14.1.12) (respectively (14.1.13)). □

We would like to remind (see also §7.3; Corollary 7.3.6).

Definition 14.1.4. The Weierstrass operator is the positive linear operator defined as

$$(W_nf)(t):=\sqrt{\frac{n}{\pi}}\cdot\int_{-\infty}^{\infty}f(x)\cdot e^{-n(t-x)^2}dx, \quad \text{all } n\in\mathbb{N},$$

where $f\in C_B(\mathbb{R})$.

One has $W_nf \stackrel{u}{\longrightarrow} f$ as $n\to\infty$.

Proposition 14.1.2. *Let $f\in C_B(\mathbb{R})$ such that $\int_{-\infty}^{\infty}|f(x)|^n dx := C_n\in(0,\infty)$; all $n\in\mathbb{N}$. Consider $k\ge\lambda$ with $k,\lambda\in\mathbb{N}$.*

If

$$\lim_{n\to\infty}\left[\frac{1-\sqrt{\frac{\pi}{\lambda}}\cdot W_\lambda\left(\frac{|f|^n}{C_n},t\right)}{1-\sqrt{\pi}\cdot W_1\left(\frac{|f|^n}{C_n},t\right)}\right]=\lambda^2 \qquad (\textit{respectively } \lambda),$$

then

$$\lim_{n\to\infty}\left[\frac{1-\sqrt{\frac{\pi}{k}}\cdot W_k\left(\frac{|f|^n}{C_n},t\right)}{1-\sqrt{\pi}\cdot W_1\left(\frac{|f|^n}{C_n},t\right)}\right]=k^2 \qquad (\textit{respectively } k),$$

PROOF: Apply (14.1.9) (respectively (14.1.10)) with x replaced by $(t-x)^2$, where t is fixed. Then multiply by $|f|^n$ and integrate over $\mathbb{R}$. □

CHAPTER FIFTEEN

15.1. A DISCRETE STOCHASTIC KOROVKIN TYPE CONVERGENCE THEOREM

15.1.1. Introduction

In this chapter we give a sufficient condition for the pointwise—in the first mean Korovkin property on $B_{\underline{0}}(P)$, the space of stochastic processes with real state space and countable index set P and bounded first moments.

The material of this chapter comes from Anastassiou(2) (1991).

Let $(\underline{0}, A, \tau)$ be a probability space and let P denote a fixed countable set. Consider stochastic processes X with real state space and the expectation operator

$$E(X)(t) := \int_{\underline{0}} X(t,\omega)\tau(d\omega), \qquad t \in P.$$

Define

$$B_{\underline{0}}(P) := \left\{ X : \sup_{t \in P} E|X|(t) < \infty \right\}.$$

Let $T_n : B_{\underline{0}}(P) \to B_{\underline{0}}(P)$ be any sequence of positive linear operators such that $ET_n = T_n E$, all $n = 1, 2, \ldots,$. In Theorem 15.1.1, under Korovkin type assumptions, we give a sufficient condition such that for each $X \in B_{\underline{0}}(P)$,

$$\lim_{n \to \infty} E[(T_n X)(t,\omega) - X(t,\omega)] = 0, \quad \text{for each } t \in P.$$

In Weba (1986), see Theorem 3.2, was treated the continuous case, that is, when P is an uncountable compact space. There the sufficient condition is similar to ours, however, it is produced under the additional assumption that T_n is a stochastically simple operator.

15.1.2. Results

Our main result follows

Theorem 15.1.1. *Let* $(\underline{0}, A, \tau)$ *be a probability space and* $P = \{t_1, \ldots, t_j, \ldots\}$ *be a countable set of cardinality* ≥ 2. *Consider the space of stochastic processes with real state space*

$$B_{\underline{0}}(P) := \left\{ X : \sup_{t \in P} \int_{\underline{0}} |X(t,\omega)| \tau(d\omega) < \infty \right\}$$

and the space

$$B(P) := \{ f : P \to \mathbb{R} \mid \|f\|_\infty < \infty \},$$

where

$$\|f\|_\infty := \sup_{t \in P} |f(t)|; \quad B(P) \subset B_{\underline{0}}(P).$$

Let $T_n : B_{\underline{0}}(P) \to B_{\underline{0}}(P)$ *be a sequence of positive linear operators that are E-commutative, i.e.,*

$$(E(T_n X))(t,\omega) = (T_n(EX))(t,\omega), \quad \textit{for all } (t,\omega) \in P \times \underline{0}$$

where

$$(EX)(t) := E(X(t,\omega)) := \int_{\underline{0}} X(t,\omega) \tau(d\omega)$$

is the expectation operator.

Also assume that $(T_n 1)(t,\omega) = 1$, *for all* $(t,\omega) \in P \times \underline{0}$. *For*

$$\{X_1(t,\omega), \ldots, X_k(t,\omega)\} \subset B_{\underline{0}}(P)$$

assume that

$$\lim_{n\to\infty} E[(T_n X_i)(t_j,\omega) - X_i(t_j,\omega)] = 0, \tag{15.1.1}$$

for all $t_j \in P$ *and all* $i = 1, \ldots, k$. *(I.e.,*

$$\lim_{n\to\infty} [(T_n(EX_i))(t_j) - (EX_i)(t_j)] = 0,$$

for all $t_j \in P$ *and* $i = 1, \ldots, k$.*)*

In order that

$$\lim_{n\to\infty} E[(T_n X)(t_j,\omega) - X(t_j,\omega)] = 0, \tag{15.1.2}$$

for all $t_j \in P$ and all $X \in B_{\underline{0}}(P)$, it is enough to assume that for each $t_j \in P$ there are real constants $\beta_1, \ldots, \beta_k$ such that

$$\sum_{i=1}^{k} \beta_i E[X_i(t,\omega) - X_i(t_j,\omega)] \geq 1, \quad \textit{for all } t \in P - \{t_j\}. \tag{15.1.3}$$

PROOF: If there exists $X \in B_{\underline{0}}(P)$ and $t_{j0} \in P$ such that

$$E[(T_n X)(t_{j0},\omega) - X(t_{j0},\omega)] \not\to 0,$$

then there exist a subsequence T_{λ_n} and an $\varepsilon > 0$ such that

$$|(E(T_{\lambda_n}X))(t_{j0}) - (EX)(t_{j0})| > \varepsilon, \quad \text{for all } n \geq 1.$$

By E-commutativity of T_{λ_n} we have

$$|(T_{\lambda_n}(EX))(t_{j0}) - (EX)(t_{j0})| > \varepsilon, \quad \text{for all } n \geq 1.$$

Let μ be a positive finite measure on P with $\mu(\{t\}) > 0$, for all $t \in P$. Here $B(P) \subset L_p(P,\mu)$, $1 \leq p < \infty$.

Let $f \in B(P)$, then $E(f) = f$. Hence $T_n(f) = T_n(Ef) = ET_n(f)$ and $T_n(f) \in B(P)$, that is T_n maps $B(P)$ into itself. Because each positive linear functional $T_n(\cdot, t_j)$ on $B(P)$ is bounded, by Riesz representation theorem, for the specific $j = j_0$, there exists $g_{t_{j0},n} \in L_q(P,\mu)$ where $\frac{1}{p} + \frac{1}{q} = 1$ such that

$$(T_n(f))(t_{j0}) = \int_P f(t) g_{t_{j0},n}(t)\mu(dt), \quad \text{for all } f \in B(P).$$

By $T_n(1) = 1$ and the positivity of $T_n(\cdot, t_{j0})$ one obtains

$$\int_P g_{t_{j0},n}(t)\mu(dt) = 1 \quad \text{and} \quad g_{t_{j0},n}(t) \geq 0, \quad \text{for all } t \in P.$$

Since $EX \in B(P)$, we have

$$(T_{\lambda_n}(EX))(t_{j0}) = \int_P (EX)(t) \cdot g_{t_{j0},\lambda_n}(t) \cdot \mu(dt).$$

Therefore

$$\begin{aligned}\varepsilon < |(T_{\lambda_n}(EX))(t_{j_0}) - (EX)(t_{j_0}) &= \left|\int_P (EX)(t)\cdot g_{t_{j_0},\lambda_n}(t)\cdot\mu(dt)\right.\\ &\quad \left. - \int_P (EX)(t_{j_0})\cdot g_{t_{j_0},\lambda_n}(t)\cdot\mu(dt)\right|\\ &= \left|\int_{P-\{t_{j_0}\}}[(EX)(t)-(EX)(t_{j_0})]\cdot g_{t_{j_0},\lambda_n}(t)\cdot\mu(dt)\right|\\ &\le \|EX-(EX)(t_{j_0})\|_\infty\cdot\left(\int_{P-\{t_{j_0}\}} g_{t_{j_0},\lambda_n}(t)\mu(dt)\right),\end{aligned}$$

so that

$$\int_{P-\{t_{j_0}\}} g_{t_{j_0},\lambda_n}(t)\mu(dt) > \frac{\varepsilon}{\|EX-(EX)(t_{j_0})\|_\infty} =: \delta > 0, \quad \text{for all } n \ge 1.$$

There cannot be real constants $\beta_1,\ldots,\beta_k$ with

$$\sum_{i=1}^k \beta_i E[X_i(t,\omega)-X_i(t_{j_0},\omega)] \ge 1, \quad \text{for all } t\in P-\{t_{j_0}\}.$$

Since, otherwise, we would have

$$\sum_{i=1}^k \beta_i E[X_i(t,\omega)-X_i(t_{j_0},\omega)]\cdot g_{t_{j_0},\lambda_n}(t) \ge g_{t_{j_0},\lambda_n}(t), \quad \text{for all } t\in P-\{t_{j_0}\}$$

and thus

$$\begin{aligned}\sum_{i=1}^k \beta_i\cdot\int_{P-\{t_{j_0}\}}&[(EX_i)(t)-(EX_i)(t_{j_0})]\cdot g_{t_{j_0},\lambda_n}(t)\cdot\mu(dt)\\ &\ge \int_{P-\{t_{j_0}\}} g_{t_{j_0},\lambda_n}(t)\cdot\mu(dt) > \delta.\end{aligned}$$

(Note that

$$(T_{\lambda_n}(EX_i))(t_{j_0}) = \int_P (EX_i)(t)\cdot g_{t_{j_0},\lambda_n}(t)\cdot\mu(dt), \quad i=1,\ldots,k.)$$

However from the assumptions of the theorem, we have

$$\lim_{n\to\infty}(T_{\lambda_n}(EX_i))(t_{j_0}) = (EX_i)(t_{j_0}), \quad \text{all } i=1,\ldots,k.$$

Hence

$$0 = \lim_{n\to\infty}\left(\sum_{i=1}^k \beta_i[(T_{\lambda_n}(EX_i))(t_{j_0})-(EX_i)(t_{j_0})]\right) > \delta.$$

Thus $\delta < 0$, contradicting $\delta > 0$. □

To show that the assumptions of Theorem 15.1.1 are not void and they are powerful, we present

Example 15.1.1. (i) Consider the probability space $([-a,a], \mathcal{B}, \frac{\lambda}{2a})$, where $a > 0$, $\mathcal{B}$ the Borel σ-algebra on $[-a,a]$, λ the Lebesgue measure on $[-a,a]$. Since $\frac{\lambda}{2a}([-a,a]) = 1$, $\frac{\lambda}{2a}$ is a probability measure on $[-a,a]$. Let also $P = \{\pm 1, \pm 2, \ldots, \pm T\}$ be a finite set of integers. That is here $\omega \in \underline{\Omega} = [-a,a]$ and $t \in P$.

Consider the sequence of operators

$$T_n\colon B_{\underline{\Omega}}(P) \to B_{\underline{\Omega}}(P)$$

such that

$$(T_n X)(t,\omega) = X(t,\omega)(1 - e^{-n|t|}) + X(-t,\omega)e^{-n|t|}, \quad \text{for all } n \geq 1. \tag{15.1.4}$$

If $X \geq 0$ then $T_n X \geq 0$, that is T_n is a positive operator, furthermore $T_n(1) = 1$, for all $n \geq 1$. It is obvious that T_n is linear.

Observe that

$$(E(T_n X))(t,\omega) = (EX)(t) \cdot (1 - e^{-n|t|}) + (EX)(-t) \cdot e^{-n|t|} = (T_n(EX))(t,\omega),$$

i.e., $ET_n = T_n E$, that is T_n is E-commutative for all $n \geq 1$. Therefore T_n fulfills the assumptions of Theorem 15.1.1.

From

$$(E(T_n X))(t) = (EX)(t) \cdot (1 - e^{-n|t|}) + (EX)(-t) \cdot e^{-n|t|},$$

it is clear that

$$\lim_{n\to\infty} E[(T_n X)(t,\omega) - X(t,\omega)] = 0,$$

for all $t \in P$ and all $X \in B_{\underline{\Omega}}(P)$. Thus T_n fulfills the conclusion of Theorem 15.1.1.

(ii) Continuing in the setting of part (i): Let $X_1(t,\omega) = 1$, $X_2(t,\omega) = 2t|\omega|/a$ and $X_3(t,\omega) = 3t^2\omega^2/a^2$. Then $(EX_1)(t) = 1$, $(EX_2)(t) = t$ and $(EX_3)(t) = t^2$. It is obvious that $X_1, X_2, X_3 \in B_{\underline{0}}(P)$. We would like to find $\beta_1, \beta_2, \beta_3$ such that

$$\sum_{i=1}^{3} \beta_i[(EX_i)(t) - (EX_i)(t_j)] \geq 1, \quad \text{for all } t \in P - \{t_j\}.$$

For that we can pick β_1 an arbitrary real number, $\beta_2 = -2t_j$ and $\beta_3 = 1$. We get

$$\beta_1(1-1) + (-2t_j)(t - t_j) + (t^2 - t_j^2) = (t - t_j)^2 \geq 1,$$

for all $t \in P - \{t_j\}$. Hence X_i, $i = 1, 2, 3$ fulfill the sufficient condition of Theorem 15.1.1.

Trivially $T_n X_i = X_i$, giving us $ET_n X_i = EX_i$, for $i = 1, 3$. And

$$(T_n X_2)(t, \omega) = X_2(t, \omega)(1 - e^{-n|t|}) + X_2(-t, \omega) \cdot e^{-n|t|},$$

implying

$$(E(T_n X_2))(t) = t(1 - 2e^{-n|t|}).$$

It is clear that

$$\lim_{n \to \infty} (E(T_n X_2))(t) = (EX_2)(t).$$

We have seen how X_i, $i = 1, 2, 3$ fulfill the assumptions of Theorem 15.1.1.

Index

Adams rule, 107
admissible pair of hyperplanes, 48
concave operator, 265
converge weakly, 205
convex moment methods, 59
convolution operator, positive, linear, 296
convolution type operators, non-positive, 303
E-commutative, 314
exponential operator, 269
φ-decomposable, 372
Fourier–Stieltjes coefficients, 380
generalized reduced K-functional, 357
hyperbolic coefficients, 382
Jefferson rounding, 100
Kantorovich radius, 73
Laplace transform, 384
Levy distance, 123
Levy radius, 123
Markov kernel, 60
Markov transformation, 60
modified modulus of continuity, 226
modulus of continuity, 208
modulus of smoothness, 304
moment, 38
moment condition, 42
moment point, 43
moment space, 42
Newton–Puiseux diagram, 146
optimal distance method, 43
optimal ratio method, 52, 56
positive linear functional, 201
positive linear operator, 199
Prokhorov distance (metric), 150
Prokhorov radius, 150
V-representation, 51
W-representation, 51
Tchebycheff system, 129
trigonometric Prokhorov radius, 170
Trotter operator, 371
Truncated exponential p.d.f., 292
two sides-truncated normal p.d.f., 293

LIST OF SYMBOLS

$B(\cdot, r)$ ball of radius r

$\int$ integral sign

μ, v measures (positive)

f function

$\mu(f) = \int f \, d\mu$

$L(\cdot), L(\cdot \mid f) = \inf_{\mu} \mu(f)$

$U(\cdot), U(\cdot \mid f) = \sup_{\mu} \mu(f)$

conv convex hull

convc convex cone

$\mathrm{int}(A), A^0$ interior of set A

$Bd(A)$ boundary of A

$\emptyset$ empty set

$\cup$ union

$\cap$ intersection

$\exists$ there exists

$\overline{A}$, cl A closure of A

X_S characteristic function on S

$L_S(\cdot) = \inf \mu(S)$

$U_S(\cdot) = \sup \mu(S)$

$m_S(x_0)$ set of convex type probability measures

F, F_X cumulative distribution function

δ_α Dirac (unit) measure

E expectation

Var variance

$\mathbb{N}$ natural numbers

$\mathbb{Z}_+$ non-negative integers

$\mathbb{R}$ real numbers

$\mathbb{R}_+$ non-negative real numbers

$\mathbb{R}^k$ k-dimensional Euclidean space

$(V, \|\cdot\|)$ real normed vector space

$[\cdot]$ integral part of number

$\lceil\cdot\rceil$ ceiling of number

$[x]_c$ p. 80

$$[x]_+ := \begin{cases} x, & x \geq 0 \\ 0, & x < 0 \end{cases}$$

$\overset{\frown}{AB}$ arc(AB)

$\overline{AB}$ line segment (AB)

$A\overset{\Delta}{B}C$ triangle (ABC)

$\varepsilon \ll 1$ ε much smaller than 1

$\rightrightarrows, \Rightarrow$ weak convergence

ω_1 first modulus of continuity

$\overline{\omega}_1$ modified first modulus of continuity

ω_r rth modulus of smoothness, $r \in \mathbb{N}$

$\omega_r(t, f; C_B)$ p. 371

$K_r(t, f, C_B)$ p. 371

$\|\cdot\|_L$ Lipschitz seminorm p. 206

$\|\cdot\|_{BL}$ p. 206

$C(X)$ continuous functions on X

$B(X)$ bounded functions on X

$C_B(\mathbb{R})$ bounded continuous real valued functions on $\mathbb{R}$

$C^n(Q)$ n-times continuously differentiable functions on $Q \subset \mathbb{R}^k$

C_B^r p. 370

$L^1(\underline{0}, A, P)$ p. 314

$C_{\underline{0}}([a,b])$ p. 314

$B_{\underline{0}}([a,b])$ p. 314

$C_{\underline{0}}(Q)$ p. 322

$B_{\underline{0}}(Q)$ p. 322

$C_{\underline{0}}^{n}([a,b])$ p. 314

$C_{\underline{0}}^{n}(Q)$ p. 323

$X(t,w)$ stochastic process

$\text{Lip}(\alpha, r, C_B)$ p. 371

REFERENCES

1. Akhiezer, N.I. and Krein, M. (1962) *Some Questions in the Theory of Moments*, American Mathematical Society, Providence, RI.

2. Akhiezer, N.I. (1965) *The Classical Moment Problem*, Hafner Publ. Co., New York.

3. Alfsen, E.M. (1971) *Compact Convex Sets and Boundary Integrals*, Springer-Verlag, Berlin-New York.

4. Amir, D. and Ziegler, Z. (1978) Korovkin shadows and Korovkin systems in $C(S)$-spaces, *J. Math. Anal. Appl.* **62**, 640–675.

5. Anastassiou, G.A. (1983) An improved general stochastic inequality, *Bull. of the Greek Math. Soc.*, **24**, 1–11.

6. Anastassiou, G.A. (1984) A Study of Positive Linear Operators by the Method of Moments, Ph.D. Thesis, University of Rochester.

7. Anastassiou, G.A. (1) (1985) A study of positive linear operators by the method of moments, one dimensional case, *J. Approx. Theory*, **45**, No. 3, 247–270.

8. Anastassiou, G.A. (2) (1985) A "K-attainable" inequality related to the convergence of positive linear operators, *J. Approx. Theory*, **44**, 380–383.

9. Anastassiou, G.A. (3) (1985) Miscellaneous sharp inequalities and Korovkin type convergence theorems involving sequences of probability measures, *J. Approx. Theory*, **44**, No. 4, 384–390.

10. Anastassiou, G.A. (1) (1986) Korovkin type inequalities in real normed vector spaces, *Approx. Theory Appl.*, No. 2, **2**, 39–53.

11. Anastassiou, G.A. (2) (1986) Korovkin inequalities in real normed vector spaces, Inter. Conf. Approx. Th., Proceedings, "Approx. Th. V," pp. 235–238, Academic Press.

12. Anastassiou, G.A. (3) (1986) Multi-dimensional quantitative results for probability measures approximating the unit measure, *Approx. Theory Appl.*, **2**, 93–103.

13. Anastassiou, G.A. (1) (1987) The Levy radius of a set of probability measures satisfying basic moment conditions involving $\{t, t^2\}$, *Constructive Approx. Journal*, **3**, 257–263.

14. Anastassiou, G.A. (2) (1987) The Levy radius of a set of probability measures satisfying moment conditions involving a Tchebycheff system of functions, *Approx. Th. & Appl.*, Vol. 3, No. 2–3, 50–65.

15. Anastassiou, G.A. (3) (1987) On the degree of weak convergence of a sequence of finite measures to the unit measure under convexity, *J. Approx. Th.*, Vol. 51, No. 4, 333–349.

16. Anastassiou, G.A. (1987) A comparison of Prohorov and Lipschitz metrics, *Bulletin of Greek Math. Soc.*, **28**, A, 1–5.

17. Anastassiou, G.A. (1988) The rate of weak convergence of convex type positive finite measures, *J. Math. Anal. and Appl*, Vol. 136, No. 1, 229–248.

18. Anastassiou, G.A. (1988) A generalized K-attainable inequality related to the weak convergence of probability measures, M.S.U. Report Series 1988-1.

19. Anastassiou, G.A. and Rachev, S.T. (1988) How precise is the approximation of a random queue by means of deterministic queueing models, M.S.U. Report Series 1988-27.

20. Anastassiou, G.A. (1) (1989) Smooth rate of weak convergence of convex type positive finite measures, *J. Math. Anal. and Appl.* **141**, 491–508.

21. Anastassiou, G.A. (2) (1989) Sharp inequalities for convolution type operators, *J. Approx. Theory*, **58**, 259–266.

22. Anastassiou, G.A. (3) (1989) Rate of convergence of non-positive generalized convolution type operators, *J. Math. Anal. and Appl.*, **142**, 441–451.

23. Anastassiou, G.A. (4) (1989) Rate of convergence of positive linear operators using an extended complete Tchebycheff system, *Journal Approx. Theory*, **59**, 125–149.

24. Anastassiou, G.A. and S.T. Rachev (1989) Approximation of a random queue by means of deterministic queueing models, *Approximation Theory VI*, Vol. I, pp. 9–11, Academic Press.

25. Anastassiou, G.A. (1) (1990) Weak convergence and the Prokhorov radius for probability measures, *Approx. Opt. & Comp.: Th. & Appl.*, A.G. Law & C.L. Wang (eds.), North Holland, pp. 47–48.

26. Anastassiou, G.A. (2) (1990) On a discrete Korovkin theorem, *Journal Approx. Theory*, **61**, 384–386, correction.

27. Anastassiou, G.A. (1) (1991) Korovkin inequalities for stochastic processes, *J. Math. Anal. & Appl.*, Vol. 157, No. 2, 366–384.

28. Anastassiou, G.A. (2) (1991) A discrete stochastic Korovkin theorem, *Int. J. Math. & Math. Sci.*, Vol. 14, No. 4, 679–682.

29. Anastassiou, G.A. (1992) Weak convergence and the Prokhorov radius, *J. Math. Anal. & Appl.*, Vol. 163, No. 2, 541–558.

30. Anastassiou, G.A. and Rachev, S.T. (1992) Moment problems and their applications to characterization of stochastic processes, queueing theory, and rounding problem, Proceedings 6th S.E.A. Meeting, Memphis State University, "Approximation Theory," pp. 1–77, Marcel Dekker, New York.

31. Apostol, T.M. (1969) *Mathematical Analysis*, Addison-Wesley, London.

32. Arens, R. (1950) Representation of functionals by integrals, *Duke Math. J.*, **17**, 499–506.

33. Arens, R. (1963) Representation of moments by integrals, *Illinois J. Math.*, **7**, 609–614.

34. Atzmon, A. (1975) *A moment problem for positive measures on the unit disc*, *Pacific J. Math.*, **59**, 317–325.

35. Balinski, M.L. and Young, H.P. (1982) *Fair Representation: Meeting the Ideal of One Man, One Vote*, Yale University Press, New Haven.

36. Bartoszynski, R. (1961) A characterization of the weak convergence of measures, *Annals of Math. Stat.*, **32**, 561–567.

37. Basu, S.K. and Simons, G. (1983) Moment spaces for IFR distributions, applications, and related material, *Contributions to Statistics: Essay in Honor of Norman L. Johnson*, P.K. Sen (Ed.), North-Holland Publishing Company, pp. 27–46.

38. Bauer, H. (1978) Approximation and abstract boundaries, *Ann. Math. Monthly*, **85**, 632–647.

39. Berg, C. and Maserick, P.H. (1982) Polynomially positive definite sequences, *Math. Ann.* **259** 487–495.

40. Berg, C. (1987) The multidimensional moment problem and semigroups, *Proc. Symposia Appl. Math.*, **37**, 110–124.

41. Bernshtein, S.N. (1969) *Collected Works*, Vol. 4, Nauka, Moscow (in Russian).

42. Billingsley, P. (1968) *Convergence of Probability Measures*, John Wiley, New York.

43. Billingsley, P. (1979) *Probability and Measure*, John Wiley & Sons, New York.

44. Bloom, W.R. and Sussich, J.F. (1982) The degree of approximation by positive linear operators on compact connected Abelian groups, *J. Austral. Math. Soc. A*, **33**, 364–373.

45. Borovkov, A.A. (1984) *Asymptotic Methods in Queueing Theory*, John Wiley & Sons, New York, Toronto.

46. Butzer, P.L. and Berens, H. (1967) *Semi-Groups of Operators and Approximation*, Springer-Verlag, Berlin.

47. Butzer, P.L. and Oberdörster, W. (1975) Linear functionals defined on various spaces of continuous functions on $\mathbb{R}$, *J. Approx. Th.*, **13**, 451–469.

48. Butzer, P.L. and Oberdörster, W. (1975), Darstellungssätze für Beschränkte Lineare funktionale in zusammenhang mit Hausdorff, Stieltjes – und Hamburger – Moment problemen, *Forschungsberichte des Landes Nordrhein-Westfalen*, Nr. 2515, Westdeutscher Verlag Opladen, iii + 47pp.

49. Butzer, P.L. and Hahn, L. (1) (1978) General theorems on rates of convergence in distribution of random variables, I: General limit theorems, *J. Multivariate Anal.*, **8**, 181–201.

50. Butzer, P.L. and Hahn, L. (2) (1978) General theorems on rates of convergence in distribution of random variables II: Applications to the stable limit laws and weak law of large numbers, *J. Multivariate Anal.*, **8**, 202–221.

51. Cambanis, S., Simons, G. and Stolt, W. (1976) Inequalities for $Ek(X,Y)$ when the marginals are fixed, *Z. Wahrsch. Verw. Geb.*, **36**, 4, 285–294.

52. Cambanis, S. and Simons, G. (1982) Probability and expectation inequalities, *Z. Wahrsch. Verw. Geb.*, **59**, 1–25.

53. Cartan, H. (1971) *Differential Calculus*, Hermann, Paris.

54. Censor, E. (1971) Quantitative results for positive linear approximation operators, *J. Approx. Theory*, **4**, 442–450.

55. Chow, Y.S. and Teicher, H. (1978) *Probability Theory*, Springer-Verlag, New York.

56. De Acosta, A. (1982) Invariance principles in probability for triangle arrays of B-valued random vectors and some applications, *Ann. Probab.*, **10**, 346–373.

57. DeVore, R.A. (1972) *The Approximation of Continuous Function by Positive Linear Operators*, Lecture Notes in Mathematics, Vol. 293, Springer-Verlag, Berlin/New York.

58. Diaconis, P. and Freedman, D. (1979) On rounding percentages, *Journal of the American Statistical Assoc.*, **74**, 359–364.

59. Ditzian, Z. (1975) Convergence of sequences of linear positive operators: Remarks and applications, *J. Approx. Theory*, **14**, 296–301.

60. Dobrushin, R.L. (1970) Prescribing a system of random variables by conditional distributions, *Theory Prob. Appl.*, **15**, 458–486.

61. Donner, K. (1982) *Extension of Positive Operators and Korovkin Theorems*, Lecture Notes in Mathematics, Vol. 904, Springer-Verlag, Berlin/New York.

62. Dowson, C.D. and Landau, B.U. (1982) The Frechet distance between multivariate normal distributions, *J. Multivar. Anal.*, **12**, 450–455.

63. Dudley, R.M. (1966) Convergence of Baire measures, Studia Math., XXVII, **3**, 251–268.

64. Dudley, R.M. (1968) Distances of probability measures and random variables, *Ann. Math. Statist.*, **40**, 40–50.

65. Dudley, R.M. (1972) Speeds of metric probability convergence, *Z. Wahrscheinlichkeitsth.*, **22**, 323–332.

66. Dudley, R.M. (1976) *Probabilities and Metrics: Convergence of laws on metric spaces with a view to statistical testing*, Aarhus Univ. Mat. Inst. Lecture Note Series, No. 45, Aarhus.

67. Dudley, R.M. (1989) *Real Analysis and Probability*, Wadsworth & Brooks/Cole, Pacific Grove, California.

68. Dunford, N. and Schwartz, J. (1957) *Linear Operators, Part I*, Interscience, New York.

69. Fan, Ky (1959) *Convex Sets and Their Applications*, Argonne National Laboratory.

70. Feller, W. (1966), *An Introduction to Probability Theory and Its Applications*, Vol. II, Wiley, New York.

71. Givens, C.R. and Shortt, R.M. (1984) A class of Wasserstein metrics for probability distributions, *Michigan Math. J.*, **31**, 231–240.

72. Glicksberg, I. (1952) The representation of functionals by integrals, *Duke Math. J.*, **19**, 253–261.

73. Gonska, H.H. (1) (1983) On approximation in spaces of continuous functions, *Bull. Austral. Math. Soc.*, **28**, 411–432.

74. Gonska, H.H. (2) (1983) On approximation of continuously differentiable functions by positive linear operators, *Bull. Austral. Math. Soc.*, **27**, 73–81.

75. Gonska, H.H. (1984) On approximation in $C(X)$, constructive theory of functions, in Proc. Int. Conf., Varna, Bulgaria 1984 (B. Sendov et al., Eds.), pp. 364–369, Sofia Publishing House, Bulgarian Academy of Sciences.

76. Gordon, W.J. and Wixom, J.A. (1978) Shepard's method of metric interpolation to bivariate and multivariate interpolation, *Math. Comp.*, **32**, 253–264.

77. Grigorevski, N.N. and Shiganov, I.S. (1976) On certain modifications of Dudley's metric, *Zapiskli Nauchnich Sem.*, LOMI, **41**, 17–27, in Russian.

78. Hahn, L. (1982) Stochastic methods in connection with approximation theorems for positive linear operators, *Pacific J. Math.*, **101** (2), 307–319.

79. Hamburger, H. (1920) Über eine Erweiterung des Stieltjes' Schen Momenten problems, *Math. Ann.*, **81**, 235–319; **82** (1921), 120–164, 168–187.

80. Haneveld, W.K.K. (1985) *Duality in Stochastic Linear and Dynamic Programming*, Centrum voor Wiskunde en Informatica, Amsterdam.

81. Hausdorff, F. (1921), Summations methoden und moment folgen I, II, *Math. Z.*, **9**, 74–109, 280–299.

82. Hausdorff, F. (1923), Momentenprobleme für ein endliches intervall, *Math. Z.*, **16**, 220–248.

83. Hewitt, E. and Stromberg, K. (1965) *Real and Abstract Analysis*, Springer-Verlag, New York/Berlin.

84. Hildebrandt, T.H. (1932) On the moment problem for a finite interval, *Bull. Amer. Math. Soc.*, **38**, 269–270.

85. Hildebrandt, T.H. and Schoenberg, I.J. (1933) On linear functional operations and the moment problem for a finite interval in one or several dimensions, *Ann. of Math.*, **34**, 317–328.

86. Hille, E. (1973) *Analytic Function Theory*, Vol. II, Chelsea, New York.

87. Hoeffding, W. (1955) The extrema of the expected value of a function of independent random variables. *Annals of Mathematical Statistics*, **26**, 268–275.

88. Hoeffding, W. and Shrikhande, S.S. (1956) Bounds for the distribution function of a sum of independent, identically distributed random variables, *Annals of Mathematical Statistics*, **27**, 439–449.

89. Jakimovski, A. and Ramanujan, M.S. (1964) A uniform approximation theorem and its application to moment problems, *Math. Z.*, **84**, 143–153.

90. Junggeburth, J., Scherer, K. and Trebels, W. (1973) *Zur besten Approximation auf Banachräumen mit Anwendungen auf ganze Funktionen*, Forschungsberichte des Landes Nordrhein-Westfalen Nr. 2311, 51–75.

91. Kalashnikov, V.V. and Rachev, S.T. (1984) Characterization problems in queueing and their stability, In: *Stability Problems for Stochastic Models*, Proceedings Moscow, VNIISI, 1984, pp. 49–86 (in Russian); English Transl., J. Soviet Math., Vol. 35, No. 2, pp. 2336–2360.

92. Kalashnikov, V.V. and Rachev, S.T. (1985) Characterization problems in queueing and their stability, *Adv. Appl. Prob.*, **17**, 868–886.

93. Kalashnikov, V.V. and Rachev, S.T. (1990) *Mathematical Methods for Construction of Stochastic Queueing Models*, Wadsworth, Brooks/Cole.

94. Kantorovich, L.V. (1942) On the translocation of masses, *C.R. Acad. Sci. URSS*, (N.S.) **37**, 199–201.

95. Kantorovich, L.V. and Rubinshtein, G.Sh. (1957) On a function space and some extremum problems, *Dokl. Akad. Nauk SSSR*, **115**, 6, 1058–1061. (In Russian).

96. Kantorovich, L.V. and Rubinshtein, G.Sh. (1958) On the space of completely additive functions, *Vestnik LGU, Ser. Mat., Mekh. i. Astron.*, **7**, 2, 52–59. (In Russian).

97. Karlin, S. and Shapley, L.S. (1953) *Geometry of Moment Spaces*, Memoirs **12**, Amer. Math. Soc., Providence.

98. Karlin, S. and Studden, W.J. (1966) *Tchebycheff Systems: With Applications in Analysis and Statistics*, Interscience, New York.

99. Kellerer, H.G. (1984) Duality theorems and probability metrics, In: Proc. 7th Brasov Conf. 1982, pp. 211–220, Bucuresti.

100. Kellerer, H.G. (1984) Duality theorems for marginal problems, *Zeitschrift fur Wahrscheinlichkeitstheorie*, **67**, 399–432.

101. Kemperman, J.H.B. (1965) On the sharpness of Tchebycheff type inequalities, *Indag. Math.*, **27**, 554–571; 572–587; 588–601.

102. Kemperman, J.H.B. (1968) The general moment problem, a geometric approach, *Ann. Math. Stat.*, **39**, 93–122.

103. Kemperman, J.H.B. (1971) *Moment Problems with Convexity Conditions, Optimizing Methods in Statistics* (J.S. Rustagi, Ed.), pp. 115–178, Academic Press, New York.

104. Kemperman, J.H.B. (1972) On a class of moment problems, *Proceedings Sixth Berkeley Symposium on Mathematical Statistics and Probability*, **2**, pp. 101–126.

105. Kemperman, J.H.B. (1977) On the FKG-inequality for measures on a partially ordered space, *Proc. Nederl. Akad. Wet.*, **80**, 313–331.

106. Kemperman, J.H.B. (1) (1983) *On the role of duality in the theory of moments. Semi-infinite programming and applications*, Lecture Notes in Economics and Mathematical Systems, **215**, Springer, Berlin, 63–92.

107. Kemperman, J.H.B. (2) (1983) Moment Theory, Class Notes, Springer Semester, 1983, University of Rochester, New York.

108. Kemperman, J.H.B. (1987) Geometry of the moment problem, *Proceedings of Symposia in Applied Mathematics*, **27**, A.M.S., 16–53.

109. King, J.P. (1975) Probability and positive linear operators, *Rev. Roumaine Math. Pures Appl.*, **20**, No. 3, 325–327.

110. Korovkin, P.P. (1958) An asymptotic property of positive methods of summation of Fourier series and best approximations of functions of class Z_2, by linear positive polynomial operators, (Russian) *Uspekhi Mat. Nauk*, **13** (6)(84), 99–103.

111. Korovkin, P.P. (1960) *Linear Operators and Approximation Theory*, Hindustan Publ. Corp., Delhi, India.

112. Krein, M.G. (1951) The ideas of P.L. Cebysev and A.A. Markov in the theory of limiting values of integrals and their further development, *Uspehi Mat. Nauk N.S.*, **6**, 3–120.

113. Krein, M. and Nudel'man, A. (1977) *The Markov Moment Problem and Extremal Problems*, Transl. Math. Monographs **50**, American Math. Soc., Providence.

114. Kruskal, W.M. (1958) Ordinal measures of association, *J. Amer. Statist. Assoc.*, **53**, 814–861.

115. Kuznezova-Sholpo, I. and Rachev, S.T. (1987) Explicit solutions of moment problems, Technical Report No. 197, Center for Stochastic Processes, Univ. of North Carolina at Chapel Hill.

116. Landau, H., ed., (1987) *Moments in Mathematics*, Proc. Symposia Appl. Math. **37**, Amer. Math. Soc., Providence.

117. Levin, V.L. and Milyutin, A.A. (1978) The problem of mass transfer with a discontinuous cost function and a mass statement of the duality problem for convex external problems, *Russian Math. Surveys*, 34/3, 1–78.

118. Lorentz, G.G. (1953) *Bernstein Polynomials*, Univ. of Toronto Press, Toronto.

119. Lorentz, G.G. (1966) *Approximation of Functions*, Holt, Rinehart & Winston, New York.

120. Lorentz, G.G. (1972) Korovkin Sets, Lecture Notes, Sept. 1972, U.C. Riverside, Center for Numerical Analysis, University of Texas at Austin.

121. Mamedov, R.A. (1959), D.A.N., USSR, **128**, p. 674.

122. Mises, R. von (1939) The limits of a distribution function if two expected values are given. *Ann. Math. Statist.*, **10**, 99–104.

123. Mond, B. (1976) On the degree of approximation by linear positive operators, *J. Approx. Theory*, **18**, 304–306.

124. Mond, B. and Vasudevan, R. (1980) On approximation by linear positive operators, *J. Approx. Theory*, **30**, 334–336.

125. Mosteller, F., Youtz C. and Zahn, D. (1967) The distribution of sums of rounded percentages, *Demography*, **4**, 850–858.

126. Mulholland, H.P. and Rogers, C.A. (1958) Representation theorems for distribution functions, *Proc. London Math. Soc.*, **8**, 177–223.

127. McGregor, J. (1980) Solvability criteria for certain N-dimensional moment problems, *J. Approx. Th.*, **30**, 315–333.

128. Nishishiraho, T. (1983) The rate of convergence of positive linear approximation process, in *Approximation Theory, IV* (C.K. Chui, L.L. Schumaker, and J.D. Ward, Eds.), pp. 635–641, Proc. Int. Symp. Approx. Th., College Station, 1983, Academic Press, New York.

129. Nishishiraho, T. (1986) The degree of approximation by iterations of positive linear operators, in *Approximation Theory V* (C.K. Chui, L.L. Schumaker, and J.D. Ward, Eds.), pp. 507–510, Int. Symp. Approx. Th., College Station, 1986, Academic Press, New York.

130. Olkin, I. and Pukelheim, F. (1982) The distance between two random vectors with given dispersion matrices, *Linear Algebra and Appl.*, **48**, 257–263.

131. Peetre, J. (1963) *A Theory of Interpolation of Normed Spaces*, Notes Universidade de Brasilia.

132. Jiménez Pozo, M.A. (1974) Sur les opérateurs lineaires positifs et la methode des fonctions tests, *C.R. Acad. Sci. Paris Sér. I. Math.*, **278**, 149–152.

133. Jiménez Pozo, M. A. (1978) On the problem of the convergence of a sequence of linear operators, *Moscow Univ., Math. Bull.*, **33**(4), 1–8.

134. Jiménez Pozo, M.A. (1981) Quantitative theorems of Korovkin type in bounded function spaces, Constructive theory of functions, in Proc. Int. Conf. Varna, Bulgaria (D. Vacov, Ed.), pp. 488–494, Sofia Publishing House, Bulgarian Academic of Sciences.

135. Putinar, M. (1988) A two-dimensional moment problem, *J. Funct. Anal.*, **80**, 1–8.

136. Rachev, S.T. (1981) Minimal metrics in a space of random vectors with fixed one-dimensional marginal distributions, in *Stability Problems for Stochastic Models*, Proceedings, Moscow VNIISI, 1981, pp. 112–128 (Russian); English transl., J. Soviet Math., Vol. 34, No. 2, 1986, pp. 1542–1555.

137. Rachev, S.T. (1982) Minimal metrics in the minimal variables space, *Pub. Inst. Statist. Univ. Paris*, v. XXVII, fasc. I, pp. 27–47.

138. Rachev, S.T. (1984) On a problem of Dudley, *Soviet Math. Dokl.*, v. 29, No. 2, 162–164.

139. Rachev, S.T. (1984) On a class of minimal functionals in a space of probability measures, *Theor. Probab. Appl.*, v. 29, 41–49.

140. Rachev, S.T. (1984) The Monge–Kantorovich mass transfer problem and its stochastic applications, *Theory of Prob. Appl.*, **29**, 647–676.

141. Rachev, S.T. (1985) Extreme functionals in the space of probability measures, In: Proc. Stability problems for stochastic models, 1984, pp. 320–348, Lecture Notes in Math., 1155, Springer-Verlag, Berlin, New York.

142. Rachev, S.T. (1) (1989) The problem of stability in queueing theory, *Queueing Systems Theory and Applications*, **4**, 287–318.

143. Rachev, S.T. and Shortt, R.M. (2) (1989) Classification problem for probability metrics, *Contemporary Mathematics*, A.M.S. **96**, 221–262.

144. Rachev, S.T. (1991) *Probability Metrics and the Stability of Stochastic Models*, Wiley, England.

145. Rao, R.R. (1962) Relations between weak and uniform convergence of measures with applications, *Ann. Math. Statist.*, **33**, 659–680.

146. Reidemeister, K. (1921) Über die singulären Randpunkte eines konvexen Körpers, *Math. Ann.*, **83**, 116–118.

147. Richter, H. (1957) Parameterfreie Abschätzung und Realisierung von Erwartungswerten, *Blätter der Deutschen Gesellschaft für Versicherungsmathematik*, **3**, 147–161.

148. Riesz, F. (1911) Sur certaines systèmes singuliers d'équations intégrales, *Ann. Sci. Ecole Norm. Sup.*, **28**, 33–62.

149. Rogosinsky, W.W. (1958) Moments of non-negative mass, *Proceedings of Royal Society London*, Ser. A. **245**, 1–27.

150. Rogosinsky, W.W. (1962) Non-negative linear functionals, moment problems, and extremum problems in polynomial spaces, *Studies in mathematical analysis and related topics*, 316–324, Stanford Univ. Press.

151. Ruschendorf, L. (1985) The Wasserstein distance and approximation theorems, *Z. Wahrscheinlichkeitstheorie verw. Gebiete*, **70**, 117–129.

152. Sarason, D. (1987) *Moment problems and operators in Hilbert space*, Moments in Math., Proc. Symposia Applied Math. **37**, 54–70, Amer. Math. Soc., Providence.

153. Saskin, J.A. (1966) Korovkin systems in spaces of continuous functions, *Amer. Math. Soc. Transl. Anal. Ser. 2*, **54**, 125–144.

154. Schmüdgen, K. (1991) The K-moment problem for semi-algebraic sets, *Math. Ann.*, **289**, 203–206.

155. Schumaker, L.L (1981) *Spline Functions. Basic Theory*, John Wiley & Sons, New York.

156. Schurer, F. and Steutel, F.W. (1979) On the exact degree of approximation of Bernstein operators on $C([0,1]^2)$. *Multivariate Approx. Theory*, edited by W. Schempp and K. Zeller, Birkhaüser-Verlag, 413–435.

157. Schurer, F. and Steutel, F.W. (1980) Exact degrees of approximation for two-dimensional Bernstein operators on C^1, in *Symposium Volume, Approximation Theory III* (E.W. Cheney, Ed.), pp. 823–829, Academic Press, New York/London.

158. Selberg, H.L. (1940) Zwei Ungleichungen zur Ergänzung des Tchebycheffschen Lemmas, *Skand. Aktuarietids Krift*, **23**, 121–125.

159. Shepard, D. (1968) A two-dimensional interpolation function for irregularly-spaced data, *Proc. 1968 ACM National Conference*, 517–524.

160. Shisha, O. and Mond, B. (1) (1968) The degree of convergence of sequences of linear positive operators, *Nat. Acad. of Sci. U.S.*, **60**, 1196–1200.

161. Shisha, O. and Mond, B. (2) (1968) The degree of approximation to periodic functions by linear positive operators, *J. Approx. Theory*, **1**, 335–339.

162. Shohat, J.A. and Tamarkin, J.D. (1943) *The Problems of Moments*, American Math. Soc., Providence.

163. Singh, S.P. (1981) On the degree of approximation by Szász operators, *Bull. Austral. Math. Soc.*, **24**, 221–225.

164. Stark, E.L. (1972) An extension of a theorem of P.P. Korovkin to singular integrals with not necessarily positive kernels, *Akad. Van. Wetensc. Proc. A Math. Sci.*, **75**, 227–235.

165. Szulga, A. (1978) On the Wasserstein metric, In: *Translations of the 8th Prague Conference on Information Theory, Statistical Decision Functions, and Random Processes*, Prague, 1978, v. B. Akademia, Praha, pp. 267–273.

166. Szulga, A. (1982) On minimal metrics in the space of random variables, *Theory Prob. Appl.*, **27**, 424–430.

167. Timan, A. (1963) *The Theory of Approximation of Functions of a Real Variable*, (translation from Russian), Pergamon, Oxford/New York.

168. Trotter, H.F. (1959) An elementary proof of the central limit theorem, *Arch. Math.*, (Basel) **10**, 226–234.

169. Valentine, F.A. (1964) *Convex Sets*, McGraw-Hill, New York.

170. Weba, M. (1986) Korovkin systems of stochastic processes, *Mathematische Zeitschrift*, **192**, 73–80.

171. Weba, M. (1990) A quantitative Korovkin theorem for random functions with multivariate domains, *J. Approx. Theory*, **61**, 74–87.

172. Widder, D.V. (1928) A generalization of Taylor's series, *Trans. A.M.S.*, **30**, 126–154.

173. Zolotarev, V.M. (1976) Metric distances in spaces of random variables and their distributions, *Math. USSR Sbornik*, **30**, 3, 373–401.

174. Zolotarev, V.M. (1983) Probability metrics, *Theory Prob. Appl.*, **28**, 278–302.

175. Zolotarev, V.M. (1986) *Contemporary Theory of Summation of Independent Random Variables*, Nauka, Moscow (in Russian).